METHODS IN MOLECULAR BIOLOGY

Series Editor

John M. Walker

School of Life and Medical Sciences

University of Hertfordshire

Hatfield, Hertfordshire, AL10 9AB, UK

For further volumes:
http://www.springer.com/series/7651

Polyamines

Methods and Protocols

Edited by

Rubén Alcázar and Antonio F. Tiburcio

Department of Biology, Healthcare and Environment, Faculty of Pharmacy and Food Sciences, Section of Plant Physiology, University of Barcelona, Spain

Humana Press

Editors
Rubén Alcázar
Department of Biology,
Healthcare and Environment,
Faculty of Pharmacy and Food Sciences,
Section of Plant Physiology
University of Barcelona
Spain

Antonio F. Tiburcio
Department of Biology, Healthcare and Environment,
Faculty of Pharmacy and Food Sciences,
Section of Plant Physiology
University of Barcelona
Spain

ISSN 1064-3745 ISSN 1940-6029 (electronic)
Methods in Molecular Biology
ISBN 978-1-4939-7397-2 (hardcover) ISBN 978-1-4939-7398-9 (eBook)
ISBN 978-1-4939-8473-2 (softcover)
DOI 10.1007/978-1-4939-7398-9

Library of Congress Control Number: 2017956111

Printed on acid-free paper

This Humana Press imprint is published by Springer Nature
The registered company is Springer Science+Business Media LLC
The registered company address is: 233 Spring Street, New York, NY 10013, U.S.A.

Preface

The history of polyamine (PA) research starts more than 300 years ago. In 1678, Antoni van Leeuwenhoek noted the deposition of stellate crystals in aging human semen. More than 200 years later, the basic constituent of such crystals was named spermine, but it was not until the mid-1920s that its correct chemical composition and structure was determined. Spermidine was also discovered and named at this time. In 1971, Seymor Cohen [1] addressed the biological relevance of these compounds and stimulated PA research in many areas, including plant physiology. Among the provocative generalizations in his book [1] was the observation that modern biochemistry was concerned mainly with anions and generally tended to neglect cations, of which PAs are one of the cell's major organic representatives.

Since then, pioneering works in the PA field started in a limited number of research laboratories studying plant physiology and biochemistry, microbiology, cancer research and chemistry, among others. Research on this topic expanded during the last decades and nowadays, polyamines are recognized to participate in relevant biological functions in plants, microorganisms, and animals.

Protocols for PA research are continuously evolving, and we attempted to provide a comprehensive collection of updated protocols by key researchers in different fields. In addition, we provide the reader with a selection of excellent reviews introducing the different topics of polyamine research covered by the protocols (Chapters 1–4, 36–37, and 40).

The book covers the quantification of different polyamines and conjugates (Chapters 5–10), polyamine and related enzymatic activities (Chapters 11–18), subcellular localization studies (Chapters 19–20), transport (Chapters 21–23), DNA methylation (Chapter 24), ODC regulation (Chapters 25–26), modulation of protein synthesis and post-translational modifications (Chapters 27–28), genetic and phenotyping analyses (Chapters 29–33), determination of ROS (Chapter 34), genome-wide association mapping (Chapter 35), and polyamine applications and cancer (Chapters 36–40). We believe that this book contains most of the essential protocols for polyamine research and will be a helpful manual in research laboratories.

The editors would like to express their gratitude to all authors who have greatly contributed to this book, which we hope will stimulate research on the fascinating topic of polyamines.

Barcelona, Spain

Rubén Alcázar
Antonio F. Tiburcio

Reference

1. Cohen SS (1971) Introduction to the Polyamines. Prentice-Hall, Englewood Cliffs, NJ.

Contents

Contributors

PATRICIA AGUDELO-ROMERO • *Australian Research Council (ARC) Centre of Excellence in Plant Energy Biology, The University of Western Australia, Perth, WA, Australia*

RUBÉN ALCÁZAR • *Department of Biology, Healthcare and Environment, Faculty of Pharmacy and Food Sciences, Section of Plant Physiology, University of Barcelona, Barcelona, Spain*

I. ALOISI • *Department of Biological, Geological and Environmental Sciences, University of Bologna, Bologna, Italy*

RICCARDO ANGELINI • *Department of Science, University "Roma Tre", Rome, Italy*

CHRYSTALLA ANTONIOU • *Department of Agricultural Sciences, Biotechnology and Food Science, Cyprus University of Technology, Lemesos, Cyprus*

KOSTADIN E. ATANASOV • *Department of Biology, Healthcare and Environment, Faculty of Pharmacy and Food Sciences, Section of Plant Physiology, University of Barcelona, Barcelona, Spain*

LUIS BARBOZA-BARQUERO • *CIGRAS, Universidad de Costa Rica, Sede Rodrigo Facio, San Jose, Costa Rica*

JAUME BASTIDA • *Department of Biology, Healthcare and Environment, Faculty of Pharmacy and Food Sciences, Section of Plant Physiology, University of Barcelona, Barcelona, Spain*

FULLER W. BAZER • *Department of Animal Science, Texas A&M University, College Station, TX, USA*

THOMAS BERBERICH • *Laboratory Center, Senckenberg Biodiversity and Climate Research Centre (BiK-F), Frankfurt am Main, Germany*

MARTA BITRIÁN • *Department of Biology, Healthcare and Environment, Faculty of Pharmacy and Food Sciences, Section of Plant Physiology, University of Barcelona, Barcelona, Spain*

P.L.R. BONNER • *School of Science and Technology, Nottingham Trent University, Nottingham, UK*

N. BRIGLIA • *UNIBAS—Dipartimento delle Culture Europee e del Mediterraneo: Architettura, Ambiente, Patrimoni Culturali, Matera, Italy*

G. CAI • *Department of Life Sciences, University of Siena, Siena, Italy*

PEDRO CARRASCO • *Departament de Bioquímica i Biologia Molecular, Facultat de Ciències Biològiques, Universitat de València, València, Spain; Estructura de Recerca Interdisciplinar en Biotecnologia i Biomedicina (ERI BIOTECMED), Universitat de València, València, Spain*

F. CELLINI • *ALSIA—Metapontum Agrobios Research Center, Metaponto (MT), Italy*

ALESSANDRA CONA • *Department of Science, University "Roma Tre", Rome, Italy*

ZHAOLAI DAI • *State Key Laboratory of Animal Nutrition, China Agricultural University, Beijing, China*

ELISABETTA DAMIANI • *School of Medicine, Medical Sciences and Nutrition, University of Aberdeen, Aberdeen, UK; Department of Life and Environmental Sciences, Polytechnic University of Marche, Ancona, Italy*

S. DEL DUCA • *Department of Biological, Geological and Environmental Sciences, University of Bologna, Bologna, Italy*

PAUL ESKER • *Department of Plant Pathology and Environmental Microbiology, Penn State University, University Park, PA, USA*

Gabriele Favero • *Department of Chemistry and Drug Technologies, Sapienza University of Rome, Rome, Italy*
Andreas Finke • *Max Planck Institute for Plant Breeding Research, Cologne, Germany*
Ana Margarida Fortes • *Biosystems & Integrative Sciences Institute (BioISI), Faculdade de Ciências, Universidade de Lisboa, Lisbon, Portugal*
Vasileios Fotopoulos • *Department of Agricultural Sciences, Biotechnology and Food Science, Cyprus University of Technology, Lemesos, Cyprus*
Shinsuke Fujiwara • *Department of Bioscience, Graduate School of Science and Technology, Kwansei-Gakuin University, Sanda, Hyogo, Japan; Research Center for Intelligent, Kwansei-Gakuin University, Sanda, Hyogo, Japan*
Wakao Fukuda • *Department of Biotechnology, College of Life Sciences, Ritsumeikan University, Kusatsu, Shiga, Japan*
Andrés Gárriz • *Instituto de Investigaciones Biotecnológicas-Instituto Tecnológico Chascomús, Universidad Nacional de General San Martín-Consejo Nacional de Investigaciones Científicas y Técnicas (IIB-INTECH/UNSAM-CONICET), Chascomús, Buenos Aires, Argentina*
Daniela Gödderz • *Deallus Consulting Ltd, London, UK*
Egli C. Georgiadou • *Department of Agricultural Sciences, Biotechnology and Food Science, Cyprus University of Technology, Lemesos, Cyprus*
Nicole M. Gibbs • *Laboratory of Genetics, University of Wisconsin-Madison, Madison, WI, USA*
Ryota Hidese • *Department of Bioscience, Graduate School of Science and Technology, Kwansei-Gakuin University, Sanda, Hyogo, Japan*
Yongqing Hou • *Hubei Key Laboratory of Animal Nutrition and Feed Science, Hubei Collaborative Innovation Center for Animal Nutrition and Feed Safety, Wuhan Polytechnic University, Wuhan, China*
R. Iannacone • *ALSIA—Metapontum Agrobios Research Center, Metaponto (MT), Italy*
Kazuei Igarashi • *Amine Pharma Research Institute, Innovation Plaza at Chiba University, Chiba, Japan; Graduate School of Pharmaceutical Sciences, Chiba University, Chiba, Japan*
Jun-Ichi Kakehi • *Division of Earth, Life, and Molecular Sciences, Graduate School of Natural Science and Technology, Okayama University, Okayama, Japan*
R. J. Dohmen • *Institute for Genetics, Biocenter, University of Cologne, Cologne, Germany*
Takashi Kaneta • *Department of Chemistry, Graduate School of Natural Science and Technology, Okayama University, Okayama, Japan*
Keiko Kashiwagi • *Faculty of Pharmacy, Chiba Institute of Science, Chiba, Japan*
Leo Kurian • *Laboratory for Developmental and Regenerative RNA biology, Center of Molecular Medicine Cologne (CMMC), University of Cologne, Cologne, Germany*
Tomonobu Kusano • *Graduate School of Life Sciences, Tohoku University, Sendai, Miyagi, Japan*
Xilong Li • *Institute of Feed Science, The Chinese Academy of Agricultural Sciences, Beijing, China*
Changxin Liu • *Department of Biology, Healthcare and Environment, Faculty of Pharmacy and Food Sciences, Section of Plant Physiology, University of Barcelona, Barcelona, Spain*
Anh Mai • *Penn State Berks, Reading, PA, USA*
Santiago J. Maiale • *Instituto de Investigaciones Biotecnológicas-Instituto Tecnológico Chascomús, Universidad Nacional de General San Martín-Consejo Nacional de*

Investigaciones Científicas y Técnicas (IIB-INTECH/UNSAM-CONICET), Chascomús, Buenos Aires, Argentina

FRANCISCO MARCO • *Departament de Biologia Vegetal, Facultat de Farmàcia, Universitat de València, València, Spain; Estructura de Recerca Interdisciplinar en Biotecnologia i Biomedicina (ERI BIOTECMED), Universitat de València, València, Spain*

MARIA MARINA • *Instituto de Investigaciones Biotecnológicas-Instituto Tecnológico Chascomús, Universidad Nacional de General San Martín-Consejo Nacional de Investigaciones Científicas y Técnicas (IIB-INTECH/UNSAM-CONICET), Chascomús, Buenos Aires, Argentina*

D. MARKO • *ALSIA—Metapontum Agrobios Research Center, Metaponto (MT), Italy*

PATRICK H. MASSON • *Laboratory of Genetics, University of Wisconsin-Madison, Madison, WI, USA*

FRANCO MAZZEI • *Department of Chemistry and Drug Technologies, Sapienza University of Rome, Rome, Italy*

RAKESH MINOCHA • *USDA Forest Service, Northern Research Station, Durham, NH, USA*

SUBHASH C. MINOCHA • *Department of Biological Sciences, University of New Hampshire, Durham, NH, USA*

MARIANA R. MIRANDA • *Laboratorio de Parasitología Molecular, Instituto de Investigaciones Médicas "A. Lanari", IDIM-CONICET, Universidad de Buenos Aires, Buenos Aires, Argentina*

PANAGIOTIS N. MOSCHOU • *Department of Plant Biology and Linnean Center of Plant Sciences, Swedish University of Agricultural Sciences, Uppsala, Sweden*

MASARU NIITSU • *Faculty of Pharmacy and Pharmaceutical Sciences, Josai University, Saitama, Japan*

SHANNON L. NOWOTARSKI • *Penn State Berks, Reading, PA, USA*

R. PALANIMURUGAN • *Center for Cellular and Molecular Biology (CCMB), Hyderabad, India*

KY YOUNG PARK • *Department of Biology, Sunchon National University, Chonnam, South Korea*

ALES PECINKA • *Max Planck Institute for Plant Breeding Research, Cologne, Germany*

CLAUDIO A. PEREIRA • *Laboratorio de Parasitología Molecular, Instituto de Investigaciones Médicas "A. Lanari", IDIM-CONICET, Universidad de Buenos Aires, Buenos Aires, Argentina*

A. PETROZZA • *ALSIA—Metapontum Agrobios Research Center, Metaponto (MT), Italy*

IGOR POTTOSIN • *Centro Universitario de Investigaciones Biomédicas, Universidad de Colima, Colima, México*

NARAYAN S. PUNEKAR • *Metabolism and Enzymology Laboratory, Department of Biosciences and Bioengineering, Indian Institute of Technology Bombay, Mumbai, India*

CHANTAL REIGADA • *Laboratorio de Parasitología Molecular, Instituto de Investigaciones Médicas "A. Lanari", IDIM-CONICET, Universidad de Buenos Aires, Buenos Aires, Argentina*

FERNANDO M. ROMERO • *Instituto de Investigaciones Biotecnológicas-Instituto Tecnológico Chascomús, Universidad Nacional de General San Martín-Consejo Nacional de Investigaciones Científicas y Técnicas (IIB-INTECH/UNSAM-CONICET), Chascomús, Buenos Aires, Argentina*

FRANCO R. ROSSI • *Instituto de Investigaciones Biotecnológicas-Instituto Tecnológico Chascomús, Universidad Nacional de General San Martín-Consejo Nacional de*

Investigaciones Científicas y Técnicas (IIB-INTECH/UNSAM-CONICET), Chascomús, Buenos Aires, Argentina

KALLIOPI A. ROUBELAKIS-ANGELAKIS • *Department of Biology, University of Crete, Heraklion, Greece*

LAURA VAUGHN ROUHANA • *Department of Biological Sciences, Wright State University, Dayton, OH, USA*

WILFRIED ROZHON • *Biotechnology of Horticultural Crops, Technische Universität München, Munich, Germany*

OSCAR A. RUÍZ • *Instituto de Investigaciones Biotecnológicas-Instituto Tecnológico Chascomús, Universidad Nacional de General San Martín-Consejo Nacional de Investigaciones Científicas y Técnicas (IIB-INTECH/UNSAM-CONICET), Chascomús, Buenos Aires, Argentina*

G.H.M. SAGOR • *Department of Genetics & Plant Breeding, Bangladesh Agricultural University, Mymensingh, Bangladesh*

TEJASWANI SARAGADAM • *Metabolism and Enzymology Laboratory, Department of Biosciences and Bioengineering, Indian Institute of Technology Bombay, Mumbai, India*

ANDREAS SAVVIDES • *Department of Agricultural Sciences, Biotechnology and Food Science, Cyprus University of Technology, Lemesos, Cyprus*

MELISA SAYÉ • *Laboratorio de Parasitología Molecular, Instituto de Investigaciones Médicas "A. Lanari", IDIM-CONICET, Universidad de Buenos Aires, Buenos Aires, Argentina*

JOSE M. SEGUÍ-SIMARRO • *Cell Biology Group–COMAV, Universitat Politècnica de València, Valencia, Spain*

D. SERAFINI-FRACASSINI • *Department of Biological, Geological and Environmental Sciences, University of Bologna, Bologna, Italy*

MARÍA SERRANO • *Department of Applied Biology, University Miguel Hernández, Alicante, Spain*

TSUBASA SHOJI • *Graduate School of Biological Sciences, Nara Institute of Science and Technology, Ikoma, Japan*

YOSHIAKI SUGITA • *Laboratory of Bioorganic Chemistry, Department of Pharmaceutical Sciences, Faculty of Pharmacy and Pharmaceutical Sciences, Josai University, Saitama, Japan*

S. SUMMERER • *ALSIA—Metapontum Agrobios Research Center, Metaponto (MT), Italy*

TAKU TAKAHASHI • *Division of Earth, Life, and Molecular Sciences, Graduate School of Natural Science and Technology, Okayama University, Okayama, Japan*

AYAKA TAKANO • *Division of Earth, Life, and Molecular Sciences, Graduate School of Natural Science and Technology, Okayama University, Okayama, Japan*

KOICHI TAKAO • *Laboratory of Bioorganic Chemistry, Department of Pharmaceutical Sciences, Faculty of Pharmacy and Pharmaceutical Sciences, Josai University, Saitama, Japan*

PARASKEVI TAVLADORAKI • *Department of Science, University "Roma Tre", Rome, Italy*

YUSUKE TERUI • *Faculty of Pharmacy, Chiba Institute of Science, Chiba, Japan*

ANTONIO F. TIBURCIO • *Department of Biology, Healthcare and Environment, Faculty of Pharmacy and Food Sciences, Section of Plant Physiology, University of Barcelona, Barcelona, Spain*

LAURA TORRAS-CLAVERIA • *Department of Biology, Healthcare and Environment, Faculty of Pharmacy, University of Barcelona, Barcelona, Spain*

CRISTINA TORTOLINI • *Department of Chemistry and Drug Technologies, Sapienza University of Rome, Rome, Italy*

Takeshi Uemura • *Amine Pharma Research Institute, Innovation Plaza at Chiba University, Chiba, Japan*
Daniel Valero • *Department of Food Technology, University Miguel Hernández, Alicante, Spain*
Francesc Viladomat • *Department of Biology, Healthcare and Environment, Faculty of Pharmacy and Food Sciences, Section of Plant Physiology, University of Barcelona, Barcelona, Spain*
Heather M. Wallace • *School of Medicine, Medical Sciences and Nutrition, University of Aberdeen, Aberdeen, UK*
Guoyao Wu • *Hubei Key Laboratory of Animal Nutrition and Feed Science, Hubei Collaborative Innovation Center for Animal Nutrition and Feed Safety, Wuhan Polytechnic University, Wuhan, China; State Key Laboratory of Animal Nutrition, China Agricultural University, Beijing, China; Department of Animal Science, Texas A&M University, College Station, TX, USA*
Zhenlong Wu • *State Key Laboratory of Animal Nutrition, China Agricultural University, Beijing, China*
Wegi Wuddineh • *Department of Biological Sciences, University of New Hampshire, Durham, NH, USA*
Isaac Zepeda-Jazo • *Universidad de La Ciénega del Estado de Michoacán de Ocampo, Sahuayo, Michoacán, México*

The original version of this book was revised. A correction to this book can be found at https://doi.org/10.1007/978-1-4939-7398-9_41

Chapter 1

Polyamines in the Context of Metabolic Networks

Wegi Wuddineh, Rakesh Minocha, and Subhash C. Minocha

Abstract

Polyamines (PAs) are essential biomolecules that are known to be involved in the regulation of many plant developmental and growth processes as well as their response to different environmental stimuli. Maintaining the cellular pools of PAs or their metabolic precursors and by-products is critical to accomplish their normal functions. Therefore, the titre of PAs in the cells must be under tight regulation to enable cellular PA homeostasis. Polyamine homeostasis is hence achieved by the regulation of their input into the cellular PA pool, their conversion into secondary metabolites, their transport to other issues/organs, and their catabolism or turnover. The major contributors of input to the PA pools are their in vivo biosynthesis, interconversion between different PAs, and transport from other tissues/organs; while the output or turnover of PAs is facilitated by transport, conjugation and catabolism. Polyamine metabolic pathways including the biosynthesis, catabolism/turnover and conjugation with various organic molecules have been widely studied in all kingdoms. Discoveries on the molecular transporters facilitating the intracellular and intercellular translocation of PAs have also been reported. Numerous recent studies using transgenic approaches and mutagenesis have shown that plants can tolerate quite large concentrations of PAs in the cells; even though, at times, high cellular accumulation of PAs is quite detrimental, and so is high rate of catabolism. The mechanism by which plants tolerate such large quantities of PAs is still unclear. Interestingly, enhanced PA biosynthesis via manipulation of the PA metabolic networks has been suggested to contribute directly to increased growth and improvements in plant abiotic and biotic stress responses; hence greater biomass and productivity. Genetic manipulation of the PA metabolic networks has also been shown to improve plant nitrogen assimilation capacity, which may in turn lead to enhanced carbon assimilation. These potential benefits on top of the widely accepted role of PAs in improving plants' tolerance to biotic and abiotic stressors are invaluable tools for future plant improvement strategies.

Key words Putrescine, Spermidine, Spermine, Nitrogen, NO, Glutamate, Proline, TCA, Transport

1 Introduction

Polyamines (PAs) are polycationic amines found in all living organisms, and even viruses. Major free PAs in plants include putrescine (Put), spermidine (Spd), spermine (Spm), and its structural isomer thermospermine (tSPM). Polyamines play important physiological roles in a broad range of biological functions including transcription and translation, the stabilization of nucleic acids and cell membranes, regulation of cell division and growth, organogenesis,

Rubén Alcázar and Antonio F. Tiburcio (eds.), *Polyamines: Methods and Protocols*, Methods in Molecular Biology, vol. 1694, DOI 10.1007/978-1-4939-7398-9_1, © Springer Science+Business Media LLC 2018

embryogenesis, leaf senescence, flower and fruit development as well as in responses to abiotic and biotic stressors [1–3]. Whereas some of these PA functions are common to those in animals, others are somewhat unique to plants. The cellular concentrations of PAs often fluctuate depending on the stage of plant development and growth, the type of nitrogen (N) nutrition as well as in response to various internal (i.e., developmental) and environmental signals, e.g., those induced by biotic or abiotic stressors. In general, plants have evolved robust control mechanisms for keeping PA homeostasis both to accomplish the aforementioned physiological functions and also to maintain the balance of C: N ratio in plant cells [4]. This homeostasis control is achieved through regulation of biosynthesis, interconversion among various PAs, catabolism, their conjugation with phenolic compounds and other macromolecules, and transport to other cells, tissues, and organs [5, 6]. This chapter summarizes metabolic interactions of PAs with a variety of metabolic pathways in relation to their biological significance in plants.

1.1 Polyamine Biosynthesis

One of the major players in the regulation of PA homeostasis is their biosynthesis, which in most plants occurs via two pathways, namely the arginine (Arg) decarboxylase (ADC) pathway and the ornithine (Orn) decarboxylase (ODC) pathway (Fig. 1). However, a major exception to this is the commonly used model plant *Arabidopsis thaliana*, whose genome lacks a gene encoding an ODC-like protein [8, 9]. In the ADC pathway, PA biosynthesis starts with the conversion of Arg to agmatine (Agm), then into Put, which is the precursor of all higher PAs (Fig. 1). The ODC pathway on the other hand, involves direct conversion of Orn (also an important intermediate in Arg biosynthesis) into Put. Higher PAs, namely Spd and Spm/tSpm are synthesised from Put by sequential actions of aminopropyltransferases [10], Spd synthase (SPDS), and Spm/tSpm synthase (SPMS) or tSpm synthase (tSPMS), respectively via addition of an aminopropyl moiety from decarboxylated S-adenosylmethionine (dcSAM). The enzyme SAM decarboxylase (SAMDC) produces dcSAM from SAM; this reaction is often believed to be a rate-limiting step in the biosynthesis of the higher PAs [11]. Methionine is the immediate precursor of SAM. Some plants as well as animals also produce and accumulate cadavarine (Cad), another diamine made by direct decarboxylation of Lys by Lys decarboxylase (LCD) [12] or in mammals by ODC [13]. In some plants, Cad is used for the production of speciality alkaloids [12].

1.2 Polyamine Catabolism

Catabolism of PAs is the most significant player (other than synthesis) in regulating dynamic equilibrium of cellular PA concentrations in all organisms. It also contributes to the production of several growth and development regulatory molecules in plants. The turnover of cellular PAs is rather rapid, with a half-life ($t_{1/2}$) for Put being around 6–7 h; for Spd and Spm the half-life is >20 h [14, 15]. The

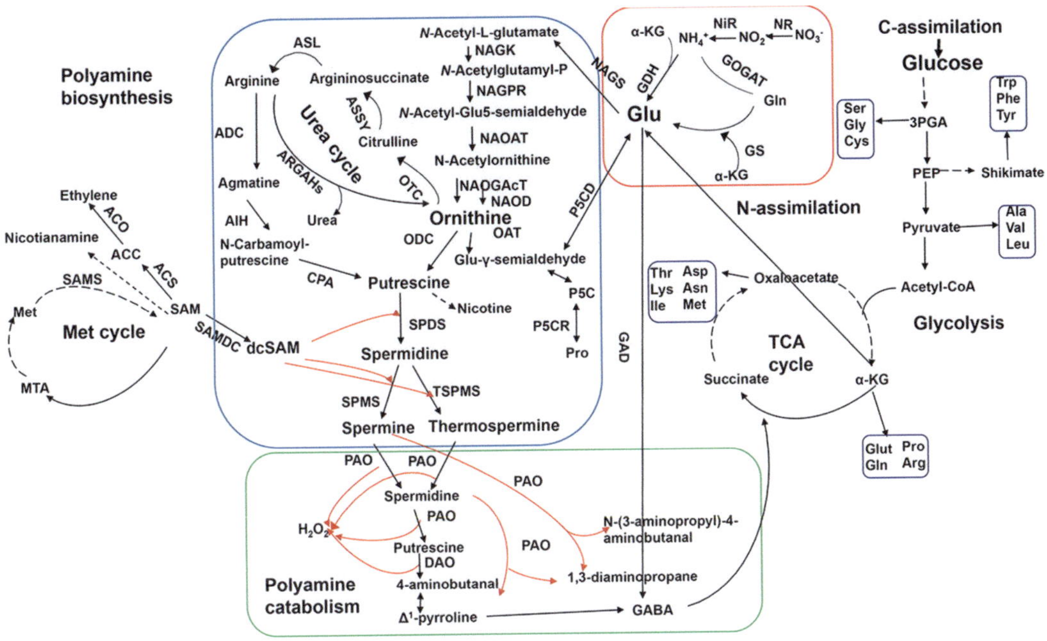

Fig. 1 Networks of polyamine metabolic pathways from N-assimilation to PA catabolism along with its interacting pathways such as TCA cycle, Glycolysis, Urea cycle, and others. Abbreviations of the enzymes and metabolites involved in the interacting pathways are as follows. *GS* Glutamine synthetase, *GOGAT* Glutamate synthase, *NAGS N*-acetylglutamate synthase, *NAGK N*-acetylglutamate kinase, *NAGPR N*-acetylglutamatyl-5-P reductase, *NAOAT N*-acetylornithine aminotransferase, *NAOGAcT N*-acetylornithine-glutamate acetyltransferase, *NAOD N*-acetylornithine deacetylase, *OTC* Ornithine transcarbamylase, *ASSY* Argininosuccinate synthase, *ASL* Argininosuccinate lyase, *CoA* Coenzyme A, *CPS* Carbamoyl phosphate synthetase, *δ-OAT* ornithine-δ-aminotransferase, *P5CR* Δ^1 pyrroline-5-carboxylate reductase, *P5CD* Δ^1 pyrroline-5-carboxylate dehydrogenase, *ACC* 1-Aminocyclopropane-1-carboxylic acid oxidase, *ADC* Arginine decarboxylase, *DAO* Diamine oxidase, *GAD* Glutamate decarboxylase, *GOGAT* Glutamate synthase, *GS* Glutamine synthetase, *LysDC* Lysine decarboxylase, *NAGK* N-Acetylglutamate kinase, *NAGPR* N-Acetylglutamate-5-P reductase, *NAGS* N-acetyl-glutamate synthase, *NAOAT* N2-acetyl-Orn aminotransferase, *NAOD* N2-acetyl-Orn deacetylase, *NIR* Nitrite reductase, *NR* Nitrate reductase, *OAT* Ornithine aminotransferase, *ODC* Ornithine decarboxylase, *OTC* Ornithine transcarbamylase, *SAMDC* S-adenosylmethionine decarboxylase, *SPDS* spermidine synthase, *SMPS* spermine synthase. (Adapted with modification from [7])

PA turnover is catalysed by copper amine oxidases (CuAOs) and the Flavin-dependent PA oxidases (PAOs) [3, 4, 16]; the former are homodimeric enzymes that catalyze breakdown of Put or Cad to γ-aminobutyric acid (GABA) via the intermediates 4-aminobutanal and Δ^1-pyrroline (Fig. 1). These enzymes can also catalyze the oxidation of Spd and Spm although with a lower affinity. On the other hand, the monomeric PAOs catabolize either Spd, Spm/tSpm (and their acetylated derivatives in animals) or back-convert them into respective lower amines (Put and Spd, respectively). Polyamine oxidation results in the formation of (a) H_2O_2, which is a signaling molecule that is perhaps a major player in PA-induced regulation of various biological processes [17, 18]; and (b) GABA, another very important molecule in numerous physiological functions in all organisms [19–21]. Besides the terminal degradation of PAs to recycle C and N, thereby bridging PA metabolism with the tricarboxylic Acid (TCA) cycle, it has now become clear that PA catabolism serves a vital role in the regulation of plant growth and development [3, 22].

1.3 Polyamine Conjugation

In plants, PAs may accumulate as free molecules or be conjugated with phenolic acids, such as hydroxycinnamic acid (HCA), coumaric acid, caffeic acid, and ferulic acid, or be bound to various macromolecules (e.g., proteins), thus significantly affecting their cellular titres. Many PA conjugates, and the enzymes catalysing their biosynthesis, have been identified in different plant species [2]. Some important PA-conjugates are the HCA amides (HCAA), also known as phenolamides, produced by hydroxycinnamoyl acylation of the mono-, di-, or tri-phenolic acid substitutions of PAs [23] by enzymes called hydroxycinnamoyl transferases belonging to the BAHD *N*-acyltransferase group [24]. Conjugation of PAs to proteins on the other hand occurs via covalent linkage to the glutamyl residues to produce either mono-γ-glutamyl-PAs or bis-γ-glutamyl-PAs; transglutaminases catalyze these reactions [25, 26]. The less polar and hydrophobic nature of PA conjugates as opposed to PAs, may favor their intercellular transport and long term storage [23]. The proportion of conjugated vs. free PAs ranges widely in different species, depending on growth and developmental status of different organs. Although some interconversion between conjugated and free PAs has also been reported [27]; it apparently is not a major contributor to the cellular PA content. The PA conjugates participate in various biological processes, e.g., pollen development, plant defense responses to pathogens and insects, and response to abiotic stresses [3, 25, 28].

1.4 Polyamine Transport

Uptake of PAs from external sources and their organellar transport contribute to the regulation of PA homeostasis in many tissues. Translocation in microorganisms is mainly facilitated by the activities of PA transporters [29]; only recently have they been reported in plants [30, 31]. The first plant PA transporter to be functionally characterized was the rice PA uptake transporter (*Os*PUT1) [31]. This protein is produced in most tissues except seeds and roots; and, was shown to be a high affinity PA transporter preferentially facilitating the import of Spd. Recent studies have identified additional proteins belonging to an L-type amino acid transporter (LAT) family of proteins in rice and Arabidopsis as transporters of PAs, and their analog paraquat (PQ—*a.k.a.* methyl viologen) [30, 32, 33]. Three of the five Arabidopsis LATs that exhibit PA transport activity displayed different subcellular localization indicating that they may be involved in different cellular activities, e.g., *At*LAT1/PUT3 responsible for resistance to PQ (e.g., RMV1 or

LAT1/PUT3) has been shown to localize in the plasma membrane and hence may be involved in the intercellular transport of PAs and PQ. *At*LAT3 and *At*LAT4, on the other hand, were shown to localize in the endoplasmic reticulum and Golgi apparatus, respectively [34]. Specific molecular functions to these putative PA transporters are only now becoming assigned, one being the Organic Cation Transporter (*At*CAT1) that is involved in Cad transport, and another *At*LAT1 in stabilizing mRNA under heat stress conditions [35, 36]. The most important missing information currently is: what signals regulate the partitioning of PAs between the cytoplasm and the vacuoles, and transport from one cell type to the other? Also, it is not clear as to whether the three common PAs are translocated to the same compartment or some are preferentially transported over the others, and perhaps into different compartments.

2 Networks of Metabolic Pathways Interacting with Polyamines and Their Metabolism

As shown in Fig. 1, PAs are synthesized from glutamate (Glu) as the starting material in plants, which is also the source of amide group for the biosynthesis of most other amino acids as well as other nitrogenous compounds in plants. The other substrate of importance for PA biosynthesis is Met, which is the primary source of SAM biosynthesis; SAM, like Glu, is also required for numerous methylation reactions in the cell. As outlined above, PAs can be accumulated in plant cells in rather high (up to mmolar) quantities, thus affecting the total N pool; their reactions with phenolic acids also contribute to the flux of C into PA conjugates. Then there are the catabolic products of PAs, including metabolites like GABA (also present in high concentrations in plants) on one side, and H_2O_2 and nitric oxide (NO) on the other. It is, therefore, not surprising that PAs play diverse physiological roles in plants, both directly and indirectly as their metabolism is intertwined in a complex network of metabolic pathways in the cell.

Some of the major interacting metabolic pathways or metabolites of considerable importance to understand the nature of regulation of these metabolic networks involving PAs include: (1) the initial N assimilation pathway (NH_4^+ and NO_3^-) leading to Glu formation; (2) the role of Glu in the biosynthesis of other amino acids by transamination reactions; (3) the biosynthetic subpathways from Glu to PAs via Orn (Glu → Orn → Put/PAs) or via Arg (Glu → Orn → Arg → Put/PAs); (4) Pro biosynthetic pathways from Glu (Glu → Pro and Orn → Pro); and (5) Arg → Orn reverse pathway involving the production of NO and urea. In addition, there is the pathway for SAM biosynthesis from Met and the metabolic pathways utilizing SAM as precursor for higher PAs, a myriad of methylation reactions that use SAM as a precursor; plus, the unique pathway of ethylene biosynthesis in plants. On the other hand, PA catabolism and its reaction products—the Reactive Oxygen Species (ROS) like H_2O_2, and GABA are major signaling molecules in plants. These reactions are then directly connected to the TCA cycle via GABA shunt. Finally, there are the PA conjugation pathways involving a variety of phenolic derivatives like HCAs and coumaric acid. The products of these pathways are involved in storage of organic N to essential signaling reactions, to interactions with cellular macromolecules ranging from DNA, RNA and proteins, and to cellular membranes. Signaling molecules derived from these pathways contribute to metabolic crosstalk with major phytohormones such as ethylene, abscisic acid (ABA), and jasmonic acid (JA). To understand the regulatory mechanism of these complex networks, it is essential to critically evaluate the individual metabolites along with the set of pathway(s) leading to the formation of individual PAs. Numerous studies have dealt with the consequences of perturbation of one subpathway on some of the related metabolic pathways in the networks, and related signaling molecules as well as the metabolites therein. A group of such interactions in plants are summarized here:

2.1 Polyamine Metabolism and Nitrogen Assimilation Pathway (NO_3/NH_4^+ → Glu)

Plants can assimilate different forms of N including inorganic forms such as NO_3^- and NH_4^+ and organic forms such as urea and amino acids. Via reduction of NO_3^-, all N assimilated by plants ends up in three main amino acids—asparagine (Asn), glutamine (Gln), and Glu. The first step in N assimilation is the conversion of NO_3^- to NH_4^+ by the successive actions of NO_3^- reductase (NR) and NO_2^- reductase (NiR) or urea to NH_4^+ by urease. Subsequently, NH_4^+ is assimilated to Glu and Gln via the Gln synthetase-Glu synthase (GS/GOGAT) pathway. The product of this pathway is Glu, which in addition to being the precursor for biosynthesis of PAs and several related metabolites like Orn, Arg, Pro and GABA, is the major donor of amine groups for most other amino acids in plants. Therefore, the form of N absorbed by plants (NO_3^- or NH_4^+) may have major effects on the metabolism of PAs.

Since cells cannot store reduced N as NH_3 or NH_4^+, PAs can be important N storage compounds in plants, their biosynthesis also performs the function of alleviating the NH_3/NH_4^+-induced toxicity in the cells [4]. In agreement with this, an increase in the cellular concentrations of Put to a larger extent, and in some cases also Spd and Spm, were reported under relatively higher NH_4^+ fertilization vs. NO_3^- feeding conditions [37]. Moreover, an increase in Put biosynthesis from radiolabeled Orn in NH_4^+-fed soybean seedlings was accompanied by reduced activities of the PA catabolic CuAOs [38]. Similarly, NH_4^+ nutrition, and also urea to a lesser extent, were found to increase Put biosynthesis leading to reduced plant growth in wheat and pepper; supplementing NH_4^+ and urea with NO_3^- ameliorated this effect.

It appears that PA biosynthesis via the Glu-Orn pathway is favoured under high NH_4^+ nutrition [39, 40]. On the other hand, studies involving poplar and red spruce cell cultures have shown that higher than optimal total N (NH_4^+ + NO_3^-) in the growth medium does not increase PA concentrations [41, 42]. However, low N concentration and lower NO_3^-: NH_4^+ ratio than the normal (2:1) ratio in the medium, led to reduced concentrations of all three major PAs [42]. Thus, the type of N fertilizer impacts the PA metabolic pathway through its effect on the rate of NH_4^+-assimilation via the GS/GOGAT pathway.

A recent study using transgenic Arabidopsis seedlings harboring a mouse *ODC* gene to channel large amounts of N assimilates to Put via Orn showed that concentrations of most amino acids and all PAs were reduced in the N deficient medium both in transgenic and wild-type seedlings [7]. High NO_3^- in the growth medium had rather negative effect on Put and Spd concentrations in transgenic seedlings probably due to depletion of the substrate Orn. The results suggest that while N availability may not be a limiting factor for PA biosynthesis, additional N and C do have positive effects on PA metabolic pathway highlighting the dynamic coordination between N and C assimilation pathways. Moreover, most reactions in the Glu → PA pathway seem to be regulated posttranscriptionally as deduced from the results of m*ODC* expressing transgenic lines showing insignificant changes in the expression of most genes coding for the PA pathway enzymes compared to the nontransgenic controls [43, 44].

Since the precursors of PA biosynthesis (Orn and Arg) are derived from Glu produced largely by the GS/GOGAT pathway, the cellular concentrations of PAs presumably depend on the concentration of Glu being made from the N assimilated via the GS/GOGAT pathway. Consistent with this is the report that inhibition of GS activity with methionine sulfoximine caused a reduction in PA concentrations in poplar cells [39]. On the other hand, studies with poplar and red spruce cell cultures also showed that higher than optimal N availability does not affect the activity of GS, whereas low N and lower NO_3^-: NH_4^+ ratio than 2:1 led to a reduction in the activity of GS [42]. Though GOGAT plays a critical role in N and C metabolism [45], its direct connection with the metabolism of PAs has not been addressed.

2.2 Pathways of Glutamate to Polyamines (Glu → Orn → Put/PAs and Glu → Orn → Arg → Put/PAs)

In addition to abiotic and biotic stresses, three approaches (i.e., chemical inhibitors, genetic mutants, and transgenic manipulations) have been used to manipulate cellular PA contents to understand the regulation of PA biosynthesis in plants. Most abiotic stress responses have shown a rapid increase in PAs, particularly Put, which is often related to increase in ADC activity with or without increase in transcription of a specific member of the ADC gene family. There are several hypotheses regarding mechanisms of signal transduction from the moment of perception of stress to the various biochemical responses; however, the signaling pathway for the induction of PA biosynthesis has not been elucidated [46]. One thing consistent in this regard is that under stress conditions, several metabolites of the PA interacting pathways (Pro, GABA, Arg, and H_2O_2—Fig. 1) are often elevated concurrent with increase in PAs. However, the physiological importance of increase in PAs is not clearly understood; neither is the role of any of the other metabolites related to these interacting pathways. Two contrasting possibilities have been repeatedly mentioned in the literature; these are a protective role vs. being the cause of stress damage. While the enzymology, and in many cases, gene expression (measurements of the transcript and analysis via promoter::reporter approach) have been variously discussed, no clear hypothesis regarding the role of increased metabolic flux into the PA pathway, and its consequences on the remainder of the pathways shown in Fig. 1 are discussed.

The use of a variety of biochemical inhibitors of the PA biosynthetic enzymes has been a common tool to lower the cellular contents of PAs to study their role in a number of growth and developmental processes in plants. Although the expected results in terms of lowering PA contents are often seen, and the specific inhibitors do lead to a better knowledge of the active pathways in various cells/tissues of plants, the complex interactions of inhibitors with other metabolic pathways (i.e., their side effects) have not been reported in most cases. The lack of this information often interferes with getting a clear understanding of the role of PAs in specific functions. Thus, the use of inhibitors is less favored than molecular approaches, which allow more precise regulation of the biosynthetic pathways.

Specific gene mutagenesis (chemical mutagen treatment or producing knockouts using T-DNA insertions) has yielded highly useful information with respect to the physiological and developmental roles of specific genes regarding the tissue- and organ-specific expression of different members of a gene family producing a functional enzyme. Once again, its usefulness in delineating metabolic consequences of manipulating PAs in an organ/tissue-specific manner with this approach has been rather limited.

Genetic engineering approach using PA metabolic pathway genes, which targets the cellular concentration of PAs as opposed to the activities of genes as the selection criterion, has been used in numerous studies both to understand PA metabolism as well as their physiological and developmental roles. Our laboratory has studied the effects of genetic manipulation of the Orn → Put step by expression of a mouse (*Mus musculus*) cDNA coding for ODC driven by either a constitutive 35S promoter or an estradiol-inducible promoter on the metabolism of PAs and related subpathways in several different plant systems including tobacco, carrot, poplar and Arabidopsis. Detailed studies on long-term, nondifferentiating, and rapidly growing cell cultures of poplar (*Populus nigra* × *maximowiczii*) and seedlings of *A. thaliana* showed several-fold increase in cellular concentration of Put but only minor changes in the concentrations of Spd and Spm as compared to nontransgenic control cells [39]. Similar results were reported earlier when m*ODC* cDNA was overexpressed in tobacco [47] and carrot [48]. Since increased Put production in m*ODC*-expressing poplar cells was due to rapid conversion of the substrate Orn into Put, it was shown to be a limiting factor in controlling overall PA production in the transgenic cells [39, 49]. However, an interesting observation is that, in spite of its low cellular concentration, large amounts of Orn (to produce up to mmolar concentrations of Put in the cells) were sustainably produced. This led to the conclusion that Orn production (i.e., the Glu → Orn part of the pathway) responds largely to its rate of utilization (i.e., Orn → Put step) with its accumulation remaining very low. In other words, the flux of Glu → Orn increased several-fold under increased Orn → Put conversion. We further observed that this change in Glu → Orn flux in these cells/plants did not require increased expression of genes whose products code for various steps in this part of the pathway (Fig. 1). Based on these studies, we postulate that Orn, which is centrally located in the Glu/Pro, Put and Arg pathways, may not only be a key regulator of the flux of Glu into these metabolites but also influence the biosynthesis of Glu via concurrent assimilation of N and C [43, 44, 50].

Since the main source of Orn in these cells is Glu, which is also a precursor for Pro and Arg production (the latter entirely via Orn), both of which concurrently accumulate in large quantities; it was postulated that the total amount of Glu production in the transgenic cells and plants must be greatly increased. Moreover, as described above, since in actively growing cells/plants, all Glu must come from N assimilation, it was suggested that the increased conversion of Orn to Put must be accompanied by increased production of Glu from assimilation of inorganic N (NH_4^+ and/or NO_3^-) from the growth medium [43]. It also follows that competition for the Orn → Put step catalysed by mODC, and for δ-Orn aminotransferase (OAT) resulting in Orn to Δ^1-pyrroline-5-carboxylate (P5C) for Pro production, must be minimal, as its inhibition did not affect Put production in poplar cells [39, 49]. Based on the rather limited effects of *mODC* expression on Spd and Spm in the presence of higher cellular Put, it has been suggested that these metabolites are subject to stricter homeostatic regulation as compared to Put.

As Glu is the primary source of the amine group for most physiologically important amino acids, the reduction in its cellular concentration due to increased flux into Orn/Put in transgenic cells resulted in significant changes in the concentration of several amino acids, as well as many related N metabolites in the mODC-transgenic cells [50]. Also, since metabolic regulatory mechanisms for various N metabolites in a cell are quite variable, and often independent; the reduction in Glu concentration did not negatively impact all amino acids. Amino acids like Gln, His, and Orn, and others that use Glu/Gln as an amine donor including members of the serine family (Ser, Gly, and Cys), shikimate family (Phe and Trp), and some of those in Asp family (Asp, Met and Lys) were found to be significantly reduced [44, 50]. In contrast, concentrations of amino acids closely associated with the PA, Pro, and Arg pathway, metabolites associated with PA catabolism (GABA, succinate, and Ala), and some amino acids like Thr, Val, and Ile were higher in the transgenic cells expressing *mODC* [44, 50].

Based on the results with poplar cells and Arabidopsis seedlings, it was postulated that increased utilization of Glu in reactions involving Orn, Pro, Arg, Put, and GABA, the increased flux of Glu into Orn/Put must be accompanied by replenishment of cellular Glu content via its increased biosynthesis directly from increased N uptake. The facts (1) that key N rich metabolites did not decrease with high flux of Orn into Put, and (2) that sustained N assimilation is intimately dependent upon C availability; it was further suggested that the transgenic cells/plants must also absorb more C from the growth medium concurrent with N. If this holds true for enhanced N assimilation by the roots from the soil and by increased C assimilation from the environment via photosynthesis; manipulation of Put biosynthesis should provide a novel approach to increase both N and C assimilation (and in turn increased biomass production) in plants. Thus, this approach would help us tackle two of our important environmental goals—namely (1) increased N use efficiency of the plants, and decreasing N fertilizer loss from soil, and (2) increasing C assimilation to help reduce atmospheric CO_2 accumulation. There is, however, an important consideration when applying this technology to design food crops vs. biomass producing nonedible crops. While in the latter, high Put/PAs can be tolerated, the PA contents in foods are known to enhance tumor growth in animals and humans (http://www.cancernetwork.com/colorectal-cancer/high-dietary-polyamines-may-foster-colorectal-adenomas). Therefore, it will be prudent to apply this approach cautiously by using tissue/organ-specific promoters in food/feed crops, where the edible part of the plant does not accumulate high concentrations of PAs.

Important prerequisites for the above chain of events (Orn depletion leading to increased flux of Glu → Orn → increased uptake of N and C) are: (1) a mechanism for cells to sense and monitor Orn concentration, and (2) a signal transduction mechanism to regulate N and C assimilation. Neither a reliable mechanism for sensing cellular Orn concentration nor the signaling pathway from Orn to increased Glu biosynthesis and uptake of extra N and C is known. Of course, there is also the need for a mechanism for increased production of Orn itself. Since Orn is an intermediate for several products of this group of interacting pathways (including Arg and Pro—Fig. 1), and its cellular concentration is rather low as compared to all end products, it is likely that stimulation of Glu biosynthesis may be affected by significant increases in the biosynthesis of any or all these metabolites. This situation raises some very interesting challenges to speculate one or more mechanisms, and then to design experiments to test them.

The results with Arabidopsis, where we used both an inducible and a constitutive promoter showed that the two systems had similar responses indicating that the role of Orn is similar under long-term as well as short-term changes in its consumption. As pointed out earlier, *A. thaliana* does not directly convert Orn to Put because it lacks a functional ODC, thus m*ODC* transgene adds a new short-cut to Put production. The inducible expression system was valuable to study the effect of short-term induction of m*ODC* (i.e., increased utilization of Orn) on the regulation of PA biosynthesis [43], which would be expected to mimic the natural conditions that plants face when they are in the phase of rapid growth or being subjected to environmental stress—abiotic or biotic.

Dalton et al. [51] have recently reported that manipulation of the Orn-Put pathway by RNAi silencing of the *ODC* gene in tobacco caused a significant reduction in the concentration of all three major PAs (Put, Spd, and Spm) along with the transcript levels of genes encoding SAMDC, SAMS, and SPDS. The transgenic plants also displayed an increase in the concentrations of downstream metabolites including Orn, Arg, Asp, Glu, and Gln. These results clearly demonstrate a coordinated regulation of this entire group of pathways (Fig. 1). The expression levels of genes coding for the alternative pathway of Put biosynthesis (ADC) was elevated, along with concentration of Arg in the RNAi plants suggesting the presence of compensation for the reduced ODC activity, thereby maintaining the baseline levels of PAs for essential metabolic functions. These observation are consistent with the postulated role of Orn in regulating the entry of Glu into the Orn/Pro/Arg/Put pathway.

On the Put catabolic side of the pathway, the turnover rate (half-life) of this diamine (which includes its conversion to Spd + Spm, catabolic breakdown, and conjugation) in both poplar and Arabidopsis was not affected by its enhanced biosynthesis [14, 49]. Likewise, the $t_{1/2}$ of Spd and Spm turnover also remained unchanged [15]. The observation that increased Put catabolism was accompanied by increased GABA accumulation in the transgenic cells/plants, suggests the absence of a feedback mechanism for GABA production from Glu by GAD. Later studies [52] support these conclusions in that there was increased Put catabolism (lower Put concentration) accompanying GABA accumulation under salinity, and a reduction in GABA concentration followed by increased Put concentrations during recovery from salinity in soybean.

2.3 Polyamines and Proline Pathways Interactions (Glu → Pro ↔ Orn ← Glu)

The first step in Pro biosynthesis from Glu is catalysed by P5C synthetase (P5CS) to form Glu-γ-semialdehyde (GSA), that subsequently cyclizes to P5C [53] (Fig. 1). In the final step, P5C reductase (P5CR) converts P5C into Pro. Alternatively, Pro can be made from Orn by OAT, which transaminates GSA joining the route shown above to Pro [53]. Like the Glu↔Pro steps, the Pro↔Orn subpathway is also reversible, using the same enzyme OAT. Proline turnover on the other hand, involves its oxidation back to Glu by Pro dehydrogenase (PRODH) and P5C dehydrogenase (P5CDH) [54]. As discussed above, since the cellular concentration of Pro does not change significantly under conditions of high Put production in the m*ODC*-transgenic cells, the reactions involving Glu↔Pro and Pro↔Glu must not be affected by the increased flux of Glu into Orn/Put. The transcriptomic analysis in poplar and Arabidopsis also show the lack of significant changes in the expression of P5CS, P5CR, OAT, and ProDH [7, 44].

2.4 Interactions Between Polyamines and S-Adenosylmethionine (SAM) Metabolism

S-Adenosylmethionine (SAM) is a common source of numerous methylation, transsulfuration, and aminopropylation reactions to various substrates such as nucleic acids, proteins, lipids and secondary metabolites. In addition, significant amounts of SAM are used in the biosyntheses of ethylene and the higher PAs (Spd, Spm and tSpm); it is also used as a substrate for nicotianamine (NA) in plants [55, 56]. Consequently, the cellular concentrations of Spd and Spm are affected by the rate of dcSAM production from SAM, the use of SAM for various methylation reactions, and the biosynthesis of ethylene [57, 58]. The production of dcSAM is regulated by SAMDC, an enzyme with some unique properties [59], e.g., (1) a long 5′ UTR with two uORFs (called "small" and "tiny"), whose translation is regulated by PAs; (2) the ribosomal binding to the small uORF further regulates SAMDC translation; (3) the last codon of small uORF is the first nucleotide of the tiny uORF; and (4) the enzyme has a rather short half-life of <30 min. This monofunctional enzyme is coded by up to five genes in *A. thaliana*, some of which are expressed constitutively, and others in a cell/tissue/organ-specific manner [60]. Most of the SAM reactions regenerate Met through the Yang cycle [61].

A study involving Arabidopsis mutants of the two genes encoding 5'methylthioadenosine[1] nucleosidase (MTN) namely *mtn1* and *mtn2* confirmed that the Met cycle is linked to the regulation of PA homeostasis [62, 63]. The concentrations of Put and Spd were higher in the *mtn1-1mtn1* double mutant with reduced MTN activity concomitant with increased concentration of MTA and SAM/dcSAM [63]; however, the production of ethylene and nicotianamine was unchanged [62]. The *mtn2*-1 and *mtn2*-2 double mutants, had only a small reduction in MTN activity; their SAM/dcSAM, MTA and PA contents were similar to wild-type plants. A mutant for both MTN genes (*mtn*1-1/*mtn*2-1) was sterile; the sterility was partially reversed by PAs [63]. Decreases in the cellular contents of Spd and Spm in one of the MTN mutants was shown to be due to MTA-mediated feedback inhibition of the activities of SPDS and SPMS [63]. These results demonstrate that MTA is a crucial regulatory metabolite acting as a link between the Yang cycle and biosynthesis of PAs, and that MTN is a crucial enzyme that regulates this metabolism. The authors suggested that key targets affected by increased MTA content were tSPM and Spd-dependent hypusine formation in the eukaryotic initiation factor elF5A.

Earlier studies in potato had shown that downregulation of SAMDC activity results in reduced concentrations of not only Spd and Spm but also Put while significantly increasing the release of ethylene resulting in abnormal plant/tuber phenotypes [64]. Studies using m*ODC* transgenic poplar cell cultures overproducing Put or in tomato fruits expressing a transgenic yeast SAMDC gene did not show notable competition between ethylene and PA biosynthetic pathways for SAM [14, 65]. Similar results (i.e., the lack of competition between PAs and ethylene for SAM) were reported in tomato during climacteric ripening [66]. Consistent with this are the results of [15], who found that elevated PAs due to transgenic expression of yeast SAMDC to channel most of the SAM to PAs did not alter the rate of ethylene synthesis in tomato fruit. All these studies together indicate that the cellular pool of SAM might be sufficient in these cells/tissues to meet the concurrent demands of the two pathways.

On the other hand, expression of sense or antisense strands of *SAMDC* cDNA for up- or down-regulation of *SAMDC* expression in Arabidopsis showed competitive relationships between the two pathways [11]. A competition between the two pathways was also observed in the developing kernels of maize [67] in response to

[1] Methylthioadenosine (MTA) is a byproduct of ethylene, PA, and nicotianamine biosynthesis

drought at grain filling stage in rice [68] and wheat [61]. Taken together, interaction between PAs and ethylene seems essential for the regulation of certain developmental processes in plants and the utilization of SAM in these pathways may be subjected to different regulatory mechanisms. Recently, a study involving miRNA396b from trifoliate orange (*Poncirus trifoliate* L.) showed that this miRNA positively affects cold tolerance via the modulation of ethylene-PA homeostasis; i.e., by inhibiting the evolution of ethylene while increasing the production of PAs [69]. The involvement of miRNA in the regulation of ethylene-PA interaction highlights the functional complexity of the PA metabolic pathway in terms of going beyond the Glu-Orn-Put-Arg-GABA-amino acids pathway.

Another common product made from SAM in plants is nicotianamine (NA), a metal chelator involved in the intra and intercellular transport and accumulation of cations, like Fe^{2+}, Fe^{3+} and Zn^{2+} [55, 56, 70, 71]. The enzyme NA synthase (NAS) trimerizes SAM to form NA. Since SAM is present in large quantities in the presence of these metal ions, it is likely to affect the availability of SAM for PA biosynthesis as well.

2.5 Polyamine Catabolism and the Regulation of PA Metabolic Pathway

Hydrogen peroxide and NO are two major signaling molecules that play critical roles in PA-induced regulation of various biological processes and responses to biotic and abiotic stresses [18]. While PAs have been advocated to protect the cells against oxidative damage by reactive oxygen species (ROS), their catabolism makes them susceptible to oxidative damage by generation of H_2O_2 and NO as well [72–74]. The formation of H_2O_2 from PA catabolism in plants is well established as opposed to PA-induced NO production; neither however is exclusively produced directly from the PAs. The catabolic breakdown of PAs generally produces H_2O_2 and 4-aminobutanal, which is the precursor of GABA (Fig. 1). Similarly, in high-Put transgenic poplar cells (from constitutive *mODC* expression), high concentration of both H_2O_2 and GABA were reported [49, 73]. The increase in H_2O_2 accumulation in transgenic cells was accompanied by increased activities of enzymes responsible for ROS scavenging, e.g., glutathione reductase (GR) and monodehydroascorbate reductase (MDHAR) [73]. Likewise, the concentration of ROS scavenging metabolites including reduced glutathione (GSH) was significantly decreased in transgenic cells indicating the creation of an elevated state of oxidative stress in the cells, which eventually led to a substantial damage to the cell membranes [73].

Nitric oxide plays a vital role in plant growth and development as well as plant immunity to pathogens. Several pathways have been suggested as routes for NO production in plants, including the metabolism of PAs [75]. Among the enzyme-based pathways that contribute to NO biosynthesis in plant cells are one involving nitrate reductase (NR) and the nitric oxide-associated-1 (NOA1) protein [76, 77]. The experimental evidence for NO production from this pathway involved the use of triple mutants of two NR structural genes (NIA1 and NIA2 and NOA1) to demonstrate the presence of an alternative pathway for NO biosynthesis [77]. Polyamine catabolism also produces NO via protein-nitrosylation [74, 78]. The latter study suggests that CuAO may be directly involved in the biosynthesis of NO from PAs. The Arabidopsis mutants of gene coding for this enzyme displayed reduced PA-induced production of NO suggesting that CuAOs might be a putative candidate that plays a role in the biosynthesis of NO from PAs. This study also reported that CuAO participates in ABA signaling. The PA-dependent signal transduction system involving NO has been suggested to regulate plant development including root growth, defense responses and abiotic stress responses. Therefore, the mechanism by which PAs regulate different responses might be via cross-linking the signaling intermediates (H_2O_2 or NO) and hormonal signaling pathways including ABA and methyl jasmonate; the latter is well known to initiate downstream response machinery of stress response.

2.6 Interaction Between Polyamines and the TCA Cycle

The TCA cycle is the primary supplier of the C backbones for assimilation of N. Biosynthesis of Glu, the primary product of N assimilation, requires α-ketoglutarate (α-KG) as a substrate for the synthesis of all amino acids in the Glu family plus the main components of the PA biosynthetic pathway including Orn and Arg. Oxaloacetate (OAA) is another TCA cycle intermediate that is important for the biosynthesis of Asp family amino acids including Met, which as discussed above, supplies dcSAM for synthesis of higher PAs. The catabolism of PAs as well as decarboxylation of Glu contribute succinate to the TCA cycle via what is known as the GABA shunt. Thus the GABA shunt lies at the interface of the C and N metabolic interactions in recycling of C and N from the PAs [75]. GABA is also an essential component of plant responses to stress conditions, and plays a signaling role in plants for other physiological processes [19–21]. Its biosynthesis occurs mainly from Glu and PA catabolism though it has been recently indicated that it could also be synthesized nonenzymatically from Pro under oxidative stress conditions. This happens when Pro acting as a hydroxyl radicals (OH) scavenger removes H from amine group leading to the spontaneous decarboxylation of Pro resulting in the formation of Δ^1-pyrroline, a precursor of GABA [79]. Besides creating a crucial link with the TCA cycle, GABA catabolism indirectly contributes to the sustainable regeneration of Glu (via supplying α-KG for its biosynthesis), and also, other amino acids [50]. These interactions between TCA cycle and both the anabolic and the catabolic pathways of PAs and amino acids signify the important contribution of these networks to maintain intracellular homeostatic equilibrium of N and C metabolites in plants.

2.7 Interaction Between Metabolic Pathways of Free and Conjugated PAs

In addition to the biosynthetic and catabolic pathways of PAs, there also are reactions of PAs with several phenolic compounds (phenolamides) in plants to produce PA conjugates, which not only produce some important secondary metabolites but also contribute to PA homeostasis [23, 80]. Additional metabolic aspects of PA conjugation are the regulation of C-rich phenolic intermediates and the end products, which play important roles in plant biotic interactions ranging from bacterial and fungal pathogenesis to herbivore avoidance [81–83]. Thus, the cellular concentrations of phenolamides and PAs may be interdependent as the biosynthesis of the former depends on the latter. Whether conjugated PAs contribute (through their reversal to free PAs) to the physiologically active cellular PA pool is still unclear. A recent study involving RNAi-mediated silencing of tobacco *ODC* gene showed a significant reduction in the concentration of phenolamides (caffeoyl-Put—CP and dicaffeoyl-Spd—DCP) along with the corresponding PAs, indicating that ODC may play an essential role in the regulation of phenolamide biosynthesis [51].

A study with Arabidopsis stamens showed that knockout of the Spd-HCA transferase (SHT), responsible for the biosynthesis of HCA conjugates of Spd did not affect the free Put, Spd and Spm concentrations, despite a drastic reduction in the concentration of HCA-bound Spd as compared to the wild-type plants [80]. The lack of increase in the substrate of SHT (Spd) in the *sht* knockouts was suggested to be due to a tight regulation of intracellular titers of free PAs. This is in line with the conclusions of our studies (cited earlier) with Arabidopsis, poplar, carrot and tobacco, which also concluded that higher PA titers in plants were more tightly regulated than Put. It was further indicated that this regulation did not operate at the level of transcription of the PA biosynthetic genes but perhaps involved posttranscriptional/translational regulation at the biochemical level.

Polyamine conjugates like CP and DCP were found to increase in response to either actual herbivore attack or simulations of it [84]. Such increases in the accumulation of PA conjugates appeared to be regulated by a transcriptional master regulator, MYB8 in *Nicotiana attenuata* based on the results from a microarray analysis, showing increased transcripts of several key enzyme genes downstream of the PA and phenylpropanoid biosynthetic pathways in a MYB8-dependent manner [85]. However, the concentrations of free PAs were not determined in this study, and hence it is difficult to discern what impact the metabolic pathway of PA-conjugates has on the metabolism of its free PA counterpart.

2.8 Interaction Between PA Metabolic Pathway and Plant Hormones

As much as the PAs are associated with stress response, ABA is implicated as one of the major players in plant response to stress. Coordination between PAs and ABA in the regulation of response to environmental stressors has been reported by several studies; however, little details are available on the molecular mechanism of their interactions and crosstalk in most cases. The involvement of ABA in transcriptional upregulation of genes coding for *SPDS1*, *SPMS* and *ADC2* in Arabidopsis has been suggested concurrent with changes in Put concentration under drought conditions [86]. Conversely, Cuevas et al. [87] showed that Put can modulate transcriptional upregulation of an ABA biosynthetic gene, *9-cis-epoxycarotenoid dioxygenase-3* (*NCED3*) and other ABA-regulated genes thereby increasing the concentration of ABA during cold stress. These findings signify the existence of positive feedback loop between ABA and Put in response to cold stress [2]. Furthermore, ABA signaling has also been shown to induce accumulation of PAs as well as its catabolism, thereby increasing the rate of H_2O_2 generation leading to activated stress responses in grape [88]. A crosstalk between NO and ABA has been suggested to play a role during Arabidopsis seed germination with endogenous increase in NO ameliorating the inhibitory effect of ABA on seed germination via induction of its catabolism [89]. Therefore, the molecular mechanism of PA-induced regulation of stress responses may likely involve its interaction with ABA via the signaling molecules H_2O_2 and NO.

Other phytohormones, such as auxins and cytokinins, have also been suggested to interact with PAs to regulate certain physiological functions. Previous studies suggested that auxin could stimulate PA biosynthesis especially during early root initiation stage [90, 91]. The involvement of PAs in modulation of cytokinin-regulated physiological phenomena via its effect on the expression of cytokinin response genes is also known [92, 93].

3 Methods Used for Studying PA Interacting Pathways and the Future

It is apparent from the above discussion that PAs themselves and their metabolic pathway substrates and products interact with a wide network of cellular metabolic pathways, such as the TCA cycle, GABA shunt, N assimilation, modulation of phenolics through conjugation and binding, modulating SAM synthesis and its utilization, signaling pathways involving phytohormones, stress and senescence related metabolism, and oxidative stress pathways to facilitate the regulation of various biological functions in which they play vital roles. Due to their accumulation in relatively large quantities (up to mmolar in concentrations), PAs are also capable of affecting total soluble N balance in the cells; i.e., at times they act as N storage compounds and at other times, they release the stored N.

These interactions may be manifested by changes in the transcription of genes that code for PA metabolic enzymes, the level of active proteins/enzymes, and the substrates used for PA biosynthesis. They also influence directly (through hypusilation of elF5) or indirectly (by interactions with mRNA) the process of translation, including that of their own biosynthetic enzyme (e.g., interaction with the uORF in the 5′UTR of SAMDC). To this end, some of the recent studies have highlighted substantial changes in the transcriptome and metabolome of the interacting pathways, resulting from genetic perturbation of a single step in the PA biosynthetic pathway [7, 44]. The reverse is also true where "Omics" studies targeted at other genotypic or phenotypic variations in plants have revealed significant changes in plant PA metabolism where the transcripts of PA metabolic genes, as well as precursors and intermediates of the PA metabolic pathway were related to specific growth and development phenomena being investigated. It is time that we use the combined approaches of next-generation sequencing (NGS) of DNA and RNA, and biochemical analyses using tools of proteomics and metabolomics (GC-MS-MS, LC-MS-MS, NMR, stable- and radio-isotope labeling, etc.) to unravel biochemical mechanisms of the complex cellular responses accompanying these alterations in the PA metabolic network. These approaches can help discern changes in the rates of biosynthesis of thousands of genes and metabolites of not only the interacting pathways but also other previously considered unrelated pathways; and hence, may lead to the discovery of some novel relationships among PA and other pathways.

While the mutants and constitutive transgene expression approaches have been commonly used, and provide excellent tools to delineate metabolic interactions of the PA metabolic pathways and other pathways, a major drawback of these two approaches is that they don't often mimic the real-life dynamic and transient changes in cellular metabolite concentrations during short-term responses of plants to internal (e.g., growth, development, and differentiation), and to external stimuli (abiotic and biotic stresses, diurnal fluctuations of C and N availability, etc.). In this regard, studies involving well-designed regulated promoters (inducible, cell, tissue, and organ specific, and developmentally regulated) must be planned in conjunction with the modern tools of transcriptomic, proteomic, and metabolomics analyses. Furthermore, the aspects of PA metabolism related to the flux rates of various metabolites in the PA associated pathways, have received minimal attention. These studies capable of monitoring small changes in N- and C-rich metabolites together would contribute immensely to our understanding of the flow of various metabolites and intermediates into different interactive and branched chain pathways where a metabolite (e.g., Glu, Orn, Arg, and SAM) serves as a common precursor for several pathways that produce products with either competing (e.g., SAM for PAs and ethylene) or complementary biological functions (e.g., H_2O_2, ABA, NO). These advances, combined with improvements in bioinformatics tools for analysis of large datasets, could be utilized to better understand the complexities of PA metabolic networks.

References

1. Kusano T, Berberich T, Tateda C, Takahashi Y (2008) Polyamines: essential factors for growth and survival. Planta 228:367–381
2. Alcázar R, Altabella T, Marco F, Bortolotti C, Reymond M, Koncz C, Carrasco P, Tiburcio AF (2010) Polyamines: molecules with regulatory functions in plant abiotic stress tolerance. Planta 231:1237–1249
3. Tiburcio AF, Altabella T, Bitrión M, Alcázar R (2014) The roles of polyamines during the lifespan of plants: from development to stress. Planta 240:1–18
4. Moschou PN, Wu J, Cona A, Tavladoraki P, Angelini R, Roubelakis-Angelakis KA (2012) The polyamines and their catabolic products are significant players in the turnover of nitrogenous molecules in plants. J Exp Bot 63:5003–5015
5. Moschou PN, Paschalidis KA, Roubelakis-Angelakis KA (2008) Plant polyamine catabolism: the state of the art. Plant Signal Behav 3:1061–1066
6. Bitrión M, Zarza X, Altabella T, Tiburcio AF, Alcózar R (2012) Polyamines under abiotic stress: metabolic crossroads and hormonal crosstalks in plants. Metabolites 2:516–528
7. Majumdar R, Barchi B, Turlapati SA, Gagne M, Minocha R, Long S, Minocha SC (2016) Glutamate, ornithine, arginine, proline, and polyamine metabolic interactions: the pathway is regulated at the post-transcriptional level. Front Plant Sci 7:78
8. Hanfrey C, Sommer S, Mayer MJ, Burtin D, Michael AJ (2001) Arabidopsis polyamine biosynthesis: absence of ornithine decarboxylase and the mechanism of arginine decarboxylase activity. Plant J 27:551–560
9. Alcázar R, Tiburcio AF (2016) Polyamines in stress protection: applications in agriculture. In: Abiotic stress response in plants. Wiley-VCH Verlag GmbH & Co, KGaA, pp 411–422
10. Shao L, Majumdar R, Minocha S (2012) Profiling the aminopropyltransferases in plants: their structure, expression and manipulation. Amino Acids 42:813–830
11. Hu WW, Gong H, Pua EC (2005) The pivotal roles of the plant S-adenosylmethionine decarboxylase 5′ untranslated leader sequence in regulation of gene expression at the transcriptional and posttranscriptional levels. Plant Physiol Biochem 138:276–286
12. Bunsupa S, Katayama K, Ikeura E, Oikawa A, Toyooka K, Saito K, Yamazaki M (2012) Lysine decarboxylase catalyzes the first step of quinolizidine alkaloid biosynthesis and coevolved with alkaloid production in leguminosae. Plant Cell 24:1202–1216
13. Persson L (1977) Evidence of decarboxylation of lysine by mammalian ornithine decarboxylase. Acta Physiol Scand 100:424–429
14. Quan Y, Minocha R, Minocha SC (2002) Genetic manipulation of polyamine metabolism in poplar II: effects on ethylene biosynthesis. Plant Physiol Biochem 40:929–937
15. Lasanajak Y, Minocha R, Minocha SC, Goyal R, Fatima T, Handa AK, Mattoo AK (2014) Enhanced flux of substrates into polyamine biosynthesis but not ethylene in tomato fruit engineered with yeast S-adenosylmethionine decarboxylase gene. Amino Acids 46:729–742
16. Tavladoraki P, Cona A, Federico R, Tempera G, Viceconte N, Saccoccio S, Battaglia V, Toninello A, Agostinelli E (2012) Polyamine catabolism: target for antiproliferative therapies in animals and stress tolerance strategies in plants. Amino Acids 42:411–426
17. Cheeseman JM (2007) Hydrogen peroxide and plant stress: a challenging relationship. Plant Stress 1:4–15
18. Gupta K, Sengupta A, Chakraborty M, Gupta B (2016) Hydrogen peroxide and polyamines act as double edged swords in plant abiotic stress responses. Front Plant Sci 7:1343
19. Bouché N, Fromm H (2004) GABA in plants: just a metabolite? Trends Plant Sci 9:110–115
20. Bown AW, Shelp BJ (2016) Plant GABA: not just a metabolite. Trends Plant Sci 21:811–813
21. Fait A, Fromm H, Walter D, Galili G, Fernie AR (2008) Highway or byway: the metabolic role of the GABA shunt in plants. Trends Plant Sci 13:14–19
22. Tisi A, Federico R, Moreno S, Lucretti S, Moschou PN, Roubelakis-Angelakis KA, Angelini R, Cona A (2011) Perturbation of

polyamine catabolism can strongly affect root development and xylem differentiation. Plant Physiol Biochem 157:200–215

23. Bassard JE, Ullmann P, Bernier F, Werck-Reichhart D (2010) Phenolamides: bridging polyamines to the phenolic metabolism. Phytochemistry 71:1808–1824
24. Peng M, Gao Y, Chen W, Wang W, Shen S, Shi J, Wang C, Zhang Y, Zou L, Wang S, Wan J, Liu X, Gong L, Luo J (2016) Evolutionarily distinct BAHD N-acyltransferases are responsible for natural variation of aromatic amine conjugates in rice. Plant Cell 28:1533–1550
25. Aloisi I, Cai G, Serafini-Fracassini D, Del Duca S (2016) Polyamines in pollen: from microsporogenesis to fertilization. Front Plant Sci 7:155
26. Del Duca S, Serafini-Fracassini D, Cai G (2014) Senescence and programmed cell death in plants: polyamine action mediated by transglutaminase. Front Plant Sci 5:120
27. Luo J, Fuell C, Parr A, Hill L, Bailey P, Elliott K, Fairhurst SA, Martin C, Michael AJ (2009) A novel polyamine acyltransferase responsible for the accumulation of spermidine conjugates in Arabidopsis seed. Plant Cell 21:318–333
28. Tanabe K, Hojo Y, Shinya T, Galis I (2016) Molecular evidence for biochemical diversification of phenolamide biosynthesis in rice plants. J Integr Plant Biol 58:903–913
29. Igarashi K, Kashiwagi K (1999) Polyamine transport in bacteria and yeast. Biochem J 344 (Pt 3):633–642
30. Fujita M, Shinozaki K (2015) Polyamine transport systems in plants. In: Kusano T, Suzuki H (eds) Polyamines: a universal molecular nexus for growth, survival, and specialized metabolism. Springer Japan, Tokyo, pp 179–185
31. Mulangi V, Phuntumart V, Aouida M, Ramotar D, Morris P (2012) Functional analysis of OsPUT1, a rice polyamine uptake transporter. Planta 235:1–11
32. Fujita M, Fujita Y, Iuchi S, Yamada K, Kobayashi Y, Urano K, Kobayashi M, Yamaguchi-Shinozaki K, Shinozaki K (2012) Natural variation in a polyamine transporter determines paraquat tolerance in Arabidopsis. Proc Natl Acad Sci U S A 109:6343–6347
33. Fujita M, Shinozaki K (2014) Identification of polyamine transporters in plants: paraquat transport provides crucial clues. Plant Cell Physiol 55:855–861
34. Li J, Mu J, Bai J, Fu F, Zou T, An F, Zhang J, Jing H, Wang Q, Li Z, Yang S, Zuo J (2013) Paraquat resistant1, a Golgi-localized putative transporter protein, is involved in intracellular transport of paraquat. Plant Physiol 162:470–483
35. Sagor GH, Berberich T, Kojima S, Niitsu M, Kusano T (2016) Spermine modulates the expression of two probable polyamine transporter genes and determines growth responses to cadaverine in Arabidopsis. Plant Cell Rep 35:1247–1257
36. Shen Y, Ruan Q, Chai H, Yuan Y, Yang W, Chen J, Xin Z, Shi H (2016) The arabidopsis polyamine transporter LHR1/PUT3 modulates heat responsive gene expression by enhancing mRNA stability. Plant J 88:1006–1021
37. Altman A, Levin N (1993) Interactions of polyamines and nitrogen nutrition in plants. Physiol Plant 89:653–658
38. Le Rudulier D, Goas G (1977) La diamine oxydase dans les jeunes plantes de Glycine max. Phytochemistry 16:509–511
39. Bhatnagar P, Glasheen BM, Bains SK, Long SL, Minocha R, Walter C, Minocha SC (2001) Transgenic manipulation of the metabolism of polyamines in poplar cells. Plant Physiol 125:2139–2153
40. Houdusse F, Zamarreño AM, Garnica M, García-Mina J (2005) The importance of nitrate in ameliorating the effects of ammonium and urea nutrition on plant development: the relationships with free polyamines and plant proline contents. Funct Plant Biol 32:1057–1067
41. Minocha R, Lee JS, Long S, Bhatnagar P, Minocha SC (2004) Physiological responses of wild type and putrescine-overproducing transgenic cells of poplar to variations in the form and concentration of nitrogen in the medium. Tree Physiol 24:551–560
42. Serapiglia MJ, Minocha R, Minocha SC (2008) Changes in polyamines, inorganic ions and glutamine synthetase activity in response to nitrogen availability and form in red spruce (Picea Rubens). Tree Physiol 28:1793–1803
43. Majumdar R, Shao L, Minocha R, Long S, Minocha SC (2013) Ornithine: the overlooked molecule in the regulation of polyamine metabolism. Plant Cell Physiol 54:990–1004
44. Page AF, Cseke LJ, Minocha R, Turlapati SA, Podila GK, Ulanov A, Li Z, Minocha SC (2016) Genetic manipulation of putrescine biosynthesis reprograms the cellular transcriptome and the metabolome. BMC Plant Biol 16:113
45. Lu Y, Luo F, Yang M, Li X, Lian X (2011) Suppression of glutamate synthase genes significantly affects carbon and nitrogen metabolism in rice (*Oryza sativa* L.) Sci China Life Sci 54:651–663
46. Minocha R, Majumdar R, Minocha SC (2014) Polyamines and abiotic stress in plants: a

complex relationship. Front Plant Sci 5:175. doi:10.3389/fpls.2014.00175

47. DeScenzo RA, Minocha SC (1993) Modulation of cellular polyamines in tobacco by transfer and expression of mouse ornithine decarboxylase cDNA. Plant Mol Biol 22:113–127
48. Bastola DR, Minocha SC (1995) Increased putrescine biosynthesis through transfer of mouse ornithine decarboxylase cDNA in carrot promotes somatic embryogenesis. Plant Physiol 109:63–71
49. Bhatnagar P, Minocha R, Minocha SC (2002) Genetic manipulation of the metabolism of polyamines in poplar cells. The regulation of putrescine catabolism. Plant Physiol 128:1455–1469
50. Mohapatra S, Minocha R, Long S, Minocha SC (2010) Transgenic manipulation of a single polyamine in poplar cells affects the accumulation of all amino acids. Amino Acids 38:1117–1129
51. Dalton HL, Blomstedt CK, Neale AD, Gleadow R, DeBoer KD, Hamill JD (2016) Effects of down-regulating ornithine decarboxylase upon putrescine-associated metabolism and growth in *Nicotiana tabacum* L. J Exp Bot 67:3367–3381
52. Xing SG, Jun YB, Hau ZW, Liang LY (2007) Higher accumulation of gamma-aminobutyric acid induced by salt stress through stimulating the activity of diamine oxidases in *Glycine max* (L.) Merr. Roots. Plant Physiol Biochem 45:560–566
53. Szabados L, Savoure A (2010) Proline: a multifunctional amino acid. Trends Plant Sci 15:89–97
54. Ribarits A, Abdullaev A, Tashpulatov A, Richter A, Heberle-Bors E, Touraev A (2007) Two tobacco proline dehydrogenases are differentially regulated and play a role in early plant development. Planta 225:1313–1324
55. Higuchi K, Watanabe S, Takahashi M, Kawasaki S, Nakanishi H, Nishizawa NK, Mori S (2001) Nicotianamine synthase gene expression differs in barley and rice under Fe-deficient conditions. Plant J 25:159–167
56. Shojima S, Nishizawa NK, Fushiya S, Nozoe S, Kumashiro T, Nagata T, Ohata T, Mori S (1989) Biosynthesis of nicotianamine in the suspension-cultured cells of tobacco (*Nicotiana megalosiphon*). Biometals 2:142–145
57. Sauter M, Moffatt B, Saechao MC, Hell R, Wirtz M (2013) Methionine salvage and S-adenosylmethionine: essential links between sulfur, ethylene and polyamine biosynthesis. Biochem J 451:145–154
58. Harpaz-Saad S, Yoon GM, Mattoo AK, Kieber JJ (2012) The formation of ACC and competition between PAs and ethylene for SAM. Annu Plant Rev 44:53–81
59. Hanfrey C, Elliott KA, Franceschetti M, Mayer MJ, Illingworth C, Michael AJ (2005) A dual upstream open reading frame-based autoregulatory circuit controlling polyamine-responsive translation. J Biol Chem 280:39229–39237
60. Majumdar R (2011) Polyamine metabolism in Arabidopsis: transgenic manipulation and gene expression. University of New Hampshire, Durham, NH, USA
61. Yang W, Yin Y, Li Y, Cai T, Ni Y, Peng D, Wang Z (2014) Interactions between polyamines and ethylene during grain filling in wheat grown under water deficit conditions. J Plant Growth Regul 72:189–201
62. Bürstenbinder K, Waduwara I, Schoor S, Moffatt BA, Wirtz M, Minocha SC, Oppermann Y, Bouchereau A, Hell R, Sauter M (2010) Inhibition of 5′-methylthioadenosine metabolism in the Yang cycle alters polyamine levels, and impairs seedling growth and reproduction in Arabidopsis. Plant J 62:977–988
63. Waduwara-Jayabahu I, Oppermann Y, Wirtz M, Hull ZT, Schoor S, Plotnikov AN, Hell R, Sauter M, Moffatt BA (2012) Recycling of methylthioadenosine is essential for normal vascular development and reproduction in Arabidopsis. Plant Physiol Biochem 158:1728–1744
64. Kumar A, Taylor MA, Arif SAM, Davies HV (1996) Potato plants expressing antisense and sense S-adenosylmethionine decarboxylase (SAMDC) transgenes show altered levels of polyamines and ethylene: antisense plants display abnormal phenotypes. Plant J 9:147–158
65. Mehta RA, Cassol T, Li N, Ali N, Handa AK, Mattoo AK (2002) Engineered polyamine accumulation in tomato enhances phytonutrient content, juice quality, and vine life. Nat Biotechnol 20:613–618
66. Van de Poel B, Bulens I, Oppermann Y, Hertog ML, Nicolai BM, Sauter M, Geeraerd AH (2013) S-adenosyl-L-methionine usage during climacteric ripening of tomato in relation to ethylene and polyamine biosynthesis and transmethylation capacity. Physiol Plant 148:176–188
67. Feng HY, Wang ZM, Kong FN, Zhang MJ, Zhou SL (2011) Roles of carbohydrate supply and ethylene, polyamines in maize kernel set. J Integr Plant Biol 53:388–398
68. Chen T, Xu Y, Wang J, Wang Z, Yang J, Zhang J (2013) Polyamines and ethylene interact in

rice grains in response to soil drying during grain filling. J Exp Bot 64:2523–2538

69. Zhang X, Wang W, Wang M, Zhang HY, Liu JH (2016) The miR396b of *Poncirus trifoliata* functions in cold tolerance by regulating ACC oxidase gene expression and modulating ethylene-polyamine homeostasis. Plant Cell Physiol 57:1865–1878
70. von Wiren N, Klair S, Bansal S, Briat JF, Khodr H, Shioiri T, Leigh RA, Hider RC (1999) Nicotianamine chelates both FeIII and FeII. Implications for metal transport in plants. Plant Physiol Biochem 119:1107–1114
71. Shi S-Q, Shi Z, Jiang Z-P, Qi L-W, Sun X-M, Li C-X, Liu J-F, Xiao W-F, Zhang S-G (2010) Effects of exogenous GABA on gene expression of *Caragana intermedia* roots under NaCl stress: regulatory roles for H_2O_2 and ethylene production. Plant Cell Environ 33:149–162
72. Das KC, Misra HP (2004) Hydroxyl radical scavenging and singlet oxygen quenching properties of polyamines. Mol Cell Biochem 262:127–133
73. Mohapatra S, Minocha R, Long S, Minocha SC (2009) Putrescine overproduction negatively impacts the oxidative state of poplar cells in culture. Plant Physiol Biochem 47:262–271
74. Tun NN, Santa-Catarina C, Begum T, Silveira V, Handro W, Floh EI, Scherer GF (2006) Polyamines induce rapid biosynthesis of nitric oxide (NO) in *Arabidopsis thaliana* seedlings. Plant Cell Physiol 47:346–354
75. Michaeli S, Fromm H (2015) Closing the loop on the GABA shunt in plants: are GABA metabolism and signaling entwined? Front Plant Sci 6:419
76. Gupta KJ, Fernie AR, Kaiser WM, van Dongen JT (2011) On the origins of nitric oxide. Trends Plant Sci 16:160–168
77. Lozano-Juste J, León J (2010) Enhanced abscisic acid-mediated responses in nia1nia2noa1-2 triple mutant impaired in NIA/NR- and AtNOA1-dependent nitric oxide biosynthesis in Arabidopsis. Plant Physiol 152:891–903
78. Wimalasekera R, Villar C, Begum T, Scherer GF (2011) Copper amine oxidase1 (CuAO1) of Arabidopsis thaliana contributes to abscisic acid- and polyamine-induced nitric oxide biosynthesis and abscisic acid signal transduction. Mol Plant 4:663–678
79. Signorelli S, Dans PD, Coitino EL, Borsani O, Monza J (2015) Connecting proline and gamma-aminobutyric acid in stressed plants through non-enzymatic reactions. PLoS One 10:e0115349
80. Fellenberg C, Ziegler J, Handrick V, Vogt T (2012) Polyamine homeostasis in wild type and phenolamide deficient *Arabidopsis thaliana* stamens. Front Plant Sci 3:180
81. Martin-Tanguy J (1997) Conjugated polyamines and reproductive development: biochemical, molecular and physiological approaches. Physiol Plant 100:675–688
82. Takahashi Y (2016) The role of polyamines in plant disease resistance. Environ Control Biol 54:5
83. Walters DR (2000) Polyamines in plant–microbe interactions. Physiol Mol Plant Pathol 57:137–146
84. Kaur H, Heinzel N, Schottner M, Baldwin IT, Galis I (2010) R2R3-NaMYB8 regulates the accumulation of phenylpropanoid-polyamine conjugates, which are essential for local and systemic defense against insect herbivores in *Nicotiana attenuata*. Plant Physiol 152:1731–1747
85. Onkokesung N, Gaquerel E, Kotkar H, Kaur H, Baldwin IT, Galis I (2012) MYB8 controls inducible phenolamide levels by activating three novel hydroxycinnamoyl-coenzyme A: polyamine transferases in *Nicotiana attenuata*. Plant Physiol 158:389–407
86. Alcázar R, Cuevas JC, Patron M, Altabella T, Tiburcio AF (2006) Abscisic acid modulates polyamine metabolism under water stress in *Arabidopsis thaliana*. Physiol Plant 128:448–455
87. Cuevas JC, López-Cobollo R, Alcázar R, Zarza X, Koncz C, Altabella T, Salinas J, Tiburcio AF, Ferrando A (2009) Putrescine as a signal to modulate the indispensable ABA increase under cold stress. Plant Signal Behav 4:219–220
88. Toumi I, Moschou PN, Paschalidis KA, Bouamama B, Ben Salem-Fnayou A, Ghorbel AW, Mliki A, Roubelakis-Angelakis KA (2010) Abscisic acid signals reorientation of polyamine metabolism to orchestrate stress responses via the polyamine exodus pathway in grapevine. J Plant Physiol 167:519–525
89. Sanz L, Albertos P, Mateos I, Sanchez-Vicente I, Lechon T, Fernandez-Marcos M, Lorenzo O (2015) Nitric oxide (NO) and phytohormones crosstalk during early plant development. J Exp Bot 66:2857–2868
90. Heloir MC, Kevers C, Hausman JF, Gaspar T (1996) Changes in the concentrations of auxins and polyamines during rooting of in-vitro-propagated walnut shoots. Tree Physiol 16:515–519
91. Mendes AFS, Cidade LC, Otoni WC, Soares-Filho WS, Costa MGC (2011) Role of auxins,

polyamines and ethylene in root formation and growth in sweet orange. Biol Plant 55:375

92. Legocka J, Żarnowska A (2000) Role of polyamines in the cytokinin-dependent physiological processes II. Modulation of polyamine levels during cytokinin-stimulated expansion of cucumber cotyledons. Acta Physiol Plant 22:395–401

93. Anwar R, Mattoo AK, Handa AK (2015) Polyamine interactions with plant hormones: crosstalk at several levels. In: Kusano T, Suzuki H (eds) Polyamines: a universal molecular nexus for growth, survival, and specialized metabolism. Springer Japan, Tokyo, pp 267–302

Chapter 2

Molecules for Sensing Polyamines and Transducing Their Action in Plants

Tomonobu Kusano, G.H.M. Sagor, and Thomas Berberich

Abstract

Polyamines play important roles in growth, development, and adaptive responses to various stresses. In the past two decades, progress in plant polyamine research has accelerated, and the key molecules and components involved in many biological events have been identified. Recently, polyamine sensors used to detect polyamine-enriched foods and polyamines derived from degrading flesh were identified in fly and zebrafish, respectively. Work has begun to identify such molecules in plants as well. Here, we summarize the current knowledge about polyamines in plants. Furthermore, we discuss the roles of key molecules, such as calcium ions, reactive oxygen species, nitric oxide, γ-aminobutyric acid, polyamine transporters, and the mitogen-activated protein kinase cascade, from the viewpoint of polyamine action.

Key words Calcium ion, Hydrogen peroxide, Mitogen-activated protein kinase cascade, Nitric oxide, Polyamine sensor, Reactive oxygen species

1 Introduction

The polyamines (PAs) diamine putrescine (Put), triamine spermidine (Spd), and tetraamine spermine (Spm) play essential and/or critical roles in living organisms [1] including plants [2–7]. These molecules play important roles in normal physiological processes, as well as various environmental stress responses [2–7]. Another tetraamine, thermospermine (T-Spm), was initially identified in a thermophilic microorganism [8] and was shown to play an important role in stem elongation in plants [9, 10]. Another diamine, cadaverine (Cad), is a precursor of quinolizidine alkaloids [11] whose role as an effector for plant development and environmental responses has been evaluated [12, 13]. PA biosynthesis begins with arginine and ornithine, which are converted to Put by arginine decarboxylase (ADC) and ornithine decarboxylase (ODC), respectively [3–6, 14]. Spd, Spm, and T-Spm are synthesized by Spd synthase, Spm synthase, and T-Spm synthase (also known as ACAULIS5, ACL5), respectively, in the presence of another

Rubén Alcázar and Antonio F. Tiburcio (eds.), *Polyamines: Methods and Protocols*, Methods in Molecular Biology, vol. 1694, DOI 10.1007/978-1-4939-7398-9_2, © Springer Science+Business Media LLC 2018

substrate, decarboxylated S-adenosylmethionine (dcSAM), which is converted from SAM by SAM decarboxylase (SAMDC) [14]. Cad is converted from lysine by lysine decarboxylase. The catabolism of PA is governed by two enzyme families, the copper-containing amine oxidase (CuAO) and FAD-dependent polyamine oxidase (PAO) families [15–20]. The CuAO family favors diamines and triamines [17]. The PAO family is further divided into two groups based on reaction mode: one catabolizes PA via so-called terminal catabolism, and the other degrades PA via a back-conversion reaction [15, 18–20]. Several PA transporters have been identified in bacteria and yeast [21]. The first plant PA transporter gene, which was identified in *Arabidopsis thaliana*, determines sensitivity to the herbicide paraquat [22]. Other Arabidopsis and rice PA transporter genes have since been identified [23, 24]. PA levels are regulated by a complex network involving biosynthesis, catabolism, and intra- and/or inter-flux by transporters [5–7].

In this chapter, we provide an overview of the current knowledge about PA sensing molecule(s) and the molecules involved in PA-triggered responses and -signaling pathways in plants. Although the interplay between PA and ion channels (or pumps) is an important issue to consider when discussing the action of PA, we do not focus on this topic in this chapter; this issue is comprehensively summarized in other reviews [25, 26].

2 Polyamine Sensor

In mammalian systems, PA levels are monitored by an ODC-antizyme system [27, 28] and by the upstream open reading frame (uORF) of the *SAMDC* gene transcript [29]. In plants, the former system is absent [30], while the latter is conserved in a fairly complex manner [31]. PAs often function as a "social language" among species. For example, rotten flesh emits a strong smell mainly due to the presence of Cad and Put. This death-associated odor triggers avoidance behavior in zebrafish. Recently, using this model system, Hussain et al. [32] identified the high-affinity olfactory receptor, TAAR13c (trace amine-associated receptor 13c), which senses Cad and Put. The same group identified the chemosensory receptors for the taste and smell of PAs in the fly *Drosophila melanogaster* [33]. Identifying a plant PA sensor(s) represents a promising research target. An interesting finding was recently presented toward this aim [34]: Arabidopsis plants with a mutation in the nitrate transporter gene *NRT1.3* show increased resistance to PAs. This phenotype is allele-specific, as plants with a mutation in either *NRT1.1* or *NRT1.4* show normal PA sensitivity compared with the wild type. The authors suggest that PA transport or metabolism is associated with nitrate transport [34]. However, to date, no PA receptor or sensor has been identified in plants.

3 Signaling Molecules

3.1 Calcium Ions

Calcium ions (Ca^{2+}) are an important second messenger involved in various signal transduction pathways [35] including PA signaling. PA exported to the apoplast is oxidized by apoplastic CuAO and/or PAO, resulting in Ca^{2+} influx across the plasma membrane. Blocking this Ca^{2+} influx via a Ca^{2+} inhibitor impairs subsequent events such as the expression of downstream target genes [7, 36–38]. The Arabidopsis *spms acl5* double mutant, which produces null levels of Spm and T-Spm, displays abnormal, retarded growth in Ca^{2+}-depleted medium [39, 40]. In this mutant, vacuolar Ca^{2+}/H^{+} antiporter genes are highly upregulated by salt stress, suggesting that the tetraamines Spm/T-Spm are involved in Ca^{2+} dynamics, especially under abiotic stress conditions. Arabidopsis has five PAO genes, *AtPAO1* to *AtPAO5* [18–20]. AtPAO3 localizes to the peroxisome and catalyzes the back-conversion reaction of PAs [41]. Further analysis revealed that AtPAO3 functions in pollen tube growth through the activation of Ca^{2+}-permeable channels in pollen tubes [42]. H_2O_2 produced via Put catabolism by CuAO is involved in abscisic acid (ABA)-induced stomatal closure in *Vicia faba*. Ca^{2+} functions as a second messenger in this process [43]. Collectively, these findings indicate that Ca^{2+} is often associated with PA signaling.

3.2 Reactive Oxygen Species

PAs act as scavengers of reactive oxygen species (ROS). For instance, increased PA biosynthesis results in enhanced oxidative stress tolerance in rice by preventing ROS accumulation [44]. PA catabolism by CuAO and/or PAO leads to H_2O_2 production. The H_2O_2 produced by apoplastic PAO contributes to cell wall reinforcement to repair wounding sites in several plant species [15, 16, 18–20]. The versatile roles of PA-originating H_2O_2 have been summarized in several reviews [45, 46]. Tobacco plants overexpressing *ZmPAO*, a terminal catabolism-type PAO from maize, show hypersensitivity to salt stress [47]. A similar result was obtained for transgenic Arabidopsis plants overexpressing terminal catabolism-type *Citrus sinensis* PAO [48]. By contrast, transgenic tobacco plants with downregulated apoplastic *PAO* produce less H_2O_2 and exhibit less programmed cell death (PCD) than the wild type [49]. As mentioned, Arabidopsis has five *PAO* genes. AtPAO2, AtPAO3, and AtPAO4 localize to the peroxisome, while AtPAO1 and AtPAO5 localize to the cytoplasm [18–20, 50–52]. The loss-of-function *pao1 pao5* double mutant, in which cytosolic PAO activity is significantly suppressed, shows increased salt and drought stress tolerance [53]. Intriguingly, this mutant exhibits significantly reduced superoxide and H_2O_2 production and higher defense-related gene expression upon salt stress compared with the wild type [53]. In summary, whether a host plant survives or displays

PCD depends on the amounts of ROS produced by PAO-mediated PA catabolism [45, 46]. In addition to the *pao1pao5* mutant, the salt-tolerant phenotype of the loss-of-function *pao5* mutant should be noted [54]. This mutant constitutively accumulates T-Spm, as AtPAO5 is primarily involved in T-Spm catabolism [52]. Omics analysis of this mutant revealed increased ABA and jasmonic acid (JA) biosynthesis and the accumulation of compatible solutes. Furthermore, the authors showed that T-Spm modulates the expression of several target genes, including JA biosynthesis and JA-signaling genes [54]. Concerning ROS, Gemes et al. [55] recently showed that apoplastic PAO and NADPH oxidase constitute a ROS amplification module, in which high PAO levels correlate with high NADPH oxidase activity.

3.3 Nitric Oxide

Nitric oxide (NO) is a key signaling molecule involved in a variety of functions in plants, such as seed germination, root growth, respiration, and stomatal closure, as well as adaptive responses to abiotic and biotic stress [56–58]. NO is produced by at least six pathways in plants: nitric oxide synthase (NOS)-like activity, nitrate reductase (NR), xanthine oxidoreductase, plasma membrane-bound nitrite reductase, mitochondrial nitrite reductase, and the hydroxylamine- and PA-mediated pathways [57]. In 2006, Tun et al. [59] showed that exogenously applied PAs induce NO production in Arabidopsis. In the NR-deficient *nia1 nia2* double mutant, PA-induced NO production is still observed, indicating that NR is not involved in this process [60]. Instead, CuAO is involved in this process, because a mutant with a defect in *CuAO1* shows lower PA-induced NO production than the wild type [61]. This mutant shows increased ABA insensitivity and impaired ABA-induced NO production [61]. Moreover, Krasuska et al. [62] showed that crosstalk between NO and Put/Spd acts as a dormancy releasing factor in apple embryos. Exogenous Spd activates the NR and NOS pathways, resulting in increased NO production, which induces the expression of antioxidant enzyme genes and suppresses ROS production in white clover [63]. The effect of NO is often explained by protein modification, *S*-nitrosylation, and/or tyrosine nitration, which in turn affects various physiological phenomena [64, 65]. Tanou et al. [66] performed *S*-nitrosylated proteome analysis of citrus plants treated with NaCl alone and in combination with PAs, finding that tyrosine nitration is depressed by Spd or Spm, while protein *S*-nitrosylation increases in response to all PAs [66]. The biological significance of these changes in protein modification will be explored in the future.

3.4 Mitogen-Activated Protein Kinase Cascade

Tobacco mitogen-activated protein kinase 2 (NtMEK2) activates two mitogen-activated protein kinases (MAPKs): salicylic acid-induced protein kinase (SIPK) and wound-induced protein kinase (WIPK). Differential expression analysis of the downstream targets of the NtMEK2-SIPK/WIPK cascade revealed that *ADC* is one such target gene [67], indicating that this kinase cascade enhances PA biosynthesis. WIPK and SIPK activation was also detected during cryptogein (a fungal elicitor)-induced hypersensitive response (HR)-like cell death, which also involves H_2O_2 produced by PAO [38]. Furthermore, the same MAPK cascade functions in the defense response against viral pathogens in tobacco. During the HR involving a combination of tobacco mosaic virus (TMV) and tobacco plants carrying the TMV resistance gene, Spm accumulates in the apoplastic space [68]. Exogenous application of Spm to tobacco leaves, which mimics Spm accumulation in the apoplast, activates WIPK and SIPK via activation of the upstream kinase NtMEK2 [36, 37]. The authors proposed that Spm triggers mitochondrial dysfunction via Ca^{2+} channel activation and ROS generation, which leads to the activation of the MAPK cascade. One of the downstream target genes of this MAPK cascade is *NtbZIP60* [69], a tobacco ortholog of *AtbZIP60*, a key transcription factor gene of the unfolded protein response (UPR) in Arabidopsis [70]. Further analysis revealed that Spm-induced UPR requires Ca^{2+} influx to the cytoplasm and the activation of the MKK9-MPK3 (WIPK ortholog)/MPK6 (SIPK ortholog) cascade in Arabidopsis [71, 72]. This MAPK cascade may also function in the HR triggered by cucumber mosaic virus in Arabidopsis [71]. Mo et al. proposed that the MPKK1-MPK3/MPK6 cascade is involved in the Spm-signaling and camalexin (a major phytoalexin of Arabidopsis)-signaling pathways in the Arabidopsis–*Verticillium dahlia* phytopathosystem [73].

3.5 Transporter

PAs have positive or negative effects on certain sets of cation channels: detailed information about this issue is provided in other reviews [25, 26]. The first PA transporter gene identified in plants, *RMV1* (*Resistant to Methyl Viologen 1*), is a causal gene that determines the differential sensitivity to the herbicide paraquat found among various Arabidopsis accessions [22]. Since this gene presumably encodes an amino acid permease, it is also referred to as *AtLAT1* (*A. thaliana L-type amino acid transporter 1*). Arabidopsis contains four other LAT family genes, termed *AtLAT2* to *AtLAT5*; AtLAT1, AtLAT3, and AtLAT4 localize to the plasma membrane, endoplasmic reticulum, and Golgi apparatus, respectively. Independently, Mulangi et al. [23] identified PA transporter genes from rice and Arabidopsis based on a budding yeast mutant deficient in PA transport, which were designated as *PA uptake transporter* (*PUT*) genes. All *PUT* genes identified to date belong to the *LAT* family. *LAT*/*PUT* genes function in PA transport [24].

Quantitative trait locus analysis of natural variations in Cad sensitivity during root growth in various Arabidopsis accessions identified the gene *AtOCT1* (*A. thaliana Organic Cation Transporter 1*) [12]. A loss-of-function mutant of *AtOCT1* is more sensitive to Cad than the wild type, whereas Arabidopsis plants overexpressing this gene are Cad resistant. Since *AtOCT1* localizes to the plasma membrane, it is likely that this protein pumps Cad out of the cell [13]. By chance, Liu et al. [74] found that the Cad sensitivity of Arabidopsis is affected by Spm content. *AtLAT1*-overexpressing plants absorb more Cad and show increased Cad sensitivity compared with the wild type, suggesting that AtLAT1 also absorbs Cad. The authors proposed that Spm modulates the expression of *AtLAT1* and *AtOCT1*, which may alter their PA transport potential, and it determines root growth responses to Cad in Arabidopsis [75]. Other plant PA transporters and their effector molecule(s) should be identified and the crosstalk among these molecules should be explored in the future.

3.6 GABA

The GABA shunt is a pathway that produces succinate using either glutamate or α-ketoglutarate as substrate. In this pathway, GABA is synthesized from glutamate by glutamate decarboxylase (GAD) [76, 77]. The GABA shunt may help support the defense response to pathogens [76]. GABA is produced during wounding: GAD is activated and GABA is released from the damaged site. The emitted GABA functions as a signaling molecule between the plant and other organisms [77]. Another source of GABA is the amino aldehydes produced from terminal oxidation of Put by CuAO or from Spm and Spd by PAO [7, 15, 19]. Back-conversion-type PAO produces 1,3-diaminopropane from Spd and Spm, and terminal catabolism-type PAO produces 4-aminobutanal and N-(3-aminopropyl)-4-aminobutanal from Spd and Spm, respectively. CuAO converts 1,3-diaminopropane to 3-aminopropanal (APAL) and Put to 4-aminobutanal (ABAL). The oxidation of APAL and ABAL (catalyzed by members of the aldehyde dehydrogenase family) results in the biosynthesis of β-alanine and GABA, respectively [5, 15, 46]. PA degradation and GABA accumulation are tightly correlated in plants under salt [78], wounding [79], and drought stress [80]. There is direct evidence that GABA produced by aldehyde dehydrogenase 10 family members confers salt tolerance in Arabidopsis [81].

4 Crosstalk Among the Above Molecules

Pal et al. [7] proposed an effective model describing the crosstalk between PA signaling and intermediate molecules including ROS, NO, ABA, phospholipases, and NADPH oxidase, even though crosstalk among the above molecules occurs via a more complex network [7, 45]. In terms of the ROS and Ca^{2+} interaction, ROS-induced Ca^{2+} release (RICR) has been demonstrated during PA signaling [82]. Namely, the activation of Ca^{2+} channels by H_2O_2 produced by the apoplastic PAO reaction results in the entry of Ca^{2+}

into the cytoplasm. In addition, Gemes et al. reported the presence of a ROS–ROS interaction [55]. Specifically, NADPH oxidase (RBOH) interacts with apoplastic PAO, and the RBOH-PAO tandem module controls the oxidative burst during the salt response. This RBOH-PAO module functions in ROS-induced ROS amplification during PA signaling. The interplay between the NO and ROS pathways during the HR, leaf senescence, and other types of plant cell death has also been reviewed [48]; we will discuss only two examples described in this review [48]. First, during the HR, NO can affect ROS biosynthesis through the inactivation of ATRBOHD (NADPH oxidase) by *S*-nitrosylation [64]. Second, in the rice mutant *noe1* (*nitric oxide excess 1*), which produces excess NO due to a defect in the catalase gene *OsCATC*, the resulting H_2O_2 accumulation induces NO production via upregulation of *NR* [58]. If similar studies are undertaken exploring plant PAs, more will be learned about the precise roles of PAs at the molecular levels.

5 Perspectives

As mentioned, Spm-resistant Arabidopsis mutants have been isolated, and the causal gene was identified as a nitrate transporter gene [31]. This type of approach is very promising for identifying a PA sensor or receptor. Extensive screening for such PA-resistant or hypersensitive mutants, followed by map-based cloning, may help uncover the intrinsic PA sensor in plants.

To understand the role of PA at the molecular level, details about how each PA and/or metabolic intermediate(s) and end product(s) exert their effects on each molecule, as well as the crosstalk among the Ca^{2+}-, ROS-, and NO-signaling pathways and kinase pathways (such as the MAPK cascade), should be addressed. Our understanding of PA transporters in plants remains limited. Identifying additional PA transporters would stimulate further research. Revealing the effector(s) for the activity of these transporters and/or for their expression is another important research target. The roles of NO in phytohormone signaling are fairly well understood [63], and the relationships between PA and phytohormone pathways such as the ethylene, ABA, and gibberellic acid pathways have been described [5, 7]. Furthermore, the crosstalk between auxin- and the T-Spm-pathways has been clearly elucidated [83, 84]. The interaction between the PA pathway and phytohormone pathways should also be investigated in the future.

Acknowledgements

This work was financially supported by JSPS KAKENHI (no. 15K14705) to T.K.

References

1. Miller-Fleming L, Olin-Sandoval V, Campbell K, Ralser M (2015) Remaining mysteries of molecular biology: the role of polyamines in the cell. J Mol Biol 427:3389–3406
2. Kusano T, Berberich T, Tateda C, Takahashi Y (2008) Polyamines: essential factors for growth and survival. Planta 228:367–381
3. Alcazar R, Altabella T, Marco F, Bortolotti C, Reymond M, Koncz C, Carrasco P, Tiburcio AF (2010) Polyamines: molecules with regulatory functions in plant abiotic stress tolerance. Planta 231:1237–1249
4. Handa AK, Mattoo AK (2010) Differential and functional interactions emphasize the multiple roles of polyamines in plants. Plant Physiol Biochem 48:540–546
5. Tiburcio AF, Altabella T, Bitrian M, Alcazar R (2014) The roles of polyamines during the lifespan of plants: from development to stress. Planta 240:1–18
6. Minocha R, Majumdar R, Minocha SC (2014) Polyamines and abiotic stress in plants: a complex relationship. Front Plant Sci 5:175
7. Pal M, Szalai G, Janda T (2015) Speculation: polyamines are important in abiotic stress signaling. Plant Sci 237:16–23
8. Oshima T (1979) A new polyamine, thermospermine, 1,12-diamino-4,8-diazadodecane, from an extreme thermophile. J Biol Chem 254:8720–8722
9. Knott JM, Römer P, Sumper M (2007) Putative spermine synthases from *Thalassiosira pseudonana* and *Arabidopsis thaliana* synthesize thermospermine rather than spermine. FEBS Lett 581:3081–3086
10. Takano A, Kakehi J-I, Takahashi T (2012) Thermospermine is not a minor polyamine in the plant kingdom. Plant Cell Physiol 53:606–616
11. Bunsupa S, Katayama K, Ikeura E, Oikawa A, Toyooka K, Saito K, Yamazaki M (2012) Lysine decarboxylase catalyzes the first step of quinolizidine alkaloid biosynthesis and coevolved with alkaloid production in leguminosae. Plant Cell 24:1202–1216
12. Strohm AK, Vaughn LM, Masson PH (2015) Natural variation in the expression of organic cation transporter 1 affects root length responses to cadaverine in Arabidopsis. J Exp Bot 66:853–862
13. Jancewicz AL, Gibbs NM, Masson PH (2016) Cadaverine's functional role in plant development and environmental response. Front Plant Sci 7:870
14. Fuell C, Elliott KA, Hanfrey CC, Franceschetti M, Michael AJ (2010) Polyamine biosynthetic diversity in plants and algae. Plant Physiol Biochem 48:513–520
15. Cona A, Rea G, Angelini R, Federice R, Tavladoraki P (2006) Functions of amine oxidases in plant development and defence. Trends Plant Sci 11:1360–1385
16. Rea G, Metoui O, Infantino A, Federico R, Angelini R (2002) Copper amine oxidase expression in defense responses to wounding and *Ascochyta rabiei* invasion. Plant Physiol 128:865–875
17. Planas-Portell J, Gallart M, Tiburcio AF, Altabella T (2013) Copper-containing amine oxidases contribute to terminal polyamine oxidation in peroxisomes and apoplast of *Arabidopsis thaliana*. BMC Plant Biol 13:109
18. Angelini R, Cona A, Federico R, Fincato P, Tavladoraki P, Tisi A (2010) Plant amine oxidases 'on the move': an update. Plant Physiol Biochem 48:560–564
19. Kusano T, Kim DW, Liu T, Berberich T (2015) Polyamine catabolism in plants. In: Kusano T, Suzuki H (eds) Polyamine: a universal molecular nexus for growth, survival and specialised metabolism. Springer, Tokyo, pp 77–88
20. Tavladoraki P, Cona A, Angelini R (2016) Copper-containing amine oxidases and FAD-dependent polyamine oxidases are key players in plant tissue differentiation and organ development. Front Plant Sci 7:824
21. Igarashi K, Kashiwagi K (2010) Characteristics of cellular polyamine transport in prokaryotes and eukaryotes. Plant Physiol Biochem 48:506–512
22. Fujita M, Fujita Y, Iuchi S, Yamada K, Kobayashi Y, Urano K, Kobayashi M, Yamaguchi-Shinozaki K, Shinozaki K (2012) Natural variation in a polyamine transporter determines paraquat tolerance in Arabidopsis. Proc Natl Acad Sci U S A 109:6343–6347
23. Mulangi V, Phuntumart V, Aouida M, Ramotar D, Morris P (2012) Functional analysis of OsPUT1, a rice polyamine uptake transporter. Planta 235:1–11
24. Fujita M, Shinozaki K (2014) Identification of polyamine transporters in plants: paraquat transport provides crucial clues. Plant Cell Physiol 55:855–861
25. Pottosin I, Shabala S (2014) Polyamines control of cation transport across plant membrane: implications for ion homeostasis and abiotic stress signaling. Front Plant Sci 5:154

26. Pottosin I, Velarde-Buendia AM, Bose J, Zepeda-Jazo I, Shabala S, Dobrovinskaya O (2014) Cross-talk between reactive oxygen species and polyamines in regulation of ion transport across the plasma membrane: implications for plant adaptive responses. J Exp Bot 65:1271–1283
27. Kahara C (2009) Antizyme and antizyme inhibitor, a regulatory tango. Cell Mol Life Sci 66:2479–2488
28. Murai N (2015) Antizyme. In: Kusano T, Suzuki H (eds) Polyamines: a universal molecular nexus for growth, survival, and specialized metabolism. Springer, Tokyo, pp 91–99
29. Ruan H, Shantz LM, Pegg AE, Morris DR (1996) The upstream open reading frame of the mRNA encoding S-adenosylmethionine decarboxylase is a polyamine-responsive translational control element. J Biol Chem 271:29576–29582
30. Hanfrey C, Sommer S, Mayer MJ, Burtin D, Michael AJ (2001) Arabidopsis polyamine biosynthesis: absence of ornithine decarboxylase and the mechanism of arginine decarboxylase activity. Plant J 27:551–560
31. Hanfrey C, Elliott KA, Franceschetti M, Mayer MJ, Illingworth C, Michael AJ (2005) A dual upstream open reading frame-based autoregulatory circuit controlling polyamine-responsive translation. J Biol Chem 280:39229–39237
32. Hussain A, Saraiva LR, Ferrero DM, Ahuja G, Krishna VS, Liberles SD, Korsching SI (2013) High-affinity olfactory receptor for the death-associated odor cadaverine. Proc Natl Acad Sci U S A 110:19579–19584
33. Hussain A, Zhang M, Ucpunar HK, Svensson T, Quillery E, Gompel N, Ignell R, Kadow G (2016) Ionotropic chemosensory receptors mediate the taste and smell of polyamines. PLoS Biol 14:e1002454
34. Tong W, Imai A, Tabata R, Shigenobu S, Yamaguchi K, Yamada M, Hasebe M, Sawa S, Motose H, Takahashi T (2016) Polyamine resistance is increased by mutants in a nitrate transporter gene *NRT1.3* (*AtNPF6.4*) in *Arabidopsis thaliana*. Front Plant Sci 7:834
35. Tuteja N, Mahajan S (2007) Calcium signaling network in plants. Plant Signal Behav 2:79–85
36. Takahashi Y, Berberich T, Miyazaki A, Seo S, Ohashi Y, Kusano T (2003) Spermine signaling in tobacco: activation of mitogen-activated protein kinases by spermine is mediated through mitochondrial dysfunction. Plant J 36:820–829
37. Takahashi Y, Uehara Y, Berberich T, Ito A, Saitou H, Miyazaki A, Terauchi R, Kusano T (2004) A subset of the hypersensitive response marker genes including *HSR203J* is downstream target of spermine-signal transduction pathway in tobacco. Plant J 40:586–595
38. Yoda H, Hiroi Y, Sano H (2006) Polyamine oxidase is one of the key elements for oxidative burst to induce programmed cell death in tobacco cultured cells. Plant Physiol 142:193–206
39. Yamaguchi K, Takahashi Y, Berberich T, Imai A, Miyazaki A, Takahashi T, Michael A, Kusano T (2006) The polyamine spermine protects against high salt stress in *Arabidopsis thaliana*. FEBS Lett 580:6783–6788
40. Yamaguchi K, Takahashi Y, Berberich T, Imai A, Takahashi T, Michael A, Kusano T (2007) A protective role for the polyamine spermine against drought stress in *Arabidopsis*. Biochem Biophys Res Commun 352:486–490
41. Moschou PN, Sanmartin M, Andriopoulou AH, Rojo E, Sanchez-Serrano JJ, Roubelakis-Angelakis KA (2008) Bridging the gap between plant and mammalian polyamine catabolism: a novel peroxisomal polyamine oxidase responsible for a full back-conversion pathway in *Arabidopsis*. Plant Physiol 47:1845–1857
42. Wu J, Shang Z, Wu J, Jiang X, Moschou PN, Sun W, Roubelakis-Angelakis KA, Zhang S (2010) Spermidine oxidase-derived H_2O_2 regulates pollen plasma membrane hyperpolarization-activated Ca^{2+}-permeable channels and pollen tube growth. Plant J 63:1042–1053
43. An Z, Jing W, Liu Y, Zhang W (2008) Hydrogen peroxide generated by copper amine oxidase is involved in abscisic acid-induced stomatal closure in *Vicia faba*. J Exp Bot 59:815–825
44. Jang SJ, Wi SJ, Choi YJ, An G, Park KY (2012) Increased polyamine biosynthesis enhances stress tolerance by preventing the accumulation of reactive oxygen species: T-DNA mutational analysis of *Oryza sativa* lysine decarboxylase-like protein 1. Mol Cells 34:251–262
45. Gupta K, Sengupta A, Chakraborty M, Gupta B (2016) Hydrogen peroxide and polyamines at as double edged swords in plant abiotic stress responses. Front Plant Sci 7:1343
46. Moschou PN, Wu J, Cona A, Tavladoraki P, Angelini R, Roubelakis-Angelakis KA (2012) The polyamines and their catabolic products are significant players in the turnover of nitrogenous molecules in plants. J Exp Bot 63:5003–5015
47. Moschou PN, Delis ID, Paschalidis KA, Roubelakis-Angelakis KA (2008) Transgenic tobacco plants overexpressing *polyamine*

oxidase are not able to cope with oxidative burst generated by abiotic factors. Physiol Plant 133:140–156

48. Wang W, Liu J-H (2016) *CsPAO4* of *Citrus sinensis* functions in polyamine terminal catabolism and inhibits plant growth under salt stress. Sci Rep 6:31384
49. Moschou PN, Paschalidis KA, Delis ID, Andriopoulou AH, Lagiotis GD, Yakoumakis DI, Roubelakis-Angelakis KA (2008) Spermidine exodus and oxidation in the apoplast induced by abiotic stress is responsible for H_2O_2 signatures that direct tolerance responses in tobacco. Plant Cell 20:1708–1724
50. Tavladoraki P, Rossi MN, Saccuti G, Perez-Amador MA, Polticelli F, Angelini R, Federico R (2006) Heterologous expression and biochemical characterization of a polyamine oxidase from *Arabidopsis* involved in polyamine back conversion. Plant Physiol 141:1519–1532
51. Ahou A, Martignago D, Alabdallah O, Tavazza R, Stano P, Macone A, Pivato M, Masi A, Rambla JL, Vera-Sirera F, Angelini R, Federico R, Tavladoraki P (2014) A plant spermine oxidase/dehydrogenase regulated by the proteasome and polyamines. J Exp Bot 65:1585–1603
52. Kim DW, Watanabe K, Murayama C, Izawa S, Niitsu M, Michael AJ, Berberich T, Kusano T (2014) Polyamine oxidase 5 regulates *Arabidopsis thaliana* growth through a thermospermine oxidase activity. Plant Physiol 165:1575–1590
53. Sagor GHM, Zhang S, Kojima S, Simm S, Berberich T, Kusano T (2016) Reducing cytoplasmic polyamine oxidase activity in Arabidopsis increases salt and drought tolerance by reducing reactive oxygen species production and increasing defense gene expression. Front Plant Sci 7:214
54. Zarza X, Atanasov KE, Marco F, Arbona V, Carrasco P, Kopka J, Fotopoulos V, Munnik T, Gomez-Cadenas A, Tiburcio AF, Alcazar R (2016) *Polyamine oxidase 5* loss-of-function mutations in *Arabidopsis thaliana* trigger metabolic and transcriptional reprogramming and promote salt stress tolerance. Plant Cell Environ 40(4):527–542. doi:10.1111/pce.12714
55. Gemes K, Kim YJ, Park KY, Moschou PN, Andronis E, Valassaki C, Roussis A, Roubelakis-Angelakis KA (2016) An NADPH-oxidase/polyamine oxidase feedback loop controls oxidative burst under salinity. Plant Physiol 172:1418–1431
56. Baudoulin E, Hancock JT (2014) Nitric acid signaling in plants. Front Plant Sci 4:553
57. Gupta KJ, Fernie AR, Kaiser WM, van Dongen JT (2011) On the origin of nitric oxide. Trends Plant Sci 16:160–168
58. Wang Y, Loake GJ, Chu C (2013) Cross-talk of nitric oxide and reactive oxygen species in plant programed cell death. Front Plant Sci 4:314
59. Tun NN, Santa-Catarina C, Begum T, Silveira V, Handro W, Floh EI, Scherer GF (2006) Polyamines induce rapid synthesis of nitric oxide (NO) in *Arabidopsis thaliana* seedlings. Plant Cell Physiol 47:346–354
60. Wimalasekera R, Tebartz F, Scherer GF (2011) Polyamines, polyamine oxidases and nitric oxide in development, abiotic and biotic stresses. Plant Sci 181:593–603
61. Wimalasekera R, Villar C, Begum T, Scherer GF (2011) *Copper amine oxidase 1* (*CuAO1*) of *Arabidopsis thaliana* contributes to abscisic acid- and polyamine-induced nitric oxide biosynthesis and abscisic acid signal transduction. Mol Plant 4:663–678
62. Krasuska U, Ciacka K, Gniazdowska A (2016) Nitric oxide-polyamines cross-talk during dormany release and germination of apple embryos. Nitric Oxide 68:38–50. doi:10.1016/j.niox.2016.11.003
63. Peng D, Wang X, Li Z, Zhang Y, Peng Y, Li Y, He X, Zhang X, Ma X, Huang L, Yan Y (2016) NO is involved spermidine-induced drought tolerance in white clover via activation of antioxidant enzymes and genes. Protoplasma 253:1243–1254
64. Lin A, Wang Y, Tang J, Xue P, Li C, Liu L, Hu B, Yang F, Loake GJ, Chu C (2011) Nitric oxide and protein S-nitrosylation are integral to hydrogen peroxide-induced leaf cell death in rice. Plant Physiol 158:451–464
65. Paris R, Iglesias MJ, Terrile MC, Casalongue CA (2013) Function of S-nitrosylation in plant hormone networks. Front Plant Sci 4:294
66. Tanou G, Ziogas V, Belghazi M, Christou A, Filippou P, Job D, Fotopoulos V, Kojassiotis A (2014) Polyamine reprogram oxidative and nitrosative status and the proteome of citrus plants exposed to salinity stress. Plant Cell Environ 37:864–885
67. Jang EK, Min KH, Kim SH, Nam S-H, Zhang S, Kim YC, Cho BH, Yang K-Y (2009) Mitogen-activated protein kinase cascade in the signaling for polyamine biosynthesis in tobacco. Plant Cell Physiol 50:658–664
68. Yamakawa H, Kamada H, Satoh M, Ohashi Y (1998) Spermine is a salicylate-independent endogenous inducer for both tobacco acidic pathogenesis-related proteins and resistance against tobacco mosaic virus infection. Plant Physiol 118:1213–1222

69. Tateda C, Ozaki R, Onodera Y, Takahashi Y, Yamaguchi K, Berberich T, Koizumi N, Kusano T (2008) NtbZIP60, an endoplasmic reticulum-localized transcription factor, plays a role in defence response against bacterial pathogen in Nicotiana tabacum. J Plant Res 121:603–611
70. Iwata Y, Koizumi N (2012) Plant transducers of the endoplasmic reticulum unfolded protein response. Trends Plant Sci 17:720–727
71. Mitsuya Y, Takahashi Y, Berberich T, Miyazaki A, Matsumura H, Takahashi H, Terauchi R, Kusano T (2009) Spermine signaling plays a significant role in the defense response of Arabidopsis thaliana to cucumber mosaic virus. J Plant Physiol 166:626–643
72. Sagor GHM, Chawla P, Kim DW, Berberich T, Kojima S, Niitsu M, Kusano T (2015) The polyamine spermine induces the unfolded protein response via the MAPK cascade in Arabidopsis. Front Plant Sci 6:687
73. Mo H, Wang X, Zhang Y, Zhang G, Zhang J, Ma Z (2015) Cotton polyamine oxidase is required for spermine and camalexin signaling in the defence response to Verticillium dahliae. Plant J 83:962–975
74. Liu T, Dobashi H, Kim DW, Sagor GHM, Niitsu M, Berberich T, Kusano T (2014) Arabidopsis mutant plants with diverse defects in polyamine metabolism show unequal sensitivity to exogenous cadaverine probably based on their spermine content. Physiol Mol Biol Plant 20:151–159
75. Sagor GHM, Berberich T, Kojima S, Niitsu M, Kusano T (2016) Spermine modulates the expression of two probable polyamine transporter genes and determines growth responses to cadaverine in Arabidopsis. Plant Cell Rep 35:1247–1257
76. Bolton MD (2009) Primary metabolism and plant defense-fuel for the fire. Mol Plant Microbe Interact 22:487–497
77. Shelp B, Bown AW, Faure D (2006) Extracellular γ-aminobutyrate mediates communication between plants and other organisms. Plant Physiol 142:1350–1352
78. Xing SG, Jun YB, Hau ZW, Liang LY (2007) Higher accumulation of γ-aminobutyric acid induced by salt stress through stimulating the activity of diamine oxidases in Glycine max (L.) Merr. roots. Plant Physiol Biochem 45:560–566
79. Petrivalsky M, Brauner F, Luhova L, Gaqneul D, Sebela M (2007) Aminoaldehyde dehydrogenase activity during wound healing of mechanically injured pea seedlings. J Plant Physiol 164:1410–1418
80. Hatmi S, Gruau C, Trotel-Aziz P, Villaume S, Rabenoelina F, Baillieul F, Eullaffroy P, Clement C, Ferchichi A, Aziz A (2015) Drought stress tolerance in grapevine involves activation of polyamine oxidation contributing to improved immune response and low susceptibility to Botrytis cinerea. J Exp Bot 66:775–787
81. Zarei A, Trobacher CP, Shelp BJ (2016) Arabidopsis aldehyde dehydrogenase 10 family members confer salt tolerance through putrescine-derived 4-aminobutyrate (GABA) production. Sci Reports 6:35115
82. Gilroy S, Suzuki N, Miller G, Choi W-G, Toyota M, Devireddy AR, Mittler R (2014) A tidal wave of signals: calcium and ROS at the forefront of rapid systemic signaling. Trends Plant Sci 19:623–630
83. Yoshimoto K, Noutoshi Y, Hayashi K, Shirasu K, Takahashi T, Motose H (2012) A chemical biology approach reveals an opposite action between thermospermine and auxin in xylem development in Arabidopsis. Plant Cell Physiol 53:635–645
84. Baima S, Forte V, Possenti M, Penalosa A, Leoni G, Salvi S, Felici B, Ruberti I, Morelli G (2014) Negative feedback regulation of auxin signaling by ATHB8/ACL5-BUD2 transcription module. Mol Plant 7:1006–1025

Chapter 3

Polyamine Metabolism Responses to Biotic and Abiotic Stress

Fernando M. Romero, Santiago J. Maiale, Franco R. Rossi, Maria Marina, Oscar A. Ruíz, and Andrés Gárriz

Abstract

Plants have developed different strategies to cope with the environmental stresses they face during their life cycle. The responses triggered under these conditions are usually characterized by significant modifications in the metabolism of polyamines such as putrescine, spermidine, and spermine. Several works have demonstrated that a fine-tuned regulation of the enzymes involved in the biosynthesis and catabolism of polyamines leads to the increment in the concentration of these compounds. Polyamines exert different effects that could help plants to deal with stressful conditions. For instance, they interact with negatively charged macromolecules and regulate their functions, they may act as compatible osmolytes, or present antimicrobial activity against plant pathogens. In addition, they have also been proven to act as regulators of gene expression during the elicitation of stress responses. In this chapter, we reviewed the information available till date in relation to the roles played by polyamines in the responses of plants during biotic and abiotic stress.

Key words Polyamines, Microbes, Drought stress, Salt stress, Cold stress

1 Introduction

Polyamines (PAs) are aliphatic polycations that occur ubiquitously in prokaryotic and eukaryotic cells and are essential for cell growth, proliferation, and differentiation [1]. Because they are positively charged at physiological pH, PAs bind to polyanionic macromolecules like proteins, nucleic acids, and phospholipids. This ability might explain the participation of these amines in different physiological processes, such as DNA replication, RNA processing, protein synthesis and post-translational modifications [2]. The most abundant PAs are the diamine putrescine (Put), the triamine spermidine (Spd), and the tetraamine spermine (Spm), which exist in free and conjugated forms, the latter being covalently bound to small molecules and proteins. Several studies have shown that either depletion or excessive accumulation of PAs may be deleterious for

Rubén Alcázar and Antonio F. Tiburcio (eds.), *Polyamines: Methods and Protocols*, Methods in Molecular Biology, vol. 1694, DOI 10.1007/978-1-4939-7398-9_3,

cell physiology [3]. Therefore, a fine-tuned regulatory mechanism involving several processes such as synthesis, catabolism, conjugation, transport, and compartmentalization of PAs must be coordinated in order to maintain the homeostasis of these compounds under suitable levels.

The abundance of plant PAs varies with plant species and phenology [4, 5], and undergoes notable changes during plant responses to biotic and abiotic stresses. It has been demonstrated that PA metabolism is interconnected with some of the metabolic routes involved in the formation of key stress-related signaling molecules. For instance, the biosynthesis of PAs requires precursors that are also used for the synthesis of other important stress-related metabolites, such as S-adenosylmethionine (SAM, involved in the metabolism of ethylene), ornithine (necessary for the synthesis of proline), and arginine (precursor of nitric oxide) [6–8]. Moreover, a vast research in this area demonstrated that supplementation with phytohormones such as salicylic acid, jasmonic acid, abscisic acid, and ethylene evokes changes in PAs metabolism [9]. Even though there is a gap in our current knowledge of the functions of PAs in basic physiological processes, great technical advances in the field of transcriptomic and analytical chemistry have placed us on the verge of truly understanding the fundamental roles played by these compounds in the responses of plants to stress. In this chapter, we cover the information on the changes that occur in the metabolism of PA and the mechanisms underlying PA functions during the responses of plants to biotic and abiotic stresses.

2 PA Metabolism During Plant–Microbe Interactions

Notable changes in the expression of genes encoding PA biosynthetic as well as catabolic enzymes occur during the interactions established by plant and microbes, usually leading to variations in the concentration of these compounds in plant tissues [10]. Since both type of organisms are able to synthesize PAs, it might be hard to discriminate whether the increase in their concentrations is due to the metabolism of the plant or the invading microorganism. In addition, several investigations have shown that some successful pathogenic microorganisms facilitate plant invasion by taking control of the host's metabolism by the action of different effector proteins. Thus, studies on PA metabolism during plant–microbe interactions should consider that the changes in gene expression might be part of the plant defense responses induced against the pathogen, or they might also be triggered as a consequence of the pathogen's virulence mechanisms. The answer to this issue is still unclear, and probably it is not one or another but a combination of both processes. This scenario, indeed, is not only limited to pathogenic interactions, as it has been reported that the interaction of plants with beneficial microorganisms induces similar changes in the plant PA homeostasis mechanisms [4].

One of the first reports describing the alterations in PA levels during pathogenic infections was carried out by Greenland and Lewis, who reported that an increment in Spd levels occurs in barley leaves infected with the biotrophic fungus *Puccinia hordei* [11]. A comparable phenomenon was observed in the same plant species after the inoculation with *Blumeria graminis* f. sp. *hordei*, and in the interaction between wheat and *Puccinia graminis* f. sp. *tritici* [11]. These effects on PA concentrations were demonstrated to be induced after treatment with fungal elicitors, suggesting that modulation of PA metabolism might form part of a general defense mechanism [12]. The accumulation of PAs is mostly explained by the induction in the activity of the biosynthetic enzymes arginine decarboxylase (ADC) and ornithine decarboxylase (ODC) [10]. It is important to note that in the case of fungal pathogens with a biotrophic lifestyle, which require live host cells to complete their life cycles, the accumulation of PAs and the induction of gene expression were mainly located in the so-called green islands. These are areas of plant tissue surrounding the entry of the pathogen, showing high photosynthetic rates and a delay in senescence. In addition, many studies demonstrate that not only the free form, but also conjugated PAs are accumulated during the interactions of plants with pathogenic virus, fungi, and bacteria [10, 13, 14]. As these PAs contribute to the strengthening of the cell wall, thus protecting this structure against the activity of microbial hydrolytic enzymes, it has been suggested that they could play important roles in the resistance to pathogens [10].

Plant tolerance to biotic stress is often correlated with the induction of genes from the PA metabolism and the production of higher levels of PAs. This was first demonstrated with the analysis of two genotypes of barley with contrasting tolerance to *B. graminis* f. sp. *hordei* [10]. In this case, the resistant line showed a greater increase in free and conjugated forms of Put and Spd, as well as in the expression of PA biosynthetic and catabolic genes when challenged with the fungus. It has been proven recently that the accumulation of Put in tomato plants as a consequence of high NH_4^+ availability promotes tolerance to bacterial infection. This was corroborated by applying inhibitors of Put synthesis, which reverted the accumulation of Put and enhanced susceptibility [15]. The activation of PA metabolism and accumulation of Put and Spd was also observed in the resistant NN line of tobacco in response to the tobacco mosaic virus, but not in the susceptible line [16]. Similarly, the expression of thermospermine synthase (*ACL5*) in cotton, which synthesizes the spermine analogue thermospermine, was more rapidly induced in a resistant line in comparison to the susceptible one in response to *Verticillium dahliae* [17, 18]. On the other hand, plant susceptibility has been associated to repression of

PA metabolism and low levels of PAs. For instance, PA levels were reduced during the infection caused by *V. dahliae* in a susceptible cotton cultivar, whereas the expression of the PA metabolism genes remained unaltered. By contrast, the expression of *SAMDC* was induced and PA were accumulated in a tolerant cultivar during the pathogenic process [18]. The authors of this work also reported that tolerance correlated to higher expressions of *PAO*, and that this gene was significantly repressed in the susceptible cultivar during the first hours after inoculation. Altogether, gene expression analysis provides evidences that not only PA biosynthesis but also its oxidation may be essential for plant defense.

Another line of evidences demonstrating that higher PA levels are correlated to biotic stress tolerance comes from the use of genetically modified plants. In this trend, it has been shown that plants overexpressing the human *SAMDC* gene accumulate Spd and Spm and show greater tolerance to pathogens [4]. Similarly, Arabidopsis plants overexpressing the *SAMDC* gene from cotton proved to be more resistant to pathogen infection [19]. In turn, silencing of *SAMDC* or spermine synthase (*SPMS)* in cotton leads to enhanced susceptibility to *V. dahliae*, and exogenous supplementation of Spm augments the tolerance of these lines to the attack of the pathogen. These results suggest that the synthesis of Spd or Spm plays a key role in the process. Accordingly, Gonzalez et al. [20] showed that Arabidopsis plants overexpressing the *SPMS* gene were more tolerant to bacterial infections, and that mutant lines in the same gene were more susceptible, which was reverted by the application of exogenous Spm. Interestingly, a transcriptomic analysis of the Arabidopsis line overexpressing the *SPMS* gene showed the upregulation of many transcripts involved in pathogen perception and defense, including several regulatory proteins such as transcription factors and kinases [20]. Similar transcription profiles were observed in plants overexpressing *SAMDC1*, which accumulate Spm and show higher tolerances to *Pseudomonas syringae* and *Hyaloperonospora arabidopsidis* [21]. These results indicate that Spd and Spm could play important roles in the activation of essential mechanisms for plant defense. Moreover, transgenic plants overexpressing *ACL5* showed enhanced tolerance to bacterial infection. Even though this was not correlated with an increment in the levels of thermospermine (t-Spm) or any other PA, supplementation with t-Spm in Arabidopsis reduces the colonization by *Pseudomonas viridiflava* and restrict cucumber mosaic virus multiplication [22, 23], suggesting that this PA could also participate in plant defense. In this trend, it was recently shown that silencing of *ACL5* in cotton leads to a decrease in t-Spm levels, and that this line exhibited reduced resistance to *V. dahliae* [17]. Since treatment with a PAO inhibitor reduced the resistance of *ACL5* lines in Arabidopsis, it was proposed that resistance requires the oxidation of t-Spm [22].

The production of H_2O_2, a molecule playing multiple roles in plant defense, through PA oxidation and its relation to plant resistance is a topic that has been largely studied. As described above, many studies demonstrated that a remarkable increment in the activity of the PA catabolic enzymes occurs in response to biotic stress [10, 16], and that the accumulation of Spm and its oxidation at the plant apoplast induces the expression of defense-related genes. In addition, transgenic plants overexpressing PAOs show an increase in the tolerance to pathogenic infections [4, 18], and by the contrary, the use of inhibitors of these enzymes or deletion of the cognate genes strongly impairs plant resistance [18, 24]. It has been suggested that the contribution of PA oxidases to the accumulation of H_2O_2 is required for the triggering of a programmed cell death process known as hypersensitive response (HR). This reaction is usually observed in interactions established with biotrophic pathogens (those feeding on living cells), and even though it contributes to defense against these kind of microorganisms, it is known that the HR favors the attack of necrotrophic microorganisms (those killing the host cell in order to feed from them) [25]. It was recently proved that PAO and DAO activities promote defense against biotrophic or hemibiotrophic pathogens [26, 27] and by contrast, these activities favor the spread of the lesions provoked by necrotrophic pathogens [24]. Thus, it seems plausible to hypothesize that biotrophs may benefit from mechanisms that reduce PA oxidation. In this trend, Lou et al. [28] have proved that the infection of the hemibiotrophic bacteria *Pseudomonas syringae* upregulates the expression of the gene *NATA1* in Arabidopsis, which mediates the acetylation of Put. Importantly, as *nata1* mutant plants accumulate high levels of H_2O_2 and are more resistant to bacterial infection, the authors proposed that Put acetylation diminished the availability of Put for Spd and Spm synthesis, leading to a lower production of H_2O_2 by PAOs and consequently avoiding the initiation of plant defense responses. Supporting this hypothesis, both the accumulation of H_2O_2 and resistance were attenuated by the PA oxidase inhibitor guazatine, indicating that PA catabolism plays a key role in the activation of defense [28]. On the other hand, the oxidation of PAs may be required to complete the life cycle of pathogens normally considered as bio/hemibiotrophic. For instance, Jasso-Robles et al. [29] showed that inhibition of PAO activity in maize leads to a decrease in tumor size and number during the interaction with the biotrophic fungus *Ustilago maydis*. These authors hypothesized that this is due to a decrease in the production of H_2O_2 during tumor formation, a molecule important for cell elongation and cell wall maturation.

Increasing evidences seem to validate the idea that pathogens hijack the host PA metabolism in an attempt to modulate its physiology. In this way, it has been shown that the bacterium

Rhodococcus fascians produce cytokinins that are able to induce Put accumulation by activating *ADC* expression, which increase the severity of the symptoms [30]. In turn, it has been shown that the protein 10A06 secreted by the nematode *Heterodera schachtii* has the ability to interact with the product of the Spermidine synthase 2 (*SPDS2*) gene of *Arabidopsis*. Thus, expression of the 10A06 protein in Arabidopsis lead to an increase in SPDS2 synthesis, the accumulation of Spd, and the activation of PAO activity. Altogether, these changes enhance plant susceptibility to the nematode [31]. In addition, it was demonstrated that the protein C2 from the beet severe curly top virus interacts with S-adenosylmethionine decarboxylase 1 (SAMDC1) in Arabidopsis, a phenomenon contributing to increment plant susceptibility [32]. Interestingly, the interaction between proteins from the pathogen and the host PA metabolic enzymes could work in benefits for the host instead of the pathogen. For instance, Kim et al. [33] showed that the effector AvrBsT from *Xanthomonas campestris* pv *vesicatoria* interacts with the ADC enzyme from pepper, and that this interaction enhances AvrBsT-mediated cell death and plant resistance.

3 PAs and Their Roles in Plant Abiotic Stress

3.1 Drought Stress

Global climate change provokes episodes of flooding and drought in wide areas of the planet. These phenomena, particularly drought periods, impose severe limitations to agricultural productivity and threaten food security. Plants respond to environmental changes by modifying their metabolism and increasing the concentration of different molecules to cope with stress, such as PAs. There are many evidences showing the importance of the coordinated action of the synthesis and catabolism of PAs in plant adaptation and response to the lack of water [34].

In this way, the phenotypic analysis of 21 rice cultivars showed that moderated drought stress in the long-term decreases the levels of Put and Spd, whereas those of Spm are increased. However, no correlation could be uncovered between PA contents and drought tolerance [35]. This work also showed that the expression of *ADC1* was higher than other *ADC* and *ODC* genes under control conditions. Interestingly, the basal expression of *ADC1* correlated with drought tolerance, as more tolerant cultivars showed lower expression levels. In turn, expression levels of *ODC1* under drought conditions was also linearly correlated to drought tolerance, suggesting that PA biosynthesis pathways are adjusted in response to this type of stress. In addition, it was demonstrated that the levels of Spd and Spm correlated positively with grain filling rates and negatively with ethylene production [36]. Thus, it seems that elevated levels of Spd and Spm make a substantial contribution to plant tolerance during this type of stress. In this trend, the amendment of Spd in the culture medium increased the concentration of this PA in bentgrass plants under drought stress, as well as that of Put and Spm [37]. This also led to a remarkable decrease in the concentration of the oxidative stress marker malondialdehyde, superoxide radical and H_2O_2, as well as the increment in the levels of chlorophyll and antioxidant enzymes in comparison to untreated plants. The authors suggested that Spd might ameliorate the effects of drought stress by modulating the antioxidant systems. The effect of Spm supplementation was also assessed in two cultivars of maize (*Zea mays* L) subjected to drought stress [38]. This PA provoked the increment in the Spm and Spd levels in tissues from both cultivars, whilst Put concentration was raised in one genotype and diminished in the other. Besides, relative water content and RUBISCO activity were increased with the combined treatment of Spm and 24-epibrassinolide, which ameliorate drought stress better than separated treatments. Hussain et al. [39] further explored the effects of PAs by analyzing the behavior of five hybrids of maize constructed from two cultivars identified as tolerant and sensitive until 20% water holding capacity. In this case, plants treated with Put developed higher biomass and improved leaf water status under both stressed and nonstressed conditions. As a whole, these results demonstrate that PAs contribute to drought stress tolerance. However, these beneficial effects were observed only under moderated stress conditions, since PA treatment resulted deleterious under severe drought.

Most of the contribution of PAs to plant adaptation to drought stress can be explained on the basis of a modulation of the antioxidant machinery. Thus, it was recently demonstrated in a study using reciprocal-grafted, self-grafted, and ungrafted fruits of the tomato drought-tolerant cultivar Zarina and the sensitive cultivar Josefina that higher PA contents (specially Spd) stimulate catalase and superoxide dismutase activities [40]. In the same line of evidences, white clover (*Trifolium repens* L.) plants growing in 50% of relative water content and treated with Spd showed the induction of the antioxidant systems, as well as the synthesis of other defense metabolites as flavonoids and proline. Furthermore, protein biosynthesis, the ABA-responsive protein and amino acids synthesis pathways were differentially induced [41].

Several attempts have been made to generate drought-tolerant lines by overexpressing PA synthetic genes. In this trend, *Lotus tenuis* plants expressing the oat *ADC* gene under the control of the stress responsive promoter RD29A from Arabidopsis demonstrated a significant increment in Put levels, whereas the concentration of Spd and Spm remain unaltered [42]. This work reported a direct correlation between the levels of expression of *ADC* and drought tolerance. In addition, the *NCED* gene (coding for a key enzyme participating in ABA biosynthesis) was also upregulated, suggesting that the phenotype may also rely on the activation of the

ABA pathway. As a whole, research shows that PAs are key regulators of the antioxidant activity and the homeostasis of the antioxidant compounds in plants growing under drought stress.

3.2 Saline Stress

A thorough physiological and morphological characterization has been conducted to evaluate salt tolerance of 18 rice cultivars under hydroponic culture conditions. Put levels remained unaltered in most of the genotypes, but dropped significantly in those most sensitive to salt treatment. On the other hand, even though the concentration of Spd showed small changes, Spm accumulated in tolerant and sensitive genotypes cultivated in the presence of saline water [43]. Gene expression profiling demonstrated the induction of *ADC2*, *SPD/SPM2*, and *SPD/SPM3* upon stress, whereas *ADC1* and *SAMDC4* expression was repressed under the same conditions. In turn, *SAMDC1* and *SPD/SPM1* genes were constitutively expressed.

As for plants facing drought stress, the increment in PAs levels may have a critical role in the adaptation to saline conditions. Thus, the application of Spd to two chrysanthemum cultivars improved the photosynthetic parameters and prevented the damages associated to salt stress in a dose-dependent manner [44]. In addition, the activity of the antioxidant enzymes such as catalase, ascorbate peroxidase, and superoxide dismutase were increased whereas malondialdehyde contents decreased. Treated plants also showed a remarkable reduction in the accumulation of Na^+ and K^+. These results agree with works conducted in other plant species such as mung bean (*Vigna radiata* L.), soybean (*Glycine max* L.), and cucumber (*Cucumis sativus* L.) subjected to high salt treatments [45–47]. Interestingly, the inhibition of ADC activity by D-arginine suppresses salt tolerance in soybean whereas Put alleviated the negative effects of stress. In the same line of evidences, it was shown that Arabidopsis deletion lines in the *SPMS* and *tSPMS* genes accumulated higher levels of Na^+ whereas plant growth was impaired under salt conditions [48]. Moreover, European pear (*Pyrus communis* L.) lines overexpressing the apple *SPMDS1* gene accumulate Spd and show improved tolerance to osmotic stress, which was associated to the activation of antioxidant enzymes and reduction in the concentration of malondialdehyde [34].

Different works have shown that not only biosynthesis, but also catabolism of PAs is induced during plant adaptation to salt stress. For instance, salinized soybean and foxtail millet (*Setaria italica* L.) plants showed the upregulation of the catabolic oxidases DAO and PAO [46, 49]. However, the role played by the catabolism of PAs in the adaptation of plants to salinity is still known.

3.3 Cold Stress

Three terms are used to describe cold stress in plants, freezing, chilling, and suboptimal temperature. Freezing is a term regarded when the air temperature is below 0 °C, even though agrometeorological researchers used the same term for ranges between 2 and 3 °C since plants suffer damage at these temperatures. The term chilling is used for temperatures between 0 °C and the minimum temperature of growth. In turn, it is considered as suboptimal temperature the range comprehended between the minimum temperature of growth and the optimal temperature of growth [34]. Chilling occurs regularly in temperate and tropical plants, which are exceptionally subjected to freezing conditions. Plants as rice, maize, and soybean are killed at temperatures below 0 °C, whereas wheat and barley show several degrees of tolerance to freezing.

PA levels (particularly those of Spd) were incremented in seedlings of a cold-tolerant cultivar of rice subjected to chilling temperatures, which was not observed in a sensible line. In addition, a raise in the activities of ADC and SAMDC accompanied higher PA contents [50]. Inhibition of Put biosynthesis promoted cold sensitivity, indicating that this PA plays an important role in cold adaptation. Cold-tolerant tomato plants also showed a notable increment in Put levels when compared to sensible phenotypes, as long as an induction in the activities of DAO and PAO. In addition, inhibition of ADC activity reduced the tolerance to cold, which is reverted by the addition of Put [51].

Cold-tolerant cucumber plants subjected to chilling treatment displayed a raise in Spd content under cold treatment and an increased in Put levels during rewarming [52]. On the other hand, PA levels remained unaltered in sensitive cultivars. Moreover, the increments in Put and Spd were preceded by an increment in the activities of the enzymes ADC and SAMDC [52].

Mutants of Arabidopsis in *ADC1* and *ADC2* were used to investigate the specific role played by Put in plants under chilling and freezing conditions. Long-term treatments at chilling temperatures provoked a significant reduction in Put contents as well as a reduction in the expression of the gene *NCED* [53]. By contrast, plants expressing the oat *ADC* gene under the control of the RD29A promoter showed a greater tolerance to chilling temperatures in relation to WT [54].

Proteomic and transcriptomic analysis are uncovering novel roles for plant PAs in cold tolerance. For instance, proteomics has demonstrated that PA biosynthetic enzymes are differentially accumulated in tolerant soybean plants at chilling temperatures [55]. In addition, a transcriptomic analysis of two contrasting rice cultivars associated cold tolerance to transcriptional reprogramming of the PA metabolism [56]. In the same way, Dametto et al. [57] performed a transcriptomic analysis of rice seedlings under cold stress and showed increased synthesis of dehydrin, ubiquinone protein-degradation and PA biosynthesis in cold-sensitive seedlings.

PA levels constitute a good biomolecular marker that could help to predict tolerance or sensitivity to low temperatures.

Nevertheless, future research should consider different aspects related to the intensity and duration of plant exposure to cold in order to make a better description of the roles played by PAs during exposure to low temperatures.

4 Concluding Remarks

The experimental work described in this chapter clearly indicates that PAs play a remarkable role in plant responses to abiotic and biotic stresses. Thus, PA levels (both free and conjugated) tend to increase upon pathogen recognition. These compounds may act through multiple mechanisms, such as regulation of gene expression and protein stability, and the induction of defense signaling pathways. The increment in PA levels is often accompanied by the induction in their oxidation, which has been proven necessary for the triggering of defense responses. There is also a correlation between higher PA contents and improved tolerance to abiotic stress. Protection against abiotic factors mediated by PAs could rely on their ability to act as antioxidants, the capability to stabilize nucleic acids and biomembranes, regulate cytosolic pH, and on their ability to act as compatible osmolytes. Nevertheless, the precise mechanism by which PAs participate in stress responses remains to be fully understood. Several issues need to be addressed in order to have a better understanding of these processes, such as the mechanisms involved in PAs transport to different organelles and tissues in response to environmental stresses, and the signaling cross-talk between PA and phytohormone pathways. It is conceivable that a deeper exploration of this field will make great contributions to the design of better strategies to help plants to cope with environmental stresses.

Acknowledgments

This work was supported by grants of *Agencia Nacional de Promoción Científica y Tecnológica* (ANPCyT) (PICT 2011-1612, 2014-3718, 2014-3648, and 2013-0477) and *Consejo Nacional de Investigaciones Científicas y Técnicas* (CONICET) (PIP 0363 and 0980). A.G, M.M, F.R.R, S.J.M and O.A.R are members of the Research Career of *Consejo Nacional de Investigaciones Científicas y Técnicas* (CONICET). FMR is postdoctoral fellow of ANPCyT.

References

1. Michael AJ (2016) Polyamines in eukaryotes, bacteria and archaea. J Biol Chem 291:14896–14903
2. Igarashi K, Kashiwagi K (2000) Polyamines: mysterious modulators of cellular functions. Biochem Biophys Res Commun 271:559–564

3. Poulin R, Coward JK, Lakanen JR, Pegg AE (1993) Enhancement of the spermidine uptake system and lethal effects of spermidine over-accumulation in ornithine decarboxylase-overproducing L1210 cells under hyposmotic stress. J Biol Chem 268:4690–4698
4. Jiménez-Bremont JF, Marina M, Guerrero-González ML, Rossi FR, Sánchez-Rangel D, Rodríguez-Kessler M, Ruiz OA, Gárriz A (2014) Physiological and molecular implications of plant polyamine metabolism during biotic interactions. Front Plant Sci 5:95
5. Alcazar R, Altabella T, Marco F, Bortolotti C, Reymond M, Koncz C, Carrasco P, Tiburcio AF (2010) Polyamines: molecules with regulatory functions in plant abiotic stress tolerance. Planta 231:1237–1249
6. Pandey S, Ranade SA, Nagar PK, Kumar N (2000) Role of polyamines and ethylene as modulators of plant senescence. J Biosci 25:291–299
7. Urano K, Maruyama K, Ogata Y, Morishita Y, Takeda M, Sakurai N, Suzuki H, Saito K, Shibata D, Kobayashi M, Yamaguchi-Shinozaki K, Shinozaki K (2009) Characterization of the ABA-regulated global responses to dehydration in *Arabidopsis* by metabolomics. Plant J 57:1065–1078
8. Yamasaki H, Cohen MF (2006) NO signal at the crossroads: polyamine-induced nitric oxide synthesis in plants? Trends Plant Sci 11:522–524
9. Tiburcio AF, Altabella T, Bitrián M, Alcázar R (2014) The roles of polyamines during the lifespan of plants: from development to stress. Planta 240:1–18
10. Walters DR (2003) Polyamines and plant disease. Phytochemistry 64:97–107
11. Walters DR (2000) Polyamines in plant-microbe interactions. Physiol Mol Plant Pathol 57:137–146
12. Broetto F, Marchese JA, Leonardo M, Regina M (2005) Fungal elicitor- mediated changes in polyamine content, phenylalanine-ammonia lyase and peroxidase activities in bean cell culture. Gen Appl Plant Physiol 31:235–246
13. Rodríguez-Kessler M, Ruiz OA, Maiale S, Ruiz-Herrera J, Jiménez-Bremont JF (2008) Polyamine metabolism in maize tumors induced by *Ustilago maydis*. Plant Physiol Biochem 46:805–814
14. Rossi FR, Marina M, Pieckenstain FL (2015) Role of arginine decarboxylase (ADC) in *Arabidopsis thaliana* defence against the pathogenic bacterium *Pseudomonas viridiflava*. Plant Biol 17:831–839
15. Fernández-Crespo E, Scalschi L, Llorens E, García-Agustín P, Camañes G (2015) $NH4^+$ protects tomato plants against *Pseudomonas syringae* by activation of systemic acquired acclimation. J Exp Bot 66:6777–6790
16. Marini F, Betti L, Scaramagli S, Biondi S, Torrigiani P (2001) Polyamine metabolism is upregulated in response to tobacco mosaic virus in hypersensitive, but not in susceptible, tobacco. New Phytol 149:301–309
17. Mo H, Wang X, Zhang Y, Yang J, Ma Z (2015) Cotton ACAULIS5 is involved in stem elongation and the plant defense response to *Verticillium dahliae* through thermospermine alteration. Plant Cell Rep 34:1975–1985
18. Mo H, Wang X, Zhang Y, Zhang G, Zhang J, Ma Z (2015) Cotton polyamine oxidase is required for spermine and camalexin signalling in the defence response to *Verticillium dahliae*. Plant J 83:962–975
19. Mo H-J, Sun Y-X, Zhu X-L, Wang X-F, Zhang Y, Yang J, Yan G-J, Ma Z-Y (2016) Cotton S-adenosylmethionine decarboxylase-mediated spermine biosynthesis is required for salicylic acid- and leucine-correlated signaling in the defense response to *Verticillium dahliae*. Planta 243:1023–1039
20. Gonzalez ME, Marco F, Minguet EG, Carrasco-Sorli P, Blazquez MA, Carbonell J, Ruiz OA, Pieckenstain FL (2011) Perturbation of spermine synthase gene expression and transcript profiling provide new insights on the role of the tetraamine spermine in *Arabidopsis* defense against *Pseudomonas viridiflava*. Plant Physiol 156:2266–2277
21. Marco F, Busó E, Carrasco P (2014) Overexpression of SAMDC1 gene in *Arabidopsis thaliana* increases expression of defense-related genes as well as resistance to *Pseudomonas syringae* and *Hyaloperonospora arabidopsidis*. Front Plant Sci 5:115
22. Marina M, Sirera FV, Rambla JL, Gonzalez ME, Blázquez MA, Carbonell J, Pieckenstain FL, Ruiz OA (2013) Thermospermine catabolism increases *Arabidopsis thaliana* resistance to *Pseudomonas viridiflava*. J Exp Bot 64:1393–1402
23. Sagor GHM, Takahashi H, Niitsu M, Takahashi Y, Berberich T, Kusano T (2012) Exogenous thermospermine has an activity to induce a subset of the defense genes and restrict cucumber mosaic virus multiplication in *Arabidopsis thaliana*. Plant Cell Rep 31:1227–1232
24. Marina M, Maiale SJ, Rossi FR, Romero FM, Rivas EI, Garriz A, Ruiz OA, Pieckenstain FL (2008) Apoplastic polyamine oxidation plays different roles in local responses of tobacco to

infection by the necrotrophic fungus *Sclerotinia sclerotiorum* and the biotrophic bacterium *Pseudomonas viridiflava*. Plant Physiol 147:2164–2178

25. Govrin EM, Levine A (2000) The hypersensitive response facilitates plant infection by the necrotrophic pathogen *Botrytis cinerea*. Curr Biol 10:751–757
26. Moschou PN, Sarris PF, Skandalis N, Andriopoulou AH, Paschalidis KA, Panopoulos NJ, Roubelakis-Angelakis KA (2009) Engineered polyamine catabolism preinduces tolerance of tobacco to bacteria and oomycetes. Plant Physiol 149:1970–1981
27. Yoda H, Fujimura K, Takahashi H, Munemura I, Uchimiya H, Sano H (2009) Polyamines as a common source of hydrogen peroxide in host- and nonhost hypersensitive response during pathogen infection. Plant Mol Biol 70:103–112
28. Lou Y-R, Bor M, Yan J, Preuss AS, Jander G (2016) Arabidopsis NATA1 acetylates putrescine and decreases defense-related hydrogen peroxide accumulation. Plant Physiol 171:1443–1455
29. Jasso-Robles FI, Jiménez-Bremont JF, Becerra-Flora A, Juárez-Montiel M, Gonzalez ME, Pieckenstain FL, García de la Cruz RF, Rodríguez-Kessler M (2016) Inhibition of polyamine oxidase activity affects tumor development during the maize-*Ustilago maydis* interaction. Plant Physiol Biochem 102:115–124
30. Stes E, Biondi S, Holsters M, Vereecke D (2011) Bacterial and plant signal integration via D3-type cyclins enhances symptom development in the *Arabidopsis-Rhodococcus fascians* interaction. Plant Physiol 156:712–725
31. Hewezi T, Howe PJ, Maier TR, Hussey RS, Mitchum MG, Davis EL, Baum TJ (2010) *Arabidopsis* spermidine synthase is targeted by an effector protein of the cyst nematode *Heterodera schachtii*. Plant Physiol 152:968–984
32. Zhang Z, Chen H, Huang X, Xia R, Zhao Q, Lai J, Teng K, Li Y, Liang L, Du Q, Zhou X, Guo H, Xie Q (2011) BSCTV C2 attenuates the degradation of SAMDC1 to suppress DNA methylation-mediated gene silencing in *Arabidopsis*. Plant Cell 23:273–288
33. Kim NH, Kim BS, Hwang BK (2013) Pepper arginine decarboxylase is required for polyamine and γ-aminobutyric acid signaling in cell death and defense response. Plant Physiol 162:2067–2083
34. Menéndez AB, Rodriguez AA, Maiale SJ, Rodriguez-Kessler M, Jimenez-Bremont JF, Ruiz OA (2013) Polyamines contribution to the improvement of crop plants tolerance to abiotic stress. In: Tuteja N, Gill SS (eds) Crop improvement under adverse conditions. Springer New York, New York, NY, pp 113–136
35. Do PT, Degenkolbe T, Erban A, Heyer AG, Kopka J, Köhl KI, Hincha DK, Zuther E (2013) Dissecting rice polyamine metabolism under controlled long-term drought stress. PLoS One 8(4):e60325
36. Yang W, Yin Y, Li Y, Cai T, Ni Y, Peng D, Wang Z (2014) Interactions between polyamines and ethylene during grain filling in wheat grown under water deficit conditions. Plant Growth Regul 72:189–201
37. Li Z, Zhou H, Peng Y, Zhang X, Ma X, Huang L, Yan Y (2015) Exogenously applied spermidine improves drought tolerance in creeping bentgrass associated with changes in antioxidant defense, endogenous polyamines and phytohormones. Plant Growth Regul 76:71–82
38. Talaat NB, Shawky BT (2016) Dual application of 24-epibrassinolide and spermine confers drought stress tolerance in maize (*Zea mays* L.) by modulating polyamine and protein metabolism. J Plant Growth Regul 35:518–533
39. Hussain S, Farooq M, Wahid M, Wahid A (2013) Seed priming with putrescine improves the drought resistance of maize hybrids. Int J Agric Biol 15:1349–1353
40. Sánchez-Rodríguez E, Romero L, Ruiz JM (2016) Accumulation of free polyamines enhances the antioxidant response in fruits of grafted tomato plants under water stress. J Plant Physiol 190:72–78
41. Li Z, Zhang Y, Xu Y, Zhang X, Peng Y, Ma X, Huang L, Yan Y (2016) Physiological and iTRAQ-based proteomic analyses reveal the function of spermidine on improving drought tolerance in white clover. J Proteome Res 15:1563–1579
42. Espasandin FD, Maiale SJ, Calzadilla P, Ruiz OA, Sansberro PA (2014) Transcriptional regulation of 9-cis-epoxycarotenoid dioxygenase (NCED) gene by putrescine accumulation positively modulates ABA synthesis and drought tolerance in *Lotus tenuis* plants. Plant Physiol Biochem 76:29–35
43. Do PT, Drechsel O, Heyer AG, Hincha DK, Zuther E (2014) Changes in free polyamine levels, expression of polyamine biosynthesis genes, and performance of rice cultivars under salt stress: a comparison with responses to drought. Front Plant Sci 5:182
44. Zhang N, Shi X, Guan Z, Zhao S, Zhang F, Chen S, Fang W, Chen F (2016) Treatment

with spermidine protects chrysanthemum seedlings against salinity stress damage. Plant Physiol Biochem 105:260–270

45. Nahar K, Hasanuzzaman M, Rahman A, Alam MM, Mahmud J-A, Suzuki T, Fujita M (2016) Polyamines confer salt tolerance in mung bean (*Vigna radiata* L.) by reducing sodium uptake, improving nutrient homeostasis, antioxidant defense, and methylglyoxal detoxification systems. Front Plant Sci 7:1104
46. Zhang G-w, Xu S-c, Hu Q-z, Mao W-h, Gong Y-m (2014) Putrescine plays a positive role in salt-tolerance mechanisms by reducing oxidative damage in roots of vegetable soybean. J Integr Agric 13:349–357
47. Yuan Y, Zhong M, Shu S, Du N, Sun J, Guo S (2016) Proteomic and physiological analyses reveal putrescine responses in roots of cucumber stressed by NaCl. Front Plant Sci 7:1035
48. Alet AI, Sánchez DH, Cuevas JC, Marina M, Carrasco P, Altabella T, Tiburcio AF, Ruiz OA (2012) New insights into the role of spermine in *Arabidopsis thaliana* under long-term salt stress. Plant Sci 182:94–100
49. Sudhakar C, Veeranagamallaiah G, Nareshkumar A, Sudhakarbabu O, Sivakumar M, Pandurangaiah M, Kiranmai K, Lokesh U (2015) Polyamine metabolism influences antioxidant defense mechanism in foxtail millet (*Setaria italica* L.) cultivars with different salinity tolerance. Plant Cell Rep 34:141–156
50. Pillai MA, Akiyama T (2004) Differential expression of an S-adenosyl-L-methionine decarboxylase gene involved in polyamine biosynthesis under low temperature stress in japonica and indica rice genotypes. Mol Genet Genomics 271:141–149
51. Song Y, Diao Q, Qi H (2015) Polyamine metabolism and biosynthetic genes expression in tomato (*Lycopersicon esculentum* Mill.) seedlings during cold acclimation. Plant Growth Regul 75:21–32
52. Shen W, Nada K, Tachibana S (2000) Involvement of polyamines in the chilling tolerance of cucumber cultivars. Plant Physiol 124:431–440
53. Cuevas JC, López-Cobollo R, Alcázar R, Zarza X, Koncz C, Altabella T, Salinas J, Tiburcio AF, Ferrando A (2008) Putrescine is involved in *Arabidopsis* freezing tolerance and cold acclimation by regulating abscisic acid levels in response to low temperature. Plant Physiol 148:1094–1105
54. Alet AI, Sanchez DH, Cuevas JC, del Valle S, Altabella T, Tiburcio AF, Marco F, Ferrando A, Espasandín FD, González ME, Carrasco P, Ruiz OA (2011) Putrescine accumulation in *Arabidopsis thaliana* transgenic lines enhances tolerance to dehydration and freezing stress. Plant Signal Behav 6:278–286
55. Tian X, Liu Y, Huang Z, Duan H, Tong J, He X, Gu W, Ma H, Xiao L (2015) Comparative proteomic analysis of seedling leaves of cold-tolerant and -sensitive spring soybean cultivars. Mol Biol Rep 42:581–601
56. Yang Y-W, Chen H-C, Jen W-F, Liu L-Y, Chang M-C (2015) Comparative transcriptome analysis of shoots and roots of TNG67 and TCN1 rice seedlings under cold stress and following subsequent recovery: insights into metabolic pathways, phytohormones, and transcription factors. PLoS One 10:e0131391
57. Dametto A, Sperotto RA, Adamski JM, Blasi ÉAR, Cargnelutti D, de Oliveira LFV, Ricachenevsky FK, Fregonezi JN, Mariath JEA, da Cruz RP, Margis R, Fett JP (2015) Cold tolerance in rice germinating seeds revealed by deep RNAseq analysis of contrasting indica genotypes. Plant Sci 238:1–12

Chapter 4

Thermospermine: An Evolutionarily Ancient but Functionally New Compound in Plants

Taku Takahashi

Abstract

Themospermine is a structural isomer of spermine and is present in some bacteria and most of plants. An Arabidopsis mutant, *acaulis5* (*acl5*), that is defective in the biosynthesis of thermospermine displays excessive proliferation of xylem vessels with dwarfed growth. Recent studies using *acl5* and its suppressor mutants that recover the growth without thermospermine have revealed that thermospermine plays a key role in the negative control of the proliferation of xylem vessels through enhancing translation of specific mRNAs that contain a conserved upstream open-reading-frame (uORF) in the 5′ leader region.

Key words *Acaulis5*, Arabidopsis, mRNA translation, Thermospermine, uORF, Xylem

1 Distribution of Thermospermine

The *ACAULIS5* (*ACL5*) gene in *Arabidopsis thaliana* encodes a thermospermine synthase (Fig. 1) [1, 2]. The original *acl1* mutant was identified by Tsukaya et al. [3] and named after its severe dwarfism acaulis, which means stemless or nearly so in the botanical definition. The fifth mutant *acl5* also shows a severe dwarf phenotype but with excess xylem differentiation (Fig. 2) and the responsible gene *ACL5* had initially been misidentified as the gene for spermine synthase [4]. This is because polyamine analysis of bacterial cells expressing the recombinant *ACL5* gene was conducted with dansylation of the extracted polyamines, which cannot distinguish between spermine and its structural isomer thermospermine. Later, a more detailed biochemical study revealed that *ACL5* does encode a thermospermine synthase [1]. These two isomers are separable by HPLC after their benzoylation [5].

Thermospermine was first discovered in an extrathermophilic eubacterium *Thermus thermophilus* but the bacterial thermospermine synthase gene remains to be identified [6]. In *T. thermophilus*, arginine is converted to agmatine and then aminopropylated to aminopropylagmatine. Aminopropylagmatine is converted to

Rubén Alcázar and Antonio F. Tiburcio (eds.), *Polyamines: Methods and Protocols*, Methods in Molecular Biology, vol. 1694, DOI 10.1007/978-1-4939-7398-9_4, © Springer Science+Business Media LLC 2018

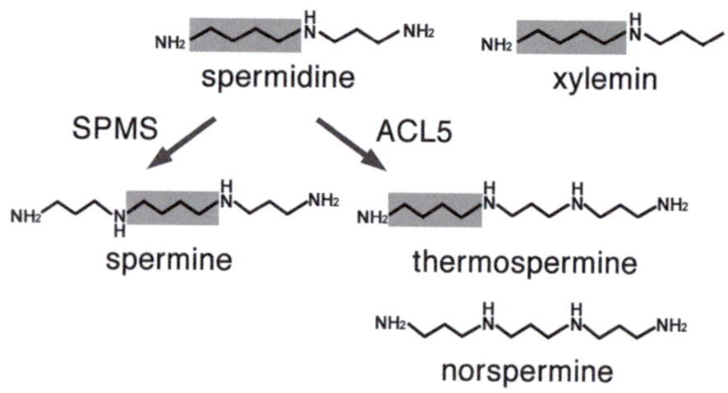

Fig. 1 Structure of major plant polyamines, norspermine, and an artificial antagonist of spermidine named xylemin. SPMS and ACL5 indicate spermine synthase and thermospermine synthase, respectively. Carbon chains of diaminobutane are *shaded*

Fig. 2 Phenotype of the *acl5* mutant of *Arabidopsis thaliana*. Thirty two-day-old flowering plants and the stem section of the wild type (*left panels*) and the *acl5* mutant (*right panels*) are shown

spermidine by agmatinase and further aminopropylated to spermine. Probably because of the reduced substrate specificity of aminopropyltransferase and agmatinase, aminopropylagmatine may be also aminopropylated to aminopropylaminopropylagmatine and then converted to thermospermine by agmatinase [7]. In eukaryotes, *ACL5* homologs have been detected in all plants so far investigated, suggesting that thermospermine is widely present in the plant kingdom [8]. Except some insects including a digger wasp, *Philanthus triangulum*, whose venom contains a thermospermine-bound toxin, philanthotoxin 433 [9], most animals and fungi may not have thermospermine. In contrast, spermine is present in most animals, fungi, seed plants, and some algae [10] but not in other algae and mosses (Table 1). While eukaryotic spermine synthase genes may have evolved from spermidine synthase genes independently in animal, plant, and fungal kingdoms, plant thermospermine synthase genes namely *ACL5* orthologs may have been obtained by a lateral trajectory from the

Table 1
Distribution of spermine and thermospermine

Domain	Organism	Tspm	Spm	Reference
Bacteria				
	Escherichia coli	–	–	[11]
	Thermoanaerobacter ethanolicus	–	+	[12]
	Thermus thermophilus	+	+	[13]
Archaea				
	Staphylothermus hellenicus	+	+	[14]
Eukaryota				
(Amoebozoa)	*Acanthamoeba culbertsoni*	–	–	[15]
(Fungi)	*Saccharomyces cerevisiae*	–	+	[16]
(Animalia)	*Caenorhabditis elegans*	–	–	[17]
	Drosophila melanogaster	–	+	[18]
	Homo sapiens	–	+	[11]
(Chromista)	*Thalassiosira pseudonana*	+	–	[1]
	Phytophthora infestans	+	–	[8]
(Plantae)	*Chlorogonium capillatum*	+	+	[10]
	Physcomitrella patens	+	–	[8]
	Selaginella lepidophylla	+	+	[19]
	Ginkgo biloba	+	+	[8]
	Oryza sativa	+	+	[5]
	Arabidopsis thaliana	+	+	[2]

cyanobacterial ancestor of the chloroplast to the host algal nucleus [20] or, alternatively, may have been acquired by the eukaryotic ancestor of the plant lineage [21].

2 Physiological Functions of Thermospermine

In the stem of the Arabidopsis *acl5* mutant, xylem vessels are over-proliferated and pith cells are reduced (Fig. 2), suggesting that thermospermine is required for repressing the differentiation of xylem vessels, which are destined to die, and the dwarf phenotype may be attributed to the reduction of the number of living cells that contribute to stem elongation [2]. In contrast, the *spms* mutant of Arabidopsis, which is defective in spermine synthase, is wild type in appearance [22]. The *acl5 spms* double mutant shows the dwarf phenotype indistinguishable from *acl5*, indicating that spermine is not essential for normal growth. Some studies suggest that spermine plays a role in stress responses such as salinity and drought as putrescine does while spermine catabolism is proposed to be a key source of hydrogen peroxide in stress responses [23]. The *SPMS* gene is expressed ubiquitously in most tissues but the *ACL5* gene is expressed only in xylem precursor cells [24, 25]. The total amount of thermospermine in seedlings is several-fold lower than that of spermine. The structural difference between thermospermine and spermine is that the former contains a tandem array of the aminopropyl chain (NC3NC3) but the latter does not. We thus examined the effect of norspermine which contains three continuous arrays of the aminopropyl chain (NC3NC3NC3) on the stem growth of *acl5* by its application to the shoot tip and confirmed that norspermine can substitute for thermospermine as a plant growth regulator [26]. On the other hand, norspermidine, a triamine also containing a tandem array of the aminopropyl chain (NC3NC3), cannot rescue the growth of *acl5*. Thus, the structure of tetraamines containing the tandem array of the aminopropyl chain may be important for the action of thermospermine in xylem differentiation. Exogenous supply of high concentrations of thermospermine to wild-type seedlings specifically represses the differentiation of xylem vessels in the root and severely inhibits the formation of lateral roots and the shoot growth [27]. Thermospermine and norspermine might be applicable as an herbicide.

Plant vascular formation is triggered by auxin. *ACL5* expression is also induced by auxin [4]. Furthermore, by screening for compounds that can modulate xylem differentiation in the *acl5* mutant, the isooctyl ester of 2,4-dichlorophenoxyacetic acid (2,4-D IOE), was identified to enhance xylem vessel differentiation remarkably in *acl5* but not in wild-type seedlings [28]. 2,4-D IOE is a pro-drug of a synthetic auxin 2,4-D and the effect on xylem differentiation in *acl5* is suppressed by the anti-auxin, *p*-chlorophenoxyisobutyric acid (PCIB) and also by thermospermine. This suggests that the auxin signaling leads to both xylem differentiation and thermospermine biosynthesis and the former is normally limited by thermospermine signaling but can be continually stimulated by an exogenously supplied auxin in the absence of thermospermine. The opposite action of thermospermine and auxin may be required to fine-tune the timing and spatial pattern of xylem vessel differentiation [29].

Expression of *ACL5* is decreased by exogenous thermospermine while it is increased in the *acl5* mutant, suggesting a negative feedback control of thermospermine biosynthesis [2]. The Arabidopsis genome has four genes encoding S-adenosyl methionine decarboxylase (*AdoMetDC* or *SAMDC*), which provides an aminopropyl donor for the synthesis of polyamines and the vascular-specific *SAMDC4* is involved in thermospermine biosynthesis. Indeed, a loss-of-function mutant of *SAMDC4* shows *acl5*-like dwarf phenotype [30]. *SAMDC4* expression is also negatively regulated by thermospermine [31].

Thermospermine is also implicated in biotic and abiotic stress responses. In Arabidopsis, thermospermine enhances expression of a subset of pathogenesis-related genes, which are upregulated during cucumber mosaic virus (CMV)-triggered hypersensitive response, and can repress CMV multiplication [32]. Exogenous supply of thermospermine as well as ectopic *ACL5* expression confers increased resistance to the biotrophic bacterium *Pseudomonas viridiflava* while *acl5* mutants are less resistant than the wild type [33]. In cotton, expression of *GhACL5* is induced upon treatment with the fungal pathogen *Verticillium dahliae*, its gene silencing enhances the susceptibility to *V. dahliae* infection, and the fungal growth is effectively inhibited by treatment with thermospermine [34]. On the other hand, the *acl5* mutant of Arabidopsis has been also shown to be more sensitive than wild-type plants to high salinity [35]. Thermospermine is catabolized by the vascular-specific polyamine oxidase encoded by *PAO5* in Arabidopsis. In accordance with the salt-sensitive phenotype of *acl5*, the loss-of-function mutant of *PAO5*, which has an increased level of thermospermine, shows increased salt tolerance [36].

3 Molecular Mechanism of the Action of Thermospermine

How does thermospermine act in the control of xylem differentiation in plants? The study on the mode of action of thermospermine has been progressed by genetic approaches with the isolation of suppressor mutants of *acl5* that restore the dwarf phenotype without thermospermine. The first clue, *SAC51*, was identified from a dominant suppressor of *acl5*, *sac51-d*, and encodes a basic helix-loop-helix (bHLH) protein. *sac51-d* has a premature termination codon in one of five upstream open-reading frames (uORFs) that is conserved among the *SAC51* family [37]. uORFs are present in approximately 40% of eukaryotic mRNAs and conserved ones are generally involved in the negative control of main ORF translation [38]. The conserved uORF of *SAC51* was also shown to have an inhibitory effect on main ORF translation [37]. A detailed study revealed that the 5′ leader sequence of the *SAC51* mRNA is responsive to thermospermine and the SAC51 protein may be overproduced in *sac51-d* [2]. Moreover, three dominant suppressors, *sac52-d*, *sac53-d*, and *sac56-d*, have been shown to contain a point mutation in genes encoding a ribosomal protein L10 (RPL10), a Receptor for activated C kinase 1 (RACK1) which is also known as a ribosomal component, and RPL4, respectively [39, 40]. These mutants also enhance translation of the *SAC51* main ORF without thermospermine. These results suggest that thermospermine acts in cancelling the inhibitory effect of the conserved uORF of the *SAC51* mRNA on translation of the main ORF. Given the fact that polyamines mainly exist as a polyamine–RNA complex, thermospermine might have an effect on the secondary structure of the *SAC51* mRNA or the ribosomal RNA thereby leading scanning ribosomes to the start codon of the main ORF (Fig. 3). This uORF is highly conserved in all four members of the *SAC51* family in Arabidopsis. Dominant mutations that cause an amino acid substitution in the corresponding uORF of *SACL1* and *SACL3* have also been shown to suppress the dwarf phenotype of *acl5* [41]. However, thermospermine enhances main ORF translation of *SAC51* and *SACL1* but not of *SACL2* and *SACL3* [27]. A detailed comparison of 5′ leader sequences of these mRNAs revealed that an additional AUG codon is present in the same reading frame as the conserved uORF of *SAC51* and *SACL1*, suggesting that the additional short uORF conserved in *SAC51* and *SACL1* might be indeed involved in the response to thermospermine [42]. Both

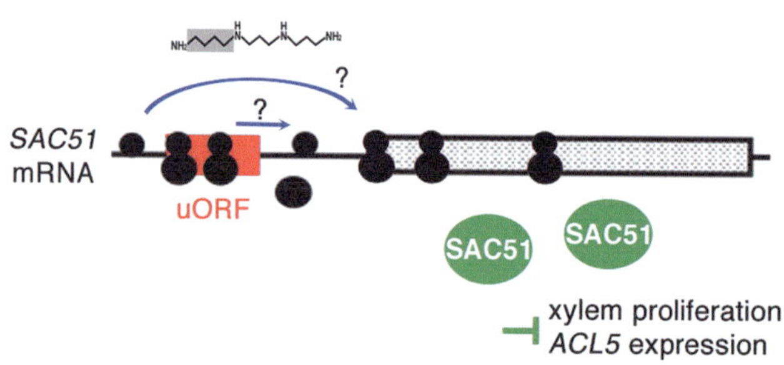

Fig. 3 A hypothetical model for the action of thermospermine in enhancing the translation of the *SAC51* main ORF. A conserved uORF and the main ORF of the *SAC51* mRNA are shown in *red* and *dotted boxes*, respectively. *Black circles* indicate small and large subunits of the ribosome. The difference in gene expression patterns between *ACL5* and *SAC51* suggests a non-cell-autonomous function of thermospermine as a mobile plant growth regulator

ACL5 and *SACL3* are expressed in xylem precursor cells under the control of the LHW-TMO5 and LHW-T5L1 bHLH heterodimeric transcription factors [43] while *SAC51* and *SACL1* are expressed throughout vascular tissues and in phloem precursor cells in the root, respectively [41]. It is thus possible that thermospermine synthesized in xylem precursor cells moves to adjacent cells and enhances translation of *SAC51* and *SACL1*. Such a non-cell-autonomous function could define thermospermine as a plant hormone.

4 Future Perspective

The regulatory targets of thermospermine that have been identified so far are only *SAC51* and *SACL1*. Given the distribution of thermospermine in plants and bacteria, there could be more targets whose translation is regulated by thermospermine. Further identification of these targets would help to clear the precise mode of action of thermospermine in translation of specific mRNAs. We have found that an antagonist of spermidine (Fig. 1) can act as an inhibitor of thermospermine biosynthesis and result in excessive xylem differentiation, a phenocopy of *acl5* in Arabidopsis, and named it xylemin for its xylem-inducing effect [44]. Simultaneous treatment of tobacco seedlings with both xylemin and 2,4-D IOE causes a drastic induction of xylem vessels in leaves. Xylemin could be expected to serve as a useful compound for the control of xylem induction and woody biomass production and also for the study of the function of thermospermine in other organisms than vascular plants.

Acknowledgement

This work was supported in part by Grants-in-Aid for Scientific Research [No. 16H0124518] from the Japan Society for the Promotion of Science (JSPS) to T.T.

References

1. Knott JM, Römer P, Sumper M (2007) Putative spermine synthases from *Thalassiosira pseudonana* and *Arabidopsis thaliana* synthesize thermospermine rather than spermine. FEBS Lett 581:3081–3086
2. Kakehi JI, Kuwashiro Y, Niitsu M, Takahashi T (2008) Thermospermine is required for stem elongation in *Arabidopsis thaliana*. Plant Cell Physiol 49:1342–1349
3. Tsukaya H, Naito S, Rédei GP, Komeda Y (1993) A new class of mutations in *Arabidopsis thaliana*, *acaulis1*, affecting the development of both inflorescences and leaves. Development 118:751–764
4. Hanzawa Y, Takahashi T, Michael AJ, Burtin D, Long D, Pineiro M et al (2000) *ACAULIS5*, an Arabidopsis gene required for stem elongation, encodes a spermine synthase. EMBO J 19:4248–4256

5. Naka Y, Watanabe K, Sagor GH, Niitsu M, Pillai MA, Kusano T et al (2010) Quantitative analysis of plant polyamines including thermospermine during growth and salinity stress. Plant Physiol Biochem 48:527–533
6. Oshima T (2010) Enigmas of biosyntheses of unusual polyamines in an extreme thermophile, *Thermus thermophilus*. Plant Physiol Biochem 48:521–526
7. Oshima T, Moriya T, Terui Y (2011) Identification, chemical synthesis, and biological functions of unusual polyamines produced by extreme thermophiles. Methods Mol Biol 720:81–111
8. Takano A, Kakehi JI, Takahashi T (2012) Thermospermine is not a minor polyamine in the plant kingdom. Plant Cell Physiol 53:606–616
9. Eldefrawi AT, Eldefrawi ME, Konno K, Mansour NA, Nakanishi K, Oltz E, Usherwood PN (1988) Structure and synthesis of a potent glutamate receptor antagonist in wasp venom. Proc Natl Acad Sci U S A 85:4910–4913
10. Hamana K, Niitsu M, Hayashi H (2013) Occurrence of homospermidine and thermospermine as a cellular polyamine in unicellular chlorophyte and multicellular charophyte green algae. J Gen Appl Microbiol 59:313–319
11. Pegg AE, Michael AJ (2010) Spermine synthase. Cell Mol Life Sci 67:113–121
12. Hosoya R, Hamana K, Niitsu M, Itoh T (2004) Polyamine analysis for chemotaxonomy of thermophilic eubacteria: polyamine distribution profiles within the orders *Aquificales*, *Thermotogales*, *Thermodesulfobacteriales*, *Thermales*, *Thermoanaerobacteriales*, *Clostridiales* and *Bacillales*. J Gen Appl Microbiol 50:271–287
13. Terui Y, Ohnuma M, Hiraga K, Kawashima E, Oshima T (2005) Stabilization of nucleic acids by unusual polyamines produced by an extreme thermophile, *Thermus thermophilus*. Biochem J 388:427–433
14. Hamana K, Tanaka T, Hosoya R, Niitsu M, Itoh T (2003) Cellular polyamines of the acidophilic, thermophilic and thermoacidophilic archaebacteria, *Acidilobus*, *Ferroplasma*, *Pyrobaculum*, *Pyrococcus*, *Staphylothermus*, *Thermococcus*, *Thermodiscus* and *Vulcanisaeta*. J Gen Appl Microbiol 49:287–293
15. Kishore P, Wittich RM, Walter RD, Shukla OP (1993) Polyamine metabolism in *Acanthamoeba culbertsoni*. Microbios 73:7–21
16. Chattopadhyay MK, Tabor CW, Tabor H (2003) Spermidine but not spermine is essential for hypusine biosynthesis and growth in *Saccharomyces cerevisiae*: spermine is converted to spermidine in vivo by the FMS1-amine oxidase. Proc Natl Acad Sci U S A 100:13869–13874
17. MacRae M, Kramer DL, Coffino P (1998) Developmental effect of polyamine depletion in *Caenorhabditis elegans*. Biochem J 333:309–315
18. Callaerts P, Geuns J, De Loof A (1992) Polyamine changes during early development of *Drosophila melanogaster*. J Insect Physiol 38:751–758
19. Sagor GH, Inoue M, Kim DW, Kojima S, Niitsu M, Berberich T, Kusano T (2015) The polyamine oxidase from lycophyte *Selaginella lepidophylla* (SelPAO5), unlike that of angiosperms, back-converts thermospermine to norspermidine. FEBS Lett 589:3071–3078
20. Minguet EG, Vera-Sirera F, Marina A, Carbonell J, Blázquez MA (2008) Evolutionary diversification in polyamine biosynthesis. Mol Biol Evol 25:2119–2128
21. Fuell C, Elliott KA, Hanfrey CC, Franceschetti M, Michael AJ (2010) Polyamine biosynthetic diversity in plants and algae. Plant Physiol Biochem 48:513–520
22. Imai A, Akiyama T, Kato T, Sato S, Tabata S, Yamamoto KT et al (2004) Spermine is not essential for survival of Arabidopsis. FEBS Lett 556:148–152
23. Mitsuya Y, Takahashi Y, Berberich T, Miyazaki A, Matsumura H, Takahashi H et al (2009) Spermine signaling plays a significant role in the defense response of *Arabidopsis thaliana* to cucumber mosaic virus. J Plant Physiol 166:626–643
24. Clay NK, Nelson T (2005) Arabidopsis *thickvein* mutation affects vein thickness and organ vascularization, and resides in a provascular cell-specific spermine synthase involved in vein definition and in polar auxin transport. Plant Physiol 138:767–777
25. Muñiz L, Minguet EG, Singh SK, Pesquet E, Vera-Sirera F, Moreau-Courtois CL et al (2008) *ACAULIS5* controls Arabidopsis xylem specification through the prevention of premature cell death. Development 135:2573–2582
26. Kakehi JI, Kuwashiro Y, Motose H, Igarashi K, Takahashi T (2010) Norspermine substitutes for thermospermine in the control of stem elongation in *Arabidopsis thaliana*. FEBS Lett 584:3042–3046
27. Cai Q, Fukushima H, Yamamoto M, Ishii N, Sakamoto T, Kurata T et al (2016) The *SAC51* family plays a central role in thermospermine responses in Arabidopsis. Plant Cell Physiol 57:1583–1592

28. Yoshimoto K, Noutoshi Y, Hayashi K, Shirasu K, Takahashi T, Motose H (2012) A chemical biology approach reveals an opposite action between thermospermine and auxin in xylem development in *Arabidopsis thaliana*. Plant Cell Physiol 53:635–645
29. Yoshimoto K, Noutoshi Y, Hayashi K, Shirasu K, Takahashi T, Motose H (2012) Thermospermine suppresses auxin-inducible xylem differentiation in *Arabidopsis thaliana*. Plant Signal Behav 7:937–939
30. Ge C, Cui X, Wang Y, Hu Y, Fu Z, Zhang D et al (2006) *BUD2,* encoding an S-adenosylmethionine decarboxylase, is required for Arabidopsis growth and development. Cell Res 16:446–456
31. Tong W, Yoshimoto K, Kakehi JI, Motose H, Niitsu M, Takahashi T (2014) Thermospermine modulates expression of auxin-related genes in Arabidopsis. Front Plant Sci 5:94
32. Sagor GH, Takahashi H, Niitsu M, Takahashi Y, Berberich T, Kusano T (2012) Exogenous thermospermine has an activity to induce a subset of the defense genes and restrict cucumber mosaic virus multiplication in *Arabidopsis thaliana*. Plant Cell Rep 31:1227–1232
33. Marina M, Sirera FV, Rambla JL, Gonzalez ME, Blázquez MA, Carbonell J et al (2013) Thermospermine catabolism increases *Arabidopsis thaliana* resistance to *Pseudomonas viridiflava*. J Exp Bot 64:1393–1402
34. Mo H, Wang X, Zhang Y, Yang J, Ma Z (2015) Cotton *ACAULIS5* is involved in stem elongation and the plant defense response to *Verticillium dahliae* through thermospermine alteration. Plant Cell Rep 34:1975–1985
35. Alet AI, Sánchez DH, Cuevas JC, Marina M, Carrasco P, Altabella T et al (2012) New insights into the role of spermine in *Arabidopsis thaliana* under long-term salt stress. Plant Sci 182:94–100
36. Zarza X, Atanasov KE, Marco F, Arbona V, Carrasco P, Kopka J et al (2016) *Polyamine oxidase 5* loss-of-function mutations in *Arabidopsis thaliana* trigger metabolic and transcriptional reprogramming and promote salt stress tolerance. Plant Cell Environ 40 (4):527–542. doi:10.1111/pce.12714
37. Imai A, Hanzawa Y, Komura M, Yamamoto KT, Komeda Y, Takahashi T (2006) The dwarf phenotype of the Arabidopsis *acl5-1* mutant is suppressed by a mutation in an upstream ORF of a bHLH gene. Development 133:3575–3585
38. Wethmar K, Smink JJ, Leutz A (2010) Upstream open reading frames: molecular switches in (patho)physiology. Bioessays 32:885–893
39. Imai A, Komura M, Kawano E, Kuwashiro Y, Takahashi T (2008) A semi-dominant mutation in the ribosomal protein L10 gene suppresses the dwarf phenotype of the *acl5* mutant in Arabidopsis. Plant J 56:881–890
40. Kakehi JI, Kawano E, Yoshimoto K, Cai Q, Imai A, Takahashi T (2015) Mutations in ribosomal proteins, RPL4 and RACK1, suppress the phenotype of a thermospermine-deficient mutant of *Arabidopsis thaliana*. PLoS One 27: e0117309
41. Vera-Sirera F, De Rybel B, Úrbez C, Kouklas E, Pesquera M, Álvarez-Mahecha JC et al (2015) A bHLH-based feedback loop restricts vascular cell proliferation in plants. Dev Cell 35:432–443
42. Yamamoto M, Takahashi T (2017) Thermospermine enhances translation of *SAC51* and *SACL1* in Arabidopsis. Plant Signal Behav 12 (1):e1276685. doi:10.1080/15592324.2016.1276685
43. Katayama H, Iwamoto K, Kariya Y, Asakawa T, Kan T, Fukuda H et al (2015) A negative feedback loop controlling bHLH complexes is involved in vascular cell division and differentiation in the root apical meristem. Curr Biol 25:3144–3150
44. Yoshimoto K, Takamura H, Kadota I, Motose H, Takahashi T (2016) Chemical control of xylem differentiation by thermospermine, xylemin, and auxin. Sci Rep 6:21487

Chapter 5

Determination of Polyamines by Capillary Electrophoresis Using Salicylaldehyde-5-Sulfonate as a Derivatizing Reagent

Takashi Kaneta

Abstract

Here we describe a protocol for the determination of polyamines using capillary electrophoresis with ultraviolet absorbance detection. Aliphatic polyamines were derivatized with salicylaldehyde-5-sulfonate (SAS) which formed Schiff base with amino groups, resulting in anionic derivatives. The derivatization of polyamines, including putrescine (PUT), cadaverine (CAD), spermidine (SPD), and spermine (SPM), was conducted in 10 mM HEPES buffer with pH 7.8 containing 5 mM SAS in a mixed solvent of water and ethanol. The SAS derivatives were separated using a background electrolyte composed of 10 mM phosphate buffer with pH 7.8. Calibration curves were linear over a concentration range of 20–200 μM for CAD, PUT, SPD, and SPM, and the limits of detection (LOD) were several μM for all polyamines. Furthermore, solid phase extraction was coupled with the CE method to improve the LOD to sub-μM levels, and the calibration curves were linear over a concentration range of 1–20 μM for CAD, PUT, and SPD.

Key words Capillary electrophoresis, Polyamines, Salicylaldehyde-5-sulfonate

1 Introduction

Polyamines, which are found in biological cells including plants and animals, play important roles in the biological systems. Many kinds of polyamines are known to function in biosynthesis and metabolism, e.g., putrescine is synthesized from arginine in microorganisms and plants, and is converted to spermidine and spermine in metabolism [1, 2].

These polyamines must be determined to understand their functions in biological systems, so it is essential to employ accurate and precise analytical techniques. High-performance liquid chromatography (HPLC) is frequently employed for the separation and determination of polyamines using derivatizing reagents such as dansyl chloride and *o*-phthalaldehyde (OPA) [3, 4]. Recently, mass spectrometry (MS) is combined with HPLC because of informative results [5], although the instrument is relatively expensive.

Rubén Alcázar and Antonio F. Tiburcio (eds.), *Polyamines: Methods and Protocols*, Methods in Molecular Biology, vol. 1694, DOI 10.1007/978-1-4939-7398-9_5, © Springer Science+Business Media LLC 2018

Capillary electrophoresis (CE) is also a complementary separation technique since it requires only a small amount of the sample for separation and determination. In fact, CE permits the determination of analytes in several μL of a sample using an automated sampler which is generally included as standard equipment for commercial CE instruments. Because of low consumption of the samples and separation medium, CE has been applied to the determination of polyamines using several labeling reagents [6–11]. Salicylaldehyde-5-sulfonate (SAS) is also a useful labeling reagent for primary amines in CE analysis, as being reported in previous publications [12, 13]. When using SAS as the labeling reagent of primary amines, electrically neutral amines are converted to anionic derivatives due to the negative charge of SAS. An advantage of SAS in CE analysis of polyamines is that the apparent electrophoretic mobility of SAS is much smaller than those of SAS-polyamine conjugates, resulting in complete separation of them from excess amounts of SAS, which is contained in sample solutions.

Here, we describe the protocol for determining polyamines using pre-column derivatization with SAS. The protocol includes synthesis of SAS, derivatization of polyamines with SAS, CE separation of the polyamine derivatives, and coupling of the CE method with preconcentration using solid phase extraction [14].

2 Materials

Prepare all solutions using ultrapure water (*see* **Note 1**) and analytical-grade reagents. Prepare and store all reagents at room temperature except for polyamine solutions, which must be stored at 4 °C or −20 °C. Diligently follow all waste disposal regulations when disposing waste materials. Filter all solutions with 0.2-μm membrane filters before use (*see* **Note 2**).

2.1 Synthesis of SAS

1. 3 M Sulfuric acid in water.
2. 20% Na_2CO_3: Dissolve 20 g of Na_2CO_3 in 100 mL of water.
3. Aniline.
4. Salycylaldehyde.
5. Hexane, ethanol and acetone.

2.2 Derivatization

1. 1 M NaOH solution in water.
2. HEPES buffer: 0.1 M HEPES–NaOH, pH 7.8 (*see* **Note 3**).
3. 0.01 M Putrescine dihydrochloride (PUT) solution in water. Store at 4 °C.
4. 0.01 M Cadaverine dihydrochloride (CAD) solution in water. Store at 4 °C.

5. 0.01 M Spermidine (SPD) solution in ethanol. Store at −20 °C.
6. 0.01 M Spermine tetrahydrochloride (SPM) solution in water. Store at −20 °C.
7. 1 mM Polyamine mixture: 1 mM PUT, 1 mM CAD, 1 mM SPD, 1 mM SPM solution in water.
8. SAS solution: 50 mM SAS in 0.1 M HEPES buffer.

2.3 Capillary Electrophoresis

1. Phosphate buffer (background electrolyte): 100 mL 0.1 M Na_2HPO_4, 10 mL of 0.1 M KH_2PO_4, pH 7.8.

2.4 Solid Phase Extraction

1. 0.1 M Sodium n-dodecylbenzene sulfonate (DBS) in water.
2. 1 M acetic acid solution.
3. 0.1 M acetic acid–NaOH, pH 5.0.
4. Standard solutions of polyamines: 0.1, 0.25, 0.5, 1, or 2 mM of each polyamine, 5% perchloric acid in water (*see* **Note 4**).
5. Solid phase extraction cartridge: HyperSep C18 cartridge (Thermo Fisher Scientific, 100 mg, 1 mL).

3 Methods

3.1 Synthesis of SAS

The derivatization reagent SAS was synthesized according to [12, 13, 15, 16].

3.1.1 Synthesis of N-Phenylsalicylaldimine

1. Mix 20 mL of methanol, 9.4 mL (103 mmol) of aniline and 10.8 mL (103 mmol) of salicylaldehyde in a 50-mL beaker (*see* **Note 5**).
2. Stir the solution overnight.
3. Transfer the solution into a separation funnel. Collect the lower phase into a 50-mL beaker, resulting in a yellow solid (*see* **Note 6**).
4. Put the beaker in a vacuum desiccator to dry the solid.
5. Recrystallize the solid by adding 100 mL of hexane.
6. Filter the solid using a filter paper (*see* **Note 7**).

3.1.2 Synthesis of N-Phenylsalicylaldimine-5-Sulfonic Acid

1. Mix 12.6 g of N-phenylsalicylaldimine with 35 mL of concentrated sulfuric acid.
2. Stir the solution in a water bath at 100 °C for 2.5 h.
3. Add the solution gently to 35 mL cold water with stirring (*see* **Note 8**).
4. Filter the precipitate with a Buchner funnel.

5. Recrystallize the precipitate in 70 mL 3 M sulfuric acid. Filter the precipitate and then wash with cold water, ethanol, and acetone.
6. Dry the precipitate in a vacuum desiccator (*see* **Note 9**).

3.1.3 Synthesis of Sodium SAS Monohydrate

1. Mix 1 g of N-phenylsalicylaldimine-5-sulfonic with 10 mL 20% (w/v) Na_2CO_3.
2. Boil the solution for 2.5 h. Meanwhile, add hot water to keep the volume constant (*see* **Note 10**).
3. Reduce the volume to 16 mL by evaporation.
4. Cool the solution and adjust the pH to 5 with acetic acid (*see* **Note 11**).
5. Add an equal volume of ethanol, and then cool to 0 °C in an ice bath to precipitate the product.
6. Filter the precipitate, and then wash with ethanol. (*see* **Note 12**).
7. Dry the precipitate in a vacuum desiccator (*see* **Note 13**).

3.2 CE System

Use a commercially available CE system equipped with a spectrophotometric detector (Model ^{3D}CE, (Agilent Technologies, CA, USA) is employed in our study). Electropherograms were recorded via Hewlett-Packard ChemStation software, which permits the automatic measurement of peak height, peak area, and migration time.

3.3 Electrophoresis

1. Keep the temperature of the CE system at 20 °C. Set the detection wavelength at 240 nm.
2. Condition the separation capillary by rinsing with 0.1 M NaOH for 5 min, deionized water for 5 min, and the background electrolyte for 5 min, sequentially.
3. Take 50 μL of 1 mM polyamine mixture, 50 μL of 50 mM SAS solution, 100 μL of 0.1 M HEPES buffer, 200 μL of water, and 600 μL of ethanol into a sample vial for electrophoresis and put in the vial tray of the CE system (*see* **Note 14**). React for 40 min in the vial tray to produce the SAS derivatives of the polyamines (*see* **Note 15**) (Fig. 1).
4. Inject the solution of the SAS derivatives into the capillary by pressure at 50 mbar for 5 s.
5. Apply a constant potential of +30 kV to the inlet side of the separation capillary and run for 10 min.
6. Save the data as a file and converted it to an ASCII file. The ASCII file can be processed with Microsoft Excel to draw the electropherogram (Fig. 2).
7. Flush the capillary with 0.1 M NaOH and fill it with water after the experiment.

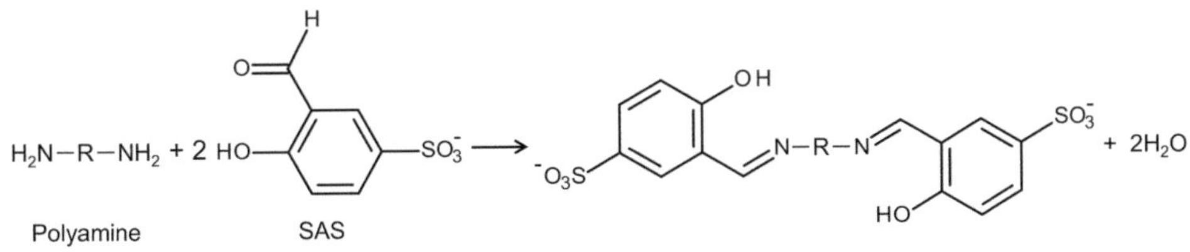

Fig. 1 Reaction of a polyamine with SAS

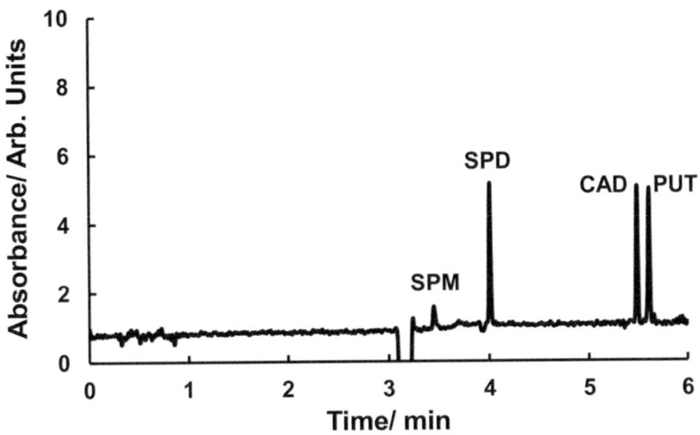

Fig. 2 Typical electropherogram of a polyamine mixture. Sample solution; 50 μM CAD, PUT, SPD, SPM, and 5 mM SAS in 10 mM HEPES buffer (pH 7.8) containing 60% ethanol; background electrolyte for electrophoresis, 10 mM phosphate (pH 7.8)

3.4 Calibration Curve

1. Take an appropriate volume (25, 50, 100, or 200 μL) of 1 mM polyamine mixture, 50 μL of 50 mM SAS solution, 100 μL of 0.1 M HEPES buffer, 50 μL of water, and 600 μL of ethanol into a sample vial to prepare a standard solution of SAS derivatives with 25, 50, 100, or 200 μM (*see* **Note 16**).
2. Put the sample vials in the vial tray of the CE system. React for 40 min in the vial tray to produce the SAS derivatives of the polyamines.
3. Measure the peak areas and plot values against polyamines concentrations (*see* **Note 17**).

3.5 Solid Phase Extraction

1. Take 3 mL of a standard solution of polyamines. Adjust the pH to 5 by adding solid potassium hydrogen carbonate and cool to 0 °C (*see* **Note 18**).
2. Filter the cooled solution to remove the precipitate of potassium perchlorate by a disposable syringe filter (*see* **Note 19**). Transfer the filtrate to a 5-mL beaker.
3. Add 600 μL of 0.1 M acetate buffer and 60 μL of 0.1 M DBS into the 5-mL beaker.

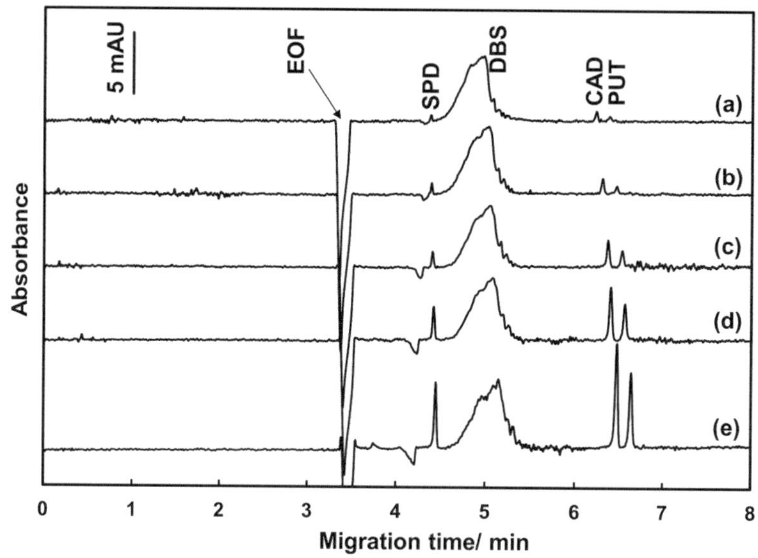

Fig. 3 Electropherograms of SAS-derivatives after preconcentration by SPE. Background electrolyte for electrophoresis, 10 mM phosphate (pH 7.8). Sample solutions; (**a**) 1 μM CAD, PUT, SPD, and SPM, (**b**) 2.5 μM CAD, PUT, SPD, and SPM, (**c**) 5 μM CAD, PUT, SPD, and SPM, (**d**) 10 μM CAD, PUT, SPD, and SPM, (**e**) 20 μM CAD, PUT, SPD. All solutions also contained 5 mM SAS, 10 mM HEPES buffer (pH 7.8), and 90% ethanol

4. Set the HyperSep C18 cartridge to a bell jar of a vacuum filtration system.
5. Put a vial in the bell jar to collect eluent.
6. Condition the HyperSep C18 cartridge with 1 mL of acetone, 1 mL of methanol, 1 mL of water, and 1 mL of acetate buffer successively.
7. Pour the polyamine solution into the HyperSep C18 cartridge. Wash the HyperSep C18 cartridge with 1 mL of water.
8. Replace the vial in the bell jar with a vacant vial.
9. Pour 270 μL of ethanol to elute polyamine-DBS ion pair.
10. Add 30 μL of 50 mM SAS into the collected ethanol solution.
11. Put the vial in the vial tray of the CE system and left for 40 min to react polyamines with SAS.
12. Start electrophoretic runs (*see* **Notes 20** and **21**) (Fig. 3).

4 Notes

1. Filter pure water before the use for the preparation of solutions. A disposable syringe filter is usually employed for rapid filtration.
2. The filtration helps to remove particulate matters in solutions since they may appear as spike noises in the electropherogram and may clog the capillary.

3. HEPES buffer is more suitable than phosphate buffer since phosphate salts form precipitation at the ethanol concentration of more than 60%.
4. Perchloric acid was added because polyamines are frequently extracted from plant samples with 5% perchloric acid.
5. The solution turns the color to yellow and then separates into two phases.
6. Solid appears immediately when the solution flows away from the separation funnel.
7. Filter paper of Whatman No. 5 is suitable for the filtration. The yield is around 62% (12.6 g).
8. Yellow precipitation appears while adding the solution to cold water.
9. The yield is around 30% (~4 g).
10. Pour hot water regularly to keep the volume constant.
11. Measure with pH test strips.
12. Sodium SAS monohydrate is white while N-phenylsalicylaldimine-5-sulfonic acid is yellow.
13. The yield is around 56% (0.56 g).
14. Cool the sample tray to 20 °C if the CE system equipped a thermostat system. The optimal temperature for the derivatization is 20 °C.
15. The sample contains 50 μM of PUT, CAD, SPD, and SPM, 50 mM of SAS, 10 mM of HEPES buffer (pH 7.8), and 60% of ethanol.
16. All solutions must contain 60% of ethanol. The concentration of ethanol must be constant for all solutions since ethanol content influences the stability of the SAS derivatives.
17. Correlation coefficients (R^2) were 0.9959 for PUT, 0.9986 for CAD, 0.9993 for SPD, and 0.9518 for SPM.
18. We do not need to cool the solution if the sample contains no perchlorate ion.
19. Use the sample solution directly without filtration when it contains no precipitate of $KClO_4$.
20. No peak for SPM appears in the electropherograms since SPM is not retained in the HyperSep C18 cartridge.
21. Although a broad peak for DBS appears between SPD and CAD in the electropherograms, it is completely separated from the SAS-polyamine derivatives.

References

1. Jänne J, Pösö H, Raina A (1978) Polyamines in rapid growth and cancer. Biochim Biophys Acta 473:241–293
2. Luk GD, Casero RA (1987) Polyamines in normal and cancer cells. Adv Enzym Regul 26:91–105
3. Busto O, Valero Y, Guasch J, Borrull F (1994) Solid phase extraction applied to the determination of biogenic amines in wines by HPLC. Chromatographia 38:571–578
4. Busto O, Mestres M, Guasch J, Borrull F (1995) Determination of biogenic amines in wine after clean-up by solid-phase extraction. Chromatographia 40:404–410
5. Saccani G, Tanzi E, Pastore P, Cavalli S, Reyd M (2005) Determination of biogenic amines in fresh and processed meat by suppressed ion chromatography-mass spectrometry using a cation-exchange column. J Chromatogr A 1082:43–50
6. Mattusch J, Huhn G, Wennrich R (1995) Sensitive laser induced fluorescence detection of polyamine-fluoresceinisothiocyanate-derivatives after capillary zone electrophoretic separation. Fresenius J Anal Chem 351:732–738
7. Legaz ME, Vicente C, Pedrosa MM (1998) Separation of tosylated polyamines by high-performance capillary zone electrophoresis. J Chromatogr A 823:511–521
8. Paproski RE, Roy KI, Lucy CA (2002) Selective fluorometric detection of polyamines using micellar electrokinetic chromatography with laser-induced fluorescence detection. J Chromatogr A 946:265–273
9. Liu G, Chen J, Ma Y (2004) Simultaneous determination of catecholamines and polyamines in PC-12 cell extracts by micellar electrokinetic capillary chromatography with ultraviolet absorbance detection. J Chromatogr B805:281–288
10. Zhang L-Y, Tang X-C, Sun M-X (2005) Simultaneous determination of histamine and polyamines by capillary zone electrophoresis with 4-fluor-7-nitro-2,1,3-benzoxadiazole derivatization and fluorescence detection. J Chromatogr B 820:211–219
11. Fu N-N, Zhang H-S, Ma M, Wang H (2007) Quantification of polyamines in human erythrocytes using a new near-infrared cyanine 1-(epsilon-succinimidyl-hexanoate)-1′-methyl-3,3,3′,3′-tetramethyl-indocarbocyanine-5,5′-disulfonate potassium with CE-LIF detection. Electrophoresis 28:822–829
12. Driouich R, Takayanagi T, Oshima M, Motomizu S (2001) Separation and determination of n-alkylamines and histamine by capillary zone electrophoresis using salicylaldehyde-5-sulfonate as a derivatizing reagent. J Chromatogr A 934:95–103
13. Driouich R, Takayanagi T, Oshima M, Motomizu S (2003) Investigation of salicylaldehyde-5-sulfonate as a precolumn derivatizing agent for the determination of n-alkane diamines, lysine, diaminopimelic acid, and isoniazid by capillary zone electrophoresis. J Pharm Biomed Anal 30:1523–1530
14. Inoue G, Kaneta T, Takayanagi T, Kakehi J, Motose H, Takahashi T (2013) Determination of polyamines in *Arabidopsis Thaliana* by capillary electrophoresis using salicylaldehyde-5-sulfonate as a derivatizing reagent. Anal Methods 5:2854–2859
15. Berry KJ, Moya F, Murray KS, van den Bergen AMB, West BO (1982)Water-solublecobalt(II) complexes ofNN'-substitutedbis(salicylaldimine-5-sulphonic acids).Oxygen-carrying-properties and conversion into cobalt(III) organometallic compounds. J Chem Soc Dalton Trans 1982:109–116
16. Baugh LS, Sissano JA (2002) Polymerization of methyl methacrylate and other polar monomers with alkylaluminum initiators bearing bidentate and tridentate N- and O-donor ligands. J Polym Sci Part A Polym Chem 40:1633–1651

Chapter 6

Detection of Thermospermine and Spermine by HPLC in Plants

Taku Takahashi, Ayaka Takano, and Jun-Ichi Kakehi

Abstract

Thermospermine, a structural isomer of spermine, is widely spread in the plant kingdom and has recently been shown to play a key role in the repression of xylem differentiation in vascular plants. However, a standard high-performance liquid chromatography (HPLC) protocol for detecting polyamines as their dansyl derivative cannot distinguish themospermine from spermine. These isomers become separated from each other after benzoylation. In this chapter, we describe a simple protocol for extraction, benzoylation, and HPLC detection of thermospermine and spermine with other polyamines from plant material.

Key words *Arabidopsis*, Benzoylation, HPLC, Spermine, Thermospermine

1 Introduction

Polyamines such as putrescine, cadaverine, spermidine, and spermine can be easily detected by a rapid and simple HPLC method that involves their labeling with dansyl chloride [1]. However, one disadvantage of this method is that spermine and its structural isomer, thermospermine, cannot be distinguished from each other. Themospermine is present in some bacteria and most of plants [2–4]. As described in Chapter 4, it is only in the recent years that the key and specific role of thermospermine in the repression of xylem differentiation in vascular plants has been found [5] and the field of study on plant thermospermine still remains largely unexplored. Studies that report on the polyamine content in plant tissue often overlook the difference between spermine and thermospermine because they use the conventional sensitive method using dansyl chloride for detecting polyamines [6]. A study by Knott et al. clearly showed using gas chromatography-mass spectrometry that the *ACAULIS5* (*ACL5*) gene of *Arabidopsis thaliana* does not encode a spermine synthase but a thermospermine synthase [7]. We have shown by thin-layer chromatography (TLC) that wild-type seedlings of *Arabidopsis* contain both spermine and

Rubén Alcázar and Antonio F. Tiburcio (eds.), *Polyamines: Methods and Protocols*, Methods in Molecular Biology, vol. 1694, DOI 10.1007/978-1-4939-7398-9_6, © Springer Science+Business Media LLC 2018

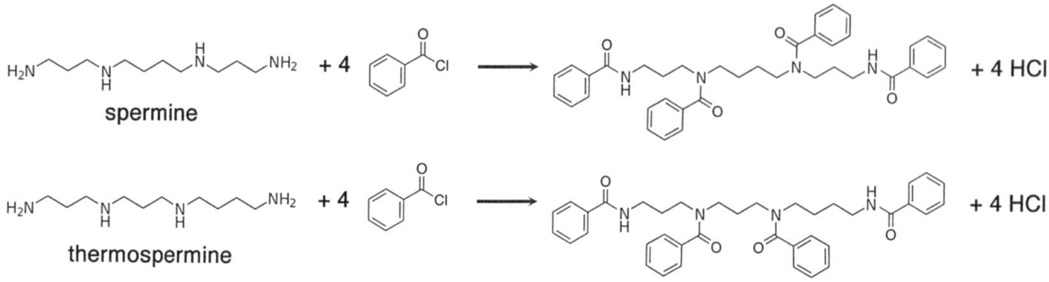

Fig. 1 The reaction between benzoyl chloride and spermine or thermospermine under alkaline conditions

thermospermine but because of the low sensitivity of the TLC assay, the seedlings had to be incubated with the substrate spermidine before polyamine extraction [5]. Another sensitive method had been developed for the analysis of polyamines in higher plant extracts based on reverse-phase HPLC of their benzoyl derivatives [8, 9]. Benzoylation of polyamines is a substitution reaction of active hydrogen atoms of amino groups with benzoyl groups (Fig. 1) [10]. Thermospermine and spermine have also been shown to be relatively easily distinguishable in HPLC analysis after their benzoylation [11]. Based on the reported procedure, we describe here a simple protocol for extraction, benzoylation, and HPLC detection of polyamines from *Arabidopsis* seedlings. The protocol is also applicable to extracts from other plant material and bacterial or yeast cells.

2 Materials

Most reagents are common in molecular biology laboratories and may be purchased from a preferred supplier. Prepare all solutions using ultrapure water with a sensitivity of 18 MΩ cm at 25 °C and analytical-grade reagents. Prepare and store all reagents at room temperature unless otherwise indicated.

2.1 Sample Preparation

1. Liquid nitrogen.
2. Mortar and pestle.
3. Perchloric acid (PCA): Prepare 5% (v/v) solution in water and store at 4 °C.
4. 2 N NaOH.
5. Benzoyl chloride.
6. Saturated NaCl: Add NaCl in small quantities at a time to approximately 80 mL water in a glass beaker, with stirring and heating, until no further NaCl will dissolve.
7. Diethyl ether.
8. Methanol.

2.2 Reverse Phase HPLC

1. HPLC system (Agilent 1120 Compact LC).
2. Reverse phase column; TSKgel ODS-80Ts with 4.6 mm ID × 250 mm length and 5 μm particle size (Toso, Tokyo, Japan).
3. HPLC-grade acetonitrile: Prepare 42% (v/v) solution in water (*see* **Note 1**).

2.3 Polyamines

Hydrochloride salts of putrescine, spermidine, and spermine, are available from several chemical companies, i.e. Sigma-Aldrich. Thermospermine tetrahydrochloride can be purchased from Santa Cruz Biotechnology, Inc.

3 Methods

3.1 Extraction and Benzoylation of Plant Polyamines

1. Harvest 0.5 g fresh weight of young seedlings of *Arabidopsis thaliana* and extract polyamines by grinding the sample with mortar and pestle in liquid nitrogen (*see* **Notes 2–4**).
2. Suspend the powder in 5-fold volumes (2.5 mL) of 5% (v/v) cold PCA in a 15 mL falcon tube, incubate on ice for 1 h, and centrifuge at 3000 × *g* for 10 min at 4 °C (*see* **Note 5**).
3. Attach the syringe filter of 0.2 μm pore size to a 5-mL syringe. After centrifugation, transfer the supernatant into the syringe and press the plunger gently to push sample through the filter into a new 15 mL falcon tube.
4. To 2 mL of the extract, add 1 mL 2 N NaOH for neutralization and mix by vortex.
5. Add 10 μL of benzoyl chloride, mix, and incubate for 20 min at room temperature.
6. Add 2 mL saturated NaCl, mix, add 2 mL diethyl ether and mix by vigorous shaking for 10–15 s (*see* **Note 6**).
7. Centrifuge at 3000 × *g* for 10 min at 4 °C and transfer the upper ether layer into a 2 mL new tube.
8. Place the sample tube on ice, evaporate to dryness under vacuum, and suspend the residue in 50 μL of methanol. Store the sample at −20 °C until use (*see* **Note 7**).

3.2 HPLC Analysis

1. Set up the HPLC system equipped with a TSKgel ODS-80Ts column (4.6 mm × 250 mm) that is equilibrated with 42% (v/v) acetonitrile. Set the flow rate at 0.5 mL/min.
2. Start the run with 10 μL injection of standard mixture. Standard solution contains 0.5 nmol each of polyamines. Monitor the UV absorbance of benzoylated polyamines at 254 nm. The HPLC standards are putrescine, 7 min; spermidine, 13.5 min; thermospermine 23 min; spermine, 25 min (Fig. 2a).

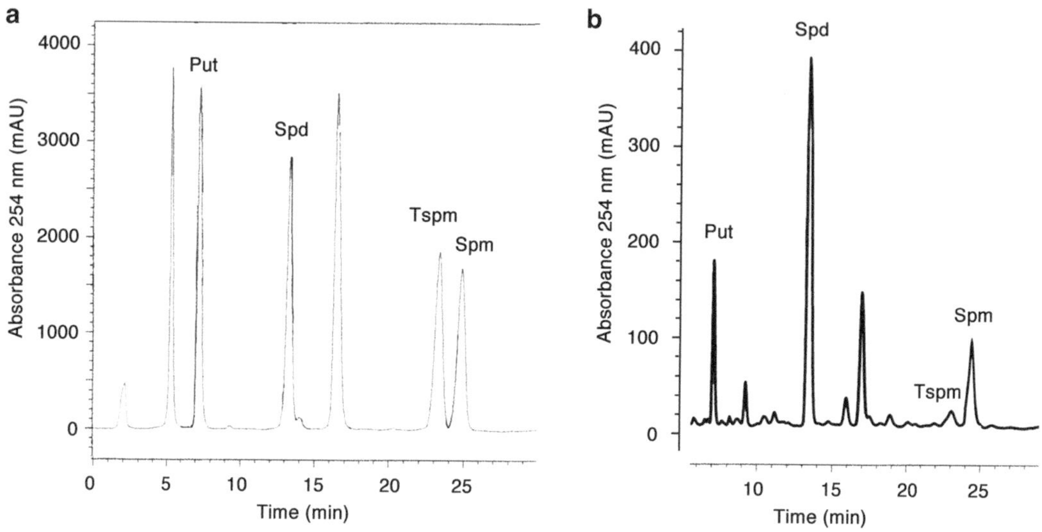

Fig. 2 HPLC profile of benzoylated polyamines-putrescine (Put), spermidine (Spd), thermospermine (Tspm), and spermine (Spm) of standards (**a**) and those prepared from wild-type *Arabidopsis* seedlings (**b**)

3. Start the sample analysis when the chromatogram of the standards is satisfactory (*see* **Note 8**). An example of the chromatogram of plant polyamines is shown in Fig. 2b.
4. Continue to assay samples.

4 Notes

1. Sonication of acetonitrile solutions for 10 min before use is preferable to remove air bubbles. Failure to degas the mobile phase may result in spurious peaks throughout the chromatogram.
2. This protocol can be adopted for other plant material and bacterial or yeast cell cultures. For extraction from cultured cells, collect the cells by centrifuging 50 mL of the cell suspension at 3000 × *g* for 10 min at 4 °C, wash twice with 0.8% (w/v) NaCl, resuspend with 5% (v/v) cold PCA, sonicate eight times, each for 30 s with 30-s intervals on ice, and place on ice for 1 h. Proceed as in **step 3** of Subheading 3.1 but filter twice.
3. Frozen tissues can be stored at −80 °C for 1–2 months.
4. If the study is intended to detect weak activity of spermine or thermospermine synthase in vivo, it might be preferable to add the substrate spermidine, i.e. at 0.1–1 mM final concentration for 24 h before extraction, into culture media.
5. For preparation of standard mixture, add 2.5 mL of 5% (v/v) cold PCA to 10 μL solution containing 1 mM each of polyamines and follow the steps thereafter.

6. Diethyl ether is highly volatile and its vapors are extremely flammable. It should be used in a fume hood.
7. Unlike dansyl derivatives, the benzoyl derivatives are not light sensitive and are stable in methanol for several months at −20 °C [8].
8. One cycle of the run usually takes up to 2 h to complete but can be shortened by increasing the acetonitrile concentration gradually from 42% to 100% in 3 min, eluting for 20 min, and decreasing down to 42% in 3 min in the intervals between polyamine separation.

Acknowledgment

This work was supported in part by Grants-in-Aid for Scientific Research [No. 16H0124518] from the Japan Society for the Promotion of Science (JSPS) to T.T.

References

1. Marcé M, Brown DS, Capell T, Figueras X, Tiburcio AF (1995) Rapid high-performance liquid chromatographic method for the quantitation of polyamines as their dansyl derivatives: application to plant and animal tissues. J Chromatogr B Biomed Appl 666:329–335
2. Minguet EG, Vera-Sirera F, Marina A, Carbonell J, Blázquez MA (2008) Evolutionary diversification in polyamine biosynthesis. Mol Biol Evol 25:2119–2128
3. Fuell C, Elliott KA, Hanfrey CC, Franceschetti M, Michael AJ (2010) Polyamine biosynthetic diversity in plants and algae. Plant Physiol Biochem 48:513–520
4. Takano A, Kakehi JI, Takahashi T (2012) Thermospermine is not a minor polyamine in the plant kingdom. Plant Cell Physiol 53:606–616
5. Kakehi JI, Kuwashiro Y, Niitsu M, Takahashi T (2008) Thermospermine is required for stem elongation in *Arabidopsis thaliana*. Plant Cell Physiol 49:1342–1349
6. Hanzawa Y, Takahashi T, Michael AJ, Burtin D, Long D, Pineiro M et al (2000) *ACAULIS5*, an *Arabidopsis* gene required for stem elongation, encodes a spermine synthase. EMBO J 19:4248–4256
7. Knott JM, Römer P, Sumper M (2007) Putative spermine synthases from *Thalassiosira pseudonana* and *Arabidopsis thaliana* synthesize thermospermine rather than spermine. FEBS Lett 581:3081–3086
8. Flores HE, Galston AW (1982) Analysis of polyamines in higher plants by high performance liquid chromatography. Plant Physiol 69:701–706
9. Slocum RD, Flores HE, Galston AW, Einstein LH (1989) Improved method for HPLC analysis of polyamines, agmatine and aromatic monoamines in plant tissue. Plant Physiol 89:512–517
10. Redmond JW, Tseng A (1979) High-pressure liquid chromatographic determination of putrescine, cadaverine, spermidine and spermine. J Chromatogr 170:479–481
11. Naka Y, Watanabe K, Sagor GH, Niitsu M, Pillai MA, Kusano T et al (2010) Quantitative analysis of plant polyamines including thermospermine during growth and salinity stress. Plant Physiol Biochem 48:527–533

Chapter 7

Development of Amine-Oxidase-Based Biosensors for Spermine and Spermidine Analysis

Cristina Tortolini, Gabriele Favero, and Franco Mazzei

Abstract

In this work a detailed description of the development of amine oxidase-based electrochemical biosensors for the selective determination of the biogenic amines is presented. The enzymes required for this operation are Polyamine Oxidase (PAO) and Spermine Oxidase (SMO) which are physically entrapped in poly(vinyl alcohol) bearing styrylpyridinium groups (PVA-SbQ), a photo-cross-linkable gel, onto screen printed electrode (SPE) surface. The developed biosensors are deeply characterized in the analysis of biogenic amines by using flow injection amperometric (FIA) technique. The enzymatic electrodes are characterized by good sensitivity, long-term stability, and reproducibility. To test the feasibility of the developed biosensors in the analysis of real matrices, they are used for the analysis of blood samples. The results obtained are in good agreement with those obtained with the GC-MS reference method.

Key words Enzymatic electrode, Polyamine oxidase, Spermine oxidase, Biogenic amines, Clinical and biomedical analyses

1 Introduction

Biogenic amines (BAs) are nitrogenous organic compounds characterized by a low molecular weight, produced by the decarboxylation of amino acids in animals, plants, and microorganisms. BAs can be divided into aliphatic (putrescine, cadaverine, spermidine, spermine), heterocyclic (histamine, tryptamine), and aromatic (tyramine, phenylethylamine) [1, 2]. More recently, it has become apparent that polyamines may also have useful medical applications. Under normal conditions, levels of polyamines in physiological fluids are low whereas they may rise remarkably in the presence of several neoplastic diseases. In this framework, they have been considered as markers of tumor growth (putrescine concentrations) and cell turnover (spermidine levels) [3–6].

Conversely, high levels of BAs can also be found in processed food as evidence of an ongoing putrefaction process, thus representing a marker for food freshness [7–9]. In fact, polyamines are

Rubén Alcázar and Antonio F. Tiburcio (eds.), *Polyamines: Methods and Protocols*, Methods in Molecular Biology, vol. 1694, DOI 10.1007/978-1-4939-7398-9_7,

related to the quality of the food products as they may yield carcinogenic compounds, especially when also nitrites are present [1, 10, 11]. In the last years, a growing demand for quick, cheap, and disposable methods both as point of care testing (POCT) and for the quality and safety assessment of food arose [12, 13]. The traditional analytical techniques are time-consuming, laborious, and require well-trained operators. In this respect, electrochemical biosensors for their peculiar characteristics: simplicity, selectivity, sensitivity, cost-effectiveness, and miniaturizability, could play an important role.

In this work a detailed description of the development of amine oxidase-based electrochemical biosensors for the selective determination of the biogenic amines is presented. The use of amine oxidases with various amperometric transducers has been described in the literature in the recent years. Amine oxidases [1, 14, 15] are a class of enzymes catalyzing, in the presence of molecular oxygen as an electron acceptor, the oxidative deamination of primary amines, diamines, and substituted amines in aldehydes, with the formation of ammonia and hydrogen peroxide. Hydrogen peroxide appears to be the most appropriate electrochemical-measuring analyte as it offers accuracy and reproducibility [1, 12]. One of the main problems affecting the use of amine oxidase first-generation-based biosensors is bound to the direct electrochemical detection of hydrogen peroxide. In fact, this measurement must be performed at high applied potential (generally +700 mV) where other electroactive substances can be oxidized interfering in the measurements [16]. To solve this problem, the use of electrochemical mediators reducing the applied potential measurement value [16–19] has been performed. In this work, is reported the development of SMO-based biosensor for the selective and sensitive detection of spermine (Spm) [20, 21], and of a PAO-based biosensor for the measurement of the total concentration of Spermine and Spermidine (Spmd) substrates [22]. PAO and SMO are immobilized onto the Prussian Blue (PB) modified electrode by means of photo-cross-linkable poly(vinyl alcohol) with styrylpyridinium groups (PVA-SbQ) onto the electrode surface. PB is used as redox mediator for measuring the enzymatically generated hydrogen peroxide at low applied potential (−100 mV vs. Ag/AgCl) reducing electrochemical interferences [16, 21, 22]. The developed biosensors are also employed in the analysis of Spm and Spmd in blood samples comparing the results obtained with those achieved with the GC-MS reference method.

2 Materials

Prepare all the solutions using high-purity deionized water (resistance 18 MΩ cm at 25 °C; TOC < 10 μg L^{-1}). All other chemicals are of analytical grade.

2.1 Stock Solutions of Substrates

1. PAO (polyamine oxidase) biosensor stock solutions of Spermine (Spm) and Spermidine (Spmd) are prepared in phosphate buffer (PBS) 50 mM pH 5.8, with 100 mM KCl.
2. SMO (spermine oxidase) biosensor stock solutions of Spm are prepared in PBS 50 mM pH 8.0, with 100 mM KCl.

2.2 Blood Samples

Blood samples are treated with 4 volumes of 8% ice-cold perchloric acid, mixed thoroughly for 5 min and centrifuged at 5000 × *g* for 20 min. The supernatant is used for the analysis.

3 Methods

Carry out all procedures at room temperature unless otherwise specified.

3.1 Prussian Blue Electrochemical Deposition

1. Prepare a solution containing 0.1 M KCl and 0.1 M HCl as a supporting electrolyte, 1.9 mM of $K_3[Fe(CN)_6]$ and 1.9 mM of $FeCl_3$.
2. Put a graphite screen-printed electrode (GP-SPE) into the solution prepared above.
3. Perform the electrodeposition of the redox mediator Prussian Blue (PB) in the potential range 0.35 to −0.1 V vs Ag/AgCl at scan rate of 50 mVs^{-1}, for 10 cycles.
4. Activate the mediator employing the same supporting electrolyte solution, in the potential range 0.35 to −0.1 V vs Ag/AgCl at scan rate of 50 mVs^{-1}, for 40 cycles.

3.2 Enzyme Immobilization

1. Prepare the enzyme/PVA-SbQ solution: a known aliquot (6 μL) of the enzymatic solution (PAO or SMO) are mixed with an aliquot (8 μL) of PVA-SbQ solution (*see* **Note 1**).
2. Homogenize the resulting enzyme/PVA-SbQ solution by vortex mixing (*see* **Note 2**), and homogeneously spread 4 μL of this solution onto the SPE working electrode surface (*see* **Note 3**).
3. Maintain the enzyme-SPEs surfaces for 20 min under a UV lamp at room temperature allowing the immobilization of the enzyme by photopolymerization (*see* **Note 4**).

3.3 Electrochemical Measurements: Amperometric Measurements

1. A μAutolab type III potenstiotat with a conventional three-electrodes configuration with a microliter flow cell connected to a flow injection valve and a Gilson's Minipuls 3 peristaltic pump (*see* **Note 5**) are employed for the amperometric detection. For the measurements, GP-SPEs are used (*see* **Notes 6** and 7).
2. The calibration plot is obtained by adding several aliquots of the substrate standard solution at different concentrations into the buffer solution.

3.4 Characterization of PAO and SMO Biosensors

1. Increasing concentrations of substrate solutions are injected into the FIA valve and send to the microliter flow-cell with the enzymatic SPE. The measurement is performed acquiring the maximum of the signal recorded for each standard solution (*see* **Note 8**).
2. Data obtained are elaborated to calculate the main kinetic parameters by modeling the experimental results using the classical Michaelis-Menten equation with Prism 5.04 software from GraphPad Software, Inc. (USA).

3.5 Analysis of Real Samples by Electrochemical Measurements

1. The analysis of blood samples are performed in the same condition early described. The samples, appropriately diluted, according the linear range of the calibration plot, are injected in the FIA system (Fig. 1).

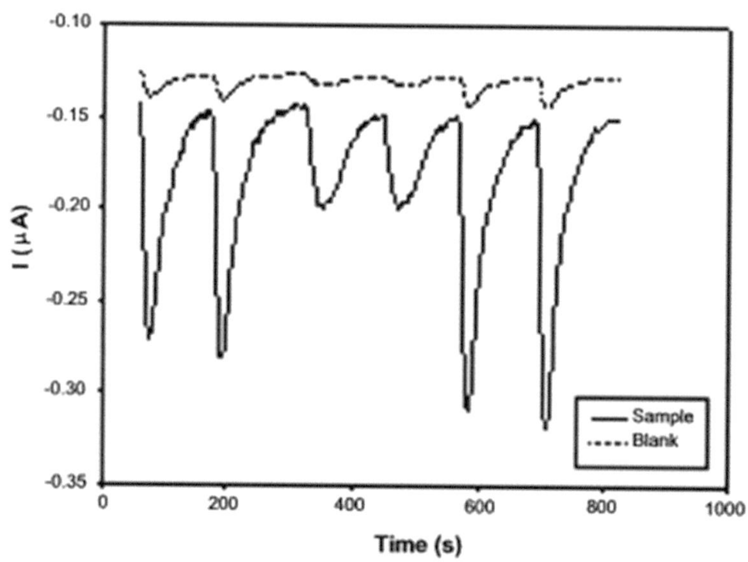

Fig. 1 Amperometric current response under FIA conditions using SMO-SPE biosensor in the analysis of different concentration of Spm in blood samples

4 Notes

1. 50 mg of PVA-SbQ are solubilized in 400 μL distillated water.
2. Few minutes of mixing are enough to produce a homogenous solution.
3. The final activity of the enzyme immobilized onto the electrode surface was about 0.16 U for PAO and 0.05 U for SMO biosensor, respectively.
4. After photopolymerization the biosensors are stored at 4 °C.
5. Flow rate of the peristaltic pump is 0.716 mL/min.
6. GP-SPE has working electrode of graphite with a surface diameter of 4 mm, carbon as counter electrode and silver as reference one. They are employed at potential of +700 mV vs Ag/AgCl.
7. PB-modified GP-SPEs are employed at potential of −100 mV vs Ag/AgCl.
8. The maximum value of the signal peak is referred to the baseline obtained by the buffer solution injection.

References

1. Di Fusco M, Federico R, Boffi A, Macone A, Favero G, Mazzei F (2011) Characterization and application of a diamine oxidase from Lathyrus sativus as component of an electrochemical biosensor for the determination of biogenic amines in wine and beer. Anal Bioanal Chem 401:707–716
2. Karovicova J, Kohajdova Z (2005) Review: biogenic amines in food. Chem Pap 59:70–79
3. Bachrach U (2004) Polyamines and cancer: minireview article. Amino Acids 26:307–309
4. Russel DH (1977) Clinical relevance of polyamines as biochemical markers of tumor kinetics. Clin Chem 23:22–27
5. Fujita K, Nagatsu T, Maruta K, Ito M, Senba H, Miki K (1976) Urinary putrescine, spermidine and spermine in human blood and solid cancers and in an experimental gastric tumor of rats. Cancer Res 36:1320–1324
6. Cohen LF, Lundgren DW, Farrell PM (1976) Distribution of spermidine and spermine in blood from cystic fibrosis patients and control subjects. Blood 48:469–475
7. Armagan Ö (2007) A review: current analytical methods for determination of biogenic amines in food. Food Chem 103:1475–1486
8. Kivirand K, Rinken T (2011) Biosensors for biogenic amines: the present state of art minireview. Anal Lett 44:2821–2833
9. Ruiz-Capillas C, Jimènez-Colmenero F (2004) Biogenic amines in meat and meat products. Crit Rev Food Sci Nutr 44:489–499
10. Muresan L, Valera RR, Frèbort I, Popescu IC, Csöregi E (2008) Amine oxidase amperometric biosensor coupled to liquid chromatography for biogenic amine determination. Microchim Acta 163:219–225
11. Jae-Ick L, Young-Wan K (2013) Characterization of amine oxidases from Arthrobacter aurescens and application for determination of biogenic amines. World J Microbiol Biotechnol 29:673–682
12. Henao-Escobar W, Domìnguez-Renedo O, Alonso-Lomillo MA, Arcos-Martinez MJ (2013) A screen-printed disposable biosensor for selective determination of putrescine. Microchim Acta 180:687–693
13. Rackus DG, Shamsi MH, Wheeler AR (2015) Electrochemistry, biosensors and microfluidics: a convergence of fields. Chem Soc Rev 44:5320–5340
14. Shanmugam S, Thandavan K, Gandhi S, Sethuraman S, Rayappan JB, Krishnan UM (2011) Development and evaluation of a highly sensitive rapid response enzymatic nanointerfaced biosensor for detection of putrescine. Analyst 136:5234–5240
15. Alonso-Lomillo MA, Domínguez-Renedo O, Matos P, Arcos-Martínez MJ (2010)

Disposable biosensors for determination of biogenic amines. Anal Chim Acta 665:26–31

16. Ricci F, Palleschi G (2005) Sensor and biosensor preparation, optimisation and applications of Prussian blue modified electrodes. Biosens Bioelectron 21:389–407
17. Karyakin AA, Karyakina EE (1999) Prussian blue-based artificial peroxidase as a transducer for hydrogen peroxide detection. Application to biosensors. Sensors Actuators B 57:268–273
18. Chaubey A, Malhotra BD (2002) Review. Mediated biosensors. Biosens Bioelectron 17:441–456
19. Dou Y, Haswell S, Greenman J, Wadhawan J (2009) Immobilized anthraquinone for redox mediation of horseradish peroxidase for hydrogen peroxide sensing. Electrochem Commun 11:1976–1981
20. Cona A, Rea G, Angelini R, Federico R, Tavladoraki P (2006) Functions of amine oxidases in plant development and defence. Trends Plant Sci 11:80–88
21. Federico R, Cona A, Angelini R, Schininà ME, Giartosio A (1990) Characterization of maize polyamine oxidase. Phytochemistry 29 (8):2411–2414
22. Cervelli M, Polticelli F, Federico R, Mariottini P (2003) Heterologous expression and characterization of mouse spermine oxidase. J Biol Chem 278(7):5271–5276

Chapter 8

Identification of Branched-Chain Polyamines in Hyperthermophiles

Ryota Hidese, Wakao Fukuda, Masaru Niitsu, and Shinsuke Fujiwara

Abstract

Thermophiles are organisms that grow optimally at temperatures higher than 55 °C. They contain two types of unusual longer/branched-chain polyamines in addition to common polyamines such as spermidine and putrescine. These unusual polyamines contribute to the survival of hyperthermophiles at high temperatures. Recently, the novel aminopropyltransferase BpsA was found to be responsible for the biosynthesis of branched-chain polyamines in the hyperthermophilic archaeon *Thermococcus kodakarensis*, which contains N^4-bis(aminopropyl)spermidine as the major polyamine. This compound is synthesized by the sequential addition of decarboxylated *S*-adenosylmethionine (dcSAM) aminopropyl groups to spermidine via the bifunctional catalytic action of BpsA. In this chapter, methods for the extraction and identification of branched-chain polyamines are presented, along with methods for the production and characterization of recombinant *T. kodakarensis* BpsA as a model aminopropyltransferase.

Key words Hyperthermophile, Archaea, Branched-chain polyamine, Branched-chain polyamine synthase, *Thermococcus kodakarensis*, N^4-bis(aminopropyl)spermidine [3(3)(3)4]

1 Introduction

Thermophilic microorganisms, which grow at temperatures above 55 °C, inhabit high-temperature environments such as hot springs, terrestrial solfatara, deep-sea hydrothermal vents, and composting organic matter. Those that thrive optimally at temperatures higher than 80 °C, known as hyperthermophiles, include representatives from archaea and bacteria domains [1]. One of the unique features of (hyper)thermophiles is the presence of longer- and branched-chain polyamines (Fig. 1) [2–9]. Long linear polyamines such as caldopentamine (3333) and caldohexamine (33333), the names of which are abbreviated based on the number of methylene (CH_2) chain units between NH_2, NH, N or N^+, are found in thermophilic archaea (both in crenarchaeota and euryarchaeota) and bacteria, while branched-chain polyamines such as N^4-aminopropylnorspermidine {3(3)3}, N^4-aminopropylspermidine {3(3)4}, tetrakis-(3-

Rubén Alcázar and Antonio F. Tiburcio (eds.), *Polyamines: Methods and Protocols*, Methods in Molecular Biology, vol. 1694, DOI 10.1007/978-1-4939-7398-9_8, © Springer Science+Business Media LLC 2018

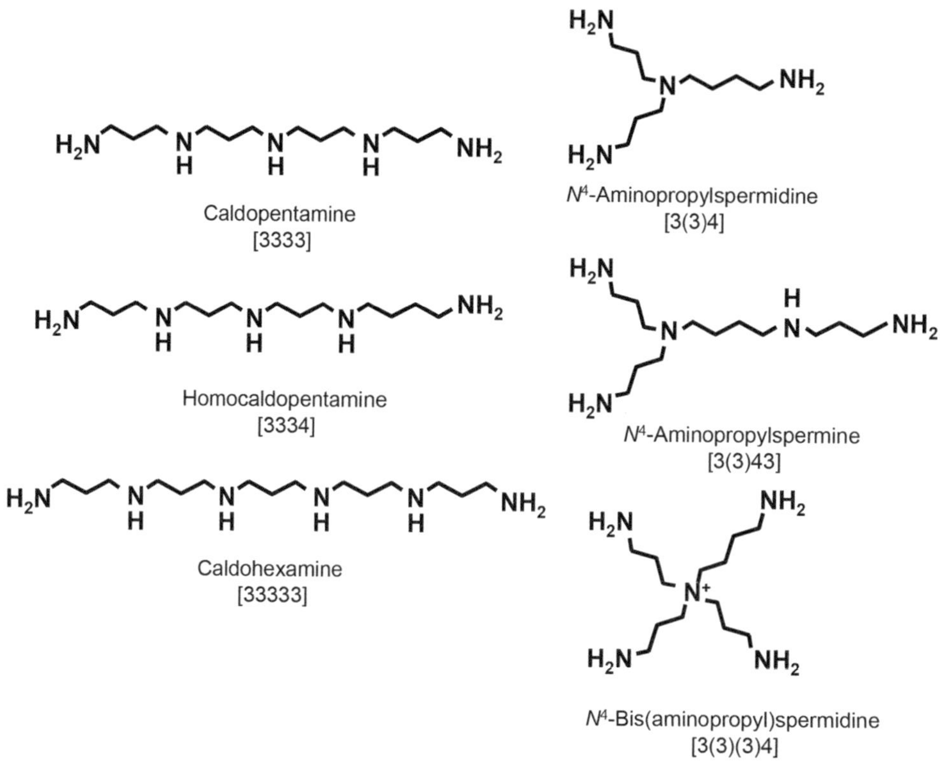

Fig. 1 Major polyamines found in (hyper)thermophile

aminopropyl)ammonium {3(3)(3)3} and N^4-bis(aminopropyl) spermidine {3(3)(3)4} are mainly found in euryarchaeotal hyperthermophiles belonging to the genera *Methanocaldococcus*, *Archaeoglobus* and *Thermococcus* [9] that grow at temperatures above 80 °C, and thermophilic bacteria of the phyla *Aquificae* and *Thermus-Deinococcus.* Because the relative amount of these longer- and branched-chain polyamines in (hyper)thermophiles is significantly increased with increasing growth temperature, these unique polyamines are believed to greatly contribute to microbial survival at high temperatures [3, 10]. Indeed, an in vitro study indicated that long-chain and branched-chain polyamines effectively stabilize DNA and RNA, respectively [11].

In plants and some bacteria, arginine decarboxylase catalyzes the decarboxylation of L-arginine to produce agmatine, which is subsequently converted to putrescine by agmatine ureohydrolase or a combination of agmatine iminohydrolase and *N*-carbamoylputrescine amidohydrolase. Aminopropyltransferase catalyzes the transfer of the aminopropyl group of decarboxylated *S*-adenosylmethionine (dcSAM) to putrescine, resulting in the production of spermidine [12, 13]. By contrast, a thermophilic bacterium *Thermus thermophilus* and the hyperthermophilic archaeon *Thermococcus kodakarenis* possess a unique polyamine biosynthetic pathway in which spermidine is synthesized from agmatine via N^1-aminopropylagmatine [10, 14]. A unique aminopropyltransferase responsible for the biosynthesis of branched-chain polyamines was identified in *T. kodakarensis* [15]. This novel enzyme, termed BpsA, appears to catalyze sequential aminopropyl transfer reactions from dcSAM to the initial substrate spermidine, resulting in the production of N^4-bis(aminopropyl)spermidine {3(3)(3)4} via the intermediate N^4-aminopropylspermidine {3(3)4}. The primary structure of BpsA is distinct from that of known aminopropyltransferases including spermidine synthase [15]. Consistent with the distribution of branched-chain polyamines in (hyper)thermophiles, BpsA orthologs are highly conserved in both (hyper)thermophilic bacteria and euryarchaeotal archaea (Fig. 2). Branched-chain polyamines are one of the key molecules for survival of these organisms at high temperatures. However, the exact physiological roles of branched-chain polyamines remain unknown. In this chapter, we present methods for the extraction and identification of branched-chain polyamines, and for enzyme assays to characterize branched-chain polyamine synthases.

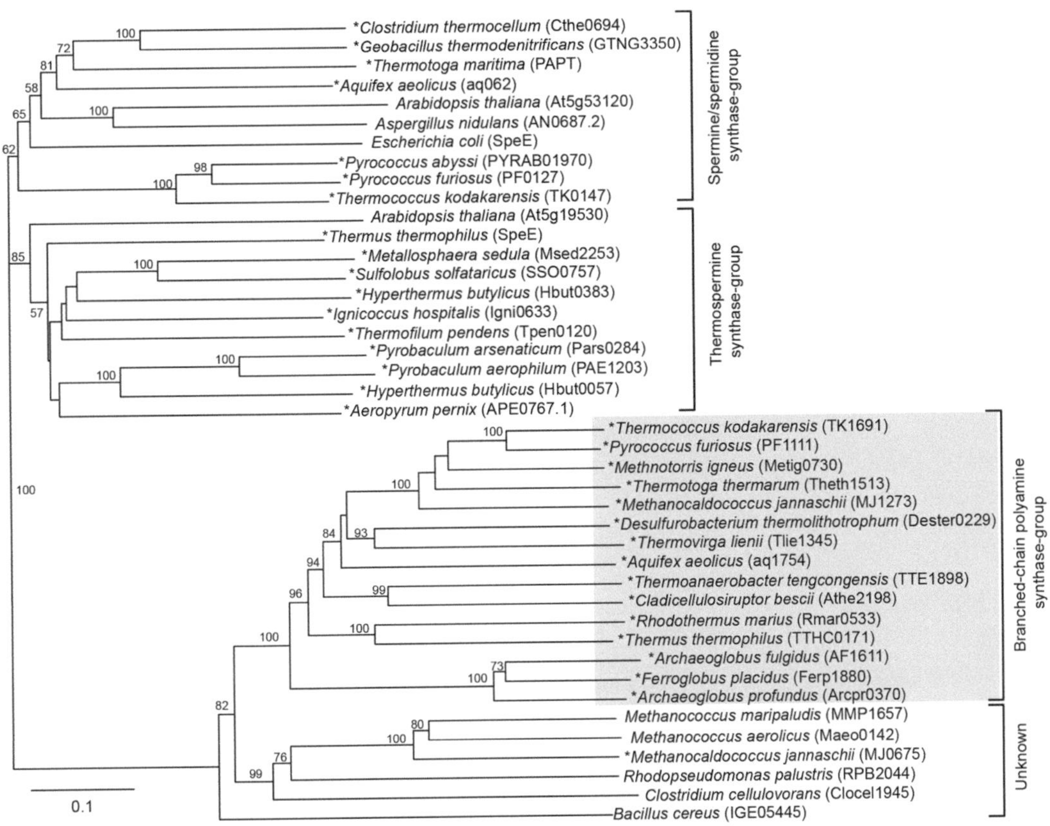

Fig. 2 Phylogenetic tree of aminopropyltransferases involved in polyamine synthesis. The scale bar represents one substitution per 10 amino acids. Bootstrap values of ~50 to ~100 trials are shown. BpsA orthologs are shown in the shaded square. *Asterisk* indicates (hyper)thermophile

2 Materials

2.1 Media

1. 15 ml glass test tubes.
2. 0.2 μm syringe filters.
3. Constant temperature incubator (~30–100 °C).
4. Anaerobic chamber (TABAI anaerobic incubator EAN-140, Espec corp., Osaka, Japan).
5. 50 ml, 100 ml, 1 l PYREX round media storage bottles.
6. 5.0% $Na_2S{\cdot}9H_2O$.
7. 100 mg/ml Ampicillin.
8. 50 mg/ml Kanamycin.
9. 4× Artificial sea water (ASW) solution (per l): 80 g NaCl, 12 g $MgCl_2{\cdot}6H_2O$, 24 g $MgSO_4{\cdot}7H_2O$, 4 g $(NH_4)SO_4$, 0.8 g $NaHCO_3$, 1.2 g $CaCl_2{\cdot}2H_2O$, 2 g KCl, 1.68 g KH_2PO_4, 0.2 g NaBr, 0.08 g $SrCl_2{\cdot}6H_2O$, 0.04 g $Fe(NH_4)$ citrate.
10. 0.8× ASW.
11. ASW-YT-S^0 broth (per l): 250 ml 4× ASW solution, 5 g tryptone, 5 g yeast extract, 5 g elemental sulphur (S^0), 1 ml 5.0% $Na_2S{\cdot}9H_2O$, pH to 7.2 with NaOH (*see* **Note 1**).
12. Luria-Bertani broth (per l): 10 g Tryptone, 5 g yeast extract, 10 g NaCl, pH to 7.2 with NaOH.
13. Luria-Bertani agar plates (per l): 10 g Tryptone, 5 g yeast extract, 10 g NaCl, 20 g agar, pH to 7.2 with NaOH.

2.2 Polyamine Extraction

1. 1% perchloric acid (PCA).
2. *T. kodakarensis* cells [16].
3. 1.5 ml, 15 ml, 50 ml plastic tubes.
4. 1 ml plastic syringes.
5. Ultrasonicator (Branson Sonifier model 250, SciQuip, Shrewsbury, UK).
6. 0.45 μm syringe filters (Millex-LH filter, Merck-Millipore, Billerica, MA).

2.3 High Pressure Liquid Chromatography (HPLC) Analysis

1. High pressure liquid chromatography (HPLC) system (Model GL7700, GL science Inc., Tokyo, Japan; *see* Fig. 3).
2. CK-10S cation exchange column (6.0 mm diameter × 50 mm length, Mitsubishi chemicals Co., Tokyo, Japan). The column contained sulfonated polystyrene divinylbenzene copolymer resin particles with an average diameter of 11 μm.
3. Mobile phase solution: 100 mM potassium citrate monohydrate, 2.0 M KCl, 5% 2-propanol, 0.03% Brij35 (polyoxyethylene lauryl ether). The pH was adjusted to 3.2 with HCl.

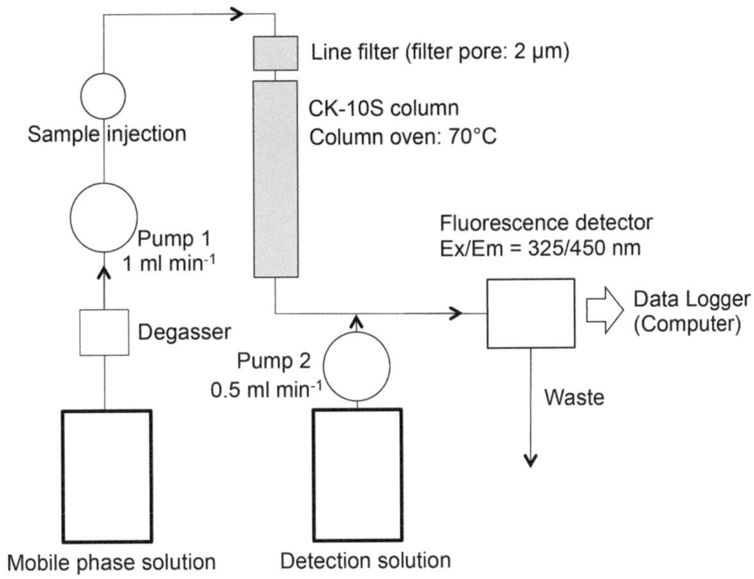

Fig. 3 Schematic diagram of HPLC system for polyamine analysis

4. Detection solution: 400 mM boric acid, 400 mM NaOH, 0.06% Brij35, 7.5 mM *o*-phthalaldehyde, 1% ethanol, 28 mM 2-mercaptoethanol (*see* **Note 2**).
5. 0.22 μm bottle top filters.

2.4 Gas Chromatography (GC) and GC-Mass Spectrometry (MS) Analyses

1. Dowex 50W-X8 (Wako pure chemical industries, Ltd., Osaka, Japan).
2. 1 M, 6 M HCl.
3. Rotary evaporator.
4. 7 ml PYREX test tubes with screw caps.
5. N_2 gas.
6. Vacuum evaporator (Model RMC-24, Koike Precision instruments, Hyogo, Japan).
7. ≥99% Heptafluorobutyric anhydride.
8. ≥99% Acetonitrile.
9. ≥99.5% Diethyl ether.
10. 0.5 M Na_2CO_3.
11. pH test paper.
12. Gas chromatography (GC) system (Model GC-17A, Shimadzu, Kyoto, Japan).
13. InertCap 1MS (0.32 mm inner diameter × 30 mm length, GL Science Inc., Tokyo, Japan).
14. GC-mass spectrometry (MS) system (JMS-700, JEOL Ltd., Tokyo, Japan).

2.5 Preparation of Recombinant BpsA

1. pET28a(+) expression vector (Merck-Millipore, Billerica, MA).
2. Thermal cycler.
3. *T. kodakarensis* genomic DNA (*see* **Note 3**).
4. Phenol:chloroform:isoamyl alcohol solution (PCI) (25:24:1).
5. Gel and PCR purification kit.
6. Agarose.
7. 50× TAE (per l): 242 g TRIS-Base, 57.1 ml glacial acetic acid, 18.6 g EDTA-2Na. Store at room temperature. For 1× TAE, dilute 1:50 from a 50× TAE buffer stock.
8. UV transilluminator.
9. Agarose gel electrophoresis unit.
10. 0.5 μg/ml Ethidium bromide (EtBr).
11. DNA Ligation kit (2× ligation mix, Nippongene, Tokyo, Japan).
12. Shaking incubator.
13. *Escherichia coli* BL21-CodonPlus (DE3)-RIL cells (Agilent technology, Santa Clara, CA).
14. *E. coli* DH5α.
15. Restriction endonuclease *Nde*I.
16. Restriction endonuclease *Eco*RI.
17. DNA sequence analyser (3130 Genetic analyser, Applied Biosystems, Thermo Fisher Scientific K.K., Yokohama, Japan).
18. 1 M Isopropyl β-D-1-thiogalactopyranoside (IPTG).
19. 0.8% NaCl.
20. Ni^{2+}-nitrilotriacetic acid affinity column.
21. Dialysis membrane (10 kDa cutoff).
22. Centrifugal concentrators (15 kDa cutoff).
23. Purification buffer A: 20 mM Tris–HCl (pH 7.5), 1 mM 2-mercaptoethanol.
24. Purification buffer B: 20 mM Tris–HCl (pH 7.5), 1 mM 2-mercaptoethanol, 0.5 M imidazole.
25. Dialysis buffer: 20 mM Tris–HCl (pH 7.5).
26. 1 M Phenylmethylsulfonyl fluoride (PMSF) dissolved in ≥99.5% dimethylsulfoxide.
27. Sodium dodecyl sulphate polyacrylamide gel electrophoresis (SDS-PAGE) unit.

2.6 Enzyme Assays for Branched-Chain Polyamine Synthases

1. 10 mM Spermidine (1,8-Diamino-4-azaoctane).
2. 10 mM Spermine (1,12-Diamino-4,9-diazadodecane).
3. 1 mM Caldohexamine (*see* **Note 4**).
4. 1 mM N^4-Aminopropylspermidine (*see* **Note 5**).
5. 1 mM N^4-Bis(aminopropyl)spermidine (*see* **Note 5**).
6. 10 mM Decarboxylated-*S*-adenosylmethionine (dcSAM) (*see* **Note 6**).
7. 500 mM Piperazine-1,4-bis(2-ethanesulfonic acid) (PIPES) buffer. pH to 7.6 with NaOH.
8. Block incubator (~4–100 °C).

3 Methods

3.1 Cultivation of T. Kodakarensis

1. Autoclave 20 ml of ASW-YT broth in a 50 ml glass media bottle. After autoclaving, immediately place bottle into an anaerobic chamber to avoid oxygen dissolving. Add 0.1 g of elemental sulphur and 20 μl of 5% Na_2S solution to make ASW-YT-S^0 broth. Inoculate *T. kodakarensis* cells into 20 ml of ASW-YT-S^0 in the anaerobic chamber. Fasten the cap tightly and cultivate at 85 °C for 12 h in the incubator to prepare pre-cultures.
2. Use 8 ml of pre-cultured cells to inoculate 800 ml of ASW-YT-S^0 in 1 l bottles. Cultivate at 85 °C to late-exponential phase (6 h).
3. Cool cultures rapidly on ice.
4. Collect cells aerobically by centrifuging at 8000 × *g* for 15 min at 4 °C.
5. Wash cells twice with 20 ml of 0.8× ASW in 50 ml plastic tubes.

3.2 Preparation of Polyamine Extracts

1. Add 100 μl of 1% PCA to 0.1 g of wet *T. kodakarenis* cells (*see* **Note 7**).
2. Disrupt cells with 30 cycles of sonication (30 s pulses separated by 30 s pauses at a power output of 30 W) on ice and transfer samples to 1.5 ml tubes.
3. Centrifuge at 20,000 × *g* for 15 min at 4 °C and transfer supernatants (polyamine extracts) to new 1.5 ml tubes.
4. Pass polyamine extracts through a 0.45 μm filter using a 1 ml syringe.

3.3 Analysis of Polyamine Composition by HPLC

1. Filter mobile phases (column buffers) and detection solutions through 0.22 μm bottle top filters.
2. Equilibrate a CK-10S HPLC column with mobile phase solution at a flow rate of 1.0 ml/min. Set column oven temperature to 70 °C.

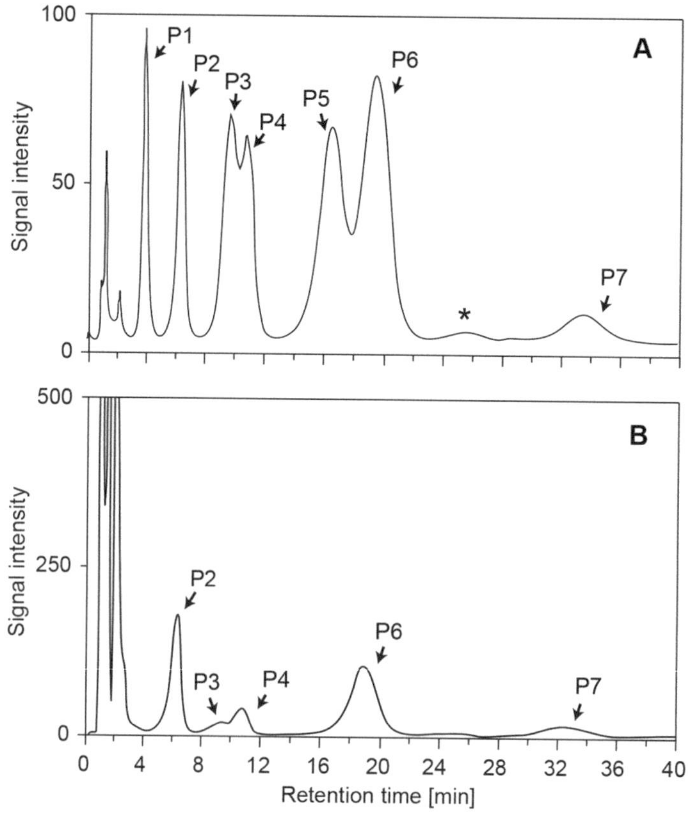

Fig. 4 Intracellular polyamines in *T. kodakarensis* analyzed by HPLC. (**a**) Peak standards of polyamines are P1, putrescine (4); p2, spermidine (34); P3, N^4-aminopropylspermidine (3(3)4); P4, spermine (343); P5, N^4-aminopropylspermine (3(3)43); P6, N^4-bis(aminopropyl)spermidine (3(3)(3)4); P7, caldohexamine (33333). The numbers in brackets represent the number of methylene CH_2 chain units between NH_2, NH, N and N^+. An *asterisk* indicates an unknown peak. (**b**) Intracellular polyamines in *T. kodakarensis* grown in ASW-YT-S^0 at 85 °C for mid-exponential phase

3. Inject 200 μl of polyamine extract (maximum sample loop volume of 100 μl) at a flow rate of 1.0 ml/min.
4. Elute polyamines and mix automatically with detection solution at a flow rate of 0.5 ml/min at 70 °C while monitoring with a fluorescence detector (excitation, 325 nm; emission, 450 nm) as shown in the schematic diagram in Fig. 3. A representative example of polyamine analysis obtained from *T. kodakarensis* is shown in Fig. 4.

3.4 Identification of Branched-Chain Polyamines by GC-MS

1. Apply 3 ml of polyamine extract to a 3 ml Dowex 50W-X8 column.
2. Wash bound polyamines with 15 ml of distilled water (five column volumes).
3. Wash bound polyamines with 15 ml of 1 M HCl (five column volumes).

4. Elute bound polyamines with 15 ml of 6 M HCl (five column volumes).
5. Dry eluted polyamines at 50 °C using a rotary evaporator.
6. Add 200 μl of distilled water to dried polyamines and transfer the polyamine solution to a PYREX test tube.
7. Evaporate polyamines using an N_2 gas blower and dryer.
8. Transfer the evaporated fraction to a vacuum desiccator and dry thoroughly for 2 h.
9. Add 200 μl each of acetonitrile and heptafluorobutyric anhydride to the completely dried polyamines, tighten screw caps and heat at 100 °C for 30 min in a dry bath.
10. Remove solvent and excess derivatisztion reagent by blowing with N_2 gas.
11. Add 0.5 ml of diethyl ether and 0.5 ml of 0.5 M Na_2CO_3 solution and stir.
12. Centrifuge at 500 × *g* for 2 min.
13. Transfer the ether phase into a sample tube and confirm alkaline pH of the water phase with pH test paper.
14. Equip GC and GC-MS with InertCap 1MS capillary columns and helium as a carrier gas at a flow rate of 1.4 ml/min. Programme the column oven from 120 to 280 °C at 16 °C/min and from 90 to 280 °C at 16 °C/min for GC and GC-MS, respectively. Use an injector temperature of 300 °C. Operate GC-MS in electron-impact mode at an ionization energy of 70 eV.
15. Apply derivatised polyamines to the GC column equipped with a flame ionisation detector coupled to the GC-MS. A representative GC-MS analysis is shown in Fig. 5 (*see* **Note 8**).

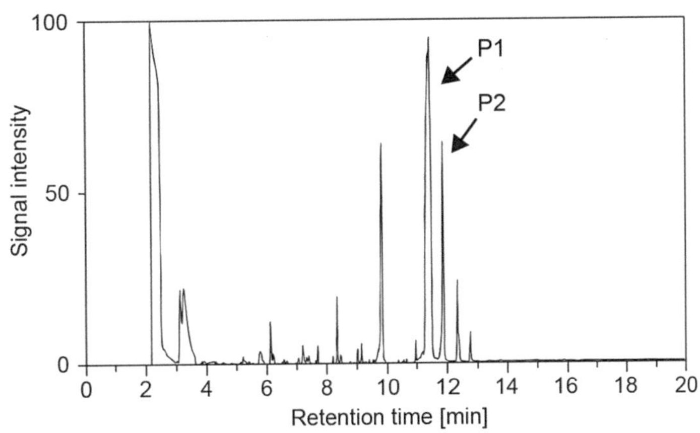

Fig. 5 GC chart of a polyamine extract of *T. kodakarensis* after derivatization to heptafluorobutyl compounds. Abbreviations: *P1* N^4-aminopropylnorspermidine (3(3)3) derivative, *P2* N^4-aminopropylspermidine (3(3)4) derivative

3.5 Cloning of Branched-Chain Polyamine Synthase Gene from T. Kodakarensis

For DNA manipulation, refer to the standard techniques described in Chapters 1, 2 and 3 in Molecular Cloning 4th edition [17]. This section describes the simple procedure for cloning *bpsA* into the expression plasmid pET28a.

1. Design and synthesize *bpsA* primers to amplify the *bpsA* (*tk1691*) gene (1056 bp). Suitable sequences for forward and reverse primers are 5′-AAAAAAACATATGATGAGGGAGA-TAATTGAGAG-3′ and 5′-AGAATTCTCAGGTAGTC-GAGCTCTCCT-3′, respectively. The underlined sequences indicate restriction endonuclease sites for *Nde*I and *Eco*RI.
2. Carry out PCR amplification of the *bpsA* gene using *T. kodakarensis* genomic DNA as a template and *bpsA* forward and reverse primers. Purify the PCR product using a standard Gel/PCR purification kit.
3. Digest the expression plasmid pET28a and the PCR product with restriction enzymes (*Nde*I and *Eco*RI in this case) in appropriate buffer according to the manufacturer's instructions (12 h at 37 °C in this case).
4. Separate the digested DNA products by gel electrophoresis using a 1% 1× TAE agarose gel. After staining with 0.5 μg/ml EtBr, cut out the fluorescent bands on a UV transilluminator and transfer each gel slice to a separate 1.5 ml tube.
5. Purify and isolate the digested DNA using a standard PCR purification kit.
6. Ligate the digested PCR product into the digested pET28a by following the manufacturer's instructions for DNA ligation.
7. Transform competent *E. coli* DH5α cells with the ligation solution. After pre-culturing transformants in LB broth for 60 min at 37 °C, grow on an LB agar plate containing 100 μg/ml kanamycin for 12 h for 37 °C.
8. A colony of transformant grown in the plate was inoculated into LB broth containing 100 μg/ml kanamycin and cultured for 12 h at 37 °C. After isolation of plasmid DNA from transformants, insertion of the *bpsA* gene was checked by digestion with *Nde*I and *Eco*RI according to the manufacturer's instructions.
9. Confirm the correct cloning of *bpsA* into pET28a (and thus generation of pET28a-*Tk*-BpsA) by nucleotide sequencing.

3.6 Preparation of Recombinant Tk-BpsA from E. coli

1. Transform competent *E. coli* BL21-CodonPlus (DE3)-RIL cells with pET28a-*Tk*-BpsA. After pre-culturing in LB broth for 60 min at 37 °C, culture on an LB agar plate containing 100 μg/ml kanamycin for 12 h at 37 °C.

2. Inoculate a colony from the LB agar plate into 5 ml of LB broth containing 100 μg/ml kanamycin and cultivate at 37 °C for 12 h with shaking.
3. Inoculate the starter culture into 500 ml of LB broth containing 100 μg/ml kanamycin and cultivate at 37 °C for 4 h with shaking.
4. Add IPTG to the final concentration of 1 mM.
5. Continue cultivation at 37 °C for 3 h to express recombinant *Tk*-BpsA.
6. Collect cells aerobically in 50 ml tubes by centrifugation at 8000 × *g* for 15 min at 4 °C.
7. Wash cells twice with 0.8% NaCl.
8. Add 10 ml of purification buffer A containing 0.1 mM PMSF to the collected cells.
9. Disrupt the cells with 40 cycles of sonication (30 s pulses separated by 30 s pauses at a power output of 30 W) on ice and transfer samples to 20 ml centrifuge tubes.
10. Centrifuge at 20,000 × *g* for 60 min at 4 °C and transfer the resultant supernatant (crude extract) to a new 15 ml tube.
11. Heat the supernatant containing 6× His-tagged *Tk*-BpsA at 80 °C for 30 min.
12. Centrifuge the heated extract at 20,000 × *g* for 30 min at 4 °C.
13. Apply the resultant supernatant to 5 ml of Ni^{2+}-nitrilotriacetic acid affinity column equilibrated in purification buffer A.
14. Wash the bound recombinant *Tk*-BpsA with 25 ml of purification buffer A (five column volumes).
15. Elute the bound recombinant *Tk*-BpsA with 10 ml of purification buffer B (two column volumes).
16. Confirm protein purity by SDS-PAGE using standard procedures [18] and dialyze the eluted recombinant *Tk*-BpsA against dialysis buffer to achieve a 10,000-fold dilution of buffer components.
17. Concentrate the recombinant *Tk*-BpsA using a centrifugal concentrator to 1 mg/ml (*see* **Note 9**).

3.7 Enzyme Assays for Branched-Chain Polyamine Synthases

1. Prepare a pre-mixture containing 500 μM dcSAM, 1 mM polyamine substrate (spermidine [34] and N^4-aminopropylspermidine [3(3)4]) and 50 mM PIPES buffer) and pre-incubate 198 μl at 80 °C for 2 min.
2. Add 2 μl of 1 mg/ml recombinant *Tk*-BpsA (final volume = 200 μl) and incubate at 80 °C for 4 min for spermidine [34] or 10 min for N^4-aminopropylspermidine [3(3)4] (*see* **Note 10**).

3. Add 2 μl of 10 M HCl to the reaction mixture and cool immediately on ice to terminate the reaction.
4. Inject 200 μl of the reaction mixture onto the HPLC (maximum sample loop volume of 100 μl) and follow the procedure described in the Subheading 3.3 above.

4 Notes

1. To confirm whether the broth is anoxic during cultivation, we recommend adding resazurin sodium salt at a final concentration of 1 mg/l to the broth before autoclaving. Pink or colourless broth indicates partial and complete anaerobic conditions, respectively.
2. Solubilize 1 g of *o*-phthalaldehyde into 10 ml of 100% ethanol. Add 0.6 g of Brij35 to the salt solution. After dissolving the Brij35 thoroughly, mix *o*-phthalaldehyde solution with the salt solution containing Brij35, then add 2 ml of 2-mercapotoethanol. The final volume is then adjusted to 1 l.
3. To prepare genomic DNA from *T. kodakarensis* cells, follow the procedure described below.
   (a) Cultivate *T. kodakarensis* KOD1 cells in 20 ml of ASW-YT-S^0 broth for 12 h at 85 °C.
   (b) Collect cells from 1.5 ml of culture in a 2.0 ml plastic tube by centrifugation (8000 × *g*, 15 min), discard the supernatant and add 50 μl of ultrapure water and 50 μl of PCI solution.
   (c) Gently mix by pipetting.
   (d) Centrifuge the PCI extracts for 10 min at 20,000 × *g* and transfer 30 μl of the upper phase to a new 1.5 ml plastic tube.
   (e) Add 3 μl of 3 M sodium acetate and 60 μl of 100% ethanol and mix thoroughly by gentle inversion.
   (f) After incubating the solution for 15 min at −20 °C, centrifuge for 15 min at 20,000 × *g*.
   (g) Discard the supernatant and add 1 ml of 70% ethanol.
   (h) Centrifuge the mixture for 5 min at 20,000 × *g*.
   (i) Discard the supernatant and dry the remaining genomic DNA fraction thoroughly.
   (j) Solubilize the genomic DNA in 15 μl of ultrapure water.
4. Caldohexamine was synthesized by following the method previously described [19]. Caldohexamine was added to the polyamine extract as an internal standard for HPLC analysis to control for extraction and separation losses.

5. N^4-Aminopropylspermidine and N^4-bis(aminopropyl)spermidine are synthesized by following the methods described previously [20].
6. dcSAM is synthesized as described previously [21], and remained stable when incubated at 80 °C for less than 15 min.
7. Trichloroacetic acid (TCA) can be used for polyamine extraction instead of PCA at the same concentration.
8. N^4-Bis(aminopropyl)spermidine is converted to N^4-aminopropylnorspermidine and N^4-aminopropylspermidine at a ratio of 3:1 during GC and GC-MS analysis after heptafluorobutyrization.
9. The prepared enzyme (1 mg/ml) dissolved in 20 mM Tris–HCl (pH 7.5) retains full activity for at least 3 months when stored at 4 °C in a refrigerator.
10. The initial reaction rate can be measured in the times stated when 0.2 μg of *Tk*-BpsA is reacted with saturated polyamine substrate and aminopropyl donor (1 mM of spermidine or 50 μM N^4-aminopropylspermidine and 500 μM dcSAM) in a 200 μl reaction volume. Furthermore, because the N^4-bis(aminopropyl)spermidine yield was less than 1% during the reaction time (4 min), the quantity of aminopropylated product N^4-bis (aminopropyl)spermidine remained relatively low and did not affect the overall kinetics of its synthesis (i.e., product inhibition is negligible under these conditions).

References

1. Mehta D, Satyanarayana T (2013) Diversity of hot environments and thermophilic microbes. In: Satyanarayana T, Littlechild J, Kawarabayasi Y (eds) Thermophilic microbes in environmental and industrial biotechnology. Springer, Heidelberg, pp 3–60. doi: 10.1007/978-94-007-5899-5_1
2. Oshima T, Kawahata S (1983) Homocaldopentamine: a new naturally occurringpentaamine.J Biochem 93:1455–1456. doi: 10.1093 oxfordjournals.jbchem.a134281
3. Oshima T, Hamasaki N, Senshu M, Kakinuma K, Kuwajima I (1987) A new naturally occurring polyamine containing a quaternary ammonium nitrogen. J Biol Chem 262:11979–11981
4. Hamana K, Niitsu M, Samejima K, Matsuzaki S (1991) Polyamine distributions in thermophilic eubacteria belonging to *Thermus* and *Acidothermus*. J Biochem 109:444–449. doi: 10.1093/oxfordjournals.jbchem.a123401
5. Hamana K, Niitsu M, Matsuzaki S, Samejima K, Igarashi Y, Kodama T (1992) Novel linear and branched polyamines in the extremely thermophilic eubacteria *Thermoleophilum*, *Bacillus*, and *Hydrogenobacter*. Biochem J 284:741–747. doi: 10.1042/bj2840741
6. Hamana K, Tanaka T, Hosoya R, Niitsu M, Itoh T (2003) Cellular polyamines of the acidophilic, thermophilic and thermoacidophilic archaebacteria, *Acidilobus*, *Ferroplasma*, *Pyrobaculum*, *Pyrococcus*, *Staphylothermus*, *Thermococcus*, *Thermodiscus*, and *Vulcanisaeta*. J Gen Appl Microbiol 49(5):287–293. doi:10.2323/jgam.49.287
7. Hosoya R, Hamana K, Niitsu M, Itoh T (2004) Polyamine analysis for chemotaxonomy of thermophilic eubacteria: polyamine distribution profiles within the orders *Aquificales*, *Thermotogales*, *Thermodesulfobacteriales*, *Thermales*, *Thermoanaerobacteriales*, *Clostridiales*, and *Bacillales*. J Gen Appl Microbiol 50 (5):271–287. doi:10.2323/jgam.50.271
8. Hamana K, Hosoya R, Itoh T (2007) Polyamine analysis of methanogens, thermophiles and extreme halophiles belonging to the

domain Archaea. J Jpn Soc Extremophiles 6:25–31. doi:10.3118/jjse.6.25

9. Hamana K, Hayashi H, Niitsu M, Itoh T (2009) Polyamine analysis of thermophilic, acidophilic, alkaliphilic and radio-tolerant bacteria belonging to the domain bacteria and methanogens, thermophiles and extreme halophiles belonging to the domain Archaea. J Jpn Soc Extremophiles 8:59–68. doi:10.2323/jgam.49.287
10. Morimoto N, Fukuda W, Nakajima N, Masuda T, Terui Y, Kanai T, Oshima T, Imanaka T, Fujiwara S (2011) Dual biosynthesis pathway for longer-chain polyamines in the hyperthermophilic archaeon *Thermococcus kodakarensis*. J Bacteriol 192:4991–5001. doi:10.1128/JB.00279-10
11. Terui Y, Ohnuma M, Hiraga K, Kawashima E, Oshima T (2005) Stabilisation of nucleic acids by unusual polyamines produced by an extreme thermophile, *Thermus thermophilus*. Biochem J 388:427–433. doi:10.1042/BJ20041778
12. Tabor CW, Tabor H (1985) Polyamines in microorganisms. Microbiol Rev 49:81–99.
13. Imai A, Matsuyama T, Hanzawa Y, Akiyama T, Tamaoki M, Saji H, Shirano Y, Kato T, Hayashi H, Shibata D, Tabata S, Komeda Y, Takahashi T (2004) Spermidine synthase genes are essential for survival of Arabidopsis. Plant Physiol 135:1565–1573. doi:10.1104/pp .104.041699
14. Ohnuma M, Terui Y, Tamakoshi M, Mitome H, Niitsu M, Samejima K, Kawashima E, Oshima T (2005) N^1-aminopropylagmatine, a new polyamine produced as a key intermediate in polyamine biosynthesis of an extreme thermophile, *Thermus thermophilus*. J Biol Chem 280:30073–30082. doi:10.1074/jbc.M413332200
15. Okada K, Hidese R, Fukuda W, Niitsu M, Takao K, Horai Y, Umezawa N, Higuchi T, Oshima T, Yoshikawa Y, Imanaka T, Fujiwara S (2014) Identification of a novel aminopropyltransferase involved in the synthesis of branched-chain polyamines in hyperthermophiles. J Bacteriol 196:1866–1876. doi:10.1128/JB.01515-14
16. Atomi H, Fukui T, Kanai T, Morikawa M, Imanaka T (2004) Description of *Thermococcus kodakaraensis* sp. nov., a well studied hyperthermophilic archaeon previously reported as *Pyrococcus* sp. KOD1. Archaea 1:263–267. doi:10.1155/2004/204953
17. Sambrook JF, Green MR (eds) (2012)-Chapter 1, 2, and 3. Molecular cloning: a laboratory manual, 4th edn. Cold Spring Harbor Laboratory Press, NewYork, pp 1–260
18. Kielkopf CL, Bauer W, Urbatsch IL (2012) Expressing cloned genes for protein production, purification, and analysis. In: Sambrook JF, Green MR (eds) Molecular cloning: a laboratory manual, 4th edn. Cold Spring Harbor Laboratory Press, NewYork, pp 1599–1610
19. Oshima T, Moriya T, Terui Y (2011) Identification, chemical synthesis, and biological functions of unusual polyamines produced by extremethermophiles. In: Pegg AE, Casero RA Jr (eds) Polyamines: methods and protocols, Methods in Mol biol, vol 720, pp 81–111. doi:10.1007/978-1-61779-034-8_5
20. Niitsu M, Sano H, Samejima K (1992) Syntheses of tertiary tetraamines and quaternary penta-amines with three and four methylene chain units. Chem Pharm Bull 40:2958–2961. doi:10.1248/cpb.40.2958
21. Dejima H, Kobayashi M, Takasaki H, Takeda N, Shirahata A, Samejima K (2003) Synthetic decarboxylated *S*-adenosyl-L-methionine as a substrate for aminopropyl transferases. Biol Pharm Bull 26:1005–1008. doi:10.1248/bpb.26.1005

Chapter 9

Analysis of Polyamines Conjugated with Hydroxycinnamoyl Acids by High-Performance Liquid Chromatography Coupled to Electrospray Ionization Tandem Mass Spectrometry

Laura Torras-Claveria, Jaume Bastida, Francesc Viladomat, and Antonio F. Tiburcio

Abstract

Polyamines conjugated with hydroxycinnamic acids are phenolic compounds, which are widespread in the plant kingdom playing important roles in development and defence responses. This chapter describes the methodology employed to analyze these phenolamides in plant material by liquid chromatography coupled to electrospray ionization tandem mass spectrometry (LC-MS-MS). These compounds are not always in sufficient concentration in plant tissues for analysis by more conventional methods such as UV detection of HPLC. Owing to their particular molecular structure, they cannot be analyzed as free polyamines. Thus, described herein is an extraction method for hydroxycinnamic acid amides in plant tissues such as leaves, and their analysis by LC-MS-MS, including identification and quantification protocols.

Key words Hydroxycinnamic acid amides (HCAA), Polyamine conjugates, LC-MS-MS, HCAA identification, HCAA quantification

1 Introduction

Plant polyamines (PAs) occur not only as free forms but also as conjugates with small molecules such as phenolic compounds, forming phenolamides, which are widespread in the plant kingdom. In this way, putrescine (Put), spermidine (Spd), spermine (Spm) and agmatine (Agm; a precursor of Put) can be conjugated with hydroxycinnamic acids (mainly *p*-coumaric, caffeic and ferulic acids) forming hydroxycinnamic acid amides (HCAA). Several saturation degrees of PA amino groups occur with mono- or polysubstitution, carrying the same or different hydroxycinnamic acids. These different combinations lead to a number of diverse HCAA [1–6]. Figure 1 shows the chemical structure of the most common HCAA occurring in plants.

Rubén Alcázar and Antonio F. Tiburcio (eds.), *Polyamines: Methods and Protocols*, Methods in Molecular Biology, vol. 1694, DOI 10.1007/978-1-4939-7398-9_9, © Springer Science+Business Media LLC 2018

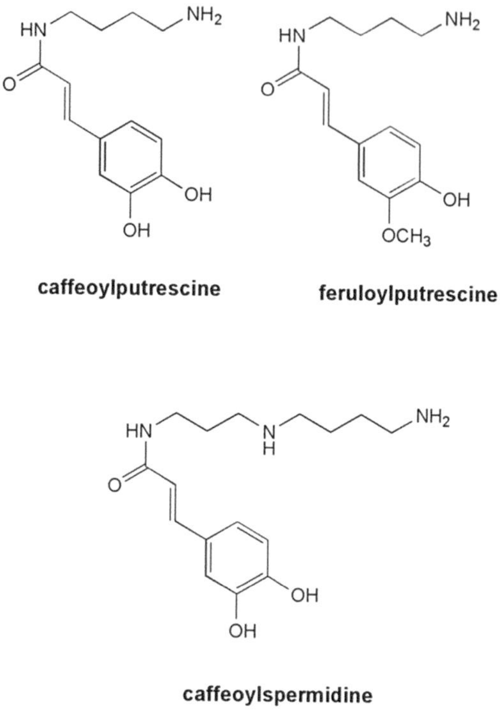

Fig. 1 Chemical structure of polyamine conjugates

Some HCAA constitutively accumulate in several organs. Thus, mono- and di-substituted putrescine (Put) and spermidine (Spd) derivatives are produced in seeds [7–9], roots [10–12], and more extensively in floral buds and reproductive organs [13-16] with proposed roles in plant growth and floral development. Tri-substituted spermidines accumulate in the pollen coat, constituting one of the major phenolic compounds in Rosaceae [17], and also in pollen of oak, sunflower, peanuts, elder, caper, hazel, *Aphelandra*, *Hippeastrum x hortorum* and *Arabidopsis* [18–25]. Several functions have been proposed for these substances such as in sporopollenin formation, pollen protection against UV radiation or pollen recognition and germination on stigma [6, 26].

There is evidence that HCAA are involved in plant defence against biotic and abiotic stresses. It has been seen that pathogen infection, wounding or elicitor treatments induce the synthesis of these compounds [27–29]. For instance, coumaroyl- and feruloyl-Agm accumulate in Arabidopsis and barley in response to pathogen attacks [30, 31]. Dicoumaroyl-caffeoyl-Spd inhibits mycelial growth of Pyrenophora and reduces powdery mildew microbial infection in barley seedlings [32]. The involvement of caffeoyl-Put and dicaffeoyl-Spd in defence against insect attack has also been reported [33]. Furthermore, an antiviral effect on tobacco mosaic virus (TMV) multiplication has been proposed for

N-acylated-Put derivatives [14, 34]. HCAA have also been suggested to play a role in the tolerance of plants to water, cold and oxidative stresses, boron deficiency, wounding and UV-B and UV-C radiations [35–37].

The analysis of HCAA was first performed by HPLC-UV techniques. However, the tissues used for the identification of these compounds were pollen or pistils, where HCAA are found in a high concentration. Initially, these compounds were also considered as phylogenetic markers because they were not detected in some species, although eventually their presence was confirmed in all the species analyzed. The lack of detection of HCAA was due to their low concentration in the plant and limited sensitivity of the analytical techniques employed at that time. The concentration of HCAA can vary depending on the organ analyzed, the stage of plant development, the age of the plant and the environmental growth conditions [38–41].

With the development of mass spectrometry techniques and improvement in sensitivity, LC-ESI-MS-MS and high resolution LC-Q-TOF-MS have been widely used to detect and quantify HCAA [4, 27, 28, 42–45]. These techniques provide robust results, which are similar when experiments are repeated under the same or similar conditions on different occasions, and with accuracies of 10 ppm or better, thus providing useful information for molecular formula assignment [46]. Furthermore, concentrations can be estimated with external calibration curves by reference to authentic standards such as Spm [43].

A mass spectrometer consists of three separate units with three functions: ion generation in an ionization chamber, ion separation in the mass analyzer, and ion detection in the detection device. In the electrospray ionization (ESI) analysis, the sample is dissolved in an organic or aqueous solvent, and then conducted through a capillary to the ionization chamber where a high voltage is applied. The result is the formation of an aerosol of charged droplets, which are conducted to the mass detector [47].

In the following protocol we describe the procedure for extraction, purification, and analysis of HCAA by an LC-MS-MS technique, using tobacco and Arabidopsis leaves as plant materials.

2 Materials

2.1 Plant Materials

Tobacco (*Nicotiana tabacum* L. cv. Wisconsin) and Arabidopsis (*Arabidopsis thaliana* Col-0) leaves were used as plant materials.

2.2 Materials for Processing Plant Samples

1. Liquid nitrogen.
2. Ceramic pestle and mortar.
3. 15 mL centrifuge tubes.

4. MeOH (HPLC grade).
5. MiliQ water.
6. HCOOH.
7. AcOH.
8. Ultrasound bath.
9. Centrifuge for tubes of 15 mL.
10. Vacuum concentration system.
11. 0.45 μm polytetrafluoroethylene (PTFE) filter.
12. 2 mL HPLC opaque crystal vials.

2.3 Equipment

1. LC-MS-MS system with ESI ionization.
2. LC system: An Agilent 1100 quaternary pump with an auto-sampler and UV detector at λ 280 nm.
3. MS system: API 3000 triple quadrupole coupled on line with an LC system.

3 Methods

3.1 Extraction of HCAA

1. Place 1 g of plant tissue in a mortar and grind the tissue in liquid N_2 (*see* **Note 1**).
2. Immediately extract the powdered sample in 3.5 mL of extraction mixture ($MeOH/H_2O$ (3% HCOOH) (1:1) for 30 min in a sonication bath.
3. Centrifuge the mixture for 20 min at 15,000 × *g*.
4. Keep the supernatant and evaporate the solvent under vacuum at 30 °C.
5. Redissolve the dried residue in 2.5 mL of 0.05% AcOH solution.
6. Filter the sample through a 0.45 μm PTFE filter.
7. Place the filtered extracts in an opaque crystal 2 mL HPLC vial (*see* **Note 1**).

3.2 LC-MS-MS

3.2.1 LC Conditions

1. Solvent A: H_2O, 0.05% AcOH.
2. Solvent B: Methanol (MeOH).
3. Gradient: Increasing linear of solvent B (v/v) (*t* min, %B): (0, 0), (25, 34), (30, 54), (35, 80), (37, 100), (40, 100), (43, 0), (50, 0).
4. Flow rate: 400 μL/min.
5. Injection volume: 25 μL.

3.2.2 MS Conditions

1. Operation mode: positive.
2. Capillary voltage: 3500 V.

3. Nebulizer gas: N_2, 10 (arbitrary units).
4. Curtain gas: N_2, 12 (arbitrary units).
5. Collision gas: N_2, 4 (arbitrary units).
6. Focusing potential: 200 V.
7. Entrance potential: 10 V.
8. Drying gas: N_2, heated to 400 °C and introduced at flow rate of 8000 cm^3/min.
9. Declustering potential (DP) and collision energy (CE): After analyzing HCAA standards in different DP and CE, the optimal conditions for a correct fragmentation of HCAA were DP = 20 and CE = 20 (*see* **Note 2**).
10. Type of experiments performed: Precursor Ion Scan, Product Ion Scan, Neutral Loss Scan and Multiple Reaction Monitoring.

3.2.3 MS Experiments

1. Precursor Ion Scan. In this type of experiment the quadrupole 1 (*Q1*) scans over all possible precursors of the selected ion in quadrupole 3 (*Q3*).
2. Product Ion Scan. In this experiment selected precursors in *Q1* are fragmented in collision cell (*Q2*) and the fragments analyzed in *Q3*.
3. Neutral Loss Scan. In this experiment both *Q1* and *Q3* scan for a pair of ions with a characteristic mass difference, which corresponds to a fragment ion that is not ionized, and thus not detected by quadrupoles.
4. Multiple Reaction Monitoring (MRM). This is the most sensitive mode and provides the best specificity for a given analyte, so it is the experiment of choice for an absolute quantitative analysis and for a very low concentration analyte such as HCAA. In this experiment there is a selection of predefined fragment ion pairs. The ion pairs checked in MRM are chosen based on the pair of ions that have previously shown the highest intensity in Product Ion and Precursor Ion Scan experiments with optimal DP and CE of HCAA standards.

3.3 HCAA Identification

3.3.1 Analysis of Standards

1. Dissolve the HCAA standards in 0.05% AcOH in a 100 ppm concentration.
2. Analyze the HCAA standards by LC-MS-MS following the method and parameters described in Subheading 3.2.
3. Experiments performed: Product Ion Scan, Precursor Ion Scan, Neutral Loss Scan.
4. Assay different DP and CE and select the best option for these kind of compounds. In our equipment the best option was DP 20 and CE 20 (*see* **Note 2**).

Table 1
Pair of HCAA ions of more intensity selected for MRM experiments

Compound	Parent ion $[M+H]^+$	Fragment ion
Feruloylspermidine	322	177
Coumaroylspermidine	292	147
Caffeoylspermidine	308	163
Cinnamoylspermidine	276	131
Sinapoylspermidine	352	207
5-Hydroxyferuloylspermidine	338	193
Feruloylputrescine	265	177
Coumaroylputrescneine	235	147
Caffeoylputrescine	251	163
Cinnamoylputrescine	219	131
Sinapoylputrescine	295	207
5-Hydroxyferuloylputrescine	281	193

5. Select the ion pair with the highest intensity for performing MRM experiments (Table 1).
6. Perform MRM experiment with the ion pair selected for all HCAA standards to verify the method and check the retention time.

3.3.2 Analysis of Samples

1. Analyze plant extracts by LC-MS-MS (*see* Subheading 3.2).
2. Experiments performed: Product Ion Scan, Precursor Ion Scan and Neutral Loss Scan of the selected ions.
3. Perform MRM experiment with the pair of ions selected in each HCAA standard. (*See* Table 1 for the most common HCCA) (*see* **Note 3**.)
4. Compare mass fragments, retention times and MRM with those from authentic standards for correct identification of analytes.

3.4 HCAA Quantification

HCAA are quantified in samples with a calibration curve performed with authentic standards by MRM mode of the LC-MS-MS method.

3.4.1 Calibration Curve with Standards

1. Prepare different solutions of HCAA standards in a 0.05% AcOH in the following concentrations: 0.01, 0.05, 0.25, 0.5, 1 and 2 μg/mL.
2. Place the solutions in different HPLC opaque crystal vials ready to be analyzed by LC-MS-MS.

3. Analyze them by LC-MS-MS with the same parameters detailed in Subheading 3.2 in the MRM mode of the mass spectrometer.
4. Collect the mass areas with Analyst software (*see* **Note 4**).
5. Perform a calibration curve with Excel software.

3.4.2 Quantification

Each sample is analyzed in triplicate (corresponding to three different extractions).

1. Prepare three different vials corresponding to three separate plant extractions.
2. Analyze the three extracts by LC-MS-MS following the LC-MS-MS method described in Subheading 3.2 in MRM mode.
3. Identify the analytes to be quantified following Subheading 3.3.2.
4. Collect the analyte area with Analyst software (*see* **Note 4**).
5. Calculate the analyte concentration with the corresponding calibration curve with Excel software (*see* **Note 5**).
6. Calculate the statistical average and SD of three triplicates of the concentration with Excel software (*see* **Notes 6–8**).

4 Notes

1. HCAA are phenolic compounds, so they are very sensitive to light degradation. It is important to keep them under liquid N_2 immediately after collection, and work with opaque crystal vials throughout the analysis.
2. Be sure to assay different DP and CE parameters in Product and Precursor Ion Scan experiments in the LC-MS-MS method in order to ascertain the optimal conditions of fragmentation for the detection of parent and fragment ions prior to MRM experiments. Sometimes the lack of detection of a compound can be due to a too high or too low DP and CE intensity.
3. If different peaks of a single compound appear in MRM mode, check that both ions of the same peak belong to the same compound with Product and Precursor Ion Scan. If they belong to the same compound, the two peaks could refer to two isomers. If they belong to different compounds, check the RT with the standard to identify the correct one.
4. When collecting areas of analytes for quantification, be sure to remove electronic and chemical noise in mass spectra by setting a minimal response threshold.

5. For exact quantification of compounds, make sure that their concentration is within the range of the upper and lower limits of corresponding calibration curves.
6. Apply statistical tools such as ANOVA and post hoc test with results obtained in quantification and determine the standard deviation, coefficient of deviations and significance of differences.
7. Whenever required, perform multivariate tests such as Principal Component Analysis, *k*-Means Cluster Analysis or Hierarchical Cluster analysis.
8. Apply tools for data visualization, such as graphics or scatter-plot, to show and better appreciate the differences between samples and conditions of experiments.

Acknowledgment

The authors are thankful to the Scientific Technical Services from University of Barcelona. The research team belongs to the Natural Products Group (2016-SGR-920 Generalitat de Catalunya).

References

1. Facchini PJ, Hagel J, Zulak KG (2002) Hydroxycinnamic acid amide metabolism: physiology and biochemistry. Can J Bot 80:577–589
2. Edreva AM, Velikova VB, Tsonev TD (2007) Phenylamides in plants. Russ J Plant Physiol 54:287–301
3. Bassard JE, Ullmann P, Bernier F, Werck-Reichhart D (2010) Phenolamides: bridging polyamines to the phenolic metabolism. Phytochemistry 71:1808–1824
4. Torras-Claveria L, Jáuregui O, Codina C, Tiburcio AF, Bastida J, Viladomat F (2012) Analysis of phenolic compounds by high-performance liquid chromatography coupled to electrospray ionization tandem mass spectrometry in senescent and water-stressed tobacco. Plant Sci 182:71–78
5. Tiburcio AF, Altabella T, Bitrián M, Alcázar R (2014) The roles of polyamines during the lifespan of plants: from development to stress. Planta 240:1–18
6. Elejalde-Palmett C, de Bernonville TD, Glevarec G, Pichon O, Papon N, Courdavault V, St-Pierre B, Giglioli-Guivarc'h N, Lanoue A, Besseau S (2015) Characterization of a spermidine hydroxycinnamoyltransferase in *Malus domestica* highlights the evolutionary conservation of trihydroxycinnamoyl spermidines in pollen coat of core Eudicotyledons. J Exp Bot 66:7271–7285
7. Mbadiwe EI (1973) Caffeoylputrescine from *Pentaclethra macrophylla*. Phytochemistry 12:2546
8. Moreau RA, Nuñez A, Singh V (2001) Diferuloylputrescine and *p*-coumaroyl-feruloylputrescine, abundant polyamine conjugates in lipid extracts of maize kernels. Lipids 36:839–844
9. Luo J, Fuell C, Parr A, Hill L, Bailey P, Elliott K, Fairhurst SA, Martin C, Michael AJ (2009) A novel polyamine acyltransferase responsible for the accumulation of spermidine conjugates in *Arabidopsis* seed. Plant Cell 21:318–333
10. Yoshihara T, Katsuyoshi Y, Takamatsu S, Sakamura S (1981) A new lignan amide, grossamide, from bell pepper (*Capsicum annuum* Var. *grossurri*). Agric Biol Chem 45:2593–2598
11. Hedberg C, Hesse M, Werner C (1996) Spermine and spermidine hydroxycinnamoyl transferases in *Aphelandra tetragona*. Plant Sci 113:149–156
12. Fu XP, Wu T, Abdurahim M, Su Z, Hou XL, Aisa HA, Wu H (2008) New spermidine alkaloids from *Capparis spinosa* roots. Phytochem Lett 1:59–62

13. Buta JG, Izac RR (1972) Caffeoylputrescine in *Nicotiana tabacum*. Phytochemistry 11:1188–1189
14. Martin-Tanguy J, Martin C, Gallet M (1973) Présence de composés aromatiques liés à la putrescine dans divers *Nicotiana* virosés. C R Acad Sci Paris D 276:1433–1435
15. Cabanne F, Dalebroux MA, Martin-Tanguy J, Martin C (1981) Hydroxycinnamic acid amides and ripening to flower of *Nicotiana tabacum* var. *xanthi* n.c. Physiol Plant 53:399–404
16. Meurer B, Wray V, Grotjahn L, Wiermann R, Strack D (1986) Hydroxycinnamic acid spermidine amides from pollen of *Corylus avellana* L. Phytochemistry 25:433–435
17. Strack D, Eilert U, Wray V, Wolff J, Jaggy H (1990) Tricoumaroylspermidine in flowers of Rosaceae. Phytochemistry 29:2893–2896
18. Bokern M, Witte L, Wray V, Nimtz M, Meurer-Grimes B (1995) Trisubstituted hydroxycinnamic acid spermidines from *Quercus dentata* pollen. Phytochemistry 39:1371–1375
19. Lin S, Mullin C (1999) Lipid, polyamide, and flavonol phagostimulants for adult western corn rootworm from sunflower (*Helianthus annuus* L.) pollen. J Agric Food Chem 47:1223–1229
20. Sobolev VS, Sy A, Gloer JB (2008) Spermidine and flavonoid conjugates from peanut (*Arachis hypogaea*) flowers. J Agric Food Chem 56:2960–2969
21. Kite C, Larsson S, Veitch NC, Porter EA, Ding N, Simmonds MSJ (2013) Acyl spermidines in inflorescence extracts of elder (*Sambucus nigra* L., Adoxaceae) and elderflower drinks. J Agric Food Chem 61:3501–3508
22. Wiese S, Wubshet SG, Nielsen J, Staerk D (2013) Coupling HPLC-SPE-NMR with a microplate-based high-resolution antioxidant assay for efficient analysis of antioxidants in food–validation and proof-of-concept study with caper buds. Food Chem 141:4010–4018
23. Werner C, Hu W, Lorenzi-riatsch A, Hesse M (1995) Di-coumaroylspermidines and tricoumaroylspermidines in anthers of different species of the genus *Aphelandra*. Phytochemistry 40:461–465
24. Youhnovski N, Werner C, Hesse M (2001) *N*, *N′*,*N″*-triferuloylspermidine, a new UV absorbing polyamine derivative from pollen of *Hippeastrum* x *hortorum*. Z Naturforsch C 56:526–530
25. Grienenberger E, Besseau S, Geoffroy P, Debayle D, Heintz D, Lapierre C, Pollet B, Heitz T, Legrand M (2009) A BAHD acyltransferase is expressed in the tapetum of *Arabidopsis* anthers and is involved in the synthesis of hydroxycinnamoyl spermidines. Plant J 58:246–259
26. Fellenberg C, Vogt T (2015) Evolutionarily conserved phenylpropanoid pattern on angiosperm pollen. Trends Plant Sci 20:212–218
27. Campos L, Lisón P, López-Gresa MP, Rodrigo I, Zacarés L, Conejero V, Bellés JM (2014) Transgenic tomato plants overexpressing tyramine *N*-hydroxycinnamoyltransferase exhibit elevated hydroxycinnamic acid amide levels and enhanced resistance to *Pseudomonas syringae*. Mol Plant Microbe Interact 27:1159–1169
28. Zhang Y, Long Z, Guo Z, Wang Z, Zhang X, Ye RD, Liang X, Civelli O (2016) Hydroxycinnamic acid amides from *Scopolia tangutica* inhibit the activity of M1 muscarinic acetylcholine receptor in vitro. Fitoterapia 108:9–12
29. Takahashi T, Tong W (2015) Regulation and diversity of polyamine biosynthesis in plants. In: Kusano T, Suzuki H (eds) Polyamines. A universal molecular nexus for growth, survival, and specialized metabolism. Springer, Tokyo
30. von Ropenack E, Parr A, Schulze-Lefert P (1998) Structural analyses and dynamics of soluble and cell wall-bound phenolics in a broad spectrum resistance to the powdery mildew fungus in barley. J Biol Chem 273:9013–9022
31. Muroi A, Ishihara A, Tanaka C, Ishizuka A, Takabayashi J, Miyoshi H, Nishioka T (2009) Accumulation of hydroxycinnamic acid amides induced by pathogen infection and identification of agmatine coumaroyltransferase in *Arabidopsis thaliana*. Planta 230:517–527
32. Walters D, Meurer-Grimes B, Rovira I (2001) Antifungal activity of three spermidine conjugates. FEMS Microbiol Lett 201:255–258
33. Kaur H, Heinzel N, Schöttner M, Baldwin IT, Gális I (2010) R2R3-NaMYB8 regulates the accumulation of phenylpropanoid-polyamine conjugates, which are essential for local and systemic defense against insect herbivores in *Nicotiana attenuata*. Plant Physiol 152:1731–1747
34. Martin-Tanguy J, Martin C, Gallet M, Vernoy R (1976) Sur de puissants inhibiteurs de multiplication du virus de la mosaïque du tabac. C R Acad Sci Paris D 282:2231–2234
35. Jin S, Yoshida M, Nakajima T, Murai A (2003) Accumulation of hydroxycinnamic acid amides in winter wheat under snow. Biosci Biotechnol Biochem 67:1245–1249
36. Camacho-Cristóbal JJ, Maldonado JM, González-Fontes A (2005) Boron deficiency increases putrescine levels in tobacco plants. J Plant Physiol 162:921–928

37. Izaguirre MM, Mazza CA, Svatos A, Baldwin IT, Ballaré CL (2007) Solar ultraviolet-B radiation and insect herbivory trigger partially overlapping phenolic responses in *Nicotiana attenuata* and *Nicotiana longiflora*. Ann Bot 99:103–109
38. Meurer B, Wiermann R, Strack D (1988) Phenylpropanoid patterns in Fagales and their phylogenetic relevance. Phytochemistry 27:823–828
39. Meurer B, Wray V, Wiermann R, Strack D (1988) Hydroxycinnamic acid-spermidine amides from pollen of *Alnus glutinosa*, *Betula verrucosa* and *Pterocarya fraxinifolia*. Phytochemistry 27:839–843
40. Leubner-Metzger G, Amrhein N (1993) The distribution of hydroxycinnamoylputrescines in different organs of *Solanum tuberosum* and other solanaceous species. Phytochemistry 32:551–556
41. Panagabko C, Chenier D, Fixon-Owoo S, Atkinson JK (2000) Ion-pair HPLC determination of hydroxycinnamic acid monoconjugates of putrescine, spermidine and spermine. Phytochem Anal 11:11–17
42. López-Gresa MP, Torres C, Campos L, Lisón P, Rodrigo I, Bellés JM, Conejero V (2011) Identification of defense metabolites in tomato plants infected by the bacterial pathogen *Pseudomonas syringae*. Environ Exp Bot 74:216–228
43. Chong ESL, McGhie TK, Heyes JA, Stowell KM (2013) Metabolite profiling and quantification of phytochemicals in potato extracts using ultra-high-performance liquid chromatography-mass spectrometry. J Sci Food Agric 93:3801–3808
44. Long Z, Zhang Y, Guo Z, Wang L, Xue X, Zhang X, Wang S, Wang Z, Civelli O, Liang X (2014) Amide alkaloids from *Scopolia tangutica*. Planta Med 80:1124–1130
45. Dobritzsch M, Lübken T, Eschen-Lippold L, Gorzolka K, Blum E, Matern A, Marillonnet S, Böttcher C, Dräger B, Rosahl S (2016) MATE transporter-dependent export of hydroxycinnamic acid amides. Plant Cell 28:583–596
46. Hu Q, Noll RJ, Li H, Makarov A, Hardman M, Cooks RG (2005) The orbitrap: a new mass spectrometer. J Mass Spectrom 40:430–443
47. Sauer S, Kliem M (2010) Mass spectrometry tools for the classification and identification of bacteria. Nat Rev Microbiol 8:74–82

Chapter 10

Analysis of Glutathione in Biological Samples by HPLC Involving Pre-Column Derivatization with *o*-Phthalaldehyde

Yongqing Hou, Xilong Li, Zhaolai Dai, Zhenlong Wu, Fuller W. Bazer, and Guoyao Wu

Abstract

Glutathione (GSH) forms conjugates with polyamines in prokaryotes and eukaryotes. There is also evidence suggesting cross-talk between GSH and polyamines to regulate cellular homeostasis and function, particularly under the conditions of oxidative stress. Because of its versatile roles in cell metabolism and function, a number of high performance liquid chromatography (HPLC) methods have been developed for glutathione analysis. Here, we describe our rapid and sensitive method for the analysis of GSH and the oxidized form of glutathione (GS-SG) in animal tissues and cells by HPLC involving pre-column derivatization with *o*-phthalaldehyde (OPA). OPA reacts very rapidly (within 1 min) with S-carboxymethyl-glutathione at room temperatures (e.g., 20–25 °C) in an autosampler, and their derivatives are immediately injected into the HPLC column without any need for extraction. This method requires two simple steps (a total of 15 min) before samples are loaded into the autosampler: (a) the conversion of GS-SG into GSH by 2-mercaptoethanol; and (b) the oxidation of GSH by iodoacetic acid to yield S-carboxymethyl-glutathione. The autosampler is programmed to mix S-carboxymethyl-glutathione with OPA for 1 min to generate a highly fluorescent derivative for HPLC separation and detection (excitation wavelength 340 nm and emission wavelength 450 nm). The detection limit for GSH and GS-SG is 15 pmol/ml or 375 fmol/injection. The total time for chromatographic separation (including column regeneration) is 16 min for each sample. Our routine HPLC technique is applicable for analyses of cysteine and cystine, as well as polyamines and GSH-polyamine conjugates in biological samples.

Key words Glutathione, Cysteine, Polyamines, *o*-Phthalaldehyde, HPLC

1 Introduction

Glutathione (GSH) forms conjugates with polyamines, such as spermidine, in both prokaryotes and eukaryotes [1–4]. There is also evidence suggesting cross-talk between GSH and polyamines to regulate cellular homeostasis and function, particularly under the conditions of oxidative stress [5, 6]. Thus, GSH, which consists of L-glutamate, L-cysteine, and glycine, is the most abundant low-molecular-weight antioxidant tripeptide in cells [7]. GSH is a very special small peptide, as it is the γ-carboxyl group, rather than the

Rubén Alcázar and Antonio F. Tiburcio (eds.), *Polyamines: Methods and Protocols*, Methods in Molecular Biology, vol. 1694, DOI 10.1007/978-1-4939-7398-9_10, © Springer Science+Business Media LLC 2018

α-carboxyl group, of the L-glutamate molecule that reacts with the α-amino group of L-cysteine to form a peptidic γ-linkage, which protects GSH from hydrolysis by extracellular or intracellular peptidases. In cells, GSH peroxidase converts GSH (the reduced form) and hydrogen peroxide into oxidized glutathione (GS-SG) and water, respectively [8].

Because of its versatile roles in cell metabolism and function, a number of high-performance liquid chromatography (HPLC) methods have been developed for the analysis of both GSH and GS-SG [9–12]. A widely used technique employs the pre-column derivatization of GSH and GS-SG with dansyl chloride to form a fluorescent derivative [8–10]. Major disadvantages of this method are a prolonged period (26 h) of derivatization, technical difficulties in the online automation of the derivatization, and a need for the extraction of dansyl derivatives with chloroform before HPLC analysis [10, 11]. In the present work, we describe a rapid and sensitive method for the analysis of GSH and GS-SG by HPLC involving pre-column derivatization with *o*-phthalaldehyde (OPA) [13–16]. OPA reacts very rapidly (within 1 min) with S-carboxymethyl-glutathione at room temperatures (e.g., 20–25 °C) in an autosampler, and their derivatives are immediately injected into the HPLC column without any need for extraction [15–18]. This method requires two simple steps (a total of 15 min) before samples are loaded into the autosampler: (a) the conversion of GS-SG into GSH by 2-mercaptoethanol; and (b) the oxidation of GSH by iodoacetic acid to yield S-carboxymethyl-glutathione. The autosampler is programmed to mix S-carboxymethyl-glutathione with OPA to generate a highly fluorescent derivative for HPLC separation and detection [16–18]. Our OPA technique is applicable for analyses of cysteine [19], cystine [19], polyamines [20–22], and GSH-polyamine conjugates (e.g., GSH-spermidine [23]).

2 Materials

HPLC-grade water (H_2O) is used to prepare all reagent solutions in polypropylene tubes (for ≤50 ml solutions) or glass bottles (for >50 ml solutions), unless specified otherwise (*see* **Note 1**). All solutions are stored at 20–25 °C for use within 2 years, unless specified otherwise.

1. 1.2% Benzoic acid: Dissolve 8.4 g benzoic acid in 525 ml H_2O and add 175 ml of saturated potassium tetraborate tetrahydrate ($K_2B_4O_7 \cdot 4H_2O$).
2. 40 mM Sodium borate: Dissolve 3.051 g sodium tetraborate decahydrate ($Na_2B_4O_7 \cdot 10H_2O$) in 200 ml H_2O. 100 mM Sodium borate: Dissolve 9.55 g sodium tetraborate

decahydrate in 250 ml H_2O. 100 mM Boric acid: Dissolve 1.24 g boric acid in 200 ml H_2O.

3. 6 M HCl: Add slowly 49.1 ml of concentrated HCl (37–38%) to 50.9 ml H_2O.
4. 1.5 M Perchloric acid ($HClO_4$): Add 32.2 ml of 70% $HClO_4$ to 150 ml H_2O. Bring to a final volume of 250 ml with H_2O.
5. 2 M K_2CO_3: Dissolve 69.11 g K_2CO_3 in 150 ml H_2O. Bring to a final volume of 250 ml with H_2O.
6. *O*-Phthaldialdehyde (OPA; in a brown bottle): Dissolve 50 mg OPA in 1.25 ml methanol. Add 11.2 ml of 40 mM sodium borate buffer (pH 9.5), 50 μl of 2-mercaptoethanol, and 0.4 ml of Brij-35. Mix. This solution is stable at 4 °C for 36 h. (Turn off the light when weighing OPA.)
7. 25 mM iodoacetic acid: Dissolve 21 mg iodoacetic acid (sodium salt) in 4 ml of 40 mM sodium borate. Iodoacetic acid (12 mM): Dissolve 250 mg iodoacetic acid (sodium salt) in 100 ml H_2O. (Both solutions must be prepared fresh on the day of analysis).
8. Extraction solution (for [24] samples): Mix 50 ml of 12 mM iodoacetic acid with 50 ml of 1.5 M $HClO_4$. (This solution must be prepared fresh on the day of analysis.)
9. Glutathione, reduced (GSH, ≥98% purity) and cysteine standards (store the solutions in 0.5-ml aliquots at −80 °C). 10 mM GSH/10 mM cysteine: Dissolve 31.4 mg GSH and 12.4 mg cysteine in 10 ml of 12 mM iodoacetic acid. 100 μM GSH/100 μM cysteine: Mix 100 μl of 10 mM GSH/10 mM cysteine, 5.95 ml of the extraction solution, 1.5 ml of 2 M K_2CO_3, and 2.45 ml of 12 mM iodoacetic acid in a 15-ml polypropylene tube.
10. Oxidized glutathione (GS-SG; ≥98% purity) and cystine standards (store solutions in 0.5-ml aliquots at −80 °C). 0.4 mM GS-SG/0.4 mM Cystine standard: Dissolve 25.0 mg GS-SG and 9.71 mg cystine in 100 ml of 12 mM iodoacetic acid. 50 μM GS-SG/50 μM cystine standard: Mix 400 μl of 0.4 mM GS-SG/0.4 mM cysteine, 1.8 ml of the extraction solution, 0.5 ml of 2 M K_2CO_3, and 0.5 ml of 12 mM iodoacetic acid.
11. 28 mM 2-mercaptoethanol: Mix 20 μl of 2-mercaptoethanol with 10 ml of 40 mM sodium borate in a brown bottle. (Prepare this solution on the day of analysis.)
12. GSH/Cysteine preservation solution A: Dissolve 1 mg sodium heparin, 10 mg serine, 4.5 mg iodoacetic acid (sodium salt), and 2 ml of 100 mM sodium borate. Prepare fresh.

13. GSH/Cysteine preservation solution B (1.5 M $HClO_4$, w/v; 0.2 M boric acid): Dissolve 1.24 g boric acid in 80 ml H_2O and add 12.9 ml of 70% $HClO_4$. Bring to a final volume of 100 ml with H_2O. Store this solution at room temperature.
14. Mobile phase A (0.1 M sodium Acetate, pH 7.2): Add 27.3 g of sodium acetate-trihydrate to 1.6 l H_2O. Adjust to pH 7.2 with 96 μl of 6 M HCl. Add 180 ml methanol and 10 ml tetrahydrofuran. Bring the final volume to 2 l with H_2O. (This solution is used within one week after preparation.)
15. Millipore Water-Purifying System.
16. Waters HPLC apparatus consisting of a Model 600E Powerline multisolvent delivery system with 100-μl heads, a Model 712 WISP autosampler, a Waters 2475 Multi λ Fluorescence detector, and a Millenium-32 Workstation.
17. A Supelco C_{18} guard column (4.6 mm × 5 cm, 20–40 μm) and a Supelco C_{18} column (4.6 mm × 15 cm, 3 μm).

3 Methods

3.1 Extraction of GSH/GSSG and Cysteine/Cystine from Biological Samples

3.1.1 Tissues Other than Blood

1. Add 1.5 ml of the extraction solution to a 15-ml glass PYREX tube.
2. Grind a frozen tissue sample (50–100 mg) to powder in liquid nitrogen. Transfer the weighed powder into the glass tube.
3. Homogenize the tissue using a glass homogenizer and transfer the solution to a new 15-ml polypropylene tube.
4. Rinse the homogenizer with 1.5 ml of the extraction solution, and combine the homogenates in the 15-ml polypropylene tube.
5. Add 0.75 ml of 2 M K_2CO_3 to the polypropylene tube. Mix well.
6. Centrifuge all tubes at 600 × *g* for 5 min. Use the supernatant fluid for analysis.

3.1.2 Blood

1. Mix 0.2 ml of whole blood sample (*see* **Note 2**) with 0.2 ml of GSH/Cysteine preservation solution A in a 1.5-ml Microcentrifuge tube (*see* **Note 3**). Vortex gently for 5 s.
2. Centrifuge all tubes at 10,000 × *g* for 1 min.
3. Transfer 0.2 ml of the supernatant fluid to a new 1.5-ml Microcentrifuge tube. Add 0.1 ml of GSH/Cysteine preservation solution B to the tube.
4. Add 50 μl of 2 M K_2CO_3 to the tube. Mix well.
5. Centrifuge all tubes at 10,000 × *g* for 1 min. Use the supernatant fluid for analysis.

3.1.3 Fetal Fluid

1. Mix 0.2 ml of the fetal fluid (allantoic or amniotic fluid) with 0.2 ml of the extraction solution in a 1.5-ml Microcentrifuge tube.
2. Add 100 μl of 2 M K_2CO_3 to the tube. Mix well.
3. Centrifuge all tubes at 10,000 × g for 1 min. Use the supernatant fluid for analysis.

3.1.4 Freshly Isolated Cells

1. Add 0.2 ml of the extraction solution to ~5 × 10^6 cells in a 1.5-ml Microcentrifuge tube.
2. Homogenize cells using a small plastic homogenizer.
3. Add 0.05 ml of 2 M K_2CO_3 to the tube. Mix well.
4. Centrifuge all tubes at 10,000 × g for 1 min. Use the supernatant fluid for analysis.

3.1.5 Cells Cultured in a Petri Dish

1. Remove the culture medium to retain cells (~5 × 10^6) at the bottom of the culture dish. Rapidly wash the cells twice with phosphate-buffered saline (pH 7.4). After each wash, the solution is removed.
2. Add 0.2 ml of the extraction buffer to the cell culture dish.
3. Scrape cells from the culture dish and transfer the whole solution to a 1.5-ml tube.
4. Homogenize cells using a small plastic homogenizer.
5. Centrifuge all tubes at 10,000 × *g* for 1 min.
6. Transfer 150 μl of the supernatant fluid into a new 1.5-ml Microcentrifuge tube.
7. Add 37.5 μl of 2 M K_2CO_3 to the tube. Mix well.
8. Centrifuge all tubes at 10,000 × *g* for 1 min. Use the supernatant fluid for analysis.

3.2 HPLC Analysis of GSH/Cysteine and GS-SG/Cystine

1. For GSH/Cysteine analysis, add 50 μl of 100 μM GSH/100 μM cysteine standard (or sample) and 100 μl of 40 mM sodium borate to a 4-ml glass vial.

   For GS-SG/Cystine analysis, add 50 μl of 50 μM GS-SG/50 μM cystine standard (or sample) and 100 μl of 28 mM 2-mercaptoethanol to a 4-ml glass vial.
2. Vortex the vials for 10 s. After 10 min, add 50 μl of 25 mM iodoacetic acid to each vial. Vortex the vials for 10 s. After 5 min, add 0.1 ml of 1.2% benzoic acid and 1.4 ml of HPLC H_2O to each vial, and then place all the vials in the autosampler.
3. The autosampler is programmed to mix 25 μl of sample (or standard) with 25 μl of the OPA reagent solution (a total of 4 ml in a brown bottle) for 1 min and then deliver the derivatized solution into the HPLC column without any delay.

Table 1
Solvent gradients for the HPLC separation of GSH, GS-SG, cysteine, and cystine[a]

	Time (min)							
Mobile phase solution	**0**	**1**	**1.1**	**6.5**	**6.6**	**9**	**9.1**	**16**
A, %	97	97	86	86	0	0	97	97
B, %	3	3	14	14	100	100	3	3

[a]A combined flow rate of the HPLC is 1.1 ml/min. The mobile phase solutions A and B are degassed with helium

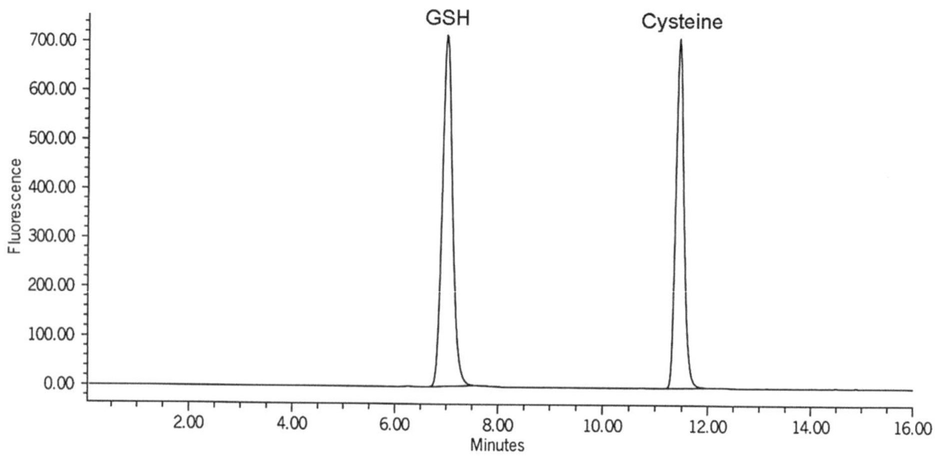

Fig. 1 A representative HPLC chromatogram for GSH and cysteine standards. Fifty microliters (50 μl) of 100 μM GSH/100 μM cysteine standards were mixed with 100 μl of 40 mM sodium borate in a 4-ml glass vial. After 10 min, 50 μl of 25 mM iodoacetic acid was added to each vial. After 5 min, 0.1 ml of 1.2% benzoic acid and 1.4 ml of HPLC H_2O were added to each vial. Twenty five microliters (25 μl) of the solution were used for derivatization with 25 μl of the *O*-phthaldialdehyde solution. The gain of the detector was 1

4. Start HPLC analysis (a total running time of 16 min including the time for HPLC column regeneration): The solvent gradients for the chromatographic separation of GSH and GS-SG are shown in Table 1.

5. Fluorescence detection (excitation wavelengths: 220 nm between 0 and 6 min, 340 nm between 6 and 12 min, and 220 nm between 12 and 16 min; emission wavelengths: 450 nm between 0 and 16 min; and gain 1) (*see* **Note 4**). GSH and cysteine standard derivatives are rapidly eluted from the HPLC column, with their retention times being approximately 6.9 and 11.4 min, respectively (Fig. 1). GS-SG and cystine standard derivatives are illustrated in Fig. 2. The detection limit for GSH, GS-SG, cysteine and cystine (signal/noise ratio > 3) is 15 pmol/ml or 375 fmol/injection. Representative HLPC chromatograms for GSH and cysteine (Fig. 3), as

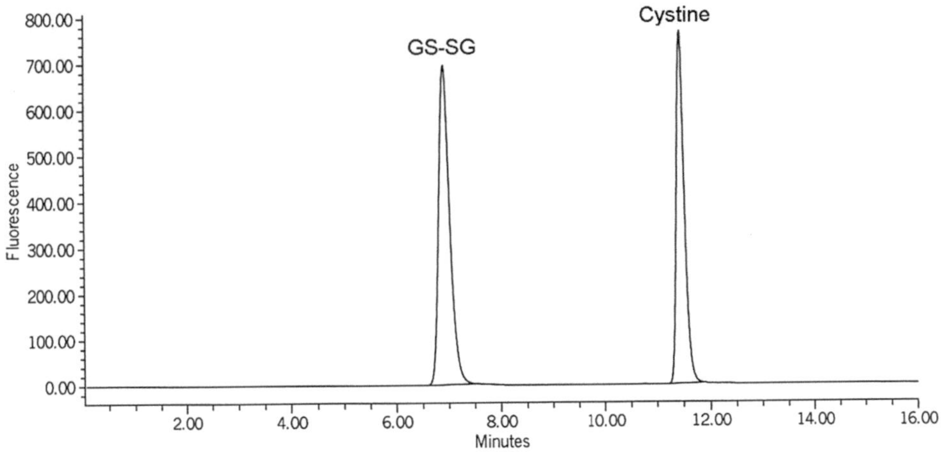

Fig. 2 A representative HPLC chromatogram for GS-SGH and cystine standards. Fifty microliters (50 μl) of 50 μM GS-SG/50 μM cystine standards were mixed with 100 μl of 40 mM sodium borate in a 4-ml glass vial. After 10 min, 50 μl of 25 mM iodoacetic acid was added to each vial. After 5 min, 0.1 ml of 1.2% benzoic acid and 1.4 ml of HPLC H_20 were added to each vial. Twenty five microliters (25 μl) of the solution were used for derivatization with 25 μl of the *O*-phthaldialdehyde solution. The gain of the detector was 1

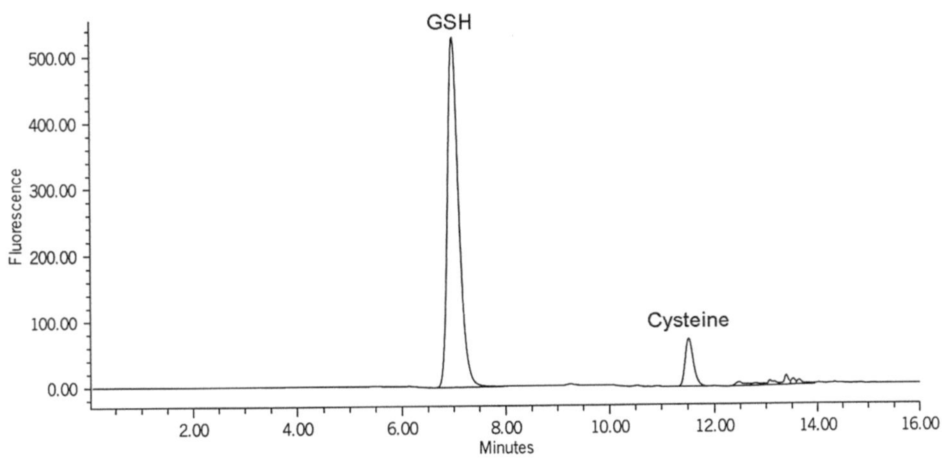

Fig. 3 A representative HPLC chromatogram for GSH and cysteine in the fetal lamb liver. A sample (50 mg) of the fetal lamb liver was homogenized in 2 ml of the extraction solution. The homogenate was neutralized with 0.5 ml of 2 M K_2CO_3, and then centrifuged at 10,000 × *g* for 1 min. Fifty microliters (50 μl) of the supernatant fluid were mixed with 100 μl of 40 mM sodium borate in a 4-ml glass vial. After 10 min, 50 μl of 25 mM iodoacetic acid was added to each vial. After 5 min, 0.1 ml of 1.2% benzoic acid and 1.4 ml of HPLC H_20 were added to each vial. Twenty five microliters (25 μl) of the solution were used for derivatization with 25 μl of the *O*-phthaldialdehyde solution. The gain of the detector was 1

well as total glutathione (GSH + ½ GS-SG) and total cysteine (cysteine + ½ cystine) (Fig. 4) in the fetal lamb liver are also shown.

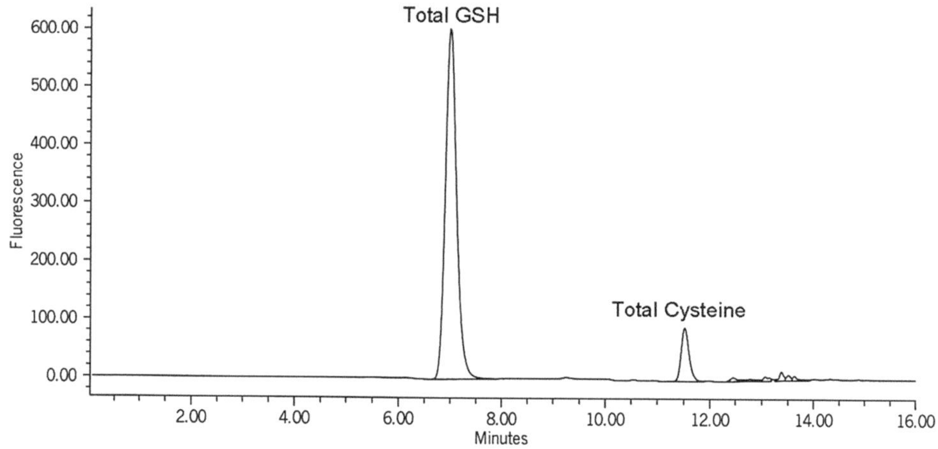

Fig. 4 A representative HPLC chromatogram for total glutathione (GSH + ½ GS-SG) and total cysteine (cysteine + ½ cystine) in liver from a fetal lamb. A sample (50 mg) of the fetal lamb liver was homogenized in 2 ml of the extraction solution. The homogenate was neutralized with 0.5 ml of 2 M K_2CO_3, and then centrifuged at 10,000 × *g* for 1 min. Fifty microliters (50 μl) of the supernatant fluid were mixed with 100 μl of 40 mM sodium borate in a 4-ml glass vial. After 10 min, 50 μl of 25 mM iodoacetic acid was added to each vial. After 5 min, 0.1 ml of 1.2% benzoic acid and 1.4 ml of HPLC H_2O were added to each vial. Twenty five microliters (25 μl) of the solution were used for derivatization with 25 μl of the *O*-phthaldialdehyde solution. The gain of the detector was 1

Table 2
Concentrations of GSH, GS-SG, cysteine and cystine in samples from animals[a]

Variable	GSH	GS-SG	Cysteine	Cystine
Ovine placentome	6.28 ± 0.17	0.34 ± 0.008	0.93 ± 0.024	0.25 ± 0.043
Fetal lamb GM	0.97 ± 0.023	0.068 ± 0.002	0.066 ± 0.002	0.014 ± 0.0004
Fetal lamb liver	4.11 ± 0.12	0.36 ± 0.010	0.52 ± 0.014	0.088 ± 0.003
Fetal lamb SI	1.71 ± 0.035	0.12 ± 0.004	0.20 ± 0.005	0.036 ± 0.001
Piglet plasma	4.58 ± 0.16	0.95 ± 0.031	4.72 ± 0.19	81.6 ± 3.15

GM Gastrocnemius muscle, *SI* small intestine

[a]Values, expressed as nmol/ml for the piglet plasma and nmol/mg wet tissue weight for the ovine tissues, are means ± SEM, $n = 6$. The ovine placentomes and fetal lamb tissues were obtained from control-fed sheep on Day 125 of gestation and then immediately placed in liquid nitrogen [25]. Blood samples from 21-day-old milk-fed pigs were processed immediately for the analysis of metabolites in plasma [17]

6. Calculation. The amount of GSH/cysteine and GS-SG/cystine in an unknown sample is calculated by the Waters Workstation on the basis of known amounts of GSH/Cysteine and GS-SG/Cystine standards (*see* **Note 5**). GS-SG = (Total glutathione − GSH)/2; and cystine = (Total cysteine − cysteine)/2. The recoveries of GSH/cysteine and GS-SG/cystine from tissues or physiological fluids are used to calculate the amounts of

the metabolites in the tissues or fluids. Table 2 shows the concentrations of GSH, GS-SG, cysteine, and cystine in: (a) the ovine placentome; (b) the liver, small intestine, and skeletal muscle of fetal lambs; and (c) plasma from piglets.

4 Notes

1. All procedures for this step and other steps of this protocol are performed at room temperatures (e.g., 20–25 °C).
2. Hemolysis should be avoided when blood samples are withdrawn. A whole blood sample is placed in a heparinized tube.
3. In the GSH/cysteine preservation solution, heparin is used to inhibit coagulation, serine borate to inhibit GSH degradation by γ-glutamyltranspeptidase, bathophenanthroline disulfonate to inhibit GSH oxidation, and iodoacetic acid to alkylate GSH and cysteine. Bathophenanthroline disulfonate sodium (BPDS; 2 mM) interferes with the reaction of cysteine with iodoacetic acid. In the presence of 12 mM iodoacetic acid, GSH and cysteine are stable in the absence of BPDS.
4. The setting of excitation wavelength at 220 nm before 6 min and after 12 min is designed to suppress fluorescence due to other amino acids that react with OPA.
5. The accuracy of the HPLC method for the analysis of GSH, cysteine, GS-SG, and cystine, as determined with known amounts of standards and expressed as the relative error [(measurement value − true value)/true value × 100%)] [26], is 1.5, 1.7, 1.6, and 1.7%, respectively, for a tissue (e.g., ovine placentome), and 1.1, 1.3, 1.2, and 1.0%, respectively for a physiological fluid (ovine fetal fluid). The precision of the HPLC method for analyses of GSH, cysteine, GS-SG, and cystine, as evaluated by the relative deviation (mean of absolute deviation/mean of replicate measurements × 100%) [26], is 1.2, 1.4, 1.3, and 1.5%, respectively, for a tissue (e.g., ovine placentome), and 0.7, 0.9, 0.8, and 0.9%, respectively for a physiological fluid (ovine fetal fluid). The rates of recovery of GSH, cysteine, GS-SG, and cystine from a tissue (e.g., ovine placentome) are 96.3, 95.6, 95.8, and 96.0%, respectively, and 98.8, 97.9, 97.3, and 98.5%, respectively, for a physiological fluid (e.g., plasma, ovine amniotic fluid, or ovine allantoic fluid).

Acknowledgments

Work in our laboratories was supported by grants from the National Natural Science Foundation of China (31572412, 31372319, 31572416), Hubei Provincial Key Project for Scientific and Technical Innovation (2014ABA022), Hubei Hundred Talent program, Natural Science Foundation of Hubei Province (2013CFA097), Agriculture and Food Research Initiative Competitive Grants (2014-67015-21770, 2015-67015-23276 and 2016-67015-24958) from the USDA National Institute of Food and Agriculture, and Texas A&M AgriLife Research (H-8200).

References

1. Dubin DT (1959) Evidence for conjugates between polyamines and glutathione in E. coli. Biochem Biophys Res Commun 1:262–265
2. Wang W, Ballatori N (1998) Endogenous glutathione conjugates: occurrence and biological functions. Pharmacol Rev 50:335–356
3. Mastri C, Thorborn DE, Davies AJ, Ariyanayagam MR, Hunter KJ (2001) Polyamine and thiol metabolism in Trypanosoma granulosum: similarities with Trypanosoma cruzi. Biochem Biophys Res Commun 282:1177–1182
4. Ariyanayagam MR, Oza SL, Mehlert A, Fairlamb AH (2003) Bis(glutathionyl)spermine and other novel trypanothione analogues in Trypanosoma cruzi. J Biol Chem 278:27612–27619
5. Kwon DH, Hekmaty S, Seecoomar G (2013) Homeostasis of glutathione is associated with polyamine-mediated β-lactam susceptibility in Acinetobacter Baumannii ATCC 19606. Antimicrob Agents Chemother 57:5457–5461
6. Ceylan S, Seidel V, Ziebart N, Berndt C, Dirdjaja N, Krauth-Siegel RL (2010) The dithiol glutaredoxins of African trypanosomes have distinct roles and are closely linked to the unique trypanothione metabolism. J Biol Chem 285:35224–35237
7. Wu G, Fang YZ, Yang S, Lupton JR, Turner ND (2004) Glutathione metabolism and its implications for health. J Nutr 134:489–492
8. Flohé L (2016) The impact of thiol peroxidases on redox regulation. Free Radic Res 50:126–142
9. Marcé M, Brown DS, Capell T, Figueras X, Tiburcio AF (1995) Rapid high-performance liquid chromatographic method for the quantitation of polyamines as their dansyl derivatives: application to plant and animal tissues. J Chromatogr B 666:329–335
10. Jones DP, Carlson JL, Samiec PS, Sternberg P, Mody VC, Reed RL, Brown LAS (1998) Glutathione measurement is human plasma: evaluation of sample collection, storage and derivatization conditions for analysis of dansyl derivatives by HPLC. Clin Chim Acta 275:175–184
11. Jobgen W, Fu WJ, Gao H, Li P, Meininger CJ, Smith SB, Spencer TE, Wu G (2009) High fat feeding and dietary L-arginine supplementation differentially regulate gene expression in rat white adipose tissue. Amino Acids 37:187–198
12. Seiler N (1986) Polyamines. J Chromatogr 379:157–176
13. Yan CC, Huxtable RJ (1995) Fluorimetric determination of monobromobimane and o-phthalaldehyde adducts of gamma-glutamylcysteine and glutathione: application to assay of gamma-glutamylcysteinyl synthetase activity and glutathione concentration in liver. J Chromatogr B 672:2172–2124
14. Michaelsen JT, Dehnert S, Giustarini D, Beckmann B, Tsikas D (2009) HPLC analysis of human erythrocytic glutathione forms using OPA and N-acetyl-cysteine ethyl ester: evidence for nitrite-induced GSH oxidation to GSSG. J Chromatogr B 877:3405–3417
15. Wu G, Meininger CJ (2008) Analysis of citrulline, arginine, and methylarginines using high-performance liquid chromatography. Methods Enzymol 440:177–189
16. Yi D, Hou YQ, Wang L, Long MH, Hu SD, Mei HM, Yan LQ, Hu CA, Wu G (2016) N-Acetylcysteine stimulates protein synthesis in enterocytes independently of glutathione synthesis. Amino Acids 48:523–533
17. Wang WW, Dai ZL, Wu ZL, Lin G, Jia SC, Hu SD, Dahanayaka S, Wu G (2014) Glycine is a nutritionally essential amino acid for maximal

growth of milk-fed young pigs. Amino Acids 46:2037–2045

18. Tekwe CD, Lei J, Yao K, Rezaei R, Li XL, Dahanayaka S, Carroll RJ, Meininger CJ, Bazer FW, Wu G (2013) Oral administration of interferon tau enhances oxidation of energy substrates and reduces adiposity in Zucker diabetic fatty rats. Biofactors 39:552–563
19. Ji Y, Wu ZL, Dai ZL, Sun KJ, Zhang Q, Wu G (2016) Excessive L-cysteine induces vacuole-like cell death by activating endoplasmic reticulum stress and mitogen-activated protein kinase signaling in intestinal porcine epithelial cells. Amino Acids 48:149–156
20. Wu G, Flynn NE, Knabe DA, Jaeger LA (2000) A cortisol surge mediates the enhanced polyamine synthesis in porcine enterocytes during weaning. Am J Physiol 279:R554–R559
21. Li H, Meininger CJ, Bazer FW, Wu G (2016) Intracellular sources of ornithine for polyamine synthesis in endothelial cells. Amino Acids 48:2401–2410
22. Dai ZL, Wu ZL, Wang JJ, Wang XQ, Jia SC, Bazer FW, Wu G (2014) Analysis of polyamines in biological samples by HPLC involving pre-column derivatization with *o*-phthalaldehyde and *N*-acetyl-L-cysteine. Amino Acids 46:1557–1564
23. Wu G, Knabe DA (1994) Free and protein-bound amino acids in sow's colostrum and milk. J Nutr 124:415–424
24. Haimeur A, Guimond C, Pilote S, Mukhopadhyay R, Rosen BP, Poulin R, Ouellette M (1999) Elevated levels of polyamines and trypanothione resulting from overexpression of the ornithine decarboxylase gene in arsenite-resistant Leishmania. Mol Microbiol 34:726–735
25. Satterfield MC, Dunlap KA, Keisler DH, Bazer FW, Wu G (2013) Arginine nutrition and fetal brown adipose tissue development in nutrient-restricted sheep. Amino Acids 45:489–499
26. Yongqing Hou, Sichao Jia, Gayan Nawaratna, Shengdi Hu, Sudath Dahanayaka, Fuller W. Bazer, Guoyao Wu, (2015) Analysis of l-homoarginine in biological samples by HPLC involving precolumn derivatization with o-phthalaldehyde and N-acetyl-l-cysteine. Amino Acids 47(9):2005-2014

Chapter 11

Determination of Arginine and Ornithine Decarboxylase Activities in Plants

Rubén Alcázar and Antonio F. Tiburcio

Abstract

In plants, putrescine is synthesized directly from the decarboxylation of ornithine and/or by the alternative arginine decarboxylase pathway. The prevalence of one or the other depends on the tissue and stress conditions. In both amino acid decarboxylation reactions, the corresponding enzymes use pyridoxal phosphate (PLP) as co-factor. PLP combines with the α-amino acid to form a Schiff base, which acts as substrate in the carboxyl group removal and CO_2 formation. We describe the methodology employed for the determination of ODC and ADC activities in plant tissues by detecting the release of (C^{14}) CO_2 using (C^{14}) labelled substrates (ornithine or arginine).

Key words Putrescine biosynthesis, Agmatine, Arginine, Ornithine, Amino acid decarboxylation

1 Introduction

Putrescine (Put) is formed directly from ornithine in a reaction catalyzed by ornithine decarboxylase (ODC; EC 4.1.117) in almost all eukaryotes [1]. However, in plants there is an alternative pathway for Put formation, in which the diamine is synthesized from arginine by the action of arginine decarboxylase (ADC; EC 4.1.1.19) followed by two successive steps catalyzed by agmatine iminohydrolase (AIH; EC 3.5.3.12) and N-carbamoylputrescine aminohydrolase (NCPAH; EC 3.5.1) [2, 3].

With the exception of Brassicaceae, where ODC is absent, *ODC* genes have been identified in several plant species [4]. Thus, in many plants, both ODC and ADC pathways co-exist and the predominance of one over the other depends on the tissue and environment. For example, the enlargement of mesocarpic cells that sustain fruit development is mainly contributed by ADC, whereas ODC is predominantly active during early fruit development when cell division occurs [5]. On the other hand, accumulation of Put under abiotic stress is often mediated by ADC [6, 7]. Contribution of ODC or ADC pathways might also be determined

Rubén Alcázar and Antonio F. Tiburcio (eds.), *Polyamines: Methods and Protocols*, Methods in Molecular Biology, vol. 1694, DOI 10.1007/978-1-4939-7398-9_11,

by their intracellular localization. Like in mammals, plant ODC has been found in the nucleus [8] whereas ADC from tobacco has been reported in the chloroplasts of photosynthetic tissues and in the nuclei of roots [9]. In oat, high ODC activity is presumably associated to nuclei in roots, whereas ADC activity is predominantly found in chloroplasts [10]. Physical separation and differential regulation of ODC and ADC pathways suggest an evolutionary specification for their functions [3].

The enzymes ODC and ADC catalyse the decarboxylation of ornithine and arginine, respectively, using pyridoxal phosphate (PLP) as a co-factor [11]. The PLP combines with the α-amino acid to form a Schiff base, which acts as the substrate in the decarboxylation (COOH group removal) reaction [12].

In the following protocol, we describe the procedure for extraction and determination of ODC and ADC activities in plant tissues by evaluating the release of (C^{14}) CO_2 using (C^{14}) labelled ornithine or arginine, respectively.

2 Materials

1. Potassium phosphate buffer pH 7.0: Mix 61.5 mL 1 M K_2HPO_4 with 38.5 mL KH_2PO_4. Dilute to 1 L with water. Autoclave and store at room temperature.
2. Protein extraction buffer: 1 mM pyridoxal phosphate, 20 mM 2-mercaptoethanol, 1 mM PMSF, protease inhibitor cocktail in 100 mM potassium phosphate buffer pH 7.0 (*see* **Note 1**).
3. 10 mM L-Arg stock (store at −20 °C).
4. 10 mM L-Orn stock (store at −20 °C).
5. 2 M NaOH.
6. Perchloric acid 10% (v/v).
7. Liquid scintillation cocktail (suitable for aqueous samples).
8. Radiolabelled substrates: Arginine L-[^{14}C(U)], L-[1-^{14}C]-Ornithine.
9. 1 mM DL-α-difluormethylarginine (DFMA) and 1 mM DL-α-difluormethylornithine (DFMO) inhibitors (*see* **Note 2**).
10. 5 × 1 cm (height × diameter) tubes with lid (*see* **Note 3**).
11. 0.5 × 0.02 cm (diameter × thickness) filter paper discs (*see* **Note 4**).
12. Scintillation counter.
13. Bradford's reagent.
14. Syringes, needles.

3 Methods

Users must get instructed about the appropriate use and disposal of radiochemicals before starting the protocol.

3.1 Protein Extraction

1. Harvest plant tissue and freeze immediately with liquid nitrogen (*see* **Note 5**).
2. Grind the tissue with the help of a mortar, pestle or disrupting device. Do not allow the samples to thaw.
3. Add 2 vol (w/v) of cold extraction buffer and homogenize the sample (*see* **Note 6**).
4. Centrifuge the extract at 12,000 × *g* for 15 min at 4 °C. Transfer the supernatant to a new tube.

3.2 Enzymatic Determination

1. Soak filter discs with 2 M NaOH and let them dry (*see* **Note 7**).
2. Hold NaOH-impregnated paper discs inside the tubes with the help of a needle (Fig. 1). Close the hub of the needle with a rubber stopper.

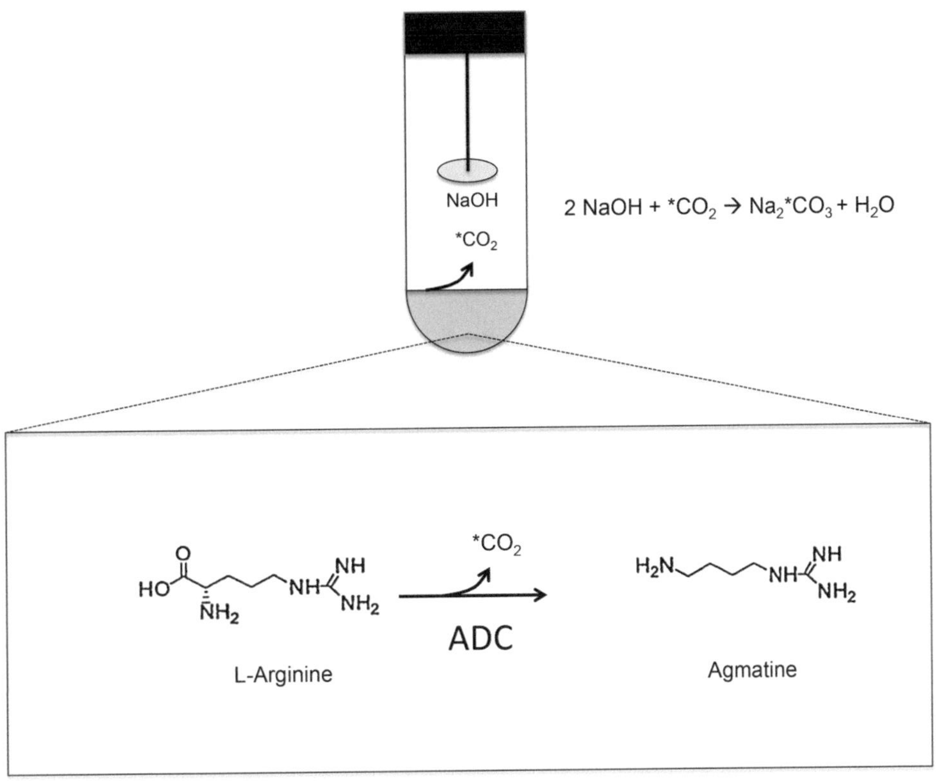

Fig. 1 Schematic representation of the device used for the enzymatic determination of ADC activity and CO_2 capture, and the catalyzed reaction by the enzyme

3. In the fume hood, pipette 100 μL of protein extract and carefully, place the extract to the bottom of the 1 × 5 cm tube.
4. Prepare a 1:5 mixture of L-[U-14C]-Arg: 10 mM L-Arg for the determination of ADC activity (Fig. 1), or L-[1-^{14}C]-Ornithine: 10 mM L-Orn for the determination of ODC activity (*see* **Note 8**).
5. Carefully, add 20 μL of the 1:5 mixture in **step 4** to each sample. Mix by pipetting up and down.
6. Close the tubes. Place a rubber stopper on top of the needle to avoid the release of CO_2 and incubate at 37 °C for 45 min (*see* **Note 9**).
7. With the help of a 1 mL-syringe, stop the reaction by adding 200 μL 20% PCA through the cannula of the needle (*see* **Note 10**). Close the tube.
8. Incubate 15 min more at room temperature to enable complete $^{14}CO_2$ capture.
9. Carefully, transfer NaOH soaked paper discs to scintillation vials containing 5 mL liquid scintillation cocktail. Mix by vortexing (*see* **Note 11**).
10. Determine the radioactive emission in a scintillation counter.

3.3 Blanks, Controls and Sample Normalization

1. Prepare blanks by adding 200 μL of extraction buffer and 20 μL of 1:5 L-[U-14C]-Arg: 10 mM L-Arg or L-[1-^{14}C]-Ornithine: 10 mM L-Orn. (*see* **Note 12**).
2. Controls for the specificity of the ADC or ODC enzymatic determination can be included by incubation of protein extracts with 1 mM DFMA or 1 mM DFMO previous to the enzymatic activity determination (*see* **Note 13**).
3. Determine protein content in the remaining supernatant of **step 4** in Subheading 3.1 using the Bradford's method and serum albumin (BSA) as the reference standard.
4. The enzyme activity is expressed in pmol CO_2/h μg of protein.

4 Notes

1. Prepare the extraction buffer before use. Read carefully safety instructions for the proper handling of 2-mercaptoethanol, PMSF and protease inhibitors.
2. DL-alpha-difluoromethylarginine (DFMA) and DL-alpha-difluoromethylornithine (DFMO) are irreversible inhibitors of ADC and ODC activities, respectively [13–15]. Alternative inhibitors to DFMA can also be used [14, 15]. Prepare stocks in water and keep at −20 °C.

3. Other plastic tubes with similar sizes might be suitable for this protocol. It is important that the lid can be perforated by the cannula of a needle. In addition, the tube lid must close tightly and the hub of the needle must be closed with a rubber stopper.
4. These discs are generated using a regular paper hole punch.
5. The amount of sample required may depend on the tissue under analysis. In our experience, at least 100 mg of plant tissue is required for the determination of ADC and ODC activities. 250 mg are optimal for most analyses. Samples must be frozen immediately to avoid that ADC activity increases due to mechanical damage induced during sampling. Perform a minimum of three to six biological replicates.
6. Keep samples on ice throughout the protocol.
7. Soaked filter discs should be prepared just before use to avoid capture of CO_2 from the atmosphere.
8. Prepare sufficient amount of fresh 1:5 mixture for all samples and distribute from the same stock.
9. Some protocols suggest shaking of the samples during the incubation. However, this may lead to accidental spilling of the radioactive mixture to the paper discs.
10. Acidification of the medium releases the remaining $^{14}CO_2$ trapped in the extraction solution.
11. Overnight incubation of discs in some scintillation cocktails provides higher and more stable counts.
12. Alternatively, blanks can also be performed using plant protein extracts that have been denatured by boiling at 95 °C 5–10 min.
13. DFMA or DFMO treated samples should exhibit a drastic drop of ADC or ODC activities. Incubation for short periods of time will also avoid conversion of radiolabelled DFMA to DFMO or arginine to ornithine through arginase activity.

References

1. Pegg AE, Casero RA Jr (2011) Current status of the polyamine research field. Methods Mol Biol 720:3–35
2. Fuell C, Elliott KA, Hanfrey CC, Franceschetti M, Michael AJ (2010) Polyamine biosynthetic diversity in plants and algae. Plant Physiol Biochem 48:513–520
3. Tiburcio AF, Altabella T, Bitrián M, Alcázar R (2014) The roles of polyamines during the lifespan of plants: from development to stress. Planta 240:1–18
4. Hanfrey C, Sommer S, Mayer MJ, Burtin D, Michael AJ (2001) Arabidopsis polyamine biosynthesis: absence of ornithine decarboxylase and the mechanism of arginine decarboxylase activity. Plant J 27:551–560
5. Carbonell J, Blázquez MA (2009) Regulatory mechanisms of polyamine biosynthesis in plants. Genes Genomics 31:107–118
6. Alcázar R, Altabella T, Marco F, Bortolotti C, Reymond M, Koncz C, Carrasco P, Tiburcio AF (2010) Polyamines: molecules with regulatory functions in plant abiotic stress tolerance. Planta 231:1237–1249
7. Bitrián M, Zarza X, Altabella T, Tiburcio AF, Alcázar R (2012) Polyamines under abiotic

stress: metabolic crossroads and hormonal crosstalks in plants. Meta 2:516–528

8. Slocum RD (1991) Tissue and subcellular localization of polyamines and enzymes of polyamine metabolism. In: Slocum RD, Flores HE (eds) Biochemistry and Physiology of Polyamines in Plants. CRC Press, Inc., Boca Raton, FL, pp 93–103
9. Bortolotti C, Cordeiro A, Alcázar R, Borrell A, Culiañez-Macià FA, Tiburcio AF, Altabella T (2004) Localization of arginine decarboxylase in tobacco plants. Physiol Plant 120:84–92
10. Borrell A, Culiañez-Macia FA, Altabella T, Besford RT, Flores D, Tiburcio AF (1995) Arginine decarboxylase is localized in chloroplasts. Plant Physiol 109:771–776
11. Lee J, Michael AJ, Martynowski D, Goldsmith EJ, Phillips MA (2007) Phylogenetic diversity and the structural basis of substrate specificity in the beta/alpha-barrel fold basic amino acid decarboxylases. J Biol Chem 282:27115–27125
12. Jordan F, Patel H (2013) Catalysis in enzymatic Decarboxylations: comparison of selected cofactor-dependent and cofactor-independent examples. ACS Catal 3:1601–1617
13. Metcalf BW, Bey P, Danzin C, Jung MJ, Casara P, Ververt JP (1978) Catalytic irreversible inhibition of mammalian ornithine decarboxylase (EC 4.1.1.17) by substrate and product analogs. J Am Chem Soc 100:2551–2553
14. Bitonti AJ, Casara PJ, McCann PP, Bey P (1987) Catalytic irreversible inhibition of bacterial and plant arginine decarboxylase activities by novel substrate and product analogues. Biochem J 242:69–74
15. McCann PP, Pegg AE, Sjodersma A (eds) (1987) Inhibition of polyamine metabolism. Academic Press, Inc., San Diego, CA

Chapter 12

Determination of S-Adenosylmethionine Decarboxylase Activity in Plants

Antonio F. Tiburcio and Rubén Alcázar

Abstract

The synthesis of spermidine, spermine and thermospermine requires the addition of aminopropyl groups from decarboxylated *S*-adenosyl-methionine (dSAM). The synthesis of dSAM is catalyzed by *S*-adenosylmethionine decarboxylase. dSAM levels are usually low, which constitutes a rate-limiting factor in the synthesis of polyamines. In this chapter, we provide a protocol for the determination of SAMDC activity in plants through the detection of radiolabelled CO_2 released during the SAMDC reaction.

Key words Spd and Spm biosynthesis, Methionine, SAM decarboxylation, SAMDC activity

1 Introduction

Once Put is produced (via ODC or ADC), the synthesis of Spd and Spm requires the addition of aminopropyl groups catalysed by aminopropyltransferases. Decarboxylated *S*-adenosyl-methionine (dSAM) is the aminopropyl donor in the synthesis of Spd and Spm, which is formed by decarboxylation of *S*-adenosylmethionine (SAM) by the action of SAM decarboxylase (SAMDC; EC 4.2.1.50). SAM is produced by the condensation of the adenosine moiety of ATP and L-methionine in a reaction catalyzed by SAM synthase (EC 2.5.1.6). Methylthioadenosine (MTA) resulting from dSAM after Spd or Spm biosynthesis is recycled through the methionine cycle to regenerate SAM. This PA metabolic pathway is employed by most organisms [1–4], and was first characterized by Herbert and Celia Tabor using *Escherichia coli* [5]. In plants, this pathway has been confirmed by examining the incorporation of $[^{14}C]$-methionine into SAM, dSAM, Spd and Spm [6, 7]. On the other hand, the 1,4-diaminobutane skeleton of Spd and Spm has been shown to be derived from $[^{14}C]$-Put (or labeled Arg or Orn incorporated via Put) [7]. dSAM titers are usually low (approx.

Rubén Alcázar and Antonio F. Tiburcio (eds.), *Polyamines: Methods and Protocols*, Methods in Molecular Biology, vol. 1694, DOI 10.1007/978-1-4939-7398-9_12, © Springer Science+Business Media LLC 2018

1–2% of the SAM content in mammals) [8], thus constituting a rate-limiting factor in the synthesis of Spd and Spm [9].

The first SAMDC characterized was from *Escherichia coli* and it requires Mg^{2+} [5]. The second characterized enzyme was from mammals, which is unaffected by Mg^{2+}, but requires Put for maximal activity [10]. Subsequent works found that Put-activated SAMDCs are widespread in fungi, protozoa and nematodes [3]. However, in plants SAMDC is not stimulated by Put or by Mg^{2+} [11–18]. It has been suggested that the lack of Put-activated SAMDCs in plants may explain the predominance of the diamine versus Spd and Spm, as compared to mammals [3].

Most of the SAMDC characterized, including those from dicot plants [13, 15], belong to a small group of decarboxylating enzymes that use a covalently bound pyruvate as a prosthetic group [3] rather than the cofactor PLP (pyridoxal 5′-phosphate) typically employed in amino acid decarboxylation reactions. The reactive carbonyl group of the pyruvate cofactor reacts with the substrate to form a Schiff base. This provides an electron sink via the amide carbonyl group of the pyruvate, and facilitates decarboxylation [3].

Isolation of genomic and cDNA sequences of SAMDC from several plant systems indicated that this enzyme is generally encoded by a small multigene family in monocot and dicots [17]. Analysis of rice, maize and *Arabidopsis thaliana* SAMDC cDNA revealed that the monocot enzyme possesses an extended C-terminus relative to dicot and human enzymes [17]. This suggests that SAMDC from monocots may have different characteristics in relation to SAMDC from dicots or mammals.

In this chapter, we describe the procedure for protein extraction and determination of SAMDC activity in oat leaves by evaluating the release of $[^{14}C]$-CO_2 using $[^{14}C]$-labelled S-adenosylmethionine (Fig. 1). However, this protocol can be used with any other plant species as starting material.

2 Materials

1. Plant material: The first leaf of 8-day-old *Avena sativa* cv. Victory (oat) seedlings was used as experimental material.
2. Potassium phosphate buffer pH 7.6: Mix 61.5 mL 1 M K_2HPO_4 with 38.5 mL KH_2PO_4. Dilute to 1 L with water. Autoclave and store at room temperature.
3. Sephadex G-25 columns (6.0 × 1.5 cm).
4. Protein extraction buffer: 1 mM pyridoxal phosphate, 20 mM 2-mercaptoethanol, 1 mM PMSF, protease inhibitor cocktail in 100 mM potassium phosphate buffer pH 7.0 (*see* **Note 1**).

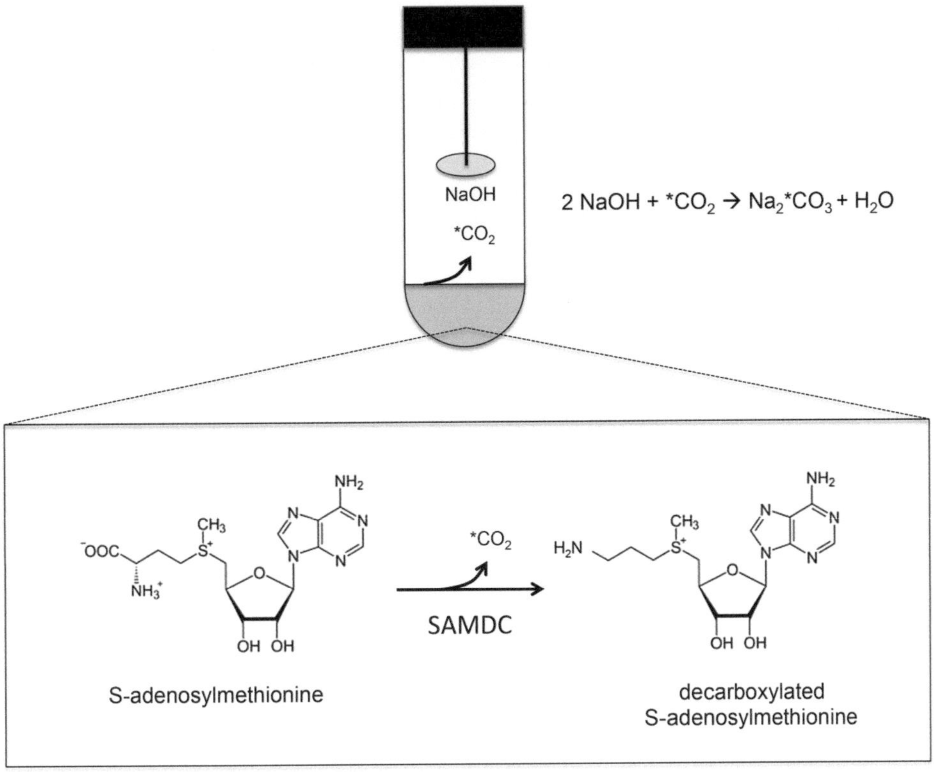

Fig. 1 Schematic representation of SAMDC activity determination

5. 2 M NaOH.
6. Perchloric acid 10% (v/v).
7. Liquid scintillation cocktail (suitable for aqueous samples).
8. Radiolabelled substrate: [1^{14}C]-SAM.
9. 10 mM $MgCl_2$.
10. 10 mM Put (store at −20 °C).
11. 10 mM Spd (store at −20 °C).
12. 10 mM Spm stock (store at −20 °C).
13. 1 mM Methylglyosal-bis-guanylhydrazone (MGMG) inhibitor (*see* **Note 2**).
14. 8.5× 1 cm (height × diameter) tubes with lid (*see* **Note 3**).
15. 0.5 cm × 0.2 mm (diameter × thickness) filter paper discs (*see* **Note 4**).
16. Scintillation counter.
17. Bradford's reagent.
18. Syringes, needles.

3 Methods

Users must get instructed about the appropriate use and disposal of radiochemicals before starting the protocol.

3.1 Protein Extraction

1. Grind the tissue in chilled mortars at a ratio of 500 mg fresh weight/mL of 100 mM potassium phosphate buffer pH 7.6 [19]
2. Centrifuge the extracts at 27,000 × *g* for 20 min at 4 °C
3. Apply 1 mL of the supernatant fraction to equilibrated Sephadex G-25 columns, and elute with phosphate buffer to yield filtrates of 1 mL (*see* **Note 5).**
4. Use these G-25 filtrates to assay SAMDC activity (*see* **Note 6**).

3.2 Enzymatic Determination

1. Soak filter discs with 2 M NaOH and let them dry (*see* **Note 7**).
2. Hold NaOH-impregnated paper discs with the help of a needle (Fig. 1). Close the hub of the needle with a rubber stopper.
3. In the fume hood, add the reaction mixture to the bottom of the tube. Mix 12 μL G-25 filtrate, 120 μL phosphate buffer (pH 7.6), 6 μL 100 μM pyridoxal phosphate (PLP) and 10 μL 20 μCi/mL of [1^{14}C]-SAM (55.3 mCi/mmol). Mix by pipetting up and down (*see* **Note 8**).
4. Close the tubes and incubate at 37 °C for 45 min (*see* **Note 9**).
5. With the help of a 1 mL-syringe, stop the reaction by adding 200 μL 20% PCA through the cannula of the needle (*see* **Note 10**). Close the tube.
6. Incubate 15 min more at room temperature to enable complete $^{14}CO_2$ capture.
7. Carefully, transfer NaOH-soaked paper discs to scintillation vials containing 5 mL liquid scintillation cocktail. Mix by vortexing (*see* **Note 11**).
8. Determine the radioactive emission in a scintillation counter (Fig. 1).

3.3 Blanks, Effectors, Inhibitors and Sample Normalization

1. Prepare blanks by adding 200 μL of extraction buffer and 10 μL 20 μCi/mL of [1^{14}C]-SAM (*see* **Note 12**).
2. Effectors: To study effects of various effectors on SAMDC enzymatic activity, G-25 filtrates were incubated with 0.5 mM $MgCl_2$ or 1 mM of each Put or Spd or Spm, previous to the enzymatic activity determination (*see* **Note 13**).
3. Inhibitors: Controls for the specificity of the SAMDC enzymatic determination were included by incubation of G-25 with 0.5 mM MGBG previous to the enzymatic activity determination (*see* **Note 14**).

4. Determine protein content using the Bradford's method and serum albumin (BSA) as the reference standard.
5. The enzyme activity is expressed in pmol CO_2/h μg of protein.

4 Notes

1. Prepare the extraction buffer before use. Read carefully safety instructions for the proper handling of 2-mercaptoethanol, PMSF and protease inhibitors.
2. MGBG is an inhibitor of SAMDC activity. Alternative inhibitors of SAMDC can also be used [1, 20]. Prepare stocks in water and keep at −20 °C.
3. Other plastic tubes with similar sizes might be suitable for this protocol. It is important that the lid can be perforated by the cannula of a needle. In addition, the tube lid must close tightly and the hub of the needle must be closed with a rubber stopper.
4. These discs are generated using a regular paper hole punch.
5. Oat leaf extracts were subjected to G-25 filtration to avoid some interferences described in crude plant extracts [12].
6. Keep samples on ice throughout the protocol.
7. Soaked filter discs should be prepared just before use to avoid capture of CO_2 from the atmosphere.
8. Unlike dicots and mammals (*see* Subheading 1), oat SAMDC activity is activated by 100 μM PLP [19]. This might be due to gene sequence differences between monocot SAMDC as compared to dicots or human enzymes [17].
9. Some protocols suggest shaking of the samples during the incubation. However, this may lead to accidental spilling of the radioactive mixture to the paper discs.
10. Acidification of the medium releases the remaining $^{14}CO_2$ trapped in the extraction solution.
11. Overnight incubation of discs in some scintillation cocktails provides higher and more stable counts.
12. Alternatively, blanks can also be performed using plant protein extracts that have been denatured by boiling at 95 °C 5–10 min.
13. It was found that oat SAMDC activity is Mg^{+2} independent and not stimulated by Put [19], such as it was observed in other plant systems and mammals (*see* Subheading 1). Addition to the reaction mixture of Spd or Spm did not increase oat SAMDC activity [19], thus ruling out a possible artifact such as decarboxylation of the substrate by polyamine oxidase-mediated H_2O_2 [12]. It was also shown that non-enzymatic

decarboxylation of SAM was not produced by addition of PLP to the reaction mixture [19] (*see* **Note 8**).

14. MGBG-treated samples exhibited a drastic drop of SAMDC activity [19], thus demonstrating the specificity of the enzymatic determination.

References

1. Slocum RD (1991) Polyamine biosynthesis in plants. In: Slocum RD, Flores HE (eds) Biochemistry and physiology of polyamines in plants. CRC Press, Boca Raton, FL, pp 23–40
2. Tiburcio AF, Altabella T, Borrell A, Masgrau C (1997) Polyamine metabolism and its regulation. Physiol Plant 100:664–674
3. Wallace HM, Fraser AV, Hughes A (2003) A perspective of polyamine metabolism. Biochem J 376:1–14
4. Pegg AE (2009) S-adenosylmethionine decarboxylase. Biochemical Society Essays Biochem 46:25–45
5. Tabor H, Rosenthal SM, Tabor CW (1958) The biosynthesis of spermidine and spermine from putrescine and methionine. J Biol Chem 233:907–914
6. Even-Chen Z, Mattoo AK, Goren R (1982) Inhibition of ethylene biosynthesis by aminoethoxyvinylglycine and by polyamines shunts label from 3,4-[C]methionine into spermidine in aged orange peel discs. Plant Physiol 69:385–388
7. Greenberg ML, Cohen SS (1985) Dicyclohexylamine-induced shift of biosynthesis from spermidine to spermine in plant protoplasts. Plant Physiol 78:568–575
8. Hibasami H, Hoffman JL, Pegg AE (1980) Decarboxylated *S*-adenosylmethionine in mammalian cells. J Biol Chem 255:6675–6678
9. Tiburcio AF, Kaur-Sawhney R, Galston AW (1990) Polyamine metabolism. In: Intermediary N metabolism. The biochemistry of plants. Academic Press, San Diego, CA, pp 238–325
10. Pegg AE, Williams-Ashman HG (1968) Stimulation of the decarboxylation of S-adenosylmethionine by putrescine in mammalian tissues. Biochem Biophys Res Commun 30:76–82
11. Suzuki Y, Hirasawa E (1980) S-adenosylmethionine decarboxylase of corn seedlings. Plant Physiol 66:1091–1094
12. Yamanoha B, Cohen SS (1985) S-adenosylmethionine decarboxylase and spermidine synthase from Chinese cabbage. Plant Physiol 178:784–790
13. Xiong H, Stanley BA, Tekwani BL, Pegg AE (1997) Processing of mammalian and plant S-adenosylmethionine decarboxylase proenzymes. J Biol Chem 272:28342–28348
14. Tassoni A, van Buuren M, Franceschetti M, Fornale S, Bagni N (2000) Polyamine content and metabolism in Arabidopsis thaliana and effect of spermidine on plant development. Plant Physiol Biochem 38:383–393
15. Bennett EM, Ekstrom JL, Pegg AE, Ealick SE (2002) Monomeric S-adenosylmethionine decarboxylase from plants provides an alternative to putrescine stimulation. Biochemistry 41:14509–14517
16. Bertoldi D, Tassoni A, Martinelli L, Bagni N (2004) Polyamines and somatic embryogenesis in two Vitis vinifera cultivars. Physiol Plant 120:657–666
17. Franceschetti M, Hanfrey C, Scaramagli S, Torrigiani P, Bagni N, Burtin D, Michael AJ (2001) Characterization of monocot and dicot plant S-adenosyl-l-methionine decarboxylase gene families including identification in the mRNA of a highly conserved pair of upstream overlapping open reading frames. Biochem J 353:403–409
18. Tassoni A, Franceschetti M, Tasco G, Casadio R, Bagni N (2007) Cloning, functional identification and structural modelling of Vitis vinifera S-adenosyl methionine decarboxylase. J Plant Physiol 164:1208–1219
19. Tiburcio AF, Kaur-Sawhney R, Galston AW (1993) Spermidine biosynthesis as affected by osmotic stress in oat leaves. Plant Growth Regul 13:103–109
20. Slocum RD, Galston AW (1987) Inhibitors of polyamine biosynthesis in plants and plant pathogenic fungi. In: McCann PP, Pegg AE, Sjodersma A (eds) Inhibition of polyamine metabolism. Academic Press Inc., San Diego, CA, pp 305–315

Chapter 13

Determination of Copper Amine Oxidase Activity in Plant Tissues

Riccardo Angelini, Alessandra Cona, and Paraskevi Tavladoraki

Abstract

Copper amine oxidases (CuAOs) involved in polyamine catabolism are emerging as physiologically relevant enzymes for their involvement in plant growth, differentiation and defence responses to biotic and abiotic stress. In this chapter, we describe two spectrophotometric and one polarographic method for determining CuAO activity in plant tissues. Some aspects related to cell wall association of apoplastic CuAOs and possible interference of plant metabolites with the enzymatic activity assays are also considered.

Key words Copper-containing amine oxidase, Flavin-containing polyamine oxidase, Peroxidase-coupled assay, Polyamines

1 Introduction

Plant polyamines are oxidatively deaminated by copper amine oxidases (CuAOs) and flavin adenine dinucleotide (FAD)-dependent polyamine oxidases (PAOs) [1–3]. Plant CuAOs are generally homodimeric enzymes containing a post-translationally biosynthesized trihydroxyphenyl alanine quinone (TPQ) cofactor and a Cu^{II} ion in each monomer which is essential for the catalytic mechanism [4, 5], while PAOs are monomeric enzymes linking a FAD molecule non covalently. These enzymes differ to each other in substrate specificity and catalytic mechanism. The apoplastic CuAOs purified from some *Fabaceae* species oxidize mainly putrescine (Put) and cadaverine (Cad), and less efficiently spermidine (Spd) and spermine (Spm), at the primary amino groups, producing ammonia, H_2O_2 and an aminoaldehyde, while three of the intracellular Arabidopsis CuAOs (AtCuAOa3, AtCuAOγ1, AtCuAOζ) are similarly efficient with both Put and Spd [3, 6]. Furthermore, a *Malus domestica* CuAO1 (MdAO1) shows preference for 1,3-diaminopropane (Dap), having no activity with Spd [7], whereas, AtCuAOζ and a *Nicotiana tabacum* CuAO (NtDAO1) oxidize also *N*-methyl-Put, though less efficiently than the non-methylated

Rubén Alcázar and Antonio F. Tiburcio (eds.), *Polyamines: Methods and Protocols*, Methods in Molecular Biology, vol. 1694, DOI 10.1007/978-1-4939-7398-9_13,

diamine [8]. In addition, a group of CuAOs shows preference for the *N*-methyl-Put and are thus named *N*-methylputrescine oxidases, whereas some plant and animal CuAOs present high catalytic activity with monoamines, such as histamine, 2-phenylethylamine, tyramine and tryptamine [3, 9]. Differently from the CuAOs, PAOs oxidize the carbon at the *endo* or *exo* side of the N^4 secondary amino group of a series of polyamines. In particular, the apoplastic PAOs oxidize the carbon at the *endo*-side of the N^4 atom of Spd and Spm producing Dap, H_2O_2, and an aminoaldehyde, whereas the intracellular PAOs oxidize the carbon at the *exo*-side of the N^4 atom of Spd or Spm, to produce Put or Spd, respectively, together with H_2O_2, and 3-aminopropanal [1, 3, 9]. Some of the intracellular PAOs are also able to oxidize N^1-acetyl Spm, as well as thermospermine (Therm-Spm) and norspermine (Nor-Spm) [3, 9]. Furthermore, the intracellular PAO5 of *A rabidopsis thaliana* (AtPAO5) was shown to have distinct properties in that it has activity mainly as a dehydrogenase rather than as an oxidase [10].

Some apoplastic CuAOs and PAOs were shown to be associated to the cell wall. Furthermore, apoplastic CuAOs and PAOs have been shown to play a key role as a source of H_2O_2, influencing cell wall architecture and maturation as well as cell wall lignification and suberisation during organ growth and wound-healing [1, 11–13]. In addition, an apoplastic Arabidopsis CuAO (AtCuAOβ; formerly ATAO1) has been shown to play a key role in Jasmonate-dependent protoxylem differentiation [3, 11] and the peroxisomal AtCuAOζ, which is expressed in guard cells, is involved in the ABA-mediated stomata closure through H_2O_2 production [14]. Other relevant physiological functions of both intracellular and apoplastic CuAOs and PAOs are associated with polyamine homeostasis, alkaloid biosynthesis, defence responses to biotic and abiotic stresses and programmed cell death [1–3].

1.1 The Catalytic Cycle

CuAO catalytic cycle can be divided into a reductive, substrate-dependent, half-reaction in which TPQ is reduced and an oxidative, oxygen-dependent, half-reaction in which reduced TPQ is re-oxidized [5]. In the case of the reductive phase catalysed by CuAO from *Lens culinaris* seedlings (LSAO) reacting with Put, a substrate Shiff-base intermediate (quinone-ketimine) involving the TPQ cofactor is formed, followed by the formation of product Shiff-base (TPQ reduction) and finally the aminoquinol intermediate with the release of the aldehyde product 4-aminobutanal [15]. The latter spontaneously cyclizes to 1-pyrroline [16]. Re-oxidation of the reduced enzyme runs through two possible routes involving: (1) O_2 binding to Cu^I-TPQ semiquinolamine radical and its reduction to superoxide anion by an inner-sphere electron transfer mechanism, or (2) O_2 binding to a hydrophobic pocket adjacent to TPQ aminoquinol and its reduction to superoxide anion via an outer-sphere electron process. In both cases a Cu^{II} – hydroperoxide iminoquinone intermediate is produced with release of H_2O_2 and ammonium upon hydrolysis [17]. Cobalt-substituted CuAO from pea seedlings shows catalytic features indicating that the former mechanism is preferred [18].

Plant CuAOs can be inactivated by long incubation time with substrates [19]. The inactivation is due to a slow reaction of the reduced protein with the turnover product H_2O_2, or the aldehyde in the case of some substrates such as Put [19].

In this chapter we describe some methods routinely used in our lab for determining CuAO activity in plant tissues. These methods will be also discussed considering association of some CuAOs to the cell wall as well as possible interference in the assays by plant metabolites or inactivation by reaction products. In particular, two spectrophotometric and one polarographic methods will be reported to cope with eventual interferences of metabolites present in plant tissues used with some of the activity assays or with necessity to determine CuAO activity in fibrous cell wall material. The spectrophotometric methods for CuAO activity determination exploit: (1) *o*-aminobenzaldehyde (OAB) which forms a yellow adduct with 1-pyrroline or 1-piperideine released from CuAO-mediated oxidation of Put or Cad, respectively, [16, 20]; (2) 3,5-dichloro-2-hydroxybenzenesulfonate (DCHBS) which, following oxidation in a peroxidase-catalysed reaction by the CuAO reaction-product H_2O_2, condenses with 4-aminoantipyrine (AAP) to form a pink adduct (Fig. 1) [16, 21–23]. The polarographic assay is based on the measurement of the level of reactant oxygen in an oxygraph equipped with a Clark-type electrode [24].

2 Materials

Prepare all buffers and solutions with deionized water using analytical grade reagents and store them at 4 °C until used. Azide is not added to buffers or reagents as it strongly inhibits CuAO or peroxidase activity, the last used in the peroxidase-coupled assay hereby described.

2.1 Extraction of Intracellular and Apoplastic CuAOs from Plant Tissues

1. Commercially available seeds are soaked in aerated tap water for 12 h, allowed to germinate at 23 °C in loam for 24 h and then grown at 25 °C in the dark or under specific dark-light cycles in a climatic chamber or greenhouse. Whenever possible, epidermis can be peeled off from stems or other organs and separately processed (*see* **Note 1**).
2. Whole organs, internodes or mechanically separated tissues are processed for obtaining total extractable and cell wall associated CuAO as reported by Angelini et al. (1990) [25]. The procedure is best suited for chickpea epicotyls and can be used

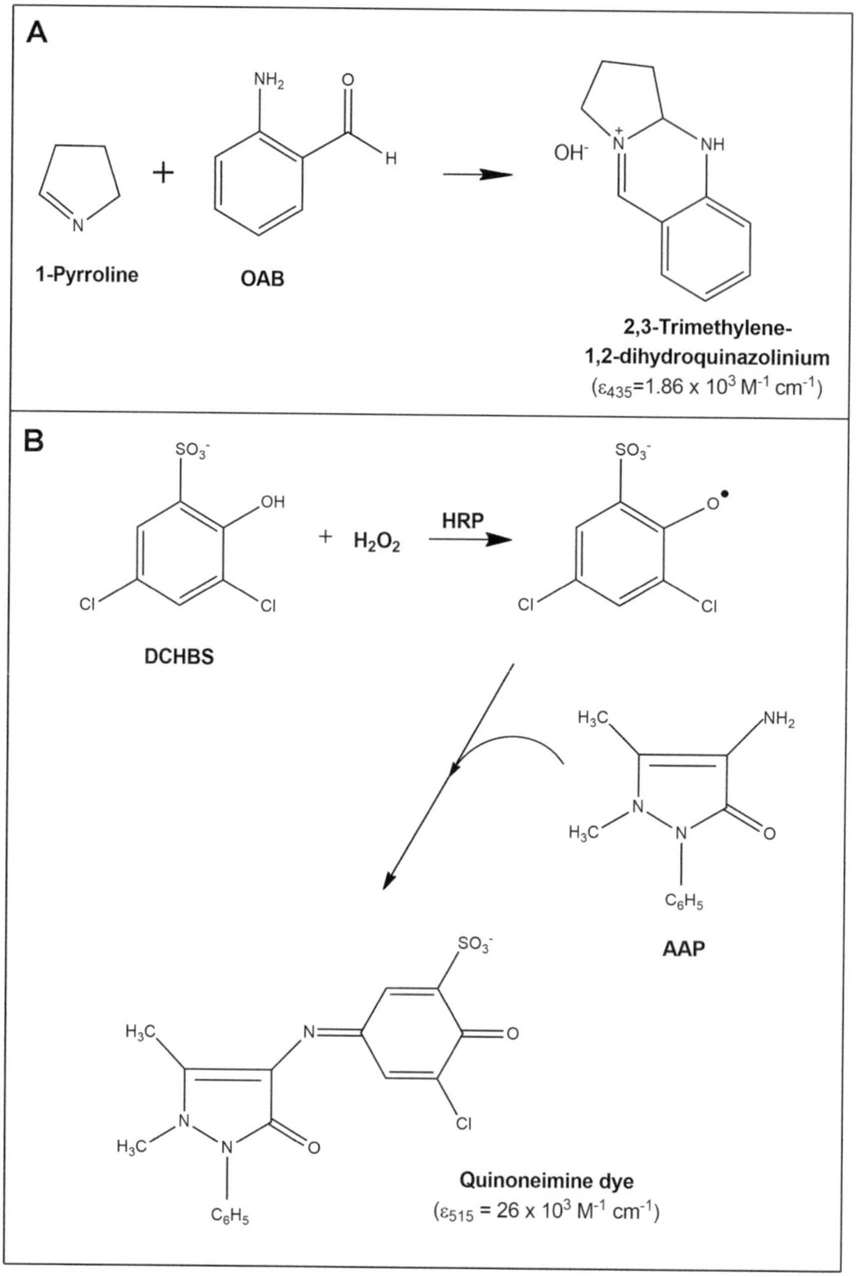

Fig. 1 Chromogenic reactions useful for CuAO activity determination. (**a**) formation of 2,3-trimethylene-1,2-dihydroquinazolinium as a result of condensation of o-aminobenzaldehyde with 1-pyrroline. The latter compound derives from spontaneous cyclisation of 4-aminobutanal produced by CuAO-catalysed oxidative deamination of Put [20]. (**b**) Formation of quinoneimine dye by condensation of AAP with DCHBS• triggered by H_2O_2-dependent peroxidase-catalysed DCHBS oxidation [16, 21, 23]

for most legumes. In detail, epidermal peels or whole organs/internodes are ground in a pre-chilled mortar (leave it on ice for the whole procedure), using 5 mL per gram of fresh tissue of 0.1 M potassium phosphate buffer, pH 6.5, containing 0.5 M

NaCl (*see* **Note 2**). The homogenate is filtered using Miracloth (EMD Millipore) or nylon cloth and then centrifuged at 12,000 × *g* for 10 min at 4 °C. The supernatant (total extractable fraction) is thus directly used for CuAO activity determination.

3. For obtaining the cell wall associated CuAO fractions (lightly and tightly bound fractions), whole organs/internodes (*see* **Note 3**) are ground as described above using 0.05 M potassium phosphate buffer, pH 6.5 (low ionic strength; buffer A). The homogenate is filtered and centrifuged as described above. The supernatant (extractable fraction lacking cell-wall bound CuAO; fraction F1) is directly used for CuAO activity determination. The fibrous residue obtained from the filtration is utilised for preparation of the cell wall fractions. This is re-suspended in an appropriate amount of buffer A (at least three volumes per starting fresh weight) in a mortar at 4 °C. The homogenate is filtered as described above and the filtrate is discarded. The procedure is repeated twice.
4. To obtain the CuAO fraction loosely bound to cell walls, the residual fibrous material is incubated in buffer A plus 1 M NaCl (buffer B) for 45 min at room temperature under constant agitation and then centrifuged at 12,000 × *g* for 10 min at 4 °C. The obtained supernatant containing the lightly bound proteins to the cell walls is dialysed against buffer A for 12 h at 4 °C (fraction F2). The remaining insoluble material is then twice washed with buffer B as described above, incubated in 0.050 M sodium acetate buffer, pH 5.0, for 10 min and then centrifuged as described above for 10 min. The supernatant is discarded and the pellet is re-suspended in a Driselase (Sigma) solution (2.5% w/v, in 0.050 M sodium acetate buffer, pH 5; 1 volume per g wet residue) for 18 h at room temperature with agitation. The suspension is then centrifuged as described above for 10 min. The supernatant represents the tightly bound cell wall fraction (fraction F3). The pellet is twice washed with buffer A as described above and then blotted dry, it represents the cell-wall residual fraction (fraction F4).
5. To isolate apoplastic CuAO isoenzymes, extracellular fluids can also be prepared [26, 27]. Briefly, 2 cm long segments (from epicotyl, hypocotyl, root) obtained by a perpendicular cut with a razor blade are incubated in ice-cold distilled water under mild agitation. Segments are then vertically packed in the barrel of a 20–50 mL syringe and washed thoroughly with cold distilled water using a circulating pump. The barrel with the packed segments is vigorously shaken down to get out excess water adhering to segments, introduced in a beaker and covered with appropriate buffer, usually 0.005–0.1 M potassium phosphate buffer, pH 7 [27]. Segments are then infiltrated

under vacuum for 5–10 min in a vacuum chamber connected to a vacuum pump (restore ambient pressure very slowly, after turning the pump off). The barrel containing the segments is centrifuged at 1000 × *g* for 5 min using a swinging-bucket rotor. Extracellular fluids are collected and stored on ice until used for CuAO activity determination. The extraction procedure may be repeated [27].

2.2 Materials for the Spectrophotometric OAB Assay

1. 10% (w/v) trichloroacetic acid (TCA) solution.
2. OAB solution (1 mg/mL). It is prepared in ethanol. Store at 4 °C.
3. 0.1 M potassium phosphate buffer, pH 7. It is prepared according to commonly used protocols by combining the appropriate amount of 1 M KH_2PO_4 and 1 M K_2HPO_4 solution (*see* for example http://cshprotocols.cshlp.org/content/2006/1/pdb.tab19). In this particular case, add 61.5 mL of 1 M K_2HPO_4 to 38.5 mL of 1 M KH_2PO_4, dilute to 1 L with distilled water.
4. 0.2 M Put (or other amine substrates) solution. It is prepared by dissolving the appropriate amount of Put hydrochloride in water.

2.3 Materials for the Spectrophotometric DHCBS/AAP Assay

1. 0.1 M potassium phosphate buffer, pH 7. It is prepared as described above.
2. Horseradish peroxidase (1 mg/mL) in water.
3. 10 mM DCHBS stock solution. It is is prepared in water and stored at 4 °C.
4. 1 mM AAP stock solution. It is prepared in water and stored at 4 °C.
5. 0.2 M Put solution. It is prepared as described above.

2.4 Materials for the Polarographic Assay

1. 0.1 M potassium phosphate buffer, pH 7. It is prepared as described above.
2. Sodium dithionite crystals.
3. Saturated potassium chloride solution.
4. Rolling paper.
5. Polytetrafluoroethylene (PTFE) membrane.
6. 0.2 M Put stock solution. It is prepared as described above.

3 Methods

This assay (*see* **Note 4**) is performed as described by Holmsted et al. (1961) [20] (Fig. 1a).

3.1 Spectrophotometric CuAO Activity Assay with OAB as a Chromogenic Reactant

1. An appropriate amount of tissue extract, or purified enzyme, is added to 4 mL of 0.1 M potassium phosphate buffer, pH 7.
2. Start the enzymatic reaction by further addition of 0.040 mL of 0.2 M Put or Cad. Catalase (0.050 mg/mL) may be added in order to reduce eventual inhibitory action of reaction product H_2O_2 on CuAO activity (*see* Subheading 1.1).
3. After incubation at 37 °C for 10–60 min, the reaction is stopped by adding 0.5 mL 10% TCA solution.
4. OAB stock solution (0.050 mL) is then added and the absorbance at 435 nm is measured in a spectrophotometer to quantify the produced chromogenic adduct ($\varepsilon_{435} = 1.86 \times 10^3\ M^{-1}\ cm^{-1}$) (*see* **Note 5**).
5. Enzyme activity can be expressed in enzyme Units (U). One U is the amount of enzyme that catalyses the reaction of 1 μmol of substrate per minute under standard conditions. Katal is also used in the literature to express enzyme activity. One katal is the amount of enzyme that converts 1 mole of substrate per second under standard conditions.

   This method, although less sensitive than the other spectrophotometric method reported in this chapter, is free from interference by plant metabolites or enzymes (catalase or peroxidase) and can be used as end-point assay after long time incubation also with fibrous cell wall material as enzyme source (in this case, centrifuge at 12,000 × *g* for 3 min, before reading at 430 nm).

3.2 Spectrophotometric Peroxidase-Coupled Assay for CuAO Activity Determination with DCHBS/AAP

This assay (*see* **Note 4**) is performed as described by Barham and Trinder (1972) [21] and modified by Smith and Barker (1988) [16] (Fig. 1b). CuAO activity is spectrophotometrically determined, following the formation of a pink adduct ($\varepsilon_{515} = 26 \times 10^3\ M^{-1}\ cm^{-1}$) as a result of the H_2O_2-dependent, peroxidase-mediated oxidation of DCHBS resulting in the formation of DCHBS radicals which condense with AAP (Fig. 1b; *see* **Note 6**). This method is more sensitive than the method exploiting OAB inasmuch as the molar extinction coefficient of the DCHBS/AAP adduct is much higher than that of the 1-pyrroline/OAB adduct. However, this assay, being coupled to peroxidase activity, is prone to interference from ascorbate [28] the content of which may greatly vary under different physiological conditions [29].

1. The assay is carried out in a reaction mixture of 1 mL containing 0.1 M phosphate buffer, pH 7.0, 0.05 mL of horseradish peroxidase stock solution, AAP (0.1 mL of 1 mM stock solution), DCHBS (0.1 mL of 10 mM stock solution) and the appropriate amount of tissue extract or purified enzyme.
2. The reaction starts by the addition of 0.01 mL of 0.2 M Put stock solution.

3.3 Determination of CuAO Activity by the Polarographic Assay

This method is based on the quantification of oxygen consumed in the re-oxidation step of CuAO-catalysed reaction in a thermostated chamber equipped with Clark type electrodes. In the device utilised (Hansatech Instruments Ltd, Norfolk, UK), Clark type electrodes are arranged in a disc, with a central platinum cathode connected through rolling paper/KCl bridge with a concentric silver anode. The platinum cathode is covered by an oxygen-permeable PTFE membrane in contact with the reaction chamber. The disc is connected to a control unit. When a potentiating voltage is applied across the two electrodes by the control unit, the platinum electrode acts as the cathode, and the silver electrode acts as the anode. Oxygen diffuses through the PTFE membrane and, provided that the appropriate difference potential is applied, it is reduced at the cathode surface so that a current flows through the circuit. Silver is oxidised at the anode. The current which is generated is proportional to the amount of oxygen reduced (Fig. 2, courtesy of Hansatech Instruments Ltd, www.hansatech-instruments.com).

1. Once calibrated the signal to oxygen content of air-saturated water at the temperature used (eventually corrected for specific buffer used) [30] and zero oxygen content (by flushing with

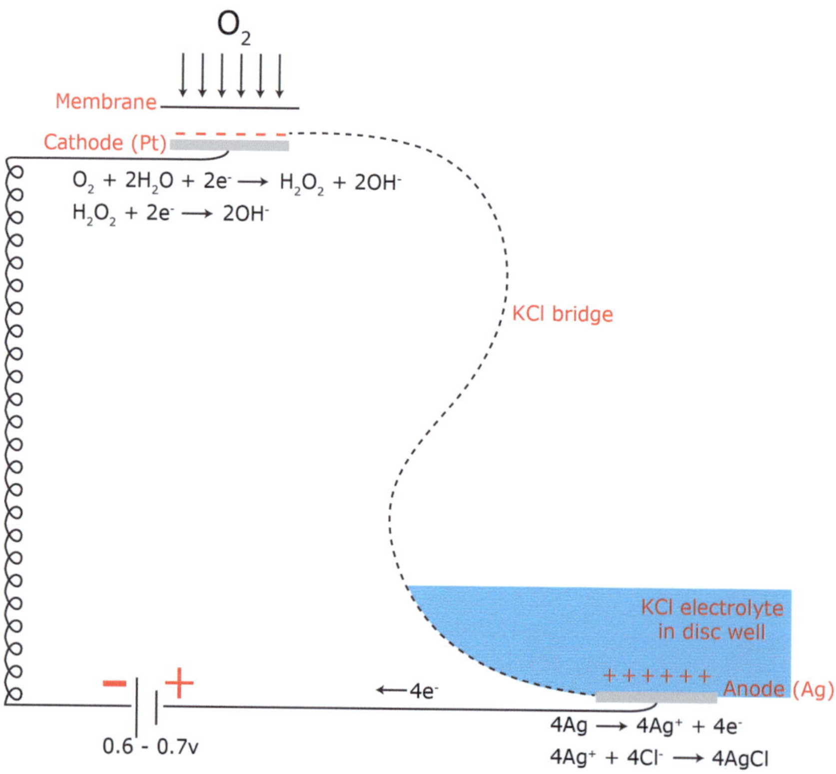

Fig. 2 Polarography principles for the determination of oxygen content in a solution. Courtesy of Hansatech Instruments Ltd. (www.hansatech-instruments.com)

N_2 or adding few crystals of sodium dithionite in the electrode chamber), oxygen consumed in the reaction is determined and so enzyme activity (*see* **Note 7**).

2. Before determining CuAO activity, carefully wash the electrode chamber with several rinses to eliminate sodium dithionite.
3. Add in the reaction chamber 1 mL 0.1 M phosphate buffer, pH 7, the appropriate amount of tissue extract or cell wall fraction and equilibrate at the temperature of the water circulating bath (usually few minutes).
4. Add 0.02 mL of 0.2 M Put solution to start the reaction.
5. If a recorder is utilised to follow the signal (current), oxygen consumed per minute is calculated considering total span and chart speed.

This method is very useful in case of necessity of CuAO activity determination in presence of fibrous cell wall material.

4 Notes

1. It has been demonstrated that CuAO accumulation and distribution in specific tissues greatly varies in different physiological conditions. In particular, CuAO activity in the epicotyl of some legume species increases in epidermal tissues upon de-etiolation, while it decreases in whole internodes [25].
2. Buffer concentration/composition and pH may need adjustments according to plant sources and CuAO characteristics.
3. The same procedure can be carried on also for epidermal peels but a considerable amount of tissue should be prepared.
4. This method is suited for CuAO activity determination only with Put or Cad as substrates inasmuch as cyclic products derived from their oxidation (1-pyrroline and 1-piperideine, respectively) are able to form adducts with o-aminobenzaldehyde [20].
5. OAB can be also added prior the incubation with the enzyme, as in the original method [20].
6. Under our experimental condition a 1:1 stoichiometric ratio exists between hydrogen peroxide reacting with peroxidase and quinoneimine dye produced.
7. The difference between the air line and the N_2 line represents the O_2 concentration of the solution contained in the electrode chamber. This can be obtained by published tables (*see* for example https://water.usgs.gov/owq/FieldManual/Chapter6/table6.2_6.pdf). A 1:1 stoichiometric ratio exists between Put oxidised and oxygen consumed.

References

1. Cona A, Rea G, Angelini R, Federico R, Tavladoraki P (2006) Functions of amine oxidases in plant development and defence. Trends Plant Sci 11:80–88
2. Tiburcio AF, Altabella T, Bitrián M, Alcázar R (2014) The roles of polyamines during the lifespan of plants: from development to stress. Planta 240:1–18
3. Tavladoraki P, Cona A, Angelini R (2016) Copper-containing amine oxidases and FAD-dependent polyamine oxidases are key players in plant tissue differentiation and organ development. Front Plant Sci 7:824
4. Padiglia A, Medda R, Pedersen JZ, Finazzi Agrò A, Lorrai A, Murgia B, Floris G (1999) Effect of metal substitution in copper amine oxidase from lentil seedlings. J Biol Inorg Chem 4:608–613
5. Medda R, Bellelli A, Peč P, Federico R, Cona A, Floris G (2009) Copper amine oxidases from plants. In: Floris G, Mondovì B (eds) Copper amine oxidases: structure, catalytic mechanism and role in pathophysiology. Taylor and Francis Group, C.R.C. Press, Boca Raton, pp 39–50
6. Planas-Portell J, Gallart M, Tiburcio AF, Altabella T (2013) Copper containing amine oxidases contribute to terminal polyamine oxidation in peroxisomes and apoplast of *Arabidopsis thaliana*. BMC Plant Biol 13:109
7. Zarei A, Trobacher CP, Cooke AR, Meyers AJ, Hall JC, Shelp BJ (2015) Apple fruit copper amine oxidase isoforms: peroxisomal MdAO1 prefers diamines as substrates, whereas extracellular MdAO2 exclusively utilizes monoamines. Plant Cell Physiol 56:137–147
8. Naconsie M, Kato K, Shoji T, Hashimoto T (2014) Molecular evolution of N-methylputrescine oxidase in tobacco. Plant Cell Physiol 55:436–444
9. Tavladoraki P, Cona A, Federico R, Tempera G, Viceconte N, Saccoccio S, Battaglia V, Toninello A, Agostinelli E (2012) Polyamine catabolism: target for antiproliferative therapies in animals and stress tolerance strategies in plants. Amino Acids 42:411–426
10. Ahou A, Martignago D, Alabdallah O, Tavazza R, Stano P, Macone A, Pivato M, Masi A, Rambla JL, Vera-Sirera F, Angelini R, Federico R, Tavladoraki P (2014) A plant spermine oxidase/dehydrogenase regulated by the proteasome and polyamines. J Exp Bot 65:1585–1603
11. Ghuge SA, Carucci A, Rodrigues-Pousada RA, Tisi A, Franchi S, Tavladoraki P, Angelini R, Cona A (2015) The apoplastic copper AMINE OXIDASE1 mediates jasmonic acid-induced protoxylem differentiation in Arabidopsis roots. Plant Physiol 168:690–707
12. Ghuge SA, Carucci A, Rodrigues-Pousada RA, Tisi A, Franchi S, Tavladoraki P, Angelini R, Cona A (2015) The MeJA-inducible copper amine oxidase AtAO1 is expressed in xylem tissue and guard cells. Plant Signal Behav 10: e1073872
13. Ghuge SA, Tisi A, Carucci A, Rodrigues-Pousada RA, Franchi S, Tavladoraki P, Angelini R, Cona A (2015) Cell wall amine oxidases: new players in root xylem differentiation under stress conditions. Plants (Basel) 4:489–504
14. Qu Y, An Z, Zhuang B, Jing W, Zhang Q, Zhang W (2014) Copper amine oxidase and phospholipase D act independently in abscisic acid (ABA)-induced stomatal closure in *Vicia faba* and Arabidopsis. J Plant Res 127:533–544
15. Bellelli A, Agro AF, Floris G, Brunori M (1991) On the mechanism and rate of substrate oxidation by amine oxidase from lentil seedlings. J Biol Chem 266:20654–20657
16. Smith TA, Barker JHA (1988) The di- and polyamine oxidase of higher plants. In: Zappia V, Pegg AE (eds) Progress in polyamine research: novel biochemical, pharmacological and clinical aspects. Plenum Press, New York, pp 573–587
17. Shepard EM, Dooley DM (2015) Inhibition and oxygen activation in copper amine oxidases. Acc Chem Res 48:1218–1226
18. Mills SA, Brown DE, Dang K, Sommer D, Bitsimis A, Nguyen J, Dooley DM (2012) Cobalt substitution supports an inner-sphere electron transfer mechanism for oxygen reduction in pea seedling amine oxidase. J Biol Inorg Chem 17:507–515
19. Pietrangeli P, Nocera S, Federico R, Mondovi B, Morpurgo L (2004) Inactivation of copper-containing amine oxidase by turnover products. Eur J Biochem 271:146–152
20. Holmsted B, Larrson L, Than R (1961) Further studies of a spectrophotometric method for the determination of diamine oxidase. Biochim Biophys Acta 48:182–186
21. Barham D, Trinder P (1972) An improved colour reagent for the determination of blood glucose by the oxidase system. Analyst 97:142–145.
22. Fossati P, Prencipe L, Berti G (1980) Use of 3, 5-dichloro-2-hydroxybenzenesulfonic acid/4-

aminophenazone chromogenic system in direct enzymic assay of uric acid in serum and urine. Clin Chem 26:227–231

23. Fraisse L, Bonnet MC, de Farcy JP, Agut C, Dersigny D, Bayolb A (2002) A colorimetric 96-well microtiter plate assay for the determination of urate oxidase activity and its kinetic parameters. Anal Biochem 309:173–179
24. Clark LC Jr (1956) Monitor and control of blood and tissue oxygen tension. Trans Am Soc Artificial Int Organs 2:41–48
25. Angelini R, Manes R, Federico R (1990) Spatial and functional correlation between diamine oxidase and peroxidase activities and their dependence upon de-etiolation and wounding in chick-pea stems. Planta 182:89–96
26. Terry ME, Bonner BA (1980) An examination of centrifugation as a method of extracting an extracellular solution from peas, and its use for the study of indoleacetic acid-induced growth. Plant Physiol 66:321–325
27. Federico R, Angelini R (1986) Occurrence of diamine oxidase in the apoplast of pea epicotyls. Planta 167:300–303
28. Martinello F, Luiz da Silva E (2006) Mechanism of ascorbic acid interference in biochemical tests that use peroxide and peroxidase to generate chromophore. Clin Chim Acta 373:108–116
29. de Pinto MC, De Gara L (2004) Changes in the ascorbate metabolism of apoplastic and symplastic spaces are associated with cell differentiation. J Exp Bot 55:2559–2569
30. Robinson J, Cooper JM (1969) Method of determining oxygen concentration in biological media, suitable for calibration of the oxygen electrode. Anal Biochem 33:390–399

Chapter 14

Determination of di−/Polyamine Oxidase Activity in Plants by an *In-Gel* Spermidine Oxidation Assay

Panagiotis N. Moschou

Abstract

Diamine and polyamine catabolism controls plant development, resistance to pathogens and stress responses. Diamine and polyamine oxidases control the catabolism of diamines and polyamines, respectively. Two major routes of di−/polyamine catabolism exist: the terminal and the interconverting. The in vitro activity of each route is assayed by the colorimetric or chemiluminescent determination of hydrogen peroxide produced by oxidation of di−/polyamine substrates. However, these assays fail to estimate activity of individual di−/polyamine oxidase isoenzymes. Herein, I describe an assay for the simultaneous *in-gel* determination of terminal and interconverting di−/polyamine oxidase isoenzyme activities.

Key words Diamine oxidases, Polyamine oxidases, Oxidation, Polyamine back-conversion, Polyamine interconversion, Hydrogen peroxide, Flavin adenine dinucleotide, Peroxisomes, Spermidine

1 Introduction

Oxidation represents the main catabolic pathway for diamines and polyamines (collectively referred as PAs). The most abundant PAs are the diamine putrescine (Put), the triamine spermidine (Spd), and the tetramines spermine (Spm) and thermospermine (t-Spm). Oxidation of PAs produces hydrogen peroxide (H_2O_2) that plays a major role in plant development and responses to abiotic and biotic stresses. The diamine oxidases (DAOs) oxidize Put and cadaverine (Cad), and with much lower affinity, Spd and Spm. The action of DAOs on Put yields pyrroline, H_2O_2 and ammonia (NH_4^+) [1, 2]. In contrast to DAOs, PA oxidases (PAOs) oxidize Spd and Spm but not Put. The apoplastic PAO catalyzes the terminal oxidation of PAs, yielding pyrroline and 1-(3-aminopropyl) pyrrollinium from Spd and Spm, respectively, along with 1,3-diaminopropane and H_2O_2. The plant intracellular (cytoplasmic or peroxisomal) PAOs produce H_2O_2, and interconvert Spm to Spd and Spd to Put [1, 3–8]. PA oxidation is mediated by multiple isoenzymes (e.g., *Arabidopsis* has at least 10 *DAO* genes [9] and five *PAO* genes,

Rubén Alcázar and Antonio F. Tiburcio (eds.), *Polyamines: Methods and Protocols*, Methods in Molecular Biology, vol. 1694, DOI 10.1007/978-1-4939-7398-9_14, © Springer Science+Business Media LLC 2018

AtPAO1-AtPAO5 [10]. DAOs and PAOs localize to the cytoplasm, apoplast or peroxisomes. The activity of each isoenzyme of DAOs and PAOs is hard to estimate with conventional in vitro assays.

Here, I describe an *in-gel* assay to estimate the activity of each of DAO and PAO isoenzymes. In this assay, isoenzymes are separated and detected in native PAGE (polyacrylamide gel electrophoresis; also, known as clear native PAGE).

2 Materials

2.1 Protein Extraction

1. Refrigerated Microcentrifuge.
2. Extraction Buffer: 0.2 M Tris–HCl (pH 8.0) prepared from stock solutions of 1 M Trizma base® base (2-Amino-2-(hydroxymethyl)-1,3-propanediol) adjusted with HCl 6 N and stored at RT, supplemented with 1 mM EDTA, 50 μM pyridoxal phosphate, 5 mM of the antioxidant dithiothreitol (DTT), 0.5 mM PMSF, 10 μM leupeptin, 10% (v/v) glycerol, 0.2% (v/v) Triton X-100, 20% (w/v) PVPP (polyvinylpolypyrrolidone). Glycerol is used for protein stabilization. Triton X-100 (polyethylene glycol *p*-(1,1,3,3-tetramethylbutyl)-phenyl ether) is a non-ionic surfactant. PVPP (polyvinyl polypyrrolidone, crospovidone, crospolividone or E1202) powder (*see* **Note 1**), is as a phenol/oxidation quencher. The protease inhibitors phenylmethylsulfonyl fluoride (PMSF) and leupeptin are dissolved in methanol at 0.5 M and in ddH_2O at 10 mM, respectively and stored at −20 °C (*see* **Note 2**); Ethylenediaminetetraacetic acid (EDTA) is dissolved in ddH_2O with pH 8.0 (*see* **Note 3**).
3. Desalting columns: Bio-gel P-6 (BioRad) (*see* **Note 4**).
4. Miracloth (Calbiochem).

2.2 Protein Quantification

1. Bovine serum albumin fraction V.
2. 20% (v/v) trichloroacetic acid.
3. Phenol Folin-Ciocalteau reagent [1 part Folin-Phenol (2×) dissolved in 1 part water].
4. Complex-forming reagent: prepare immediately before use by mixing the following stock solutions in the proportion 100:1:1 (by vol), respectively: Solution A: 2% (w/v) Na_2CO_3 in distilled water. Solution B: 1% (w/v) $CuSO_4 \cdot 5H_2O$ in distilled water. Solution C: 2% (w/v) sodium potassium tartrate in distilled water.
5. 2 N NaOH

2.3 Native Page

1. 1 mm thickness mini-Protean II gel system (BioRad) or similar.
2. 30% Acrylamide/Bis Solution, 37.5:1 stored as liquid at 4 °C. Reduce inhalation and contact hazards associated with weighing and preparing acrylamide and bis-acrylamide solutions. Precise composition and high-purity provide uniformity of gel matrices, consistent polymerization and run-to-run reproducibility.
3. 0.5 M Tris–HCl, pH 6.8 prepared from stock solutions of 1 M Trizma base® adjusted with HCl 6 N and stored at RT.
4. 1.5 M Tris–HCl, pH 8.8 prepared from stock solutions of 1 M Trizma base® adjusted with HCl 6 N and stored at RT.
5. 10% (w/v) sodium dodecyl sulphate (SDS) prepared from stock solutions of 20% SDS and stored at RT. Reduce inhalation and contact hazards associated with weighing and preparing.
6. Catalyst for polyacrylamide gel polymerization: Tetramethylethylenediamine (TEMED). Stored as liquid at 4 °C.
7. Gel polymerization agent: 10% (w/v) ammonium persulphate (APS). Powder dissolved in ddH_2O as 1 M stock solutions and stored at −20 °C.
8. Running buffer: 30.0 g of Trizma base®, 144.0 g of glycine. The pH of the buffer should be 8.3 and no pH adjustment is required.
9. Protein ladder.

2.4 Band Visualization

10. Substrate: the hydrochloride form of Spermidine (Sigma-Aldrich, St. Louis, Mo, USA) is dissolved in ddH_2O as 1 M stock solution and stored at −20 °C.
11. Reaction buffer: 50 mM Na_2HPO_4/NaH_2PO_4 prepared from stock solutions of 1 M Na_2HPO_4 and 1 M NaH_2PO_4 by mixing them to prepare the desired pH solution (e.g., pH 7.0). The final solution is dissolved 20 times to give a 200 mM solution.
12. 3,3′-diaminobenzidine (3,3′,4,4′-Biphenyltetramine, 3,3′,4,4′-Tetraaminobiphenyl or DAB) is dissolved in ddH_2O as 1 M stock solution and stored at −20 °C.
13. Guazatine acetate salt (product line PESTANAL®; Sigma-Aldrich) is dissolved in ddH_2O as 1 M stock solution and stored at −20 °C.

3 Methods

In native PAGE the protein samples are extracted and loaded directly on the gel without implementing charged dyes. The electrophoretic mobility of proteins relates to the intrinsic protein charge [11], in contrast to the charge shift technique known as blue-native PAGE. The migration distance of the proteins depends on their charge, size, and the pore size of the gel. In many cases, this method has lower resolution than blue-native PAGE, but native PAGE offers advantages whenever Coomassie dye would interfere with further assays. In addition, native PAGE is milder than blue-native PAGE and retains labile supramolecular assemblies of membrane protein complexes that are dissociated under the conditions of blue-native PAGE. This might be important for interactions required between the interacting interfaces of DAOs to retain them in their natural dimeric state.

The *in-gel* assay for DAOs/PAOs is based on oxidation of Spd that serves as a potent substrate for most DAOs and PAOs. The H_2O_2 produced by Spd-oxidation reacts with 3,3′-diaminobenzidine (DAB), forming a brownish adduct demarcating the gel regions (bands) enriched in Spd-oxidase activity. This method is also suitable for the detection of Put-oxidase or Spm-oxidase activity.

3.1 Protein Extraction

1. Tissue homogenization: grind tissue using a pre-chilled mortar and pestle and homogenize tissue in the Extraction Buffer [EB]: For each g of fresh weight, add 3–5 mL of EB.
2. Centrifuge homogenates for at least 20 min at 16,000 × *g* at 4 °C. Desalt supernatants by passing them through a Bio-gel P-6 (BioRad) column, following filtration through Miracloth. Avoid previous step if tissue purity is high.
3. Use the homogenates immediately for protein quantification and native PAGE.

3.2 Protein Quantification

Protein concentration determination: to ensure equal loading among samples in native PAGE, determine protein content by the Lowry method. This method provides decreased accuracy compared to SDS-PAGE resolution and Coomasie Brilliant Blue staining gel visualization of total proteins but is significantly faster. Lowry method is based on both the Biuret reaction, in which the peptide bonds of proteins react with copper under alkaline conditions to produce Cu^+, which reacts with the Folin reagent, and the Folin-Ciocalteau reaction, which is poorly understood but in essence phosphomolybdotungstate is reduced to heteropolymolybdenum blue by the copper-catalysed oxidation of aromatic amino acids. The reactions result in a strong blue colour. The method is sensitive in the range 0.01–1.0 mg/mL of protein.

1. In 20 μL of protein extract add 20 μL 20% (v/v) TCA.
2. Incubate samples for 30 min at 4 °C, and precipitate proteins after centrifugation at 10,000 × *g* for 20 min at RT.
3. Discard supernatant and re-dissolved protein pellet in 100 μL Solution A (*see* Subheading 2.2 and **Note 5**).

4. Prepare a mixture of Solution A + B [*see* Subheading 2.2; 10:0.2 (v/v), respectively]. Add 1 mL from Solution A + B to the samples.
5. Incubate samples for 15 min at RT.
6. Add 100 μL of Folin-Phenol (diluted with 1 volume ddH_2O), and incubate samples for 30 min at RT.
7. Read absorbance at 625 nm, after preparation of a standard curve using BSA (*see* **Notes 6** and 7).

3.3 Native Page

1. Gel casting: prepare a 7 or 10% (w/v) PAGE gel using 1 mm thickness mini-Protean II gel system or any other compatible system. Ensure that gels and buffers are SDS-free. Pre-chill gels at 4 °C (*see* **Note 8**).
2. Loading gels: use protein extracts the same day of their preparation. Load up to 37 μL (this amount needs to be tested) without adding a dye in the sample. In one slot load DNA loading dye [for 10 mL: 0.2 g (w/v) Bromophenol Blue in 6 mL 50% (v/v) and 4 mL ddH_2O water] to visualize the protein front and in an additional well load a native PAGE molecular weight marker (e.g., NativeMark™ Unstained Protein Standard). During loading keep gels chilled by placing them in an ice box.
3. Gel running: run gels at 20–40 mA using pre-chilled running buffer described in Subheading 2.3, until the protein front reaches the gel bottom. During running keep gels chilled by placing the gel tank in an ice box filled with ice.

3.4 Band Visualization

1. Pre-incubate gels for 30 min in reaction buffer (*see* Subheading 2.4) without Spd (50 mM Na_2HPO_4/NaH_2PO_4) of pH 7.0.
2. Incubate gels at RT in reaction buffer supplemented with 10 mM Spd (and/or other PAs) and 1 mM DAB (*see* **Note 9**).
3. Incubate gels until bands can be visualized.
4. To stop the reaction, rinse gels with water.
5. Results: *see* Fig. 1.

4 Notes

1. To prepare 0.5 M EDTA (pH 8.0): add 186.1 g of disodium EDTA.2H_2O to 800 mL of ddH_2O. Stir vigorously on a magnetic stirrer. Adjust the pH to 8.0 with NaOH (~20 g of NaOH pellets). Dispense into aliquots and sterilize by autoclaving. The disodium salt of EDTA will not go into solution until the pH of the solution is adjusted to ~8.0 by the addition of NaOH.
2. Toxicity of PMSF and leupeptin: avoid contact, including inhalation; wear protective clothing and use in a well-ventilated area.

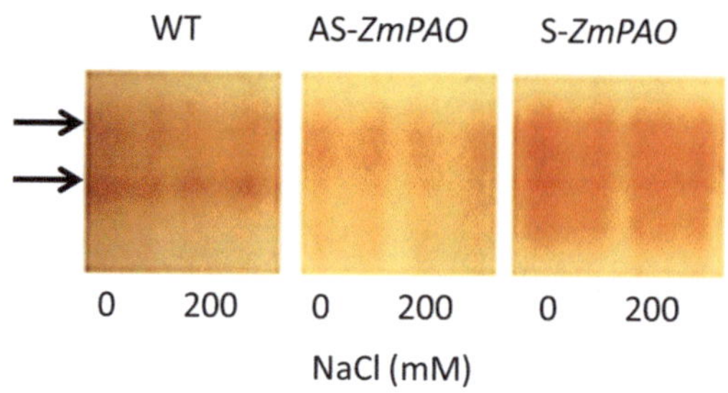

Fig. 1 Spd oxidase *in-gel* activity assay in the leaves 24 h post-salt treatment with 200 mM NaCl. *Arrowheads* indicate the major isoforms of Spd-oxidase activity in wild-type (WT) tobacco plants. Note the absence of the fast migrating Spd-oxidase isoform in antisense apoplastic PAO plants (AS-*ZmPAO*) and the increase of signal intensity in plants overexpressing apoplastic PAO (S-ZmPAO). Adapted from [12]

3. PVPP addition is necessary to decrease the content of soluble polyphenols that are abundant in tobacco, which could inhibit enzymatic assays. Moreover, polyphenols interact with proteins causing haze and precipitates. Polyphenols are common in many plant tissues and can deactivate proteins if not removed. We have tested PVPP in tobacco and grapevine.
4. Desalting columns: although they are recommended, their use it optional.
5. A 30 min incubation with periodic vortexing will increase the solubility of the pellet, thus the accuracy of the protein content determination.
6. Preparation of BSA standard curve: use a stock solution of 2 mg/mL protein in distilled water, stored at −20 °C. Prepare standards by diluting the stock solution with distilled water in a range from 0 to 2000 μg/mL.
7. Considerations when using the Lowry method: this method is 20 times more sensitive than the measurement of the UV absorption at 280 nm (which cannot be used in plant material due to the high amount of phenolics) and is several-fold more sensitive than other methods, like ninhydrin reaction. Moreover, it is simple and easy to adapt for small scale analyses. There are three major disadvantages: (1) The amount of colour varies with different proteins and is not strictly proportional to concentration. (2) The narrow pH range within which it is accurate. However, when using very small volumes of sample, changes in the pH are negligible. (3) Several compounds interfere with the Lowry method. These include some amino acid derivatives, certain buffers, drugs, lipids, sugars, salts, nucleic acids and sulphudryl reagents. Ammonium ions, zwitterionic buffers, non-ionic buffers and thiol compounds also interfere

with the Lowry method. These substances should be removed or diluted before running Lowry assays.

8. Pre-casted gels: alternatively, use Mini-PROTEAN® TGX™ gels.
9. Other additives: to distinguish between DAOs and PAOs, use the inhibitor guazatine acetate salt (product line PESTANAL®; Sigma-Aldrich), which specifically inhibits PAOs.

References

1. Moschou PN, Roubelakis-Angelakis KA (2014) Polyamines and programmed cell death. J Exp Bot 65:1285–1296. doi:10.1093/jxb/ert373
2. Moschou PN, Roubelakis-Angelakis KA (2011) Characterization, assay, and substrate specificity of plant polyamine oxidases. Methods Mol Biol 720:183–194. doi:10.1007/978-1-61779-034-8_11
3. Tavladoraki P, Rossi MN, Saccuti G, Perez-Amador MA, Polticelli F, Angelini R, Federico R (2006) Heterologous expression and biochemical characterization of a polyamine oxidase from Arabidopsis involved in polyamine back conversion. Plant Physiol 141:1519–1532. doi:10.1104/pp.106.080911
4. Kamada-Nobusada T, Hayashi M, Fukazawa M, Sakakibara H, Nishimura M (2008) A putative peroxisomal polyamine oxidase, AtPAO4, is involved in polyamine catabolism in Arabidopsis Thaliana. Plant Cell Physiol 49:1272–1282. doi:10.1093/pcp/pcn114
5. Andronis EA, Moschou PN, Toumi I, Roubelakis-Angelakis KA (2014) Peroxisomal polyamine oxidase and NADPH-oxidase cross-talk for ROS homeostasis which affects respiration rate in Arabidopsis thaliana. Front Plant Sci 5:132. doi:10.3389/fpls.2014.00132
6. Moschou P, Wu J, Cona A, Tavladoraki P, Angelini R, Roubelakis-Angelakis K (2012) The polyamines and their catabolic products are significant players in the turnover of nitrogenous molecules in plants. J Exp Bot 63:5003–5015
7. Moschou PN, Sanmartin M, Andriopoulou AH, Rojo E, Sanchez-Serrano JJ, Roubelakis-Angelakis KA (2008) Bridging the gap between plant and mammalian polyamine catabolism: a novel peroxisomal polyamine oxidase responsible for a full back-conversion pathway in Arabidopsis. Plant Physiol 147:1845–1857. doi:10.1104/pp.108.123802
8. Moschou PN, Paschalidis KA, Roubelakis-Angelakis KA (2008) Plant polyamine catabolism: the state of the art. Plant Signal Behav 3:1061–1066
9. Planas-Portell J, Gallart M, Tiburcio AF, Altabella T (2013) Copper-containing amine oxidases contribute to terminal polyamine oxidation in peroxisomes and apoplast of Arabidopsis thaliana. BMC Plant Biol 13:109. doi:10.1186/1471-2229-13-109
10. Ahou A, Martignago D, Alabdallah O, Tavazza R, Stano P, Macone A, Pivato M, Masi A, Rambla JL, Vera-Sirera F, Angelini R, Federico R, Tavladoraki P (2014) A plant spermine oxidase/dehydrogenase regulated by the proteasome and polyamines. J Exp Bot 65:1585–1603. doi:10.1093/jxb/eru016
11. Wittig I, Schagger H (2005) Advantages and limitations of clear-native PAGE. Proteomics 5:4338–4346. doi:10.1002/pmic.200500081
12. Gemes K, Jung Kim Y, Park KY, Moschou PN, Andronis E, Valassakis C, Roussis A, Roubelakis-Angelakis KA (2016) An NADPH-oxidase/polyamine oxidase feedback loop controls oxidative burst under salinity. Plant Physiol 172:1418–1431. doi:10.1104/pp.16.01118

Chapter 15

Pentamine as a Substrate for Measuring Spermine Oxidase Activity

Koichi Takao and Yoshiaki Sugita

Abstract

A method for determining spermine oxidase activity is described using the pentamine 1,16-diamino-4,8,13-triazahexadecane (3343) as the substrate, coupled with o-phthalaldehyde-post-label ion-exchange HPLC. The synthesis of 3343 is also described.

Key words Spermine oxidase, Pentamine, 1,16-diamino-4,8,13-triazahexadecane, 3343, Polyamine, Total synthesis, Assay, High-performance liquid chromatography (HPLC)

1 Introduction

The cytosolic enzyme spermine oxidase (SMO) catalyzes the direct oxidation of spermine (Spm) to spermidine (Spd), bypassing the necessity for preliminary acetylation of polyamines required by N^1-acetylpolyamine oxidase (APAO) [1, 2]. The substrate specificity of SMO appears to be limited and distinct from that of APAO: whereas APAO catalyzes a number of acetylpolyamines, to date few compounds have been reported as substrates for SMO. Two of these substrates are N^1-ethylSpm and α,ω-dimethylSpm [3–6]. In an effort to identify more substrates for SMO, a series of linear polyamines with a terminal primary amine were recently examined, including all ten pentamines composed of 3 or 4 methylene chain length, and nine pentamines were identified that showed various substrate activities releasing triamines [7]. Of these nine compounds, pentamine 3343 was shown to be superior to Spm as a substrate. Here Spd was measured using o-phthalaldehyde (OPA)-post-label ion-exchange HPLC in order to avoid crude sample-dependent error in the determination of H_2O_2 concentrations.

Rubén Alcázar and Antonio F. Tiburcio (eds.), *Polyamines: Methods and Protocols*, Methods in Molecular Biology, vol. 1694, DOI 10.1007/978-1-4939-7398-9_15,

2 Materials

Most of the reagents used are common laboratory chemicals and may be purchased from a preferred supplier. The preparation and use of the stock solutions should follow standard laboratory practice.

2.1 Synthesis of Pentamine

All organic solvents used for the synthesis are dehydrated grade.

1. Ammonia solution (28%).
2. Acetic acid (AcOH).
3. *N*-(3-Bromopropyl)phthalimide.
4. *N*-(4-Bromobutyl)phthalimide.
5. Spd.
6. NH_2NH_2 H_2O.
7. KF-Celite.
8. Pd/C.
9. Ninhydrin spray.
10. TLC plate (MERCK Silica gel 60 F254), silica gel (Wakogel C-300).
11. Teflon Millipore membrane.

2.2 Assay of SMO

1. 1.0 M Tris–HCl buffer containing 10 mM EDTA, pH 7.0. Use ultrapure-grade Tris base. Adjust to the appropriate pH at 37 °C using concentrated HCl. Store at −20 °C (*see* **Note 1**).
2. 0.36 mM pargyline solution. Store at −20 °C.
3. 5.6 mM aminoguanidine solution. Store at −20 °C.
4. 1 mM 3343 solution: 1,16-diamino-4,8,13-triazahexadecane (3343) pentahydrochloride dissolved in 0.01 M HCl. Store at −20 °C.
5. 20% trichloroacetic acid solution. Store at room temperature.

2.3 HPLC Analysis

1. HPLC system apparatus: Chromatopac C-R8A, LC-pump × 2, column oven, fluorescence detector, column packed with the cation exchange resin MCI GEL CK10S, 4.6 mmϕ × 60 mm.
2. Elution buffer: a mixture of 12.5% (v/v) methanol and 87.5% (v/v) 0.28 M sodium citrate buffer, pH 5.5, containing 2.0 M sodium chloride.
3. Post-column reagent solution: 6 mM OPA in 0.4 M potassium borate buffer, pH 10.4, containing 0.2% 2-mercaptoethanol and 0.1% Brij 35.

2.4 Preparation of Enzyme Solution (See Note 2)

1. 0.25 M sucrose-10 mM Tris–HCl buffer, pH 7.4. Use ultrapure-grade Tris base and sucrose. Adjust to the appropriate pH at 37 °C using concentrated HCl. Store at −80 °C.
2. Potter-Elvehjem homogenizer.
3. Ultracentrifuge.

3 Methods

3.1 Synthesis of 3343 [8, 9]

1. A solution of benzylamine (1 eq) and *N*-(3-bromopropyl) phthalimide (1 eq) in acetonitrile (4 mL/mmol) is refluxed for 2 h in the presence of KF-Celite (0.4 g/mmol). The reaction is monitored by TLC (*see* **Note 3**).
2. After filtration, the solvent is evaporated, then the residue is dissolved in benzene and subjected to silica gel column chromatography using a benzene:acetone (stepwisely from 50:1 to 10:1) solvent. The fractions containing *N*-(phthalimidopropyl) benzylamine are collected and the solvent evaporated.
3. *N*-(Phthalimidopropyl)benzylamine (1 eq) is similarly refluxed for 14 h in the presence of 1 eq of *N*-(4-bromobutyl)phthalimide, KF-Celite (0.4 g/mmol), and acetonitrile (4 mL/mmol).
4. After similar purification by silica gel column chromatography, the fractions containing *N*-(3-phthalimidopropyl)-*N*-(4-phthalimidobutyl)benzylamine are collected and the solvent evaporated (*see* **Note 4**).
5. *N*-(3-Phthalimidopropyl)-*N*-(4-phthalimidobutyl)benzylamine (1 eq) is dissolved in methanol (10 mL/mmol) containing NH_2NH_2 H_2O (10 eq) and refluxed for 3 h, then the solvent is evaporated. The residue is extracted with 4 M ammonia solution and $CHCl_3$, then the solvent from the $CHCl_3$ layer is evaporated to obtain N^4-benzylspd.
6. N^4-Benzylspd (1 eq), benzaldehyde (2 eq), and $MgSO_4$ (2.6 eq) are dissolved in methanol (4 mL/mmol). The solution is stirred at room temperature and $NaBH_4$ (6 eq) is carefully added to the solution on ice over a period of 1 h. Stirring is continued for another 1 h. The methanol is evaporated and the residue is extracted with Et_2O and H_2O. The Et_2O layer is washed with H_2O and the Et_2O is evaporated. The resulting oil is used as N^1,N^4,N^7-tribenzylspd.
7. A mixture of N^1,N^4,N^7-tribenzylspd (1 eq), *N*-(3-bromopropyl)phthalimide (2.1 eq), and KF-Celite (1 g/mmol) in acetonitrile (5 mL/mmol) is refluxed for 14 h.
8. After filtration, the solvent is evaporated and the residue is dissolved in benzene and subjected to silica gel column

chromatography using a benzene:acetone solvent as described above. The fractions containing N^4,N^8,N^{13}-tribenzyl-N^1,N^{16}-bis(phthaloyl)-4,8,13 triazahexadecane are collected and the solvent evaporated.

9. N^4,N^8,N^{13}-Tribenzyl-N^1,N^{16}-bis(phthaloyl)-4,8,13-triazahexadecane (1 eq) is dissolved in methanol (10 mL/mmol) containing NH_2NH_2 H_2O (10 eq), refluxed for 3 h, then the solvent is evaporated. The residue is extracted with 4 M ammonia solution and $CHCl_3$, then the solvent is evaporated from the $CHCl_3$ layer to obtain 1,16-diamino-N^4,N^8,N^{13}-tribenzyl-4,8,13 triazahexadecane.
10. 1,16-Diamino-N^4,N^8,N^{13}-tribenzyl-4,8,13-triazahexadecane (an aliquot) is hydrogenolyzed in AcOH (1 mL/mmol) at 60 °C in the presence of Pd/C. The mixture is stirred until hydrogen absorption ceases, then filtered through a Teflon Millipore membrane. The filtrate is treated with conc. HCl, evaporated to dryness, and the residue is recrystallized from aqueous EtOH. Pure 1,16-diamino-4,8,13-triazahexadecane pentahydrochloride is obtained.

3.2 Assay of SMO

All solutions should be kept at 4 °C until the start of the assay. Reaction tubes and all assay solutions should be chilled and kept on ice prior to the incubation.

1. Prepare the stock assay mixture: each 80 μL aliquot per reaction tube consists of 10 μL 1 M Tris–HCl buffer, pH 7.0 and containing 10 mM EDTA, 10 μL 0.36 mM pargyline solution, 10 μL 5.6 mM aminoguanidine solution, 25 μL 1 mM 3343 solution, and 25 μL distilled water (*see* **Note 5**).
2. Add 20 μL enzyme solution to each tube to initiate the reaction. The final reaction volume in all tubes should be 100 μL. Also include blank reactions that substitute 20 μL water or buffer for the enzyme solution.
3. 40 μL of the reaction mixture is immediately transferred to another tube containing 40 μL 20% TCA solution as the time 0 sample.
4. Incubate all the tubes at 37 °C.
5. After incubation, 40 μL of the reaction mixture is mixed with 40 μL 20% TCA solution (*see* **Note 6**).
6. The mixtures are centrifuged and the supernatants are subjected to HPLC analysis.
7. Use the product (spd from 3343) peak area to determine enzyme activity. The peak area of the time 0 sample is subtracted from each timed sample. Example data are shown in Fig. 1.

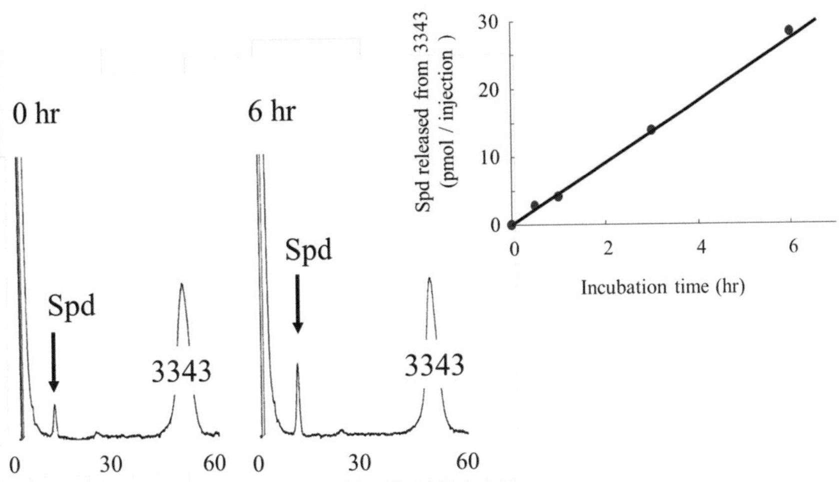

Fig. 1 Measurement of SMO activity in rat brain extract. Intrinsic Spd peak at 0 h is subtracted

3.3 HPLC Analysis

1. Equilibrate the HPLC column containing the cation exchange resin MCI GEL CK10S by pumping elution buffer at 1.0 mL/min and maintaining the column temperature at 65 °C.
2. The post-column reagent solution is mixed with the column effluent at 0.5 mL/min while maintaining the reaction oven at 65 °C.
3. The fluorescence detector is set to an excitation wavelength of 345 nm and an emission wavelength of 450 nm.
4. Start the analysis by injecting the working standard. The reaction product from 3343 is spd.
5. Start sample analysis once satisfactory chromatograms of the standard are obtained.

3.4 Preparation of Enzyme Solution

1. Homogenize a tissue sample (e.g. brain) at 4 °C with 4 volumes of 0.25 M sucrose-10 mM Tris–HCl buffer, pH 7.4, using a Potter-Elvehjem homogenizer.
2. Centrifuge the homogenate at 105,000 × *g* for 60 min at 4 °C.
3. Use the crude extract as the enzyme solution for the SMO assay.

4 Notes

1. Because the pH of Tris has a significant temperature coefficient, it is important to adjust the pH of buffers to the intended pH at the temperature at which they are to be used.
2. All preparations were conducted in a cold room (4 °C) or under cold conditions on ice.

3. Samples separated on TLC plates were detected using a fluorescence inspection lamp (254 nm) or ninhydrin spray.
4. Another synthesis method for *N*-(3-phthalimidopropyl)-*N*-(4-phthalimidobutyl)benzylamine from spd is described in the literature [10].
5. The apparent *Km* value of 45 μM 3343 was calculated as described in the literature [7].
6. The incubation time must be chosen after determining the region of apparent linearity for the enzymatic degradation of 3343.

Acknowledgements

The authors would like to thank Dr. Keijiro Samejima for his help with this manuscript.

References

1. Wang Y, Devereux W, Woster PM, Stewart TM, Hacker A, Casero RA Jr (2001) Cloning and characterization of a human polyamine oxidase that is inducible by polyamine analogue exposure. Cancer Res 61:5370–5373
2. Vujcic S, Diegelman P, Bacchi CJ, Kramer DL, Porter CW (2002) Identification and characterization of a novel flavin-containing spermine oxidase of mammalian cell origin. Biochem J 367:665–675
3. Wang Y, Murray-Stewart T, Devereux W, Hacker A, Frydman B, Woster PM, Casero RA Jr (2003) Properties of purified recombinant human polyamine oxidase, PAOh1/SMO. Biochem Biophys Res Commun 304:605–611
4. Devereux W, Wang Y, Stewart TM, Hacker A, Smith R, Frydman B, Valasinas AL, Reddy VK, Marton LJ, Ward TD, Woster PM, Casero RA (2003) Induction of the PAOh1/SMO polyamine oxidase by polyamine analogues in human lung carcinoma cells. Cancer Chemother Pharmacol 52:383–390
5. Häkkinen MR, Hyvönen MT, Auriola S, Casero RA Jr, Vepsäläinen J, Khomutov AR, Alhonen L, Keinänen TA (2010) Metabolism of *N*-alkylated spermine analogues by polyamine and spermine oxidases. Amino Acids 38:369–381
6. Hyvönen MT, Keinänen TA, Cerrada-Gimenez M, Sinervirta R, Grigorenko N, Khomutov AR, Vepsäläinen J, Alhonen L, Jänne J (2007) Role of hypusinated eukaryotic translation initiation factor 5A in polyamine depletion-induced cytostasis. J Biol Chem 282:34700–34706
7. Takao K, Shirahata A, Samejima K, Casero RA Jr, Igarashi K, Sugita Y (2013) Pentamines as substrate for human spermine oxidase. Biol Pharm Bull 36:407–411
8. Samejima K, Takeda Y, Kawase M, Okada M, Kyogoku Y (1984) Syntheses of ^{15}N-enriched polyamines. Chem Pharm Bull 32:3428–3435
9. Niitsu M, Samejima K (1986) Syntheses of a series of linear pentaamines with three and four methylene chain intervals. Chem Pharm Bull 34:1032–1038
10. Fasseur D, Lacour S, Guilard R (1998) Unequivocal synthesis of 1,9-dibenzyl-1,5,9,13-tetracyclohexadecane. Synth Commun 28:285–294

Chapter 16

Spectrophotometric Quantification of Reactive Oxygen, Nitrogen and Sulfur Species in Plant Samples

Chrystalla Antoniou, Andreas Savvides, Egli C. Georgiadou, and Vasileios Fotopoulos

Abstract

Reactive oxygen, nitrogen and sulfur species are key signalling molecules involved in multiple physiological processes that can be examined in qualitative and quantitative manners. Here, we describe simple spectrophotometric assays that allow the quantification of hydrogen peroxide, nitrite-derived nitric oxide and hydrogen sulphide from plant tissues.

Key words Reactive species quantification, Nitric oxide, Hydrogen peroxide, Hydrogen sulphide, Spectrophotometry

1 Introduction

Reactive oxygen (e.g. H_2O_2), nitrogen (e.g. NO) and sulfur (e.g. H_2S) species (RONSS), as well as polyamines (PAs; e.g. putrescine, spermidine and spermine) are all endogenous molecules involved in cell signalling and gene regulation during stress and play a pivotal role in the stress acclimation of plants [1, 2]. A considerable amount of research exists on the cross-talk between RONSS and PAs when plants are exposed to an abiotic challenge and the current knowledge is already incorporated in recent articles [2–5]. It is therefore of great importance for RONSS to be determined in PA-related studies. A number of qualitative and quantitative approaches are available. The former commonly employs fluorescent dyes which however raises questions in regard with their specificity (e.g. [6]), while the latter often requires costly approaches such as the use of commercial kits or heavily specialized equipment such as electron paramagnetic resonance (EPR) and quantum cascade lasers [7]. A commonly employed alternative is the indirect quantification of RONSS with spectrophotometric approaches that are quick, simple and cheap to carry out. A number

Rubén Alcázar and Antonio F. Tiburcio (eds.), *Polyamines: Methods and Protocols*, Methods in Molecular Biology, vol. 1694, DOI 10.1007/978-1-4939-7398-9_16, © Springer Science+Business Media LLC 2018

of protocols are available in existing literature; the present chapter describes three widely used protocols for (a) nitrite-derived NO quantification based on the Griess reaction, where sulphaniliamide and N-(1-naphthyl) ethylenediamine (NED) are used to react with NO_2 [8], (b) H_2O_2 quantification based on potassium iodide (KI) [9], and (c) H_2S quantification based on the reaction with 5,5′-Dithiobis(2-nitrobenzoic acid) (DTNB) [10].

2 Materials

Prepare all solutions using ultrapure water (prepared by purifying deionized water, to attain a sensitivity of 18 MΩ-cm at 25 °C) and analytical-grade reagents. Prepare and store all reagents at room temperature (unless indicated otherwise). Diligently follow all waste disposal regulations when disposing waste materials.

1. Ground plant tissue stored in −80 °C.
2. Pre-chilled mini-centrifuge (4 °C).
3. Microplate reader.
4. Polystyrene container with crushed ice.
5. Pre-chilled mortar and pestle with liquid nitrogen.
6. 2 mL polypropylene tubes.

2.1 Reagents for NO Quantification

1. Extraction Buffer: 50 mM acetate buffer (cool), pH 3.6, 4% (w/v) zinc acetate. To prepare 100 mL extraction buffer, dissolve 4 g of zinc acetate in 90 mL deionized water, add 285 μL glacial acetic acid and then add deionized water to reach a final volume of 100 mL (*see* **Note 1**).
2. Reaction Buffer (Griess reagent): 0.1% (w/v) naphthylethylenediamine dihydrochloride (NED), 1% (w/v) sulphanilamide in 5% (v/v) phosphoric acid. Prepare each solution separately. For NED solution add 0.1 g to 100 mL. For sulphanilamide solution, prepare 5% phosphoric acid, by adding 0.588 mL from 85% ortho-phosphoric acid to 99.422 mL deionized water and then dissolve in this solution 1 g of sulphanilamide. To create the Griess reagent mix one part of each solution. For instance, to obtain 100 mL Griess solution, 50 mL of NED solution and 50 mL of sulphanilamide solution should be mixed (*see* **Note 2**).

2.2 Reagents for H_2O_2 Quantification

1. Extraction Buffer: 0.1% (w/v) Trichloroacetic acid (TCA). Dissolve 0.1 g of TCA in 100 mL deionized water. TCA should be stored at 4 °C.
2. Reaction reagents: 10 mM potassium phosphate buffer, pH 7.0, 1 M potassium iodide (KI; stored at 4 °C). For

preparing 10 mM potassium phosphate buffer, pH 7.0, first prepare 1 M K_2HPO_4 (174.18 g/100 mL) and 1 M KH_2PO_4 (136.09 g/100 mL). Then mix 0.615 mL of 1 M K_2HPO_4, 0.385 mL of 1 M KH_2PO_4 and 90 mL of deionized water to reach a final volume of 100 mL. 1 M KI is prepared by dissolving 16.6 g in 100 mL deionized water (*see* **Note 3**).

2.3 Reagents and Equipment for H_2S Quantification

1. Extraction Buffer: 100 mM Potassium phosphate, pH 7.0 and 10 mM ethylenediaminetetraacetic acid (EDTA). Prepare 1 M K_2HPO_4 and 1 M KH_2PO_4 (as described in Subheading 2.2) and mix 6.15 mL of 1 M K_2HPO_4, 3.85 mL of 1 M KH_2PO_4 and 90 mL of deionized water to reach a final volume of 100 mL. Then dissolve 0.292 g EDTA in potassium phosphate buffer to create the extraction buffer solution.
2. Reaction reagents: 20 mM 5,5′-Dithiobis(2-nitrobenzoic acid) (DTNB; Ellman's Reagent). Dissolve 39.6 mg DTNB in 5 mL of extraction buffer (*see* **Note 4**).

3 Methods

Carry out all procedures at room temperature unless otherwise specified.

3.1 Nitrite-Derived NO Quantification

1. 100 mg of frozen plant tissue is homogenized with the addition of 1 mL extraction buffer, which is kept on ice throughout the protocol.
2. Mix well using vortex and centrifuge at 15,000 × *g* for 15 min at 4 °C.
3. Transfer the supernatant (n1) into a new tube.
4. Wash the remaining pellet with 0.5 mL extraction buffer and vortex.
5. Centrifuge at 10,000 × *g* for 10 min at 4 °C.
6. Take the supernatant (n2) and mix it with supernatant (n1) (*see* **Note 5**). Centrifuge at 10,000 × *g* for 15 min at 4 °C.
7. Transfer the clear supernatant to a new tube.
8. Add 1 mL extraction from each clear total supernatant +1 mL Griess reagent (1 part NED: 1 part sulphanilamide) (*see* **Note 6**) into 2 mL tubes.
9. Incubate at room temperature for 30 min.
10. Transfer 300 μL of the reaction mixture to 96 wells plate and read absorbance at 540 nm in a plate reader spectrometer.
11. In order to quantify the NO derivatives (NO_2^-), a standard curve is performed with known concentrations of $NaNO_2$ usually ranging from 0.050–10 μM. Results are expressed as nmole/g fresh weight. An overview of the procedure is shown in Fig. 1.

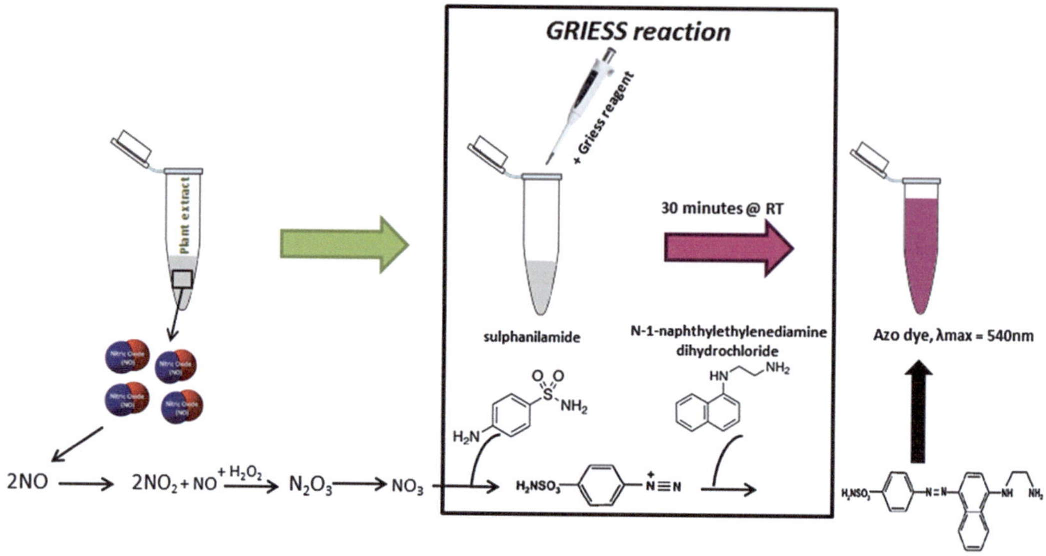

Fig. 1 Schematic representation of NO content estimation using the Griess reagent

3.2 H_2O_2 Quantification

1. 100 mg of frozen plant tissue is homogenized with the addition of 1 mL of 0.1% (w/v) TCA, which is kept on ice at all times.
2. Mix well on the vortex and centrifuge at 15,000 × *g* for 15 min at 4 °C.
3. Prepare 2 mL tubes to perform the reaction by adding 0.5 mL of the supernatant and 0.5 mL of 10 mM phosphate buffer (pH 7.0).
4. For the Blank use 0.5 mL from the 0.1% (w/v) TCA instead of the supernatant.
5. Add 1 mL of KI to initiate the reaction (as quick as possible).
6. Mix gently and leave in the dark for 2–15 min (depending on plant tissue).
7. Transfer 300 μL of the reaction mixture to 96 wells plate and read the absorbance at 390 nm in a plate reader spectrometer.
8. In order to quantify the H_2O_2, a standard curve is performed with known concentrations of H_2O_2 usually ranging from 1 to 250 μM. Results are expressed as μmol H_2O_2/g fresh weight. An overview of the procedure is shown in Fig. 2.

3.3 H_2S Quantification

1. 100 mg of frozen plant tissue is homogenized with the addition of 1 mL extraction buffer; which is kept on ice at all times.
2. Centrifuge at 15,000 × *g* for 15 min at 4 °C.
3. Transfer 100 μL of supernatant into a new tube (and keep on ice).

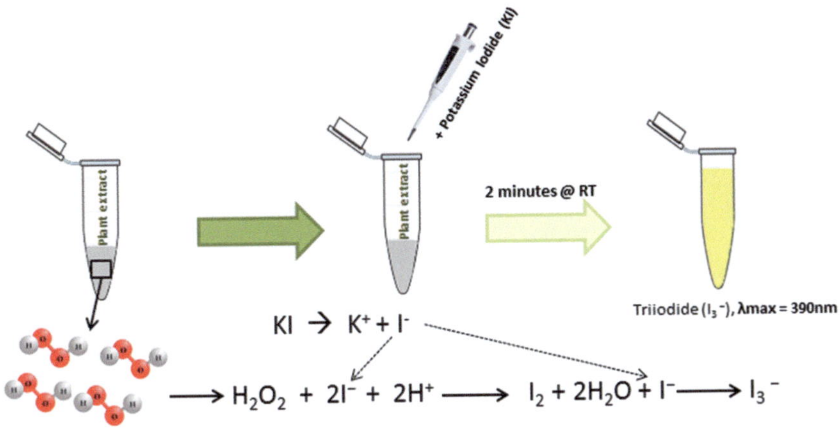

Fig. 2 Schematic representation of H_2O_2 content estimation using KI

4. Add 1880 μL extraction buffer into each tube containing 100 μL of supernatant.
5. Add 20 μL DTNB (20 mM). Total reaction volume is 2000 μL.
6. For the blank, add 1980 μL extraction buffer +20 μL DTNB (20 mM).
7. Incubate for 2 min at room temperature.
8. Transfer 300 μL of the reaction mixture to 96 wells plate and read absorbance at 412 nm in the plate reader spectrometer.
9. In order to quantify H_2S content, a standard curve is performed with known concentrations of a commonly used H_2S donor such as NaHS (sodium hydrosulphide hydrate) usually ranging from 4 to 100 μM. The results are expressed in μmole/g fresh weight. An overview of the procedure is shown in Fig. 3.

4 Notes

1. All extraction buffers should be kept cool in a container with crushed ice.
2. Store the two solutions (NED and sulphanilamide) at 4 °C with NED covered in aluminium foil as it may change colour if it is not stored protected from light.
3. It is well known that KI is liberating free iodine [11], especially in light. For this reason, it should be freshly prepared and stored in aliquots for a maximum of 1 week at 4 °C, covered with aluminium foil.
4. DTNB should be stored in the dark between 0 and 5 °C in where it can be stable for 6 months.

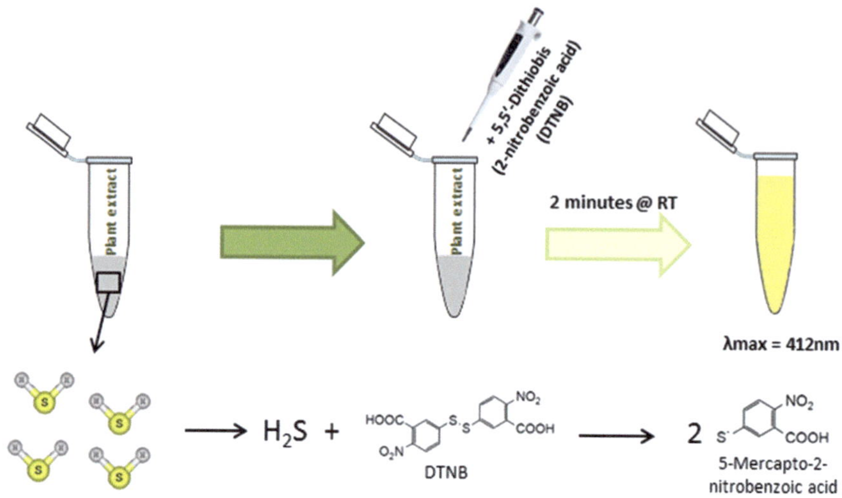

Fig. 3 Schematic representation of H_2S content estimation using DTNB

5. If your supernatant contains chlorophylls (green colour), add ~0.1 g active carbon in order to remove the pigments from the supernatant.
6. Take the Griess reagent out of the fridge to equilibrate at room temperature before the reaction takes place.

Acknowledgment

CA would like to acknowledge financial support by the Alexander S. Onassis Public Benefit Foundation. The support of Cyprus University of Technology (VF) is also gratefully acknowledged.

EG would like to acknowledge financial support by the Leventis Foundation.

References

1. Antoniou C, Savvides A, Christou A, Fotopoulos V (2016) Unravelling chemical priming machinery in plants: the role of reactive oxygen-nitrogen-sulfur species in abiotic stress tolerance enhancement. Curr Opin Plant Biol 33:101–107
2. Tiburcio AF, Altabella T, Bitrián M, Alcázar R (2014) The roles of polyamines during the lifespan of plants: from development to stress. Planta 240:1–18
3. Sequera-Mutiozabal M, Antoniou C, Tiburcio AF, Alcázar R, Fotopoulos V (2017) Polyamines: emerging hubs promoting drought and salt stress tolerance in plants. Curr Mol Biol Rep 3:28–36. doi:10.1007/s40610-017-0052-z
4. Filippou P, Antoniou C, Fotopoulos V (2013) The nitric oxide donor sodium nitroprusside regulates polyamine and proline metabolism in leaves of *Medicago truncatula* plants. Free Radic Biol Med 56:172–183
5. Minocha R, Majumdar R, Minocha SC (2014) Polyamines and abiotic stress in plants: a complex relationship. Front Plant Sci 5:175
6. Rumer S, Krischke M, Fekete A, Mueller MJ, Kaiser WM (2012) DAF fluorescence without NO: elicitor treated tobacco cells produce fluorescing DAF-derivatives not related to DAF-2 triazol. Nitric Oxide 27:123–135
7. Mur LAJ, Mandon J, Cristescu SM, Harren FJM, Prats E (2011) Methods of nitric oxide

detection in plants: a commentary. Plant Sci 181:509–519

8. Zhou B, Guo Z, Xing J, Huang B (2005) Nitric oxide is involved in abscisic acid-induced antioxidant activities in *Stylosanthes guianensis*. J Exp Bot 56:3223–3228
9. Velikova V, Yordanov I, Edreva A (2000) Oxidative stress and some antioxidant systems in acid rain-treated bean plants: protective role of exogenous polyamines. Plant Sci 151:59–66
10. Nashef AS, Osuga DT, Feeney RE (1977) Determination of hydrogen sulfide with 5,5′-dithiobis-(2-nitrobenzoic acid), N-ethylmaleimide, and parachloromercuribenzoate. Anal Biochem 79:394–405
11. Scott AH (1936) Standard iodine solutions. J Biol Chem 113:511–513

Chapter 17

Novel Route for Agmatine Catabolism in *Aspergillus niger*: 4-Guanidinobutyrase Assay

Tejaswani Saragadam and Narayan S. Punekar

Abstract

The enzyme 4-guanidinobutyrase (GBase) catalyzes the hydrolysis of 4-guanidinobutyric acid (GB) to 4-aminobutyric acid (GABA) and urea. Here we describe methods to estimate urea and GABA that were suitably adapted from the published literature. The urea is determined by colorimetric assay using modified Archibald's method. However, the low sensitivity of this method often renders it impractical to perform fine kinetic analysis. To overcome this limitation, a high sensitive method for detecting GABA is exploited that can even detect 1 μM of GABA in the assay mixture. The samples are deproteinized by perchloric acid (PCA) and potassium hydroxide treatment prior to HPLC analysis of GABA. The method involves a pre-column derivatization with *o*-phthalaldehyde (OPA) in combination with the thiol 3-mercaptopropionic acid (MPA). The fluorescent GABA derivative is then detected after reversed phase high performance liquid chromatography (RP-HPLC) using isocratic elution. The protocols described here are broadly applicable to other biological samples involving urea and GABA as metabolites.

Key words 4-Guanidinobutyrase, Agmatine, 4-Guanidinobutyric acid, GABA, Urea, *o*-Phthalaldehyde, 3-Mercaptopropionic acid, Isocratic, Reversed phase HPLC

1 Introduction

The catabolism of L-arginine through arginase is well established in *Aspergillus niger* [1]. Agmatine, a decarboxylated derivative of arginine, is a significant polyamine found in bacteria and plants. The agmatine catabolic pathway in *A. niger* was elucidated through metabolic, enzymatic, and genetic studies. This route involves GB as an intermediate metabolite and 4-guanidinobutyrase (GBase) catabolizes the hydrolysis of GB to form GABA and urea [2].

The GBase activity results in stoichiometric formation of GABA and urea; either one of them can be monitored to assay the enzyme. Urea is routinely estimated by the Archibald's method (and its variants) as described below. The method is based on the Fearon's reaction where urea (and other carbamido compounds like citrulline, methyl urea, and allantoin) reacts with dimethylglyoxime in an acidic

Rubén Alcázar and Antonio F. Tiburcio (eds.), *Polyamines: Methods and Protocols*, Methods in Molecular Biology, vol. 1694, DOI 10.1007/978-1-4939-7398-9_17, © Springer Science+Business Media LLC 2018

solution to form a yellow-colored complex; which can be measured colorimetrically [3, 4]. While the nature of this colored complex is unknown, it is proposed to consist of triazines [5] or glycolurils [6]. The lower sensitivity (detection limit is 500 μM of urea) of this method, however, limits its use in the kinetic analyses of enzymes with low K_M values.

GABA, the other product of GBase, can be monitored in many ways. A sensitive assay involves its reaction with OPA followed by detecting the fluorescent derivative upon reversed phase HPLC [7]. The OPA reacts with primary amino group of GABA in presence of thiols to form isoindole derivatives that are fluorescent. Although 2-mercaptoethanol is extensively used in combination with OPA, the method suffers because of the formation of highly unstable fluorescent adducts [8]. The corresponding thiol adduct of 3-mercaptopropionic acid (MPA) is comparatively more stable; this is attributed to the carboxylic group of MPA and was also found to enhance the fluorescence intensity of the isoindole ring [9]. The OPA exclusively reacts with GABA but not with guanidinium group of GB or carbamido group of urea; this feature is very well suited for acceptance in the GBase assay method. The other advantage with OPA is that it does not exhibit any intrinsic fluorescence [8]. The OPA-MPA method to estimate GABA is about 500 times more sensitive (*see* below) than Archibald's method of urea estimation. This feature was useful in *A. niger* GBase kinetic analysis, through access to lower range of substrate concentration for saturation; a K_M for GB of 2.7 mM could be determined (not shown).

2 Materials

1. Double distilled water (DDW): Use filtered and degassed DDW to prepare HPLC mobile phase.
2. 0.5 M Potassium phosphate buffer (pH 7.5).
3. 0.5 M Sodium phosphate buffer (pH 7.5).
4. 0.5 M Sodium acetate buffer (pH 4.0).
5. Borate buffer (pH 9.9): Prepare a 0.2 M boric acid solution in 0.2 M potassium chloride. Mix this borate solution with equal volume of 0.2 M sodium hydroxide (1:1, v/v). Adjust the final pH to 9.9.
6. 50 mM Urea.
7. 100 mM 4-Guanidinobutyric acid (GB).
8. 9.7 mM 4-Aminobutyric acid (GABA).

   Prepare all the solutions in DDW and no pH adjustment is required. Store them at −20°C and whenever needed make appropriate working solutions. All the chemicals mentioned above are from Sigma-Aldrich Co., St. Louis, MO, USA.

9. OPA: Weigh 5 mg of OPA and dissolve it in 1 mL of methanol (HPLC grade). Vortex it vigorously for 10 min to make a uniform solution (*see* **Note 1**).
10. 1.5 M Potassium hydroxide (*see* **Note 2**).
11. Acid reagent: Mix 0.5 mM $FeCl_3$ (made in DDW), 88% phosphoric acid and 98% sulfuric acid—6:3:1, v/v. Firstly weigh $FeCl_3$ and dissolve it in DDW. To this, add the acid components at regular intervals with slow stirring using a glass rod (*see* **Note 3**).
12. Color reagent: Make 3% dimethylglyoxime (from Sigma–Aldrich Co., St. Louis, MO, USA) in 98% sulfuric acid and stir it for 2–3 h until the solution becomes clear (*see* **Note 4**).
13. 11.6 N Perchloric acid (about 70%): Dilute it to 1.45 N in DDW.
14. 11.49 M MPA: Use 5 μL directly from the stock for derivatization.
15. HPLC mobile phase: To prepare 100 mL of the mobile phase, mix 30 mL of 0.05 M sodium acetate buffer (pH 4.0), 1 mL of tetrahydrofuran (THF) and 69 mL of methanol i.e., 30:1:69, v/v (*see* **Note 5**).
16. GBase source: The source of GBase is either directly from *A. niger* or from *E. coli* where the GBase cDNA is heterologously expressed. For urea estimation, both the cell-free extracts and purified GBase can be used. However, only the purified (or enriched) GBase fractions are preferred for GABA estimation. One unit of GBase is defined as the amount of enzyme that liberates 1.0 nmol of urea or GABA per min under standard assay conditions.
17. Analytical balance.
18. pH meter.
19. Vortex mixer.
20. Water bath.
21. Cooling micro-centrifuge (Eppendorf, Germany).
22. Screw capped glass tubes.
23. Sterile polypropylene tubes, micropipettes.
24. Micro-centrifuge tubes (1.5 mL).
25. Disposable tips.
26. UV-Vis spectrophotometer (from Jasco, made in Japan).
27. 0.2 μm filters (Pall Life Sciences, Ultipor®N_{66}® Nylon 6,6 Membrane).

28. Filtration unit (Merck, India).
29. Vacuum pump.
30. HPLC system: The reversed phase HPLC (RP-HPLC) system (from Jasco, made in Japan) is equipped with an isocratic pump (Jasco Pu-2080 Plus Intelligent HPLC Pump) coupled with a fluorescence detector (Jasco FP 2020 Plus Intelligent Fluorescence Detector) and an injection valve (Rheodyne 7725i, USA) with a 20 μL filling loop. For sample injection 50 μL glass syringe (Hamilton, Switzerland) is needed. The chromatographic column is made of prepacked C-18 analytical column (HiQsil™C-18 HS, made in Japan), with a particle size of 5 μm and 4.6 mm × 250 mm, ID. The parameters for fluorescence detection include an excitation and emission wavelength of 337 nm and 454 nm respectively, low sensitivity and a gain of 10×. The system is equipped with the Jasco ChromNAV software for analysis of the chromatographic peaks.

3 Methods

The procedures described below was evolved to assay GBase activity but could suitably be extended for any other system requiring urea and/or GABA estimations.

3.1 Urea Estimation

1. The standard assay mixture (200 μL) consists of GB (25 mM), 100 mM potassium phosphate buffer (pH 7.5), distilled water, and typically about 200 ng of enriched GBase protein (corresponds to 25.0 units of GBase). Prepare a master mix of 500 μL reaction to perform duplicate assays. Run an enzyme blank and substrate blank along with the enzyme reaction. Also set up for a urea standard curve simultaneously. Take urea concentrations in the range of 0.5–6.0 mM to generate the standard curve (*see* **Note 6**).
2. Pre-incubate the reaction mixture for 15 min at 37 °C before the addition of GB.
3. Initiate the reaction by adding GB and allow the reaction to proceed for 30 min at 37°C.
4. Terminate the reaction by adding 200 μL aliquots (for duplicates) of the reaction mixture to 4 mL each of the acid-reagent already dispensed in two separate screw-capped tubes.
5. Then add 200 μL of the color reagent to each tube with a dispenser. Cap all the tubes and vortex them vigorously (*see* **Note 7**).

6. Immediately place them in boiling water bath at 100°C for 20 min. The urea liberated in the reaction reacts with the color reagent to form a yellow-colored complex (*see* **Note 8**).
7. Remove the tubes from the boiling water bath and allow them to cool to room temperature (25°C). Measure the absorbance of the yellow-colored complex at 478 nm against an appropriate blank.

3.2 GABA Estimation

1. The standard assay mixture (200 μL) consists of GB (10 mM), 100 mM potassium phosphate buffer (pH 7.5), distilled water, and typically about 60 ng of enriched GBase protein (corresponds to 1.0 unit of GBase). Prepare a master mix of 700 μL reaction to perform the assay in triplicates. For kinetic studies, however, take varying concentrations of GB (0.1–20 mM) in the assay mixture with varying amounts of GBase (*see* **Note 9**).
2. Pre-incubate the reaction mixture for 15 min at 37°C before addition of the substrate GB.
3. Initiate the reaction by adding GB and allow the reaction to proceed for 15 min.
4. Terminate the reaction immediately after 15 min by adding 200 μL aliquots (for triplicates) of the reaction mixture to 50 μL of ice-cold 1.45 N PCA. Vortex the tubes vigorously and incubate them on ice for 30 min (*see* **Note 10**).
5. Centrifuge the tubes for 10 min at 4°C at 14,000 × *g*. Then carefully pipette out 200 μL of the supernatant without disturbing the denatured protein that has been precipitated (*see* **Note 11**).
6. Adjust the supernatant pH between 7.5 and 8.5 by adding 41–44 μL of 1.5 M ice cold KOH solution. Vortex the tubes vigorously and chill them on ice for 30 min (*see* **Note 12**).
7. Centrifuge the tubes for 10 min at 4°C at 14,000 × *g*. Collect the neutralized supernatant (~140 μL) and chill it on ice for additional 15 min (*see* **Note 13**).
8. Once again, spin the sample for 10 min at 4°C at 14,000 × *g*. Collect 100 μL of the supernatant which can be directly used for derivatization.
9. The derivatization procedure described here is modified from ref. 7. For this, mix 100 μL of the supernatant (from previous step) with 75 μL of 0.2 M borate buffer (pH 9.9), 20 μL of methanolic OPA (5 mg/mL) and 5 μL of 11.49 M MPA. Add the three components in sequence as mentioned. Vortex the mixture immediately and keep it for 1 min at room temperature. Inject 20 μL of the suitably diluted (typically about 20 times) derivative sample onto reversed phase HPLC column (*see* **Note 14**).

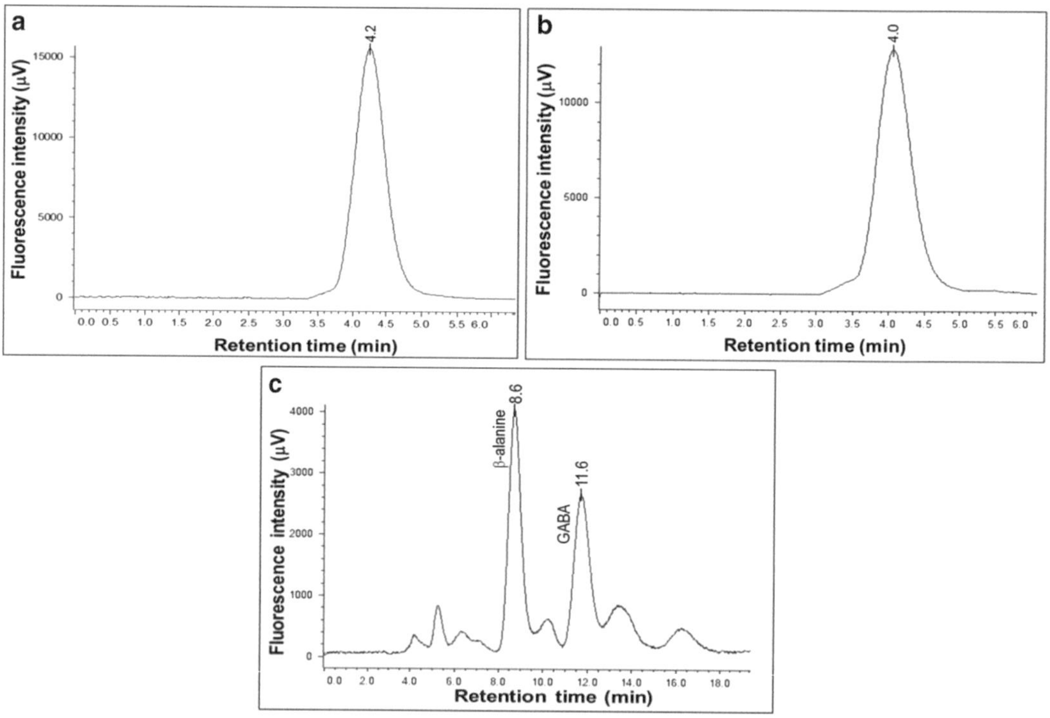

Fig. 1 Representative chromatograms of OPA-MPA derivatives. Treated samples of GABA (3.2 μM; panel **a**) and β-alanine (3.7 μM; panel **b**) were resolved by HPLC using a mobile phase with 70% methanol. Panel **c** shows the HPLC resolution of samples containing a mixture of GABA and β-alanine, albeit with a mobile phase containing 50% methanol

10. Carry out the isocratic elution at room temperature (25°C). The mobile phase (0.05 M sodium acetate, THF and methanol in the ratio of 30:1:69) should be run at a flow rate of 1 mL/min. The retention time (t_R) for GABA is 4 min. The parameters for fluorescence detection are mentioned above (see Section 30 in Materials) (*see* **Note 15**).
11. The analysis of chromatographic peaks is performed using the software Jasco ChromNAV (*see* **Note 16**).
12. Prepare GABA stock (9.7 mM) and store at −20°C; whenever needed make appropriate working solutions and treat them in the same way as an enzyme reaction mentioned above. Resolve on the HPLC column (as mentioned above) (Fig. 1, panel **a**) and generate the standard curve by plotting area under the peak against known concentrations of GABA. The linearity for GABA is over the range of 1.5–8.0 μM with a linear regression value (R^2) of >0.99 (Fig. 2). The GABA formed in the enzyme assay mixture is determined by interpolation from the standard curve. Similarly, β-alanine could also be resolved on the HPLC column with a retention time of 4 min (Fig. 1, panel **b**). The

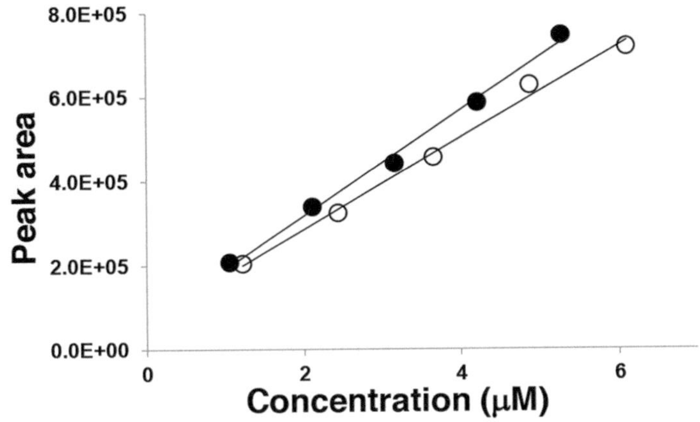

Fig. 2 Quantitation of GABA and β-alanine. Samples deproteinized with perchlorate/KOH treatment were derivatized and separated by HPLC. The GABA (•) and β-alanine (○) peak areas for each concentration were plotted

linearity for β-alanine is almost like that of GABA i.e. in the range of 1.8–9.0 μM (Fig. 2).

13. The OPA-MPA derivatization procedure is also broadly applicable and useful in estimating other amino acids like L-arginine and L-glutamic acid. The linear range for L-arginine and L-glutamic acid is about 5–30 μM and 11–50 μM, respectively. The sensitivity, specificity, and reproducibility of this method render it extremely useful in measuring compounds bearing primary amine groups.

4 Notes

1. The OPA is not readily soluble in methanol. Vortex it vigorously until the solution becomes clear. Avoid light while preparing this solution as OPA is light sensitive.
2. The KOH solution absorbs carbon dioxide on exposure to air. It should be stored in an air-tight plastic container.
3. The preparation of this acid reagent should be done on ice as it involves mixing of concentrated acids with DDW. This acid reagent is stable for at least 2 months when stored at room temperature.
4. Store the light-sensitive color reagent in an amber-colored bottle. The reagent is stable for at least a month at room temperature. Care should be taken while performing the urea assay as it involves handling of concentrated, corrosive acids.
5. Do not adjust the pH of the mobile phase after the addition of organic solvents. Mix organic solvents into acetate buffer already adjusted to pH 4.0. This helps maintain reproducibility

between experiments. Always prepare a fresh mobile phase to prevent microbial growth. All the components used to prepare HPLC mobile phase should be filtered (0.2 μm cut off membrane filters) and degassed. Allow the degassed mobile phase to reach to room temperature before use.

6. For routine GBase assays, a single confirmatory urea standard will be sufficient provided the same set of reagents is employed.
7. The assay mixture and the two reagent solutions do not mix easily. Therefore, the tubes should be vortexed vigorously for uniform color development.
8. The color reaction can also take place at room temperature albeit very slowly. But it is hastened by placing the tubes immediately for boiling. The boiling time should be kept constant (between 15 and 30 min) for consistent color development. The yellow color formed is stable for several hours.
9. Do not use potassium containing buffers in the assay mixture. Otherwise, the PCA added in the subsequent steps of GABA estimation will be titrated out and the entire deproteination procedure will require recalibration.
10. The GBase activity is maximal at pH 7.5. The drop in pH due to PCA addition results in protein denaturation and enzyme inactivation. In general, addition of 3–5% of PCA will suffice [10]. Incubation of the sample for 30 min on ice assists precipitation of denatured proteins.
11. Ensure not to disturb the precipitated protein pellet while collecting the clear supernatant. Loading particulate matter can irreversibly damage the HPLC column.
12. The potassium perchlorate precipitates around pH 6.0–7.0. However, the pH should be adjusted above 7.5 as the subsequent OPA-MPA derivatization works best at pH 9.9. Although the range of PCA and KOH volume additions are mentioned above, minor variations may have to be tried. Both these additions should be in smaller volumes to prevent sample dilution; pipetting accuracy is therefore important. As mentioned above (*see* **Note 2**), the KOH strength varies depending on its freshness or frequent use. It is critical to optimize the volumes of PCA and KOH added to adjust the pH above 7.5. Check this using a pH strip. As the KOH addition releases heat this should be done on ice to prevent GABA degradation.
13. The presence of potassium perchlorate crystals in the neutralized supernatant strongly interferes with the derivatization; additional 15 min incubation ensures its complete precipitation.
14. Prepare fresh OPA solution just before performing HPLC. The OPA derivative of GABA (and other amino acids) is not very

stable and excess OPA further destabilizes the initially formed fluorescent isoindole derivative. This can be overcome by maintaining the mole ratio of OPA to MPA in the range of 1:50 [11]. In the present protocol the OPA to MPA ratio is maintained at 1:77. Because of short stability of the derivative both the addition and vortexing steps should be performed as quickly as possible. After an optimal incubation time of 1 min the sample containing the derivative is immediately (within 5–10 s) injected on to the column. Ensure that no air bubbles enter while injecting the sample. Injecting a smaller sample volume (20 μL instead of 200 μL) helps reduce peak broadening. The sensitivity of detection is further improved when the fluorescence gain is set to 10× instead of 2× [7].

15. Including 70% methanol (and not 50% as mentioned in ref. 7) in the mobile phase significantly reduces the retention time for GABA (4 min instead of 12 min). This provides an advantage of assaying large number of GABA samples in a short time and also saves on mobile phase. However, clear resolution of two different isoindole derivatives (e.g., GABA and β-alanine) is better achieved by using 50% methanol (Fig. 1, panel **c**). In any case, ensure that the baseline is properly stabilized prior to sample injection.

16. Do not overload the column as this will greatly affect the peak symmetry. For the samples containing higher concentrations of GABA, dilute the samples in DDW before loading. The HPLC quantitation involves either measuring the peak height or area. With asymmetric and/or tailing peaks, area under the peak is normally considered and not the peak height.

References

1. Dave K, Ahuja M, Jayashri TN, Sirola RB, Punekar NS (2012) A novel selectable marker based on *Aspergillus niger* arginase expression. Enzyme Microb Technol 51:53–58
2. Kumar S, Saragadam T, Punekar NS (2015) Novel route for agmatine catabolism in *Aspergillus niger* involves 4-guanidinobutyrase. Appl Environ Microbiol 81:5593–5603
3. Archibald RM (1944) Determination of citrulline and allantoin and demonstration of citrulline in blood plasma. J Biol Chem 156:121
4. Boyde TR, Rahmatullah M (1980) Optimization of conditions for the colorimetric determination of citrulline, using diacetyl monoxime. Anal Biochem 107:424–431
5. Beale RN, Croft D (1961) A sensitive method for the colorimetric determination of urea. J Clin Pathol 14:418–424
6. Veniamin MP, Vakirtzi-Lemonias C (1970) Chemical basis of the carbamidodiacetyl micromethod for estimation of urea, citrulline, and carbamyl derivatives. Clin Chem 16:3–6
7. de Freitas Silva DM, Ferraz VP, Ribeiro AM (2009) Improved high-performance liquid chromatographic method for GABA and glutamate determination in regions of the rodent brain. J Neurosci Methods 177:289–293
8. Garcia Alvarez-Coque MC, Medina Hernandez MJ, Villanueva Camanas RM, Mongay Fernandez C (1989) Formation and instability of o-phthalaldehyde derivatives of amino acids. Anal Biochem 178:1–7
9. Kucera P, Umagat H (1984) Chemical derivatization techniques using microcolumns. In: Kucera P (ed) Microcolumn high-performance liquid chromatography, vol 28. Elsevier, Amsterdam, pp 157–160

10. Scopes RK (1994) Analysis - measurement of protein and enzyme activity. In: Cantor CR (ed) Protein purification: principles and practice, vol 3E. Springer, New York, p 58
11. Mengerink Y, Kutlan D, Toth F, Csampai A, Molnar-Perl I (2002) Advances in the evaluation of the stability and characteristics of the amino acid and amine derivatives obtained with the o-phthaldialdehyde/3-mercaptopropionic acid and o-phthaldialdehyde/N-acetyl-L-cysteine reagents. High-performance liquid chromatography-mass spectrometry study. J Chromatogr A 949:99–124

Chapter 18

Determination of Transglutaminase Activity in Plants

S. Del Duca, P.L.R. Bonner, I. Aloisi, D. Serafini-Fracassini, and G. Cai

Abstract

Transglutaminase (TGase:E.C. 2.3.2.13) catalyzes the acyl-transfer reaction between one or two primary amino groups of polyamines and protein-bound Gln residues giving rise to post-translational modifications. One increasing the positive charge on a proteins surface and the other results in the covalent crosslinking of proteins. Pioneering studies on TGase in plants started in the middle of the 1980's but the methodology designed for use with animal extracts was not directly applicable to plant extracts. Here we describe radioactive and colorimetric methods adapted to study plant TGase, as well as protocols to analyze the involvement of TGase and polyamines in the functionality of cytoskeletal proteins.

Key words Transglutaminase assay, Polyamine–protein interaction, Glutamyl-polyamines, Actin filament binding assay, Microtubule binding/motility assay

1 Introduction

Aliphatic polyamines (PAs) have been the subject of numerous investigations over many years. The historical focus of the work on PAs has been predominantly on soluble PA's in both plant and animal cells, whereas the conjoined forms of PAs has received less attention. However, due to their chemical characteristics, PAs can be covalently or non-covalently linked to several molecules (Fig. 1). Towards the end of the 1950's PAs linked by hydrogen bonds to nucleic acids attracted the attention of pioneering researchers. In the same period, other researchers found the activity of an enzyme (Transglutaminase; TGase) that was able to covalently conjugate PAs to animal proteins. In the 1980's Jack Folk and collaborators [1] published a milestone paper containing a method to quantify this enzyme-catalysed polyamine conjugation to proteins, which was later on modified by Beninati and collaborators [2]. Using this methodology, the enzyme-catalysed polyamine incorporation into proteins was also studied in plant extracts, to detect a Ca^{2+}-dependent TGase activity by its unequivocal products (the conjugated PA residues, i.e., (γ-glutamyl) –PAs, Fig. 2) in

Rubén Alcázar and Antonio F. Tiburcio (eds.), *Polyamines: Methods and Protocols*, Methods in Molecular Biology, vol. 1694, DOI 10.1007/978-1-4939-7398-9_18, © Springer Science+Business Media LLC 2018

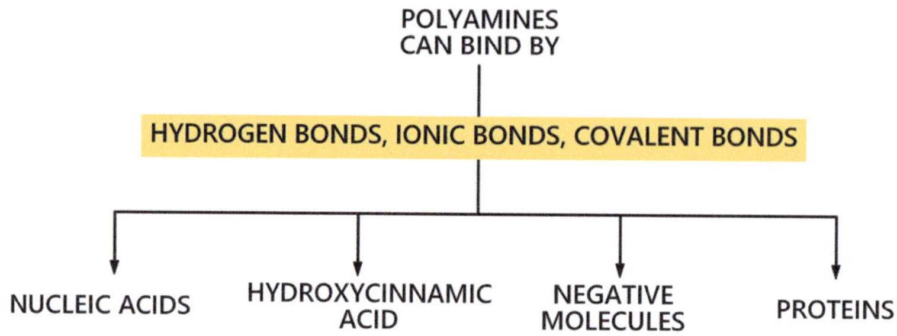

Fig. 1 PAs can be linked to several molecules with different linkages

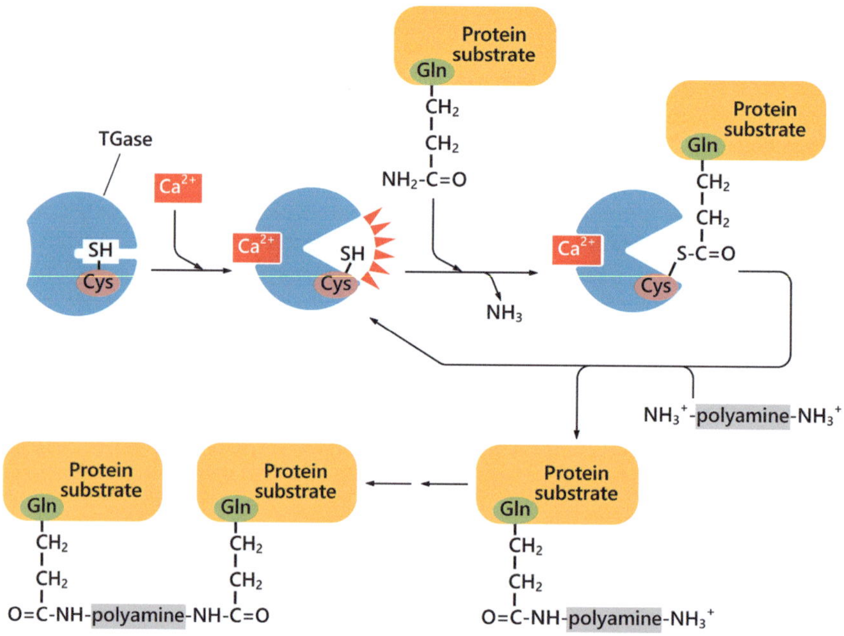

Fig. 2 The two steps of transamidase reaction of TGase. In the presence of Ca^{2+} the enzyme could change its conformation and cysteine in the active site became ready to catalize the reaction that involves the formation of a thiolester acylenzyme intermediate with the release of ammonia. Then, a molecule with a primary amine group, like a polyamine or lysine, performs a nucleophilic attack to the endo-Gln residue of the protein substrate, causing the covalent binding of polyamine to protein. Moreover, diamines and polyamines might act as a bridge in a bis-glutaminyl insert between two acceptor molecules, with the length of the bridge depending on the length of polyamine involved in the reaction

several plants in different physiological conditions. This method also reveals the metabolism of polyamines in the cell. Some problems emerged which indicated that the radioactive polyamine incorporation assay designed for animal cell extracts was not directly applicable to all plant cells extracts. The major problem was the presence of diamine oxidase activity in some plant extracts capable of incorporating radioactive putrescine into substrate protein in a calcium-independent fashion [3]. Consequently, this led to difficulties in the purification and sequencing of the plant polyamine incorporating enzyme due to the length and complexity of the method reported above to identify the enzyme activity.

Alternative assays were designed to overcome these difficulties. The use of ethyldimethylaminopropyl carboiimide(EDC)-modified casein [4] as a substrate to measure plant TGase calcium-dependent polyamine incorporation initially showed promise but was superceded by the use of biotinylated cadaverine and *N,N'*-dimethyl casein as substrates in the microtiter plate TGase assay developed by Slaughter and collaborators [5]; selective sample preparation resulted in the demonstration of calcium-dependent transglutaminase enzyme in plant tissue. Confirmation of plant TGase activity was obtained by the identification of the products of TGase protein transamidase and crosslinking activities (γ-glutamyl PA residues and ε-(γ glutamyl) lysine isodipeptide respectively) after digestion of plant protein extracts with a variety of proteolytic enzymes prior to cation exchange chromatography [4, 6].

Despite difficulties it became clear that a plant TGase enzyme was able to conjugate PAs into proteins (and form protein crosslinks), but the enzyme had a low cellular activity in most plant cells. However, exceptions to this general rule emerged with higher activities detected in apical meristematic cells, in vitro stimulated dormant or immature tissues, germinating pollen, salt-stressed unicellular algae and in the dark/light transition of photosynthetic cells/chloroplasts, i.e., during a change of developmental fate or activity. A relatively high activity was detected in chloroplasts in the presence of light [7]. Additional work based on the cross reactivity of antibodies directed to animal TGases (e.g., tissue TGase or TG2) indicated that there are more than one plant TGase enzymes which are present in different plant cell compartments [8, 9].

In animals and plants, TGase activity catalyses a two-step conjugation of PAs (or –lysine) to glutamyl derivatives of specific proteins (Fig. 2). This is referred to as transamidation (polyamine incorporation or protein crosslinking) where in the second step of the catalytic pathway an amine acts as the nucleophile. In the absence of an amine, a water molecule can act as the nucleophile resulting in the deamidation of protein-bound glutamine. Tranglutaminase PA incorporating activity and protein cross linking activity has been demonstrated in animal and plant extracts but a user friendly assay to measure TGase deamidating activity has yet to be published. The seed storage proteins of dicotyledonous plants are known to undergo deamidation prior to proteolysis. A TGase deamidation assay based on this observation is currently in preparation.There are other enzymic activities associated with TGase which have been studied in animal cells [10]; at present, they have not been studied in plants cells but the opportunity does exist for a determined researcher.

2 Materials

Prepare all solutions using ultrapure water (18.2 MΩ cm^{-1} at 25 °C) and analytical-grade reagents. Prepare and store all reagents at room temperature (unless indicated otherwise).

Plant materials concern: *Helianthus tuberosus* sprout apices, dormant tubers, isolated chloroplasts, rosaceae (*Malus, Pyrus*), *Cytrus* and *Corylus* pollens, *Arabidopsis thaliana* tissues, *Nicotiana tabacum* flowers and leaves, *Dunaliella salina* unicellular green algae, *Zea mays* calluses, leaves and isolated chloroplasts, *Oryza sativa* transgenic plants, *Cucumis sativa* isolated plastids, *Hordeum vulgaris* plastids, *Beta vulgaris, Lactuca sativa* leaves [11–13].

2.1 TGase Assay for the Incorporation of Radioactive Polyamines

1. 1.0 M Tris/HCl pH 8.2: 121.1 g of Tris in 900 ml of H_2O, adjust the pH to 8.2 with HCl and made to a final volume of 1.0 l.
2. 100 mM $CaCl_2$: 0.11 g $CaCl_2$ in 10.0 ml H_2O
3. 100 mM ethylene glycol-bis(2-aminoethylether)-*N,N,N′,N′*-tetraacetic acid (EGTA): 3.80 g EGTA in 100 ml H_2O.
4. 100 mM Dithiothreitol (DTT): 0.154 g in 10 ml H_2O.
5. 10 mM putrescine or spermidine: 0.016 g Put or 0.025 g Spd in 10 ml H_2O.
6. [1,4 (n)-^{3}H]Put (35.7 Ci/mmol, 1 mCi/ml) or [1,4 (n)-^{3}H] Spd (34.8 Ci/mmol, 1 mCi/ml); 5 μl (5 μCi) of radioactive PA are used in the assay (*see* **Note 1**).
7. Protein substrate (*N,N′*-dimethylcasein (DMC), (1 mg/ml) (*see* **Note 2**)
8. Trichloroacetic acid (TCA) 20% (w/v) containing 2 mM cold PA (Put or Spd).

2.2 Solutions for Separation of Glutamyl-Polyamines by Reversed Phase HPLC

1. 0.1 N NaOH.
2. 0.8 M morpholine acetate buffer pH 8.1.
3. Enzyme for the digestion of TGase polyamine adducts: pronase type XXI from *Streptomyces griseus*.
4. The buffers composition which are necessary to separate glutamyl-PAs by HPLC are reported in Table 1.
5. Column: Anion exchanger: DIONEX DC- 6A, 4.5 mm × 80 mm, Resin Ultropac 8 Na^+.

2.3 TGase Assay for the Incorporation of Biotin-Cadaverine

1. 100 mM Tris–HCl pH 8.5: 12.11 g of Tris in 900 ml of H_2O. Adjust the pH to 8.5 with HCl and made up to a final volume of 1 l.
2. 150 mM phosphate buffered saline (PBS)-Tween 80: 137 mM sodium chloride, 2.6 mM potassium chloride, 8.1 mM

Table 1
Buffers necessary to separate glutamyl-PAs by HPLC

Buffer	Conc (N)	pH	Na + -citrate	NaCl	Brij	phenol	Final vol
A	0.6	5.80	1.96 g	33.9 g	1 ml	1 ml	1 l
B	1.5	5.51	3.90	85.3 g	"	"	"
C	3.0	5.55	15.70	166 g	"	"	"
D	5.0	5.50	26.0	280 g	"	"	"
E	0.2	3.31	19.60	–	"	"	"
F	0.2	4.31	19.60	–	"	"	"

disodium hydrogen orthophosphate (Na_2HPO_4), 1.47 mM potassium dihydrogen orthophosphate (KH_2PO_4) and Tween 80, 0.005% (w/v). For 1.0 litre: 8.0 g of sodium chloride, 0.2 g of potassium chloride, 1.15 g of disodium hydrogen orthophosphate, 0.2 g of potassium dihydrogen orthophosphate and 500 μl of Tween 80 dissolved in 900 ml of deionized H_2O. Adjust the pH to 7.4 and make to a final volume of 1.0 l with deionized H_2O.

3. 100 mM sodium acetate pH 6.0: 8.2 g of sodium acetate in 900 ml of water. Adjust the pH value to 6.0 with acetic acid and make the volume to 1 l.
4. Assay buffer containing calcium ($CaCl_2$) positive control, 100 mM Tris–HCl pH 8.5, 6.67 mM calcium chloride, 13.3 mM DTT, 225 μM of biotin-cadaverine. Dissolve 0.30 g Tris, 0.025 g calcium chloride, 0.051 g DTT and 2.5 μg of biotin-cadaverine in 20 ml of deionized water. Adjust the pH to 8.5 with HCl and the final volume to 25 ml with water. The DTT should be added fresh prior to the assay.
5. Assay buffer containing EGTA, negative control, 100 mM Tris–HCl pH 8.5: 20 mM EGTA, 13.3 mM DTT and 225 μM of biotin-cadaverine. Dissolve 0.30 g Tris, 0.05 g EGTA, 0.051 g DTT and 2.5 μg biotin-cadaverine in 20 ml of deionized water. Adjust the pH to 8.5 with HCl and make the final volume to 25 ml with deionized H_2O. The DTT should be added fresh prior to the assay.
6. 10 mg/ml Tetramethyl benzidine (TMB): 10.0 mg TMB diluted in 1.0 ml of dimethyl sulfoxide (DMSO).
7. 3% (w/v) hydrogen peroxide: 10 μl 30% (v/v) hydrogen peroxide in 90 μl ultrapure water.
8. Development solution: 150 μl of 10 mg/ml TMB and 25 μl of 3% (v/v) hydrogen peroxide are added to 20 ml of 100 mM sodium acetate pH 6.0.

9. DMC protein substrate: 50.0 g of DMC dissolved in 500 ml of water overnight (take care this reagent is odourous) to produce a 10× stock. The DMC can be stored in aliquots at −20 °C.

2.4 TGase Assay for the Incorporation of Biotin TVQQEL Peptide into Casein (Simulating Protein Crosslinking)

The reagents for this assay are similar to the reagents above described for the biotin-cadaverine assay (Subheading 2.3). The assay buffer contains 5–20 μM biotinylated TVQQEL peptide from a 10 mg ml^{-1} stock dissolved in DMSO instead of biotin-cadaverine.DMC is replaced with bovine casein.

2.5 In vitro TGase Assay During Pollen Germination

1. Composition of pollen germination medium: 68.46 mg/ml sucrose, 300 μg/ml calcium nitrate, 20 μg/ml boric acid in ultrapure water.
2. 1 mg/ml Biotin-cadaverine stock solution.

2.6 Polyamine-Microtubule Assay

1. TGase reactivation buffer: 20 mM Tris–HCl pH 8.0 containing 1.5 mM DTT.
2. Tubulin dilution buffer (TDB): 80 mM HEPES pH 7.5 containing 1 mM EGTA, 1 mM $MgCl_2$, 2.0 mM GTP (add immediately prior to use), 20% (v/v) glycerol.
3. Cushion buffer (CB): 80 mM HEPES pH 7.5 containing 1 mM EGTA, 1 mM $MgCl_2$, 20% (w/v) sucrose.
4. Microtubule dilution buffer (MDB): 80 mM HEPES pH 7.5 containing 1.0 mM EGTA, 1 mM $MgCl_2$, 1 mM GTP (add immediately prior to use).
5. PEM buffer: 80 mM HEPES pH 7.5 containing 1.0 mM EGTA, 1.0 mM $MgCl_2$.
6. Motility buffer: 80 mM HEPES pH 7.5 containing 7.0 mM EGTA, 1 mM $MgCl_2$, 1 mM DTT, 20 μM taxol, 5 mM ATP (add immediately prior to use).
7. Perfusion chamber: made by a glass-slide and a coverslip separated by two strips of double-sided tape. Both glass-slide and coverslip must be repeatedly washed to clean their surface, they then can be stored for weeks in a beaker containing 20% (v/v) ethanol. At time of use, glass-slide and coverslips are grasped with forceps and the ethanol is removed with the flame of a Bunsen burner. After cooling, glass-slides and coverslips are assembled to form the perfusion chamber in the way indicated in Fig. 3. The volume inside the chamber should be approximately 10–15 μl.
8. PEM buffer + taxol: 80 mM HEPES pH 7.5, containing 1 mM EGTA, 1 mM $MgCl_2$, 20 μM taxol.
9. Blocking buffer: PEM buffer containing 1 mg ml^{-1} casein.

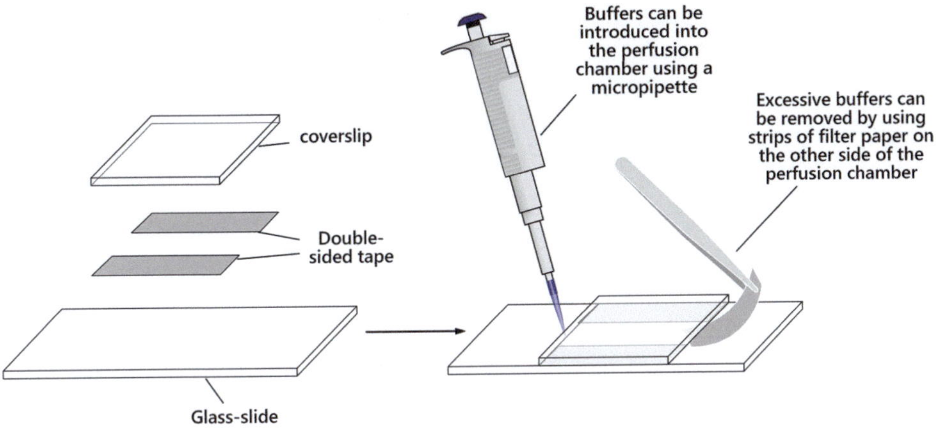

Fig. 3 Schematic representation of the assembly of a perfusion chamber using a glass-slide, a coverslip and two pieces of double-sided tape. The right part of the figure shows how to influx specific volumes of solution using a micropipette while removing previous content with filter paper

2.7 Polyamine-Actin Assay

1. TGase reactivation buffer: 20 mM Tris–HCl pH 8 containing 1.5 mM DTT
2. Buffer A (plus ATP): 5 mM Tris–HCl pH 8 containing 0.2 mM $CaCl_2$, 0.2 mM ATP (add before use).
3. 10× polymerization buffer (plus ATP): 0.5 M KCl pH 8.0, 20 mM $MgCl_2$, 10 mM ATP (add immediately before use).
4. Rhodamine-phalloidin solution: 10 μl 10× Polymerization Buffer, 25 μl of rhodamine-phalloidin stock (6.6 μM) and 65 μl of water.
5. ATPase activity assay buffer: 20 mM Pipes pH 7.0 containing 10 mM KCl, 1 mM EGTA, 2 mM $MgCl_2$, 5 mM ATP (add immediately before use).

2.8 Proteolytic Digestion of Plant Proteins for the Analysis of ε-(γ-Glutamyl) Lysine Isodipeptide Derived from Protein Crosslinks Following TGase Activity

1. 72% (w/v) TCA: 72 g of solid TCA dissolved in 100 ml distilled water.
2. 10% (w/v) TCA: 10 g of solid TCA dissolved in 100 ml distilled water.
3. 50:50 (ethanol:diethylether): 25.0 ml diethylether added to 25.0 ml ethanol.Store on ice.
4. 100 mM ammonium carbonate pH 10.0: 1.57 g ammonium carbonate dissolved in 90.0 ml of distilled water. Adjust the pH to 10.0 using concentrated sodium hydroxide. Make up to 100.0 ml with distilled water.
5. Leucine amino peptidase activation: 250 μl leucine amino peptidase (250 units), 225 μl 10 mM Tris–HCl pH 8.0, 25 μl 50 mM manganese chloride. The solution is incubated at 37°C for 2 h to activate the leucine amino peptidase.

6. Prolidase activating solution: 50 μl prolidase (100 units), 200 μl distilled water, 200 μl 10 mM Tris–HCl pH 8.0, 50 μl manganese chloride. The solution is incubated at 37°C for 2 h to activate the prolidase.
7. 0.54 M magnesium chloride: 1.09 g magnesium chloride (hexahydrate) dissolved in 10.0 ml of distilled water.
8. Chloroform:methanol:HCl (2000:1000:2): 100.0 ml chloroform, 50.0 ml methanol, 1.0 ml of concentrated HCl.

3 Methods

Plant tissue contains a variety of phenol and amine oxidase, that could interfere with the TGase assay. For some plant extract the most vulnerable assay is the radioactive polyamine incorporation assay [2, 3]; if problems are suspected and only the activity should be measured switch to either the biotin-cadaverine incorporation assay [4] or the TVQQEL peptide crosslinking assay (simulating protein crosslinking) [14] .

In all plant tissue extractions for TGase activity it is wise to plan ahead and ensure that all the reagents have been precooled. Additions, such as proteinase inhibitors, DIECA, polyvinylpolypyrrollidone, 2-mercaptoethanol and ascorbic acid are advisable. After extraction, the pH should be adjusted to between pH 8.0–9.0 before clarification at 12,000 × *g* for 30 min at 4 °C. The supernatant should then be subjected to protein precipitation using solid ammonium sulphate (0.76 g/ml). This will concentrate the protein and remove the low molecular weight contaminants. The protein ammonium sulphate mixture should be stirred and the protein precipitate collected by centrifugation 12,000 × *g* for 30 min at 4 °C. The pellet can be dissolved in a minium volume of 0.1 M Tris buffer pH 8.5 before being dialysed or desalted using size exclusion chromatography to remove excess ammonium ions. It is recommended that the extract is then put through at least one additional chromatography step such as anion exchange chromatography to separate the plant TGase from contaminating oxidase activity before being aliquoted and stored at the lowest temperature available [15].

3.1 Radioactive Assay of Transglutaminase Activity

This method (Fig. 3) can be performed (a) by supplying labelled PAs to either total plant cell extracts or isolated organelles or membranes in the test tube, or to living cells and organs (i.e., germinating pollen, unicellular algae, leaves, etc) or to in vitro coltures (preferably grown in liquid medium). (b) Alternatively, purified protein substrates, like DMC, fibronectin (FN), actin, tubulin, light harvesting complex II (LHCII) etc., or specific substrates of gplTGase (commercial Guinea pig liver TGase) such as N-benzyloxycarbonyl-L-γ-glutaminyl-L-leucine, or the histidine-tagged green fluorescent protein (His_6-X-press-GFP) a fluorescent protein derivatized with GFP, can be assayed for the plant enzymes [16]. On the other hand, if a plant substrate must be evaluated, TGases purified from either plant or animal cells can be assayed. An example of the method applied to pollen extract or entire germinating pollen is described here; some advertise for other plant samples are reported.

Reaction for transglutaminase assay occurs in 1.5 ml tubes. In details, the incubation mixture contains:

1. 15 μl solution 1.0 M Tris/HCl pH 8.2 + 3 μl solution 100 mM $CaCl_2$ (or 100 mM EGTA in case of negative control, *see* **Note 3**) + 30 μl solution 100 mM DTT + 6 μl solution 10 mM Put or Spd + 5 μl solution [1,4 (n)-^{3}H]Put (35.7 Ci/mmol, 1 mCi/ml) or [1,4 (n)-^{3}H]Spd (34.8 Ci/mmol, 1 mCi/ml) + 50 μl solution protein substrates (1 mg ml^{-1}) i.e., DMC or others (*see* **Note 2**) + ultrapure water up to 200 μl + 100 μl soln plant extract (0.5 mg ml^{-1})
2. Incubate samples for 2 h at 30 °C.
3. The reaction is stopped with 100 μl solution TCA 20% (w/v) containing cold PA (Put or Spd) 2 mM (5% (w/v) is the final concentration of TCA in order to remove trapped free PAs).
4. The mixture is stored at 4 °C for 24 h (or 2 h at −20 °C) and then centrifuged at 13,000 × *g* for 10 min at 4 °C.
5. The pellets were solubilized overnight with 0.1 M NaOH at 37 °C and then re-precipitated twice with 5% (w/v) TCA. The supernatants and final pellet, washed with anhydrous diethyl ether, were proteolytically digested as described [1, 2] with some modifications.

3.1.1 Analysis of the Products of the Enzymatic Assay

The TCA pellet obtained after the enzymatic assay can be used for different treatments, as schematically shown in Fig. 4. Direct measure of the radioactivity due to covalently conjugated polyamines: after last precipitation, the TCA pellet, re-suspended in 0.1 N NaOH, is prepared to detect the total radioactivity by a liquid scintillator counter. This simple analysis presents the possibility to erroneously detect PAs bound by a non-enzymatic activity, as it occurs when some "sticky" molecules might be present; PAs can be "glued" to these kind of molecules, severely affecting the value of results. This problem can occur especially with chlorophylls of green tissues or when phenols are present; moreover PAs can be oxidized or otherwise metabolized, thus it is important to detect glutamyl-PAs, as reported in Subheading 3.1.3. (*see* **Note 4**).

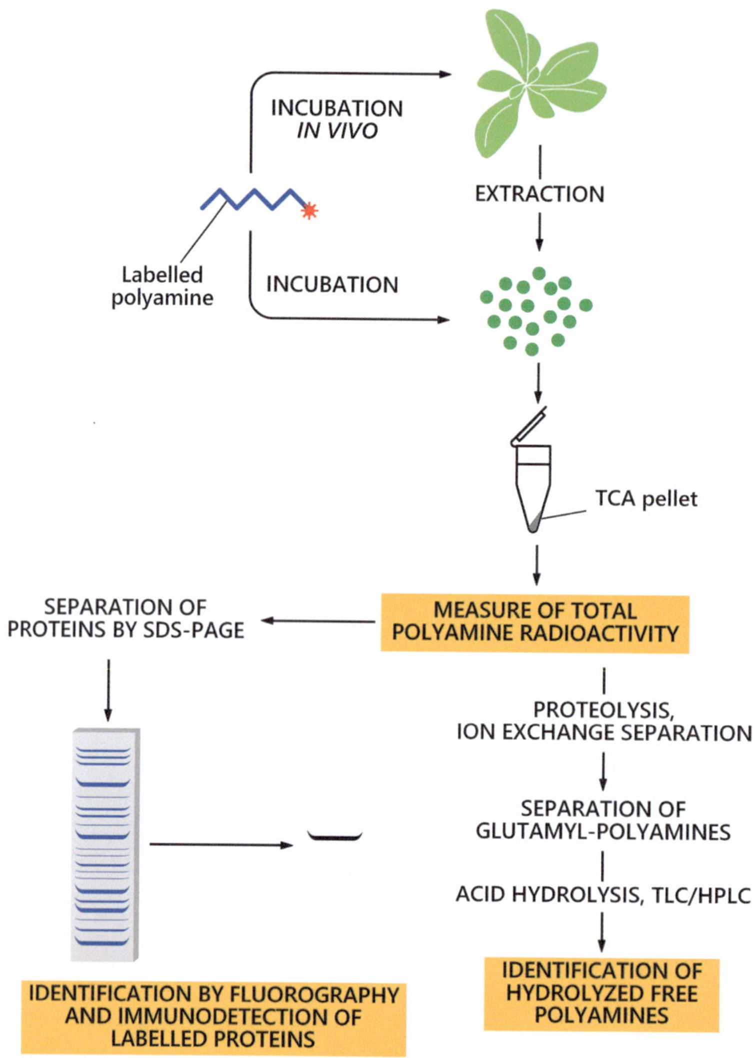

Fig. 4 Procedure that could be completely or partially performed for the study of TGase activity with radioactive PAs

3.1.2 Analysis of Proteins Conjugated to PAs Through SDS-PAGE, Immunoblot Fluorography, or Autoradiography

1. Determine the protein content using a modified bicinchoninic acid (BCA) method [17]. Use BSA as the standard protein.
2. Separate proteins through gel electrophoresis. SDS-PAGE of proteins contained in the pellet is done according to Laemmli [18] or other methods. The percentage of acrylamide (10–17%) is adjusted according to the mass of the labelled proteins.
3. Stain gels with Bio-Safe Coomassie (Bio-Rad) or other dyes. The dried gel is blotted for exposure with X-ray slides Kodak X Omat AR for 15–50 days to detect the labelled protein bands. Alternatively, the gel is exposed for the fluorography

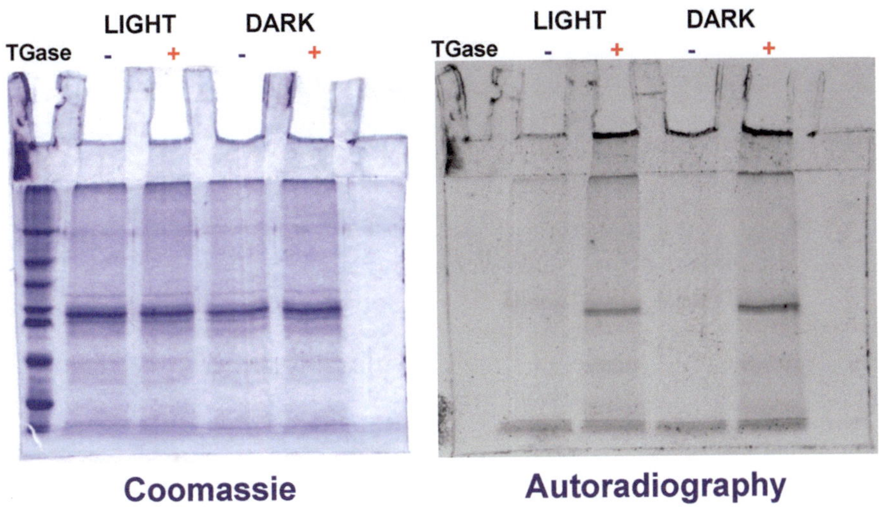

Fig. 5 SDS-PAGE separation of purified LHCII incubated with (+) and without (−) Guinea pig liver TGase (Sigma) and [^{14}C] Spm. Left: Coomassie-stained gel; right: autoradiography of the same gel after drying. Animal TGase conjugates Spm to LHCII, the main band in the gel, and is insensitive to light (third lane incubated to the light and fifth lane incubated in the dark for TGase assay). On the contrary, when chloroplast TGase and LHCII are incubated, the reaction is light-sensitive (data not shown). Bands at the bottom are due to the TGase-independent linkage of Spm with chlorophylls. In the Coomassie-stained gel, the first lane on the left contains MW standards

procedure. In Fig. 5 a Coomassie positive gel is compared with its autoradiography showing which proteins are labelled.

4. To identify the protein, the gel bands can be immunoprobed with specific antibodies. Immunoreactive bands are detected with chemiluminescent reagents available from different suppliers. Immunoblot is performed according to Towbin et al. [19]. Secondary antibodies against mouse and rabbit IgG are conjugated with horseradish peroxidase; secondary antibody to donkey IgG was conjugated with horseradish peroxidase.

3.1.3 Treatment of the Products of the Enzymatic Assay and Detection of Glutamyl-PAs

1. The TGase reaction is stopped with 5% (w/v) (final concentration) TCA also containing 2 mM unlabelled PA (i.e., Put, Spd) in order to remove trapped free polyamines; the mixture is stored at 4 °C for 24 h and then centrifuged at 13,000 × *g* for 10 min. The pellets are solubilized overnight with 0.1 M NaOH at 37 °C and then precipitated with 5% (w/v) TCA. This procedure is repeated twice. The final pellet is washed with anhydrous diethyl ether to remove TCA residues and digested with proteinases as described [1, 2] with some modifications.
2. After washing TCA pellet with anhydrous diethyl ether, diethyl ether was allowed to evaporate and the precipitate suspended in 100 μl NaOH 0.1 N and left at 37 °C for 1 h to dissolve. The liquid is neutralized with 100 μl HCl 0.1 M.

3. The proteinase solution prepared as follows: 250 mg of pronase type XXI *Streptomyces griseus* are added to 0.5 ml of 0.8 M N-ethyl morpholine acetate buffer pH 8.1. The proteinase solution (50 μl) is added to 200 μl of sample (final concentration of pronase is 100 mg ml^{-1}) and the mixture incubated at 37 °C for 48 h.
4. Following digestion, a volume of 10% (w/v) TCA equal to that of the final digestion is added and the precipitate removed by centrifugation at 13,000 × *g* for 30 min. The supernatant is extracted three times with diethyl ether and the aqueous layer evaporated under a stream of nitrogen to an appropriate volume for analysis.
5. Then the sample is solubilized in Na-citrate buffer 0.15 M at pH 2.2; 100 μl the solubilized sample is then injected onto an HPLC for ion exchange chromatographic separations of PAs and PA derivatives (on a Jasco HPLC using a column (0.4 × 8 cm) of DC-6A resin, Dionex). The anion exchanger DIONEX DC- 6A column, is set at a working temperature of 68 °C.
6. The following program of elution allows the separation of glutamyl-polyamines: Buffer E: 12 min.; Buffer F: 20 min.; Buffer A: 20 min.; Buffer B: 25 min.; Buffer C: 30 min.; Buffer D: 31 min.; Buffer E: 20 min.; Buffer F: 20 min. Flow rate: 0.5 ml/min; Fraction collected every 2 min (1 ml/tube).
7. The identity of the polyamine derivatives is determined by comparison with the corresponding retention times of glutamyl-polyamine standards prepared as reported in Table 2 [2]. Figure 6 shows an example of the separation of the Put and Spd *mono*- and *bis*- derivatives; this separation is performed after the incubation of the chloroplast extract at different pH values of the TGase assay. In these experimental conditions the *bis*-Put eluted at fraction 11, the *mono*-Put co-eluted at fraction 27–28 with *bis*-Spd and *mono*-Spd at fraction 45 (Fig. 6)
8. The identity of conjugated polyamines is determined by TLC or HPLC after release of free polyamines by acid hydrolysis (6 M HCl) of the ion exchange chromatographic fractions. (*see* **Note 5**).

3.1.4 Synthesis of Glutamyl-PAs as Reference Standards

The method to synthesize glutamyl-Put and glutamyl-Spd as product of TGase reaction, is described in Table 2. These molecules can be used as reference standards in the HPLC separation, to analyze samples tested as above reported in Subheading 3.1 (a) and/or (b) (*see* **Note 6**).

Table 2
Synthesis of glutamyl-Put and –Spd as reference standards

	Stock solutions	Final concentration	Amount (μl)	Put standard	Spd standard
1	0.5 mM Tris–HCl pH 7.4	50 mM	25	25	25
2	1.5 M NaCl	150 mM	25	25	25
3	25 mM $CaCl_2$	2.5 mM	25	25	25
4	100 mM DTT	10 mM	25	25	25
5	2% DMC 20 mg ml^{-1}	2 mg ml-1	25	25	25
6	^{3}H–put (1 μci/μl)	4 μM	10	5	–
Or	^{3}H–Spd (1 μci/μl)	4 μM	10	–	5
7*	Put 10 mM	1 mM	25 + 10 rad	–	–
Or	Spd 10 mM	1 mM	25 + 10 rad	–	–
8	GplTGase (1 mg ml^{-1})		5	5	5
9	DDW		Up to 250	115	115

Point 7 could be added to push the reaction towards the formation of *mono*-γ-glutamyl-PAs (*see* **Note 5**)

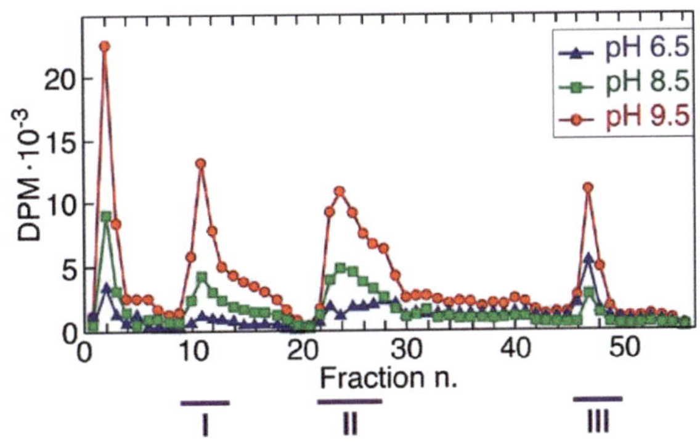

Fig. 6 Separation of glutamyl-PAs by HPLC from chloroplasts incubated with Put at different pH. Roman numbers (I, II and III) indicated the retention times range of elution for fractions of bis-γ-glutamyl Put, mono-γ-glutamyl-Put and free Put respectively. In the region II also bis-γ-glutamyl Spd could be eluted; Spd could be synthesized from Put during incubation assay. To verify if in region II is eluted mono-γ-glutamyl Put or/and bis-γ-glutamyl-Spd, it is necessary to perform an acid hydrolysis (6 M HCl) of the eluted radioactive sample in region II; then an HPLC separation of hydrolized sample will allow to identify Put or Spd

3.2 Colorimetric Assay for the Incorporation of the Biotin-Cadaverine into a Protein Substrate (Polyamine Incorporation)

1. For transglutaminase assay, the 96 wells microplate is covered for 16 h (overnight) at 4 °C with 250 μl per well of 100 mM Tris–HCl pH 8.5 containing 10 mg ml^{-1} of DMC or 5 μg ml^{-1} FN (*see* **Note 2**) well-known substrates for TGase of animal origin. After removing the protein not linked in the well, the plate is washed twice with 150 mM PBS containing 0.05% (v/v) Tween 80 and twice with ultrapure water.

2. Add 250 μl of 100 mM Tris–HCl pH 8.5 containing 3% (w/v) BSA to each well. Maintain the plate in agitation for 30 min at room temperature. This step allows an optimal saturation and, as a consequence, the reduction of the background signal.
3. The plate is washed again as above with a final washing step with 100 mM Tris–HCl pH 8.5. The last washing of each cycle is necessary to pre-equilibrate the proteins to the new assay conditions used in the following step.
4. After washing, each well is supplemented with 150 μl of 100 mM Tris–HCl pH 8.5 containing 13.3 mM DTT and 6.67 mM $CaCl_2$ or 150 μl of 100 mM Tris–HCl pH 8.5 containg 13.3 mM DTT and 20 mM EGTA for negative control. Both solutions contain 1 μl of biotin-cadaverine (1 mg ml^{-1}) per ml of buffer.
5. Then, 50 μl of the sample whose enzymatic activity needs to be tested are added in each well. This assay is necessary to check the plant TGase activity on well-known mammal TGase substrates (DMC or FN) in presence of Ca^{2+} (5 mM) or in its absence (condition obtained by 20 mM EGTA). To this aim, the same sample is assayed in triplicate (to obtain a statistical significance) in the presence or absence of calcium. A positive control is obtained incubating on DMC or on FN also the guinea pig liver (gpl) TGase at different concentrations (0.01 μg μl^{-1} e 0.02 μg μl^{-1}).
6. The activity obtained in the wells of the gpl TGase on DMC or on FN with biotin-cadaverine in presence of calcium is used to set a titration curve of the purified enzyme. On that basis, it is possible to obtain the TGase calcium-dependent activity of the sample whose putative TGase calcium-dependent activity is to be tested.
7. Plates are incubated at 37 °C for 2 h and the reaction is stopped with a washing step identical to the last described. Then, 200 μl of 100 mM Tris–HCl pH 8.5 containing 1% (w/v) BSA and extravidin peroxidase diluted 1:5000 (4 μl of extravidin peroxidase in 20 ml of buffer) are added to each well. The extravidin links to biotin residues of biotin-cavaderine amplifying the signal obtained by the conjugation of biotin-cadaverine to the glutamyl residues of protein substrate (DMC or FN used to pre-saturate the plate) catalyzed by the enzyme present in the total plant extract.
8. The plate is incubated at 37 °C for 45 min then washed as above replacing the 100 mM Tris–HCl pH 8.5 with 100 mM sodium acetate pH 6.0. The reaction is developed with 200 μl per well of 100 mM sodium acetate pH 6.0 containing 0.310 mM TMB (150 μl TMB diluted in 10 mg ml^{-1} of DMSO, in 20 ml of buffer) and 0.0045 (v/v) H_2O_2 (2.50 μl 30% H_2O_2 in 20 ml of buffer).

9. The staining development should be stopped either at 5 min or 15 min by the addition of 50 μl per well of 5 N H_2SO_4 (*see* Subheading 3.3 below). The yellow colour is relatively stable at this stage.
10. The absorbance is read at 450 nm using a microplate spectrophotometer (e.g., Titertek Multiscan ELISA). The TGase activity is obtained as ΔAbs at 450 nm, between the value obtained in presence of calcium and that obtained in presence of EGTA.
11. This value must be divided for the assay time (this represents the time the extract is in the presence of biotin-cadaverine e.g., 2 h in this procedure, not the development time) and the proteins amount present in the 50 μl of plant extract incubated in each well to obtain the value of Unit (U) of TGase specific activity.

When using this protocol consider what reported in **Note 7**.

3.3 Colorimetric Assay for the Incorporation of the Biotinylated TVQQEL Peptide into a Protein Substrate (Simulating Protein Crosslinking)

1. The method for this assay is similar to the method described above (Subheading 3.2).
2. The 96-well plates are coated with 1.0 mg ml^{-1} bovine casein and incubated overnight at 4 °C before being washed and blocked. The assay constituents include biotin TVQQEL instead of biotin-cadaverine.
3. In both the biotin-cadaverine polyamine and the biotin TVQQEL assay there is a requirement to limit the development stage. Two standard graphs should be prepared, one at a relatively high concentration of pure gplTGase (0–250 ng $well^{-1}$) developed for exactly 5 min, the other at a relatively low concentration of gplTGase (0–25 ng $well^{-1}$) developed for 15 min. The TGase used in the standard plots can be pure mammal TGase (hr or gplTGase) and the plant extracts can then be compared and quoted as mammal TGase (i.e. gplTGase) equivalents. To overcome the gradual loss in potency of stored standard mammal TGase each 96-well microplate plate should have a column with a high concentration of pure standard mammal TGase (e.g., 100 ng $well^{-1}$) and a column with a low concentration of pure standard mammal TGase (e.g., 10 ng $well^{-1}$). When the development stage of the assay is reached the buffer column, the high standard mammal TGase and the low standard mammal TGase should be developed first (the high standard for 5 min the low standard for 15 min). The plant extracts can then be developed for either 5 or 15 min depending on their potency. Comparison between extracts assayed at different times can then be adjusted depending on the potency of the standard gplTGase. This will help in the comparison of results between different microplates and different laboratories.

3.4 In vitro TGase Assay During Pollen Germination

1. Pollen was rehydrated at 30 °C and 100% relative humidity for 30 min. Pollen has been suspended in germination medium at 1 mg ml^{-1}
2. Then it was allowed to germinate into in DMC coated 96-well microtitre plates for up to 120 min, in the presence of 0.1 mM biotin-cadaverine (to check extracellular transamidase activity) supplied for the last 30 min of germination period; negative controls have been performed with TGase inhibitors (i.e., ZDON R283, iodoacetamide, cystamine) added at zero time of germination or for the last 30 min to allow pollen tube to partially grow. Pollen viability was assessed by staining with fluorescein diacetate (10 μg ml^{-1}).
3. The level of enzyme activity was expressed as Ca^{2+}-dependent increase in A_{450nm}, after subtraction of the value of the TGase inhibitor treated control. Specific activity was determined as a change in A_{450nm} of 0.1 per hour per mg of pollen protein.

3.5 TGase-Based Covalent Binding of Polyamines to Microtubules and Actin Filaments

The covalent binding assay between polyamines and tubulin/actin in the presence/absence of TGase seeks to establish whether this aspect of plant TGase activity might affect specific processes that require the full functionality of cytoskeletal proteins. In particular, here we propose to analyse their polymerization, ability to bind motor proteins, and to support the movement and activity of motor proteins. This analysis can provide information on the role of polyamines during the cellular processes that involve microtubules and actin filaments. The protocol requires an initial phase in which microtubules and actin filaments are assembled starting from the monomeric tubulin/actin, followed by the enzymatic reaction in which TGase covalently links polyamines with preformed microtubules/actin filaments. Subsequently, microtubules and actin filaments can be analyzed by means of several tests as schematically depicted in Figs. 7 and 8.

3.5.1 Method for the Polyamine-Microtubule Assay

1. Prepare TGase according to the type of enzyme you are interested in. This depends on the type of organism one is working with. If TGase comes from specific organs or tissues or cells, it must be purified according to specific protocols adapted to the particular conditions [20]. If the enzyme is purchased from companies, usually it comes in lyophilized form that must be resuspended in a minimum volume of enzyme reactivation buffer. It is advisable to leave the enzyme suspension for at least 5 min at room temperature in order to fully reactivate the enzyme. The final concentration of TGase may be critical for subsequent experiments. It is suggested to test different concentrations of enzyme in order to determine the optimum enzyme concentration for each experiment. In our hands, a working concentration of 0.01–0.1 mg ml^{-1} proved to be effective.

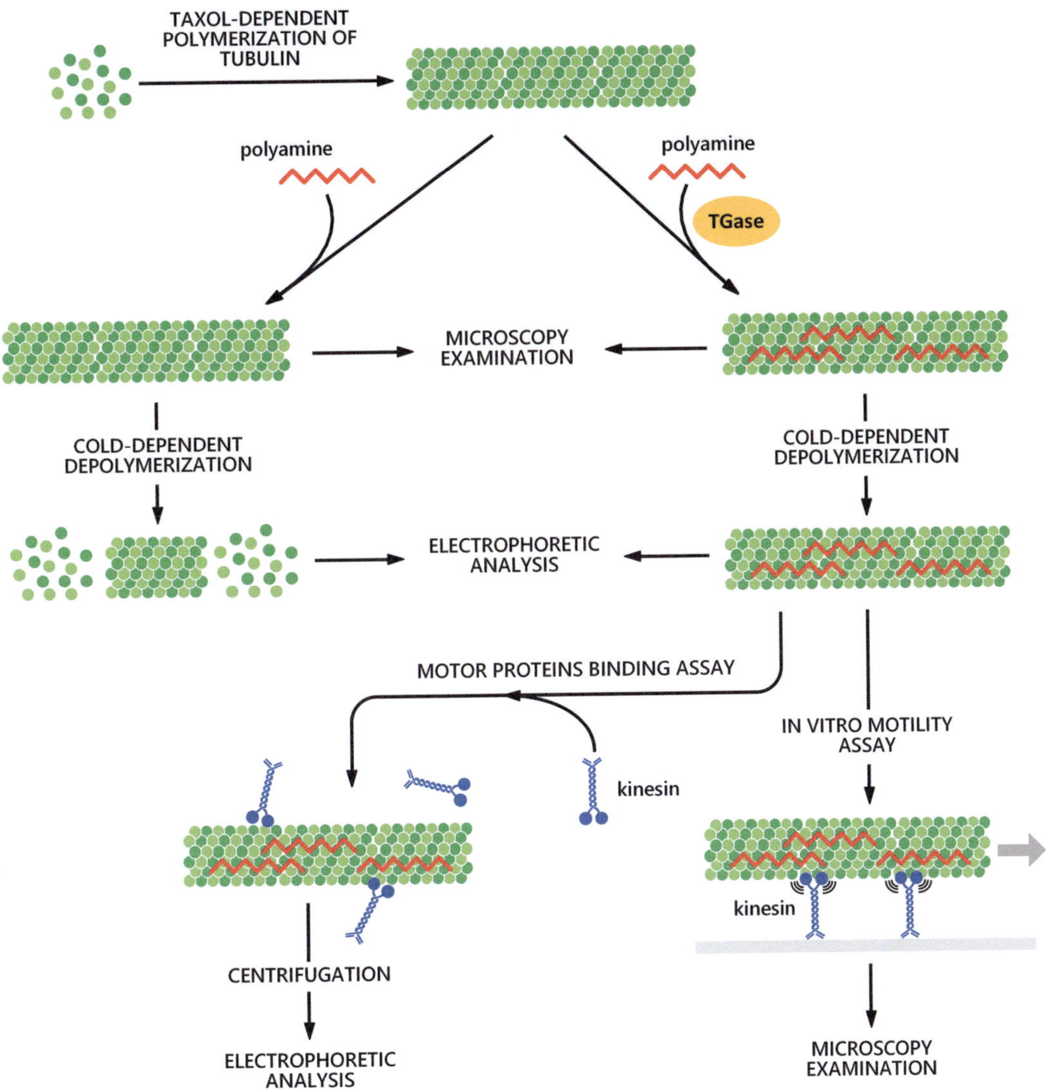

Fig. 7 Diagram of the analytical methods to test the effects of transglutaminase/polyamines on the functionality of microtubules

2. Although tubulin can be purified from different animal and plant sources [21–24], currently it can easily and cheaply be purchased from different companies. Purchased tubulin is effective and pure. Tubulin is delivered either as lyophilized or frozen at specific concentrations. Both lyophilized and frozen samples usually contain the solutes required for tubulin stability at optimal concentration. Anyway, it is suggested to check in the company website or the product brochure. For optimal stability, tubulin must be stored at −80 °C.
3. Just before starting the experiment, defrost tubulin aliquots. If tubulin is lyophilized, add the required volume of water to

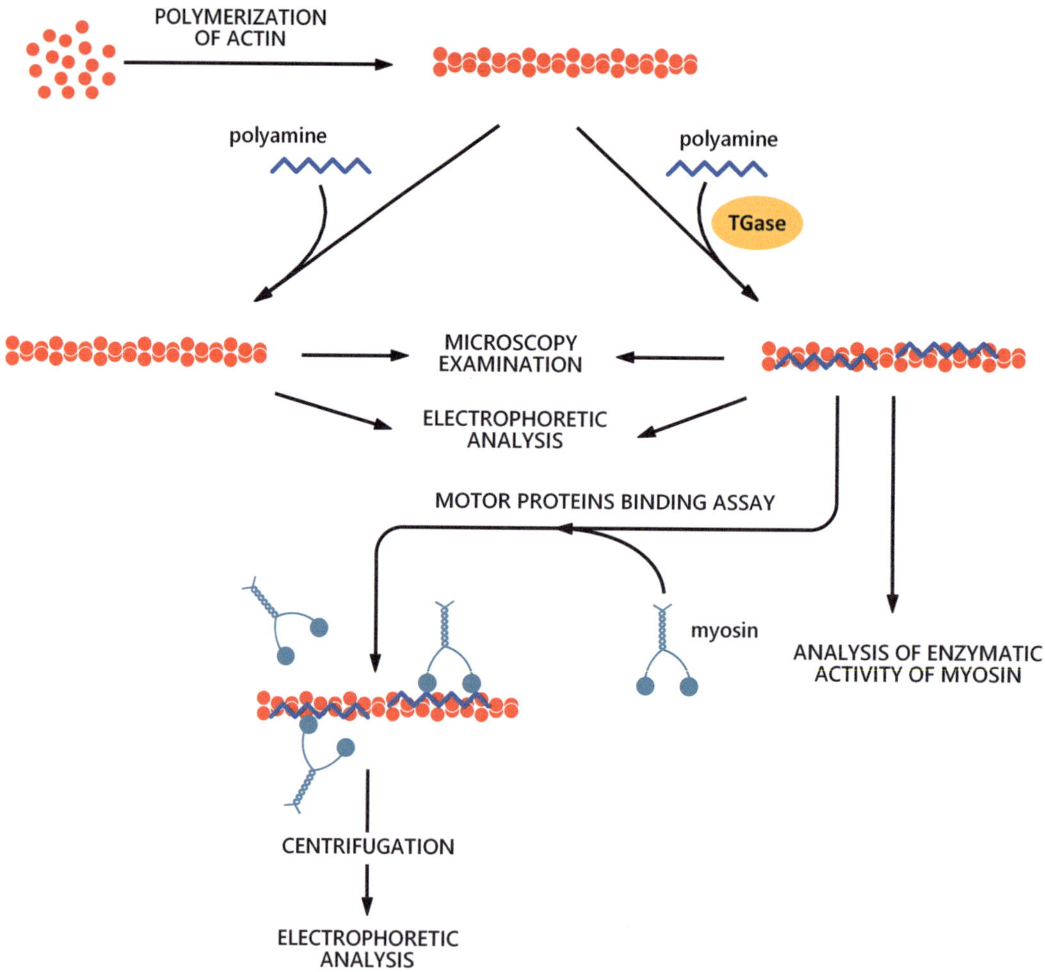

Fig. 8 Diagram of the analytical methods to test the effects of transglutaminase/polyamines on the functionality of actin filaments

adjust the protein concentration at 10 mg ml^{-1}. We usually started with 100 μl of tubulin at the concentration of 10 mg ml^{-1}. This volume and concentration is generally appropriate for a series of experiments, but the final volume must be calibrated on the specific experimental requirements. For preparation of microtubules, take 25 μl of tubulin solution (250 μg) and transfer it into a microfuge tube.

4. Add 25 μl of TDB to tubulin aliquots.The tubulin solution is now at a concentration of 5 mg ml^{-1}. Add 6.25 μl of CB into the microfuge tubes. This step initiates tubulin polymerization. In order to enhance polymerization of microtubules, incubate sample at 37 ° C for 30 min. You can do this by using a water bath set at desired temperature or more precise incubators for microfuge tubes.

5. Mix 400 μl of MDB with 32 μl of 500 μM taxol thereby making the so-called MDB-T buffer. Incubate it at 37 °C.
6. When the incubation has ended, add 400 μl of MDB-T to the microfuge tube containing newly polymerized microtubules. Mix gently and leave the samples at room temperature. Tubulin is now polymerized and at a concentration of approximately 0.5 mg ml^{-1}.
7. For the TGase-mediated incorporation of polyamines into microtubules, prepare the following reaction mixture by mixing 50 μl of microtubules (250 μg) with TGase in a ratio of 10:1 as a starting value. Please note that this ratio has to be determined experimentally on the basis of the specific TGase to be assayed. In addition, add the polyamine of your choice at a concentration of 5 mM and then add 6 mM $CaCl_2$.
8. As control, prepare a second mixture containing 50 μl of microtubule solution (250 μg), 5 mM polyamine and 6 mM $CaCl_2$.
9. Incubate both samples at 37 ° C for 2–3 h. The appropriate incubation time must be determined experimentally and this depends on the specific activity of TGase. Reactions can be stopped by adding EGTA to a final concentration of 6–7 mM. Samples can be stored at room temperature until needed.
10. For observation with a microscope, take 10 μl of TGase/polyamine-modified microtubules and perfuse it within a perfusion chamber. Wash the chamber several times with PEM buffer + taxol in order to remove unpolymerized tubulin and observe samples using a optical microscope with Köhler illumination and a DIC filter (for proper settings, you can refer to the instruction manual of your own microscope) (*see* **Note 8**). You will need specific software or hardware to remove background noise and to visualize microtubules in the perfusion chamber. For a basic idea of the required system, *see* Fig. 9. You can refer to the literature for further information on the assembly of an optical/computer system suitable for visualization of microtubules [25]. In control condition, microtubules are usually detected as single structures while in TGase/polyamine-treated samples microtubules appear as large bundles or aggregates.
11. To visualize the occurrence of TGase/polyamine-dependent aggregates, centrifuge samples for 60 min at 13,000 × *g* in a microfuge at 25 °C. Both pellets and supernatants have to be processed for SDS-PAGE analysis. The pellet can be resuspended in a small volume of PEM buffer and then added to appropriate volume of PAGE denaturing buffer [18]. Alternatively, the pellet can be directly dissolved in the denaturing buffer. The supernatant can be directly added to appropriate volumes of denaturing buffer (*see* **Note 9**). At this step, it is

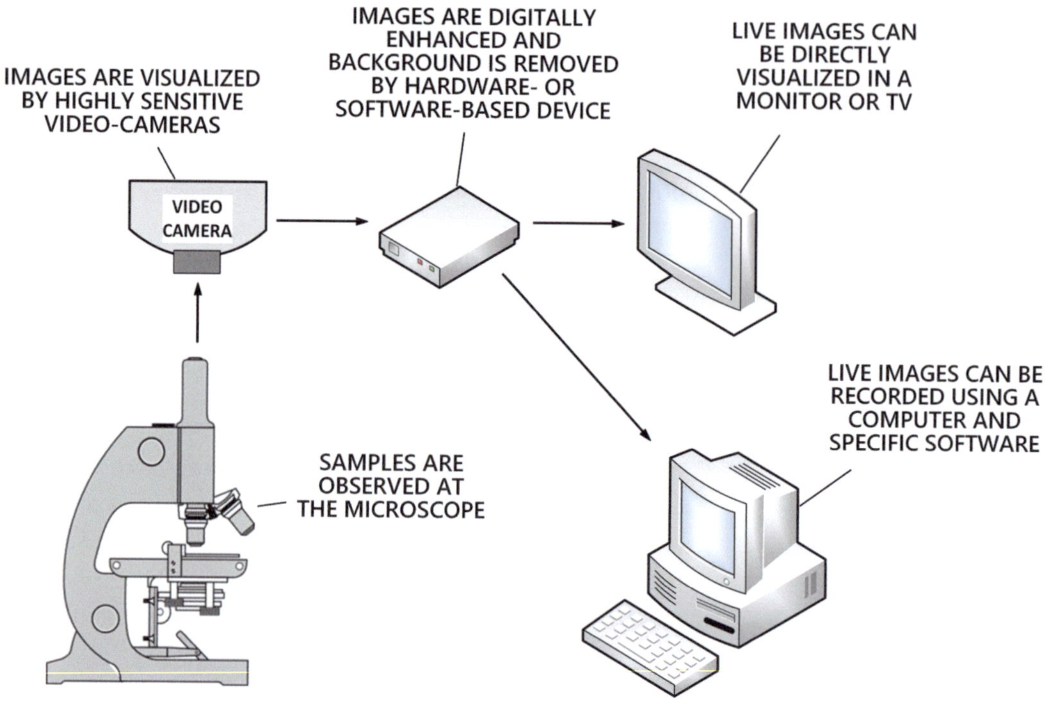

Fig. 9 Basic components of a system for viewing and recording of microtubules in a perfusion chamber. What is represented here is a relatively simple system that can be improved according to the technology offered by various companies

critical to maintain the ratio between pellets and supernatants in order to analyze corresponding volumes. Otherwise, the protein concentration has to be determined using the BCA assay [17] or commercial kits.

12. To test for temperature-dependent depolymerization, aliquots of polyamine-modified microtubules and controls can be incubated for 30 min at 4 °C. This step can be performed using a refrigerators or on ice. While this is happening acclimatize the microfuge to 4 °C.
13. At the end of incubation, proceed to the centrifugation step by centrifuging the sample at 13,000 × *g* in a microfuge for 60 min at 4 °C. Both the supernatant and pellet can be processed for SDS-PAGE by diluting with PAGE denaturing buffer (*see* **Note 10**). Again, it is important to maintain the volume ratio between samples.
14. To perform in vitro motility assay, the TGase/polyamine microtubules can be used together with a specific motor protein, such as kinesin. Purified motor proteins can be purchased from different companies. Alternatively, they can be purified from various animal or plant sources [26–28]. It is advisable to buy the purified proteins because they are normally provided at

desired working concentrations and stability. The concentration of the motor protein is not critical at this stage. However, it is important that it is maintained at a constant concentration throughout the experiments. Motor proteins purchased from a supplier are usually delivered as lyophilized proteins and must be reconstituted by adding specific volumes of water.

15. Perfuse 10 μl of the motor protein solution into a perfusion chamber, then incubate the glass-slide into a moist perfusion chamber for 5 min to allow proteins to adhere.
16. Wash out unbound motor proteins by perfusing 2–3 times with 10–20 μl of blocking buffer. This step is also used to block sites in the perfusion chamber (perform all these steps while the perfusion chamber is placed within a moist chamber in order to prevent dehydration of samples).
17. Dilute 10 μl of microtubule solution (treated or untreated with polyamines) with 40 μl of motility buffer. Add 10 μl of microtubule solution into the perfusion chamber and incubate for 1 min at room temperature.
18. Rapidly move to the microscope, which should be set up in advance. Observe the samples under Köhler illumination and using DIC filters. A 63× or 100× objective is required for optimum viewing, as well as a hardware or software device to reduce background noise (Fig. 9). A video camera is also required. Signals acquired by the video camera must be delivered to a video acquisition system, which allows recording of live video clips using dedicated software. The type of software and hardware used for video acquisition might be dependent on the model of video camera (*see* **Note 11**). It is advisable to acquire video clips at the highest video frame and at the highest frame per second.
19. The velocity of microtubule movement can be analyzed by specific software. We recommend ImageJ as a free alternative (https://imagej.nih.gov/), with the plugin "MTrackJ". The analysis allows comparison of the speed of microtubules under different experimental conditions.
20. For the microtubule binding assay with motor (or non-motor) proteins, the first step is the preparation of the test protein. Again, it is suggested to purchase purified proteins because of their standard concentration and stability. The test can be used to analyze the binding of a specific protein or of a mixture of proteins present in a specific cellular compartment. In the test, we suggest to use fixed volumes of one component (e.g., microtubules) and to vary the volume or concentration of the binding protein(s). The reaction volume can be made equivalent by the addition of the assay buffer.

21. Prepare several tubes containing 30 μl of microtubule solution (treated or untreated with polyamines) then add 2, 5, 10, 20 μl of the binding protein (as an example). Set all reaction volumes to 50 μl. If the assay involves a motor protein in the binding test, the reaction volume should also contain 10 mM ATP or 5–10 mM AMPPNP to release or enhance the binding of motor proteins to the microtubules [29].
22. Samples can be incubated for 30 min at 25 °C, then they can be centrifuged at 13,000 × *g* for 60 min at 25 °C. Both pellets and supernatants can be analyzed by SDS-PAGE .
23. For a quantitative analysis of the binding reaction, gels can be scanned and their images analyzed with dedicated software. Different equipment and software for this type of analysis are commercially available. In practical terms, this allows converting the protein band(s) of interest into a numerical values and therefore a comparable value.

3.5.2 Method for the Polyamine-Actin Assay

1. Reactivation of TGase: for procedural details, you can refer to what already stated in paragraph 1 of the previous protocol concerning the microtubule assay. Briefly, resuspend the enzyme in a minimum volume of reactivation buffer and let stand at room temperature for 5 min.
2. Thaw 1 aliquot of actin (optimum 1 mg) and dilute the actin to a concentration of 10 mg ml^{-1} by the addition of 100 μl of A-buffer. Dilute the actin to a concentration of 0.4 mg ml^{-1} with A-buffer by mixing 100 μl of actin solution with 2.4 ml of A-buffer. Incubate on ice for 1 h. Subsequently, add the polymerization inducing buffer (Polymerization Buffer 1× at final dilution) in the following ratio: 2.25 ml of actin plus 0.25 ml of 10× Polymerization Buffer. Incubate at room temperature for 1 h. The actin filaments are ready to use and can be stored at room temperature until use. It is suggested that the actin should be purchased from companies; obviously, actin can be purified from plant and animal sources and from tissues/organs selected by the operator [30].
3. The reaction mixture can be prepared as follows. Mix 625 μl of actin (≈250 μg) with TGase so that the ratio is around 10:1 (again, it is necessary to mention that the exact ratio must be determined experimentally and according to the specific conditions of the operator). Add the polyamine at the concentration of 1 mM and $CaCl_2$ at 6 mM. In parallel test tubes, mix exactly the same components but without TGase.
4. Incubate all samples at 37 °C for 2–3 h.
5. Stop the reaction by adding EGTA to the final concentration of 7 mM.

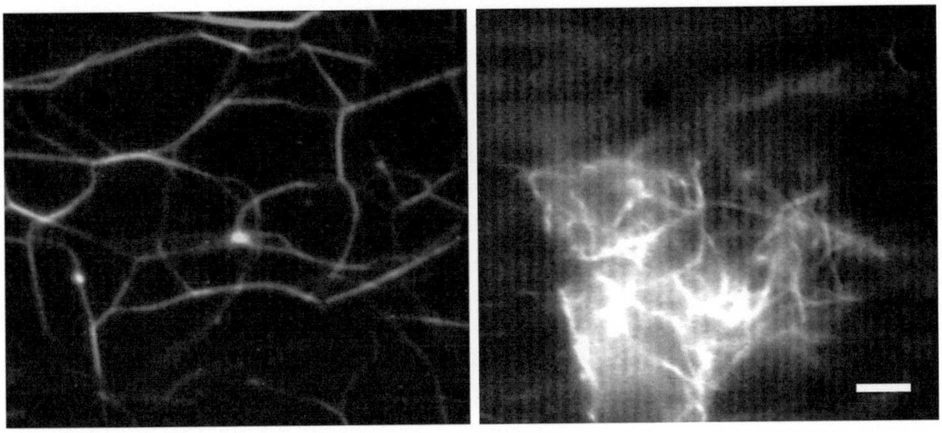

Fig. 10 Rhodamine-labelled F-actin obtained in the presence of plant TGase and Put (viewed in the rhodamine filter). On the left, Rhodamine-labelled F-actin obtained in the absence of plant TGase; on the right, Rhodamine-labelled F-actin obtained after TGase and Put treatment. Scale bar: 5 μm

6. For the observation at the microscope, perfuse 10 μl of samples within individual perfusion chambers (*see* above for their preparation) and incubate for 5 min in a moist chamber.
7. To visualize actin filaments, wash the perfusion chambers with 10–15 μl of the rhodamine-phalloidin solution to label actin filaments. After 5 min incubation, wash unbound labeling molecules and observe the samples using a fluorescent microscope (rhodamine filter). Usually, the sample incubated with TGase and polyamines shows altered actin configuration with the frequent occurrence of amorphous aggregates (Fig. 10).
8. In order to determine if TGase/polyamine alters the polymerization status of actin filaments, samples can be centrifuged in a microfuge at 13,000 × *g* for 60 min at 25 °C. Both supernatants and pellets can be processed for SDS-PAGE analysis. Again, please take care to maintaining the exact volume ratio between pellets and supernatants in order to evaluate the results correctly.
9. As stated before, gels can be scanned and images can be analyzed for densitometric values in order to evaluate statistically the percentage of actin between pellets and supernatants.
10. The binding assay to motor proteins can be performed using myosin. Actin-based motor proteins can be purchased from different companies at very good quality. Therefore, to increase reproducibility and interpretation of results, we suggest to use commercially available proteins. If myosin is delivered as lyophilized, resuspend the protein in a suitable buffer (4 mM Tris–HCl pH 7.6, 2 mM $MgCl_2$, 100 mM KCl) in order to have a concentration of about 0.5–1 mg ml^{-1}.

11. Add 10 μl of the myosin solution to 100 μl of actin filaments and incubate for 30 min at 25 °C. Centrifuge samples at 13,000 × g for 60 min at 25 °C. Both supernatants and pellets can be denatured for SDS-PAGE analysis. Again, the binding of myosin to actin filaments can be quantified by gel scanning and densitometric evaluation of the band corresponding to the myosin heavy chain (usually around 170–200 kD).
12. The activity of myosin can also be evaluated by using an ATPase assay [31]. For this analysis, one can use different volumes of the actin filament solution as prepared before. For example, 10, 20 or 50 μl of actin filaments (0.4 mg/ml, either treated or untreated with TGase/polyamines) can be mixed with 4 μg of myosin; the ATPase assay buffer can be used to bring the final volume to 100 μl.
13. Samples are incubated at 25 °C for 30 min and the reaction stopped by putting samples on ice.
14. Determine the amount of released inorganic phosphate through one of the several phosphate detection kits commercially available.

3.6 Method for the Proteolytic Digestion of Plant Proteins for the Analysis of ε-(γ-Glutamyl) Lysine Isodipeptide

1. Plant material was homogenized and protein was precipitated from the 80,000 × g supernatant by the addition of an appropriate volume of 72% (w/v) TCA to give a final concentration of 7.2% (w/v).
2. The protein was pelleted by centrifugation at 13,000 × g for 30 min at 4°C.
3. The pellet was washed once in 10% (w/v) TCA, three times in 50:50 ethanol:diethylether and three times in diethylether (each washing was followed by centrifugation at 13,000 × g for 30 min).
4. The pellets were dried by evaporation and re-dissolved in 1.0 ml of 100 mM ammonium carbonate pH 10.0. The pH of each tube was checked with pH indicator paper and adjusted to 10.0 by the addition of sodium hydroxide if necessary.
5. Subtilisin (0.1 mg) was added and incubated 16 h at 37°C The addition of subtilisin was repeated twice more over 48 h.
6. 150 μg of pronase was added to each tube as 10μl of a 15 mg ml^{-1} solution. Following a further 16 h incubation, each extract was boiled for 15 min.
7. 10 μl of activated leucine amino peptidase and activated prolidase solution were added along with a 10 μl aliquot of 0.54 M magnesium chloride. The tubes were vortexed and incubated for 24 h.
8. The pH of each digestion was then adjusted to 7.0 with 1.0 M HCl (using pH indicator paper) and 0.1 mg of

carboxypeptidase Y was added as 10 μl of a 10 mg ml^{-1} solution. Following a further 16 h incubation.

9. The digested protein solutions were placed into conical glass centrifuge tubes and added to chloroform:methanol:HCl (200:100:2) at the level of 3.6 ml to 1.0 ml of protein digestion. A further 1.2 ml of chloroform was added and the tube was vortex mixed. The tubes were centrifuged at 2500 × *g* for 5 min at 4°C. The upper methanol layer containing the liberated TG crosslinked peptide plus amino acids and the central interface containing undigested protein were carefully removed and dried.
10. The methanol phase was re-dissolved in an appropriate volume of amino acid analysis loading buffer and stored at −20°C prior to subsequent amino acid analysis using a cation exchange amino acid analyser (Biochrom UK).
11. The undigested protein was retained and quantified using the a modified BCA assay [17] to determine the efficiency of the hydrolysis.

4 Notes

1. Polyamines can be labelled in the carbon [^{14}C]-PA or in the hydrogen [^{3}H]-PA of the molecule. According to the target of the experiment, the type and position of the label might be critical. For example, if PAs are oxidized by the cell enzymes, tritium can be removed; but also carbon, if located in trimethylene moyety of Spm or Spd, can be lost if the molecule is catabolized to the respective lower PA (Spd and Put). Thus, the PA used should have the labelled carbon located in the tetramethylene part of the backbone. Alternatively, the molecule can be uniformely labelled.
2. Different protein substrates can be used to check plant TGase activity; among them the most common are: soy globulins, FN, rubisco, LHCII, actin and tubulin.
3. Plant TGase is a Ca^{2+}-dependent enzyme and thus it is inhibited by EGTA.
4. The pigments (i.e., chlorophylls) can be removed from the pellet by washing with ethanol/diethyl-ether 1:1 to remove TCA and 80% (v/v) acetone to remove chlorophylls; however other molecules to which PAs can be "glued", might be present, severely affecting the value of results.
5. *Advantages of the radioactive method.* This method has the advantage to supply the underivatized PA thus without introducing additional groups (for e.g., fluorescent groups) that can modify the hydrophylicity and the steric hindrance of the

molecule disturbing the reaction. In addition, both the terminal amino-groups are free and consequently the PA is capable to be conjugated to the glutamyl residues of a protein by only one or both amine groups, giving rise respectively to *mono*-glutamyl- or *bis*-glutamyl –PA (Fig. 2). Spd is the only asymmetric PA and thus it can give rise to two different *mono*-derivatives according if $-NH_3^+$ in position 1 or 8 is engaged. *Disadvantages of the radioactive method.* The entire procedure is very time-consuming. This method does not reveal the catalysis of glutamyl-lysine conjugates occurring via TGase reaction.

6. This method allows to synthesize either *mono*-glutamyl- or *bis*-glutamyl–PAs. To push the synthesis of *mono*-glutamyl-PA it is necessary to increase the concentration of the PA by increasing the cold reagent. Otherwise if it is necessary to synthesize preferably the *bis*-glutamyl –PA, keep the concentration of cold PA very low or do not add the cold PA into the assay.
7. Remember to warm all the reagents to 37 °C before they are used in a TGase microplate assay. Add the reducing agent (e.g., 2-mercaptoethanol or DTT) immediately before use. Close the plate with an adhesive microplate seal after the reagents have been added to the assay. During the development stage of the assay with extravidin peroxidase use exact time intervals e.g., 5 or 15 min.
8. Optical microscopy analysis of microtubules is not an easy technique to perform. It requires a very good system of optical microscopy, specifically equipped with a high-quality 63× or 100× oil-immersion objective. It also requires the expertise of trained staff. For a detailed description of the method for microscopy observation, please refer to the following textbook [25].
9. The electrophoretic analysis can be performed with standard equipments readily available in a biochemical laboratory. Generally, a 6–20% range of polyacrylamide or a linear 8% gel are generally appropriate for the analysis and visualization of TGase-dependent aggregates.
10. After staining gels with either Coomassie Blue or silver staining, the gel lanes containing the control sample and the sample treated with TGase can be compared to detect the protein complexes generated by the enzymatic reaction. The analysis may be simply visual and then descriptive, or the gel may be scanned and analyzed by specific software in order to obtain a more accurate quantitative assessment.
11. The recording of a visual field containing stationary or moving microtubules requires an adequate capturing/recording system. Nowadays, one can record images directly on a computer using video capture systems. Current videocameras directly convert the signal so that it can be immediately read and stored

as a video file using a specific software. It is important to capture the video file to the highest possible resolution and with the highest number of frames per second. The storage format can be proprietary or generic; in this case, a particular video compression codec must be installed on the computer.

References

1. Folk JE, Park MH, Chung SI, Schrode J, Lester EP, Cooper HL (1980) Polyamines as physiological substrates for transglutaminases. J Biol Chem 255:3695–3700
2. Beninati S, Martinet N, Folk JE (1988) High-performance liquid chromatographic method for the determination of ε-(γ-glutamyl)lysine and *mono*- and *bis*-γ-glutamyl derivatives of putrescine and spermidine. J Chromatogr A 443:329–335
3. Siepaio MP, Meunier JCF (1995) Diamine oxidase and transglutaminase activities in white lupin seedlings with respect to cross-linking of proteins. J Agric Food Chem 43:1151–1156
4. Lilley GR, Skill J, Griffin M, Bonner PL (1998) Detection of Ca^{2+}-dependent transglutaminase activity in root and leaf tissue of monocotyledonous and dicotyledonous plants. Plant Physiol 117:1115–1123
5. Slaughter TF, Achyuthan KE, Lai T, Greenberg CS (1992) A microtiter plate transglutaminase assay utilising 5-(Biotinamido) pentylamine as substrate. Anal Biochem 205:166–171
6. Del Duca S, Beninati S, Serafini-Fracassini D (1995) Polyamines in chloroplasts: identification of their glutamyl and acetyl derivatives. Biochem J 305:233–237
7. Del Duca S, Tidu V, Bassi R, Serafini-Fracassini D, Esposito C (1994) Identification of transglutaminase activity and its substrates in isolated chloroplast of *Helianthus tuberosus*. Planta 193:283–289
8. Della Mea M, Di Sandro A, Dondini L, Del Duca S, Vantini F, Bergamini C, Bassi R, Serafini-Fracassini D (2004) A *Zea mays* 39 kDa thylakoid transglutaminase catalyses light harvesting complex II by polyamines in a light-dependent way. Planta 219:754–764
9. Della Mea M, Serafini-Fracassini D, Del Duca S (2007) Programmed cell death: similarities and differences in animals and plants. A flower paradigm. Amino Acids 33:395–404
10. Lorand L, Graham RM (2003) Transglutaminases: crosslinking enzymes with pleiotropic functions. Nat Rev Mol Cell Biol 4:140–156
11. Aloisi I, Cai G, Serafini-Fracassini D, Del Duca S (2016a) Polyamines in pollen: from microsporogenesis to fertilization. Front Plant Sci 7:155
12. Aloisi I, Cai G, Serafini-Fracassini D, Del Duca S (2016b) Transglutaminase as polyamine mediator in plant growth and differentiation. Amino Acids 48:2467–2478
13. Signorini M, Beninati S, Bergamini C (1991) Identification of transglutaminase activity in the leaves of silver beet (*Beta vulgaris* L.) J Plant Physiol 137:547–552
14. Trigwell SM, Lynch PT, Griffin M, Hargreaves AJ, Bonner PL (2004) An improved colorimetric assay for the measurement of transglutaminase (type II) –(gamma-gltamyl) lysine cross-linking activity. Anal Biochem 330:164–166
15. Bonner PL (2007) Protein purification: the basics. Taylor and Francis, Abingdon,UK
16. Di Sandro A, Del Duca S, Verderio E, Hargreaves AJ, Scarpellini A, Cai G, Cresti M, Faleri C, Iorio RA, Hirose S, Furutani Y, Coutts IGC, Griffin M, Bonner PLR, Serafini-Fracassini D (2010) An extracellular transglutaminase is required for apple pollen tube growth. Biochem J 429:261–271
17. Brown RE, Jarvis KL, Hyland KJ (1989) Protein measurement using bicinchoninic acid: elimination of interfering substances. AnalBiochem 180:136–139
18. Laemmli UK (1970) Cleavage of structural proteins during the assembly of the head of bacteriophage T4. Nature 227:680–685
19. Towbin H, Staehelin T, Gordon J (1979) Electrophoretic transfer of proteins from polyacrylamide gels to nitrocellulose sheets: procedure and some applications. Proc.Natl.Acad.Sci. USA 76:4350–4354
20. Del Duca S, Serafini-Fracassini D, Bonner PL, Cresti M, Cai G (2009) Effects of post-translational modifications catalyzed by pollen transglutaminase on the functional properties of microtubules and actin filaments. Biochem J 418:651–664
21. Agustin C, Maria EC, Carlos AA (2013) A novel method for purification of polymerizable

tubulin with a high content of the acetylated isotype. Biochem J 449:643–648

22. Moore RC, Zhang M, Cassimeris L, Cyr RJ (1997) *In vitro* assembled plant microtubules exhibit a high state of dynamic instability. Cell Motil Cytoskeleton 38:278–286
23. Vallee RB (1982) A taxol-dependent procedure for the isolation of microtubules and microtubule-associated proteins (MAPs). J Cell Biol 92:435–442
24. Williams RC Jr, Lee JC (1982) Preparation of tubulin from brain. Methods Enzymol 85(Pt B):376–385
25. Carter NJ, Cross RA (2010) Microtubule motility assays. In: Kreitzer G, Jualin F, Espenel C (eds) Cel biology assays. Academic Press, Amsterdam, Boston
26. Meyer D, Rines DR, Kashina AS, Cole DG, Scholey JM (1998) Purification of novel kinesins from embryonic systems. Methods Enzymol 298:133–154
27. Cole DG, Scholey JM (1995) Purification of kinesin-related protein complexes from eggs and embryos. Biophysical J 68:158s–162s
28. Wagner MC, Pfister KK, Bloom GS, Brady ST (1989) Copurification of kinesin polypeptides with microtubule- stimulated mg-ATPase activity and kinetic analysis of enzymatic properties. Cell Motil Cytoskel 12:195–215
29. Ma YZ, Taylor EW (1997) Interacting head mechanism of microtubule-kinesin ATPase. J Biol Chem 272:724–730
30. Liu X, Yen L-F (1992) Purification and characterization of actin from maize pollen. Plant Physiol 99:1151–1155
31. Homma K, Saito J, Ikebe R, Ikebe M (2001) Motor function and regulation of myosin X. J Biol Chem 276:34348–34354

Chapter 19

Procedures for ADC Immunoblotting and Immunolocalization for Transmission Electron Microscopy During Organogenic Nodule Formation in Hop

Ana Margarida Fortes and Jose M. Seguí-Simarro

Abstract

Immunolocalization for transmission electron microscopy is a powerful technique to identify subcellular localization of proteins. This can be combined with molecular and physiological data in order to have a complete overview of protein function. However, optimal sample preservation is required to avoid artefacts. When using chemically fixed samples, the progressive lowering of temperature (PLT) technique is a convenient procedure to dehydrate and embed samples at low temperature, thereby preserving the antigenicity of the proteins to be detected. Despite the advantages of immunogold labelling, it is a time-consuming cell biology technique. Therefore, the quality and specificity of the antibody should be previously checked by western blot. This approach also enables to identify changes in the amount of protein under study throughout development or in response to stress conditions.

Key words Arginine decarboxylase, Hop, Immunoblotting, Immunolocalization, Transmission electron microscopy

1 Introduction

Polyamines have been implicated in a wide range of biological processes, including cell growth and division, biotic and abiotic stresses responses, morphogenesis and fruit development and ripening [1–5]. Arginine decarboxylase (ADC; EC 4.1.1.19) is one of the main enzymes controlling the biosynthesis of polyamines in plants. In a previous work, we examined the accumulation and differential subcellular location of ADC during prenodular and nodular formation in *Humulus lupulus* [3].

Hop has a long traditional use in brewing industry and an efficient tissue culture protocol for shoot regeneration through organogenic nodule formation was established for this plant [6, 7]. ADC accumulation throughout organogenic nodule formation was first examined by immunoblot. This procedure enables to evaluate the specificity of antibody before the time-consuming

Rubén Alcázar and Antonio F. Tiburcio (eds.), *Polyamines: Methods and Protocols*, Methods in Molecular Biology, vol. 1694, DOI 10.1007/978-1-4939-7398-9_19, © Springer Science+Business Media LLC 2018

immunogold labelling is carried out. Briefly, proteins are extracted and subjected to electrophoresis on a discontinuous gradient of SDS-polyacrylamide. After electrophoresis, proteins are transferred to appropriate membranes, which are incubated with the primary antibody raised against ADC and then incubated with a secondary antibody before immunodetection. One can check if unspecific bands appear or the specific signal intensity is sufficiently stronger than the background.

Following this validation, immunogold labelling can be performed to evaluate cell-specific locations of the protein and, in this way, postulate about their putative functions. Essentially, samples are fixed overnight in cold paraformaldehyde which preserves some antigenicity while preserving cell structure. Then, samples are dehydrated and embedded in appropriate resin. Semithin sections should be first obtained for preliminary histological analysis. Grids with ultrathin sections are then incubated with the same anti-ADC antibody used for the immunoblot analyses and with the secondary antibody conjugated with colloidal gold. Finally, sections are stained and can then be imaged, studied and quantified. Controls should be included in parallel, either by excluding the anti-ADC antibody or better substituting it by preimmune serum. In this chapter, we describe in detail how to perform western blotting, immunolocalization and quantification of ADC by electron microscopy in hop. This protocol is potentially applicable to the localization of ADC in other species and tissues, provided that specific parameters such as fixation times and concentrations of the anti-ADC antibody, among others, are specifically adjusted for each tissue and species.

2 Materials

This step-wise protocol was developed with internodes from in vitro micropropagated hop (*Humulus lupulus* var. Nugget) plants [6].

2.1 Solutions and Media for In Vitro Culture

1. Micropropagation media: Adams medium [8] supplemented with 5 mg/l ascorbic acid, 0.2 mg/l 6-Benzylaminopurine (BAP), 1 mg/l Indole-3-butyric acid (IBA), 20 g/l glucose and 7.8 g/l agar.
2. Organogenic nodule formation media: MS solid medium [9] supplemented with 30 mg/l cysteine, 2 mg/l BAP, 0.05 mg/l Indole-3-acetic acid (IAA), 18 g/l sucrose, and 7.8 g/l agar.

2.2 Reagents for SDS PAGE and Immunoblotting

Prepare all solutions using ultrapure water (prepared by purifying deionized water, to attain a sensitivity of 18 MΩ-cm at 25 °C) and analytical-grade reagents.

1. Extraction buffer: 20 mM Tris buffer pH 8.6, 6% β-mercaptoethanol, 1% SDS [10].
2. Miracloth (Calbiochem).
3. Acetone (Merck).
4. Loading buffer (50% glycerol, 0, 15% bromophenol blue).
5. Prestained standard proteins.
6. Solution for precipitated proteins: 100 mM Tris–HCl (pH 8), 10% SDS, 5% β-mercaptoethanol.
7. Bradford solution.
8. Resolving gel 12%: for 5 ml add 2 ml of 30% acrylamide, 1.25 resolving buffer (1, 5 M Tris–HCl pH 8.9), 1.675 ml of autoclaved water, 25 μl of 10% ammonium persulfate, 50 μl of 10% SDS and 3 μl TEMED.
9. Stacking gel ~6%: for 5 ml add 0.97 ml of 30% acrylamide, 1.25 of separation buffer (0.5 M Tris–HCl pH 7.0), 1.25 ml of autoclaved water, 25 μl of 10% ammonium persulfate, 50 μl of 10% SDS and 5 μl TEMED.
10. Minigel protein electrophoresis system.
11. Electrophoresis buffer: 25 mM Tris, 192 mM glycine, 0.1% SDS pH 8.3.
12. Trans-blot SD (semi-dry) transfer cell.
13. Transfer buffer: 25 mM Tris base, 192 mM glycine, 20% methanol.
14. PDVF membranes.
15. Phosphate-buffered saline (PBS 1×): 137 mM NaCl, 2.7 mM KCl, 10 mM Na_2HPO_4, 1.8 mM KH_2PO_4.
16. Blocking buffer: 2% powdered skimmed milk, 0.05% Tween-20 in PBS.
17. Washing buffer: 0, 2% powdered skimmed milk, 0.5% Tween-20 in PBS.
18. Anti-ADC antibody: a rabbit polyclonal antibody raised against tobacco ADC [11].
19. Alkaline phosphatase-conjugated anti-rabbit IgG.
20. Nitro blue tetrazolium (NBT)/5-bromo-4-chloro-3- indolyl phosphate (BCIP) mixture: 10 μg/ml NBT, 5 μg/ml BCIP, 100 mM $NaHCO_3$, 50 mM Na_2CO_3, 39,4 mM $MgCl_2.6H2O$.

2.3 Sample Processing for Electron Microscopy

1. Distilled water (dH_2O).
2. 4% formaldehyde in PBS (*see* **Note 1**).
3. Methanol EM grade.

4. A dissecting microscope.
5. Sterile lancet, knife and toothpicks.
6. Whatman filter paper.
7. Lowicryl K4M resin kit (Electron Microscopy Sciences, Hatfield, PA) (*see* **Note 2**).
8. UV lamp.

2.4 Sectioning and Section Staining

1. Ultramicrotome (Leica UC6, from Leica Microsystems, Vienna, Austria or similar).
2. Formvar and carbon-coated, nickel or gold EM grids (Electron Microscopy Sciences, Hatfield, PA).
3. Diamond knife (Ultra 35° from Diatome, Biel, Switzerland or similar).
4. 5% (w/v) aqueous uranyl acetate solution, pH 4–5. In a fumehood, dissolve uranyl acetate powder in dH_2O. Adjust pH with HCl. To remove precipitates, filter the solution with a 0,2 μm PTFE filter.
5. Reynolds lead citrate solution (*see* **Note 3**).

2.5 Immunogold Labeling

1. Parafilm® Laboratory Film, Bemis Flexible Packaging, Oshkosh, WI, USA.
2. Bovine serum albumin (BSA) dissolved 5% and 1% in PBS.
3. Anti-ADC antibody: a rabbit polyclonal antibody raised against tobacco ADC [11].
4. Goat anti-rabbit IgG conjugated to 10-nm colloidal gold (BBI Solutions, Cardiff, UK).

3 Methods

3.1 Plant Growth Conditions

Micropropagation of hop plants:

1. Perform every 2–3 months to glass culture tubes (13, 5 × 3 cm) with approximately 15 ml of micropropagation media per tube [6].
2. Keep at 25 °C ± 2 °C with a 16-h photoperiod (35 μmol photons·m^{-2} s^{-1}).

3.2 Internode Isolation and Induction to Organogenesis

1. Excise 6–9 mm long internodes from donor plants, wound the internodes by making 3–5 incisions throughout their length using a razor blade (wounding treatment), and inoculate them in culture flasks (9 × 8 cm) containing approximately 40 ml of culture media [6].
2. Incubate cultures at 25 °C ± 2 °C with a 16-h photoperiod (35 μmol photons·m^{-2} s^{-1}).

3. Collect material for immunoblotting and immunogold labelling experiments at the following stages:
   - (a) Stage zero: internodes at the time of excision from the donor plant.
   - (b) 12, 24 and 48 h after internode inoculation.
   - (c) 4 and 7 days after internode inoculation. At this stage, divisions in cambial and cortical cells of internodal explants can be observed.
   - (d) 15 days after internode inoculation. At this stage, several prenodular structures are formed inside the callus.
   - (e) 28 days after culture initiation when organogenic nodules are formed.
   - (f) 45 days after internode inoculation. At this stage, plantlet regeneration from organogenic nodules takes place.

3.3 Immunoblotting

1. For sample preparation, freeze samples in liquid nitrogen. Samples can be stored at this point for further use as long as they are kept at −80 °C.
2. Powder samples with a pestle and mortar and add the extraction buffer before samples unfreeze (~3–4 ml per g fresh weight). Then homogenize them in the extraction buffer.
3. Centrifuge the extract for 10 min at 10,000 × *g* and 4 °C, and filter the supernatant through Miracloth.
4. Estimate proteins using the Bradford method and BSA as standard [12].
5. Precipitate proteins overnight with four volumes of acetone (Merck) at 4 °C. Prepare proteins (30 μg per sample) in duplicate to be used for immunoblotting control. The control can also be performed in separate days.
6. Concentrate the protein by centrifugation at 15,000 × *g* for 15 min at 4 °C. Vacuum dry the pellet and resuspend it in 100 mM Tris–HCl pH 8.0, 10% SDS, 5% β-mercaptoethanol.
7. Assemble the electrophoresis cell according to the instructions provided by the manufacturer.
8. Prepare 2 × 5 ml of resolving gel, transfer it to the cell, gently overlay water or isobutanol and let it polymerize.
9. Prepare 2 × 5 ml of stacking gel. Place the comb and add water or isobutanol to the wells so they become uniformed. Avoid air bubbles.
10. Boil the samples for 5 min following the addition of 1/5 volume of loading buffer. Include pre-stained standard proteins. Centrifuge at 13,000 × *g* for 5 min at 4 °C (*see* **Note 4**). The denatured proteins (30 μg per lane) are now ready to be subjected to electrophoresis.

11. Carefully pipette the samples (*see* **Note 5**). Electrophoresis will be run on minigel system using Tris/Glycine/SDS buffer pH 8.3. The run time represents the time required for the dye front to reach the line at the bottom of the cassette. Immediately following SDS-PAGE turn off the power supply.
12. Cut the membranes to the size of the gels and immerse in methanol. Rinse twice in distilled water and once with transfer buffer.
13. Immediately assemble the transfer system according to the instructions provided by the manufacturer. Transfer the proteins separated by electrophoresis to the membranes using the Trans-blot SD semi-dry transfer cell and a freshly made buffer (25 mM Tris base, 192 mM glycine and 20% methanol).
14. Prior to immunodetection, evaluate the transfer efficiency and equal amount of total proteins with the reversible Ponceau S (Sigma) staining (*see* **Note 6**). This can be done by incubating the membrane for 5 min in 0.5% (v/v) Ponceau S in 1% glacial acetic acid, followed by washing in 1% glacial acetic acid and two washes in water.
15. For immunodetection, all the steps are carried out under gentle agitation. Membranes are first washed in PBS for 5 min and then blocked at 4 °C overnight in blocking solution (*see* **Note 7**).
16. Incubate the membrane 2 h at room temperature with a rabbit polyclonal antibody raised against ADC from tobacco, diluted 1:2000 in blocking buffer (*see* **Note 8**). Run a control by replacing the first antibody with pre-immune serum.
17. Wash the membranes three times (5–10 min each) in washing solution (*see* **Note 9**).
18. Incubate the membranes for 2 h with alkaline phosphatase-conjugated anti-rabbit IgG (Boehringer Mannheim), diluted 1:1000 in the blocking buffer.
19. Wash again three times in washing solution for 5–10 min.
20. Treat the membrane for 15–30 min with the NBT-BCIP mixture prepared in alkaline buffer as previously mentioned. Alternatively, use a purchased NBT/BCIP ready-made mix (*see* **Note 10**).
21. Wash three times in water to stop reaction. Check Fig. 1 as an example of an immunoblot obtained in hop cultures.

3.4 Processing of Organogenic Nodules for Transmission Electron Microscopy

1. Fix samples of each stage in 4% paraformaldehyde. Gently add the 4% paraformaldehyde fixative solution in excess to the sample. Keep them overnight at 4 °C (*see* **Note 11**).
2. The following day, remove the formaldehyde solution and wash the pellet with 1× PBS for 15 min on ice (three times). Store the samples in a solution of 0.1% paraformaldehyde in PBS at 4 °C until use.

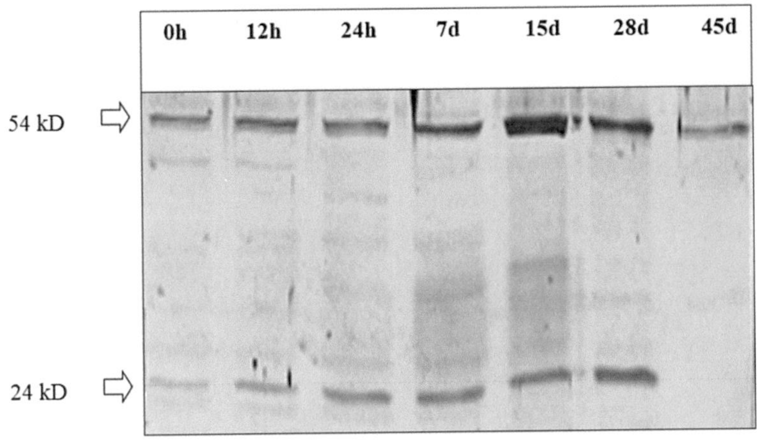

Fig. 1 Immunoblot analysis of ADC. Protein extracts corresponding to several culture stages are represented (0, 12 and 24 h, and 7, 15, 28 and 45 days). 30 μg total protein extract was loaded per lane. The 54 kD protein may be an ADC form present only in the chloroplast of photosynthetic tissues [11]. Image reproduced with permission from [3]

3. Dehydrate samples by dipping them in increasing concentrations of methanol in dH_2O as follows: 30 min in methanol 30% at 4 °C, 30 min in methanol 50% at −10 °C, 30 min in methanol 70% at −20 °C, 30 min in methanol 100% at −30 °C. Repeat the methanol 100% step a total of three times.
4. Embed samples in resin by dipping them in increasing concentrations of Lowicryl K4M in methanol at −30 °C as follows: 60 min in Lowicryl K4M 50%, 60 min in Lowicryl K4M 67%, 60 min in Lowicryl K4M 100%, and finally leave the samples in Lowicryl K4M 100% at −30 °C overnight (*see* **Note 12**).
5. The next day, move the samples to new, unused Lowicryl K4M 100% and leave them at −30 °C overnight.
6. The next day, prepare new, freshly made Lowicryl K4M, use a part of it to dip the samples on it until encapsulation, and keep the rest at −30 °C for encapsulation.
7. Transfer samples to selected capsules/molds (*see* **Note 13**) and fill with the newly made resin. In all cases, prevent contact of the resin with atmospheric oxygen as much as possible, since oxygen precludes a correct resin polymerization.
8. Polymerize the resin at least 24 h at −30 °C with a UV lamp.
9. Label the polymerized plastic resin blocks.
10. Optionally, if blocks are not to be used immediately, leave them cure close to a window, exposed to the UV radiation of sunlight for several days.
11. Store them at room temperature permanently. They are now ready for sectioning, staining and observation.

3.5 Sectioning

1. Prior to sectioning, prepare meshed nickel grids (*see* **Note 14**).
2. Trim the block to expose the sample and prepare a trapezoidal sectioning front of approximately 1 mm^2. For a correct quantification of immunogold labeling (*see* Subheading 3.7), prepare and section at least three randomly picked sample-containing plastic blocks for each stage.
3. Obtain semithin sections (0.5–2 μm) with a glass knife for preliminary observation under the light microscope.
4. Observe the semithin sections with phase contrast optics to check that the tissue is not damaged. Alternatively, if phase contrast optics is not available, sections can be stained with general stains such as toluidine blue and observed in bright field.
5. Obtain ultrathin sections (~80–100 nm) with a diamond knife. Ultrathin sections can be identified by their silver interference color when floating in the knife boat.
6. Pick sections up by touching them with the coated side of the grid.

3.6 Immunogold Labeling

1. Prepare a piece of Parafilm tape and deposit rows of three drops (20–30 μl each) of dH_2O, three drops of PBS and one drop of 5% BSA in PBS for each of the grids to be immunogold labeled.
2. Float grids carrying Lowicryl ultrathin sections sequentially on dH_2O and PBS drops, keeping the grids on each drop for few minutes, and then float them on 5% BSA in PBS drops for 5 min (*see* **Note 15**). Before floating each grid, it is important to check what the side of the grid carrying sections is, and make sure that this side is in contact with the drops.
3. Incubate sections with the primary antibody by floating each grid on a drop of anti-ADC antibody for 1 h at room temperature. For controls, incubate sections with preimmune serum diluted 1:10 in 1% BSA in PBS or only with 1% BSA in PBS, excluding the anti-ADC antibody in both cases (*see* **Note 16**).
4. Wash grids by sequentially floating them on three drops of 1% BSA in PBS.
5. Incubate sections with the secondary antibody by floating each grid on a drop of goat anti-rabbit IgG conjugated to 10 nm colloidal gold diluted 1:25 in 1% BSA in PBS for 45 min at room temperature.
6. Wash grids by sequentially floating them on three drops of 5% BSA in PBS and then three drops of dH_2O.
7. Air dry grids protected from dust with the lid of a culture dish.
8. Counterstain grids floating them on drops of 5% aqueous uranyl acetate for 7 min and lead citrate for 30 s. For uranyl

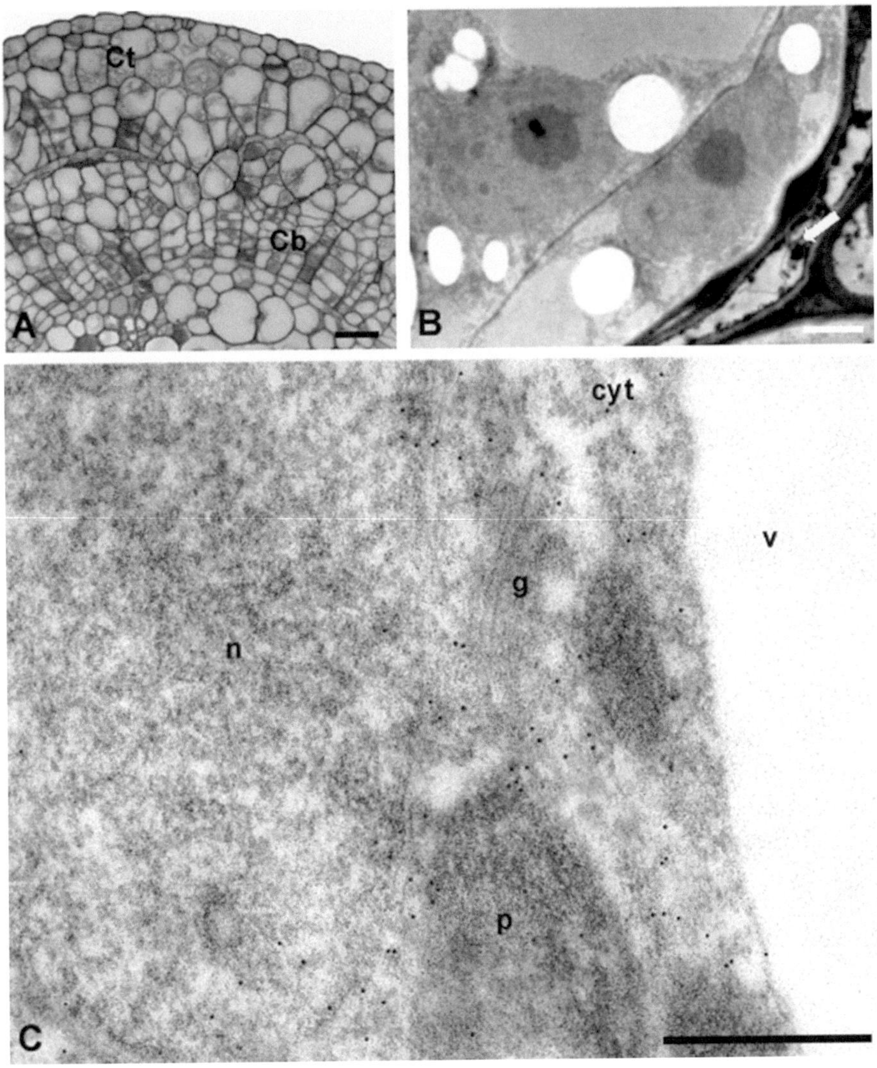

Fig. 2 ADC immunolocalization in hop, 7 days after internode in vitro inoculation. (**a**) shows a semithin section of the cortical (Ct) and cambial (Cb) internode regions. (**b**) shows an ultrathin section at low magnification, where dividing cortical cells are observed together with cells undergoing cell death (*arrow*). (**c**) shows an ultrathin section immunolabelled with anti ADC antibodies. Labeling (gold particles) is present in the cytoplasm (cyt), nucleus (n) and plastids (p), but is absent in vacuoles (v) and vesicles (g). Bars in A: 100 μm; B: 2.5 μm; C:500 nm. Image reproduced with permission from [3]

acetate, protect drops from light during incubation by covering them with an opaque lid. For lead citrate, protect drops from atmospheric oxygen by surrounding the drop-containing parafilm with NaOH pellets (*see* **Note 17**).

9. Observe sections in a transmission electron microscope. Examples of immunogold labeling detection of ADC in organogenic nodule cells can be found in Fig. 2.

3.7 Quantification of Immunogold Labeling

1. Take micrographs from the cells of interest of randomly chosen regions of the different section-containing grids corresponding to each sample.
2. The minimum number of micrographs to be used for quantification is determined using the progressive mean test [13] with minimum confidence limit of $p \leq 0.05$ (*see* **Note 18**).
3. For each micrograph, calculate the labeling density of each subcellular compartment. Labeling density is defined as the number of particles per area unit (μm^2). This number of particles is determined by hand counting particles over the compartments under study (cytoplasm, nucleus, and chloroplasts). The area occupied by each of the subcellular regions studied may be measured using a hand-made square lattice composed of 11×16 squares of 15×15 mm each, printed in transparent paper and superimposed to the image analyzed. Alternatively, the area of digital images may be computer-determined with the use of image analysis software (*see* **Note 19**). This way, areas are delimited by hand-drawing regions of interest (ROIs) with the mouse. Then, the software yields the area of each ROI.
4. As an estimation of background noise, calculate the labeling density of tissue-free resin regions and of subcellular regions where ADC is not supposed to be present, such as cell walls, nucleolus or starch deposits, for example. Ideally, background noise should never exceed 5% compared to the quantified specific signal.
5. For each stage and subcellular region, express labeling density as mean labeling density of all micrographs ± SD.
6. For paired comparisons (between the mean labeling density of the cytoplasm and background noise, for example), a *Student's t*-test, with $p \leq 0.05$ may be used to check whether differences are statistically significant.
7. For comparisons among different stages or subcellular compartments, an *analysis of variance (ANOVA)* test may be used to check out whether differences are statistically significant.

4 Notes

1. In a fumehood, prepare a solution of 4% (v/w) formaldehyde (from paraformaldehyde powder, E.M. grade) in PBS. Heat in a hot bath until the solution is transparent. Cap the erlenmeyer to avoid toxic vapors. Then put it on melting ice.
2. Prepare Lowicryl K4M by mixing the components according to manufacturer's instructions. For more information: https://www.emsdiasum.com/microscopy/.

3. High pH lead citrate solution is prepared according to the protocol initially described by [14]. Briefly, prepare 1 M lead nitrate, 1 M trisodium citrate and 1 M NaOH stock solutions. Drop by drop, add 3 ml of trisodium citrate stock to 16 ml of boiled dH_2O. Shake and then add 2 ml of lead nitrate stock. Shake until a milky solution is formed. Then add 4 ml of the NaOH stock. The solution should become transparent and precipitates disappear. Filter the solution with a 0.2 μm nylon filter, and store it until use at room temperature protected from light and atmospheric CO_2.
4. Centrifuging the samples prior to the run helps remove insoluble debris, which could produce streaks in the protein lanes.
5. Be careful when loading samples onto the gel and use appropriate long tips. Pipetting errors can affect the amount of protein in each lane. Avoid bubbles and make sure to use equal amount of protein. Excessive amount of protein may lead to saturation and incomplete transfer.
6. Make sure membrane is thoroughly wetted when beginning procedure and that remains submerged in incubation and wash buffers throughout all steps of western blotting.
7. In case incomplete blocking occurs, increase the concentration of the blocking agent or increase the duration of the blocking step (overnight at 4 °C plus an incubation at room temperature). However, overnight incubations at 4 °C should be enough to efficiently block the membrane and in certain cases even 1–2 h at room temperature.
8. Both for immunoblotting and immunolocalization procedures, the concentration of both primary and secondary antibodies must be optimized in order to obtain a high signal to noise ratio. Too much antibody can lead to unspecific binding. Also incubation periods can be extended up to overnight for the primary antibody. Before using the antibody centrifuge the antibody solution for 5 min at maximal speed in a bench microcentrifuge to remove possible debris.
9. Both for immunoblotting and immunolocalization procedures insufficient washing between incubation steps may lead to undesirable high background. The stringency of the washing procedure can be managed by increasing the duration of the washing or amount of detergent in the buffer.
10. NBT can be dissolved in 70% dimethylformamide (DMF) and BCIP in 100% DMF. The BCIP and NBT stocks will last indefinitely when stored at 4 °C or −20 °C in the dark. Rinsing the membrane strips with deionized water 2–3 times can help to reduce nonspecific binding of NBT/BCIP to the strip. Incubation period can be shortened or extended depending

on the signal strength. However, incubation period should not be too extended, otherwise inaccurate results can be obtained.

11. Due to the toxic and volatile nature of the reagents used in this procedure, all these steps must be done in a fumehood. Changes should be made as fast as possible without removing the tubes from the metal block, and then transferring the block back to the refrigerator or freezer.
12. Alternatively, a Leica EM AFS2 automatic freeze substitution system (Leica Microsystems, Vienna, Austria) can be used for the dehydration, embedding and polymerization steps. For more information: http://www.leica-microsystems.com/products/sample-preparation-for-electron-microscopy/cryo-preparation-systems/details/product/leica-em-afs2/.
13. Many different capsules and flat molds may be used for encapsulation. Most of them can be checked at (https://www.tedpella.com/Embedding_html/Embedding_Supplies_Overview.htm). The election of the best option will depend on the size, shape and nature of the sample.
14. For immunogold labeling, nickel or gold grids are recommended, because they are inert to the reagents used. Coating grids with a film helps to support sections during labeling and TEM examination. In addition to plastic films (formvar, collodion, parlodion, etc.), grids may also be coated with a carbon layer that provides additional strength and stability when exposed to the electron beam. Detailed procedures for grid coating can be found in classical transmission electron microscopy handbooks (for example, *see* [15, 16]).
15. The purpose of floating grids in dH2O and PBS drops is to hydrate the section surface. Then, 5% BSA in PBS is used to block unspecific crosslinking between the primary antibody used next and other epitopes of the section surface. If results with 5% BSA are not satisfactory, alternatives such as acetylated BSA (BSAc), fetal calf serum (FCS) or skimmed milk powder may also be tested.
16. In order to check out the specificity of immunogold labeling, two different controls may be run in parallel to the labeling assays. First, to check out the specificity of the anti-ADC antibody, it may be replaced by preimmune serum, if available, keeping the other experimental conditions unchanged. Ideally, labeling in this control should not be higher than background noise. If it is higher, results with anti-ADC should be interpreted with caution. The second control consists in excluding the anti-ADC antibody to check out the specificity of the secondary antibody conjugated to colloidal gold. Ideally, labeling in this control should not be higher than background noise.

17. NaOH pellets are used to prevent lead precipitation. When exposed to atmospheric CO_2, lead citrate may react with CO_2 producing lead carbonate, which precipitates on the sections giving rise to black dots and artefactual stains that prevent a correct visualization of cell ultrastructure. NaOH reacts with CO_2, competing for it with lead. Thus, NaOH reduces or prevents lead precipitation.
18. In short, this test is based on calculating the mean labeling density of an initial set of micrographs, and successively comparing it with the additive means of micrograph #1, #1 to #2, #1 to #3, #1 to #4... The minimum number of micrographs needed is reached when the difference between the total mean and the additive means of the last three images measured does not exceed 5%. For a more detailed explanation of the method, *see* [13].
19. An example of this type of image analysis software is Image J (https://imagej.nih.gov/ij/). This is free-licensed, easy-to-use software that includes many basic image analysis features including the definition of ROIs and calculation of their areas.

Acknowledgments

This work was supported by the grant AGL2014-55177-R from Spanish MINECO to JMSS. Funding to AMF was provided by the Portuguese Foundation for Science and Technology (SFRH/BPD/100928/2014, FCTInvestigator IF/00169/2015, PEst-OE/BIA/UI4046/2014. We also acknowledge The Society of Plant Signaling and Behavior for allowing reproduction of images.

References

1. Alcázar R, Tiburcio AF (2014) Plant polyamines in stress and development: an emerging area of research in plant sciences. Front Plant Sci 5:319
2. Minocha R, Majumdar R, Minocha SC (2014) Polyamines and abiotic stress in plants: a complex relationship. Front Plant Sci 5:175
3. Fortes AM, Costa J, Santos F, Seguí-Simarro JM, Palme K, Altabella T, Tiburcio AF, Pais MS (2011) Arginine decarboxylase expression, polyamines biosynthesis and reactive oxygen species during organogenic nodule formation in hop. Plant Signal Behav 6(2):258–269
4. Agudelo-Romero P, Bortolloti C, Pais MS, Tiburcio AF, Fortes AM (2013) Study of polyamines during grape ripening indicate an important role of polyamine catabolism. Plant Physiol Biochem 67C:105–119
5. Agudelo-Romero P, Erban A, Rego C, Carbonell-Bejerano P, Nascimento T, Sousa L, Martínez-Zapater JM, Kopka J, Fortes AM (2015) Transcriptome and metabolome reprogramming in *Vitis vinifera* cv. Trincadeira berries upon infection with *Botrytis cinerea*. J Exp Bot 66:1769–1785
6. Fortes A, Pais M (2000) Organogenesis from internode derived nodules of *Humulus lupulus* var. Nugget (Cannabinaceae): histological studies and changes in the starch content. Am J Bot 87:971–979
7. Fortes A, Santos F, Pais M (2010) Organogenic nodule formation in hop: a tool to study morphogenesis in plants with biotechnological and medicinal applications. J Biomed Biotechnol 2010:583691

8. Adams AN (1975) Elimination of viruses from the hop (*Humulus lupulus*) by heat therapy and meristem culture. J Hortic Sci 50:151–160
9. Murashige T, Skoog F (1962) A revised medium for rapid growth and bioassays with tobacco tissue cultures. Physiol Plant 15:473–497
10. Borrell A, Culianez-Macia F, Altabella T, Besford R, Flores D, Tiburcio A (1995) Arginine decarboxylase is localized in chloroplasts. Plant Physiol 109:771–776
11. Bortolotti C, Cordeiro A, Alcázar R, Borrell A, Culiañez-Macià F, Tiburcio A, Altabella T (2004) Localization of arginine decarboxylase in tobacco plants. Physiol Plant 120:84–92
12. Bradford MM (1976) A rapid and sensitive method for the quantitation of microgram quantities of protein utilizing the principle of protein-dye binding. Anal Biochem 72:248–254
13. Williams M (1977) Stereological techniques. In: Glauert AM (ed) Practical methods in electron microscopy. North Holland/American Elsevier, Amsterdam
14. Reynolds ES (1963) The use of lead citrate at high pH as an electron opaque stain in electron microscopy. J Cell Biol 17:208–212
15. Hayat MA (2000) Principles and techniques of electron microscopy: biological applications, 4th edn. Cambridge University Press, Cambridge, UK
16. Wilson AW, Robards AJ (1998) Procedures in electron microscopy. John Wiley & Sons, Sussex, UK

Chapter 20

Analysis of the Intracellular Localization of Transiently Expressed and Fluorescently Labeled Copper-Containing Amine Oxidases, Diamine Oxidase and *N*-Methylputrescine Oxidase in Tobacco, Using an *Agrobacterium* Infiltration Protocol

Tsubasa Shoji

Abstract

The intracellular localization of enzymes provides key information for understanding complex metabolic pathways. Based on enzyme localization data, the involvement of multiple organelles and the movement of metabolites between cellular compartments have been suggested for a number of pathways. Transient expression of fluorescently tagged proteins in the leaves of *Nicotiana benthamiana* through *Agrobacterium* infiltration is a simple and versatile way to examine the intracellular localization of proteins of interest. Here, this method was applied to demonstrate the peroxisomal localization of a pair of homologous copper-containing amine oxidases (CuAOs) from tobacco with distinct substrate preferences: diamine oxidase (DAO), which mediates polyamine catabolism, and *N*-methylputrescine oxidase (MPO), which is involved in nicotine biosynthesis. Our results demonstrate that the *Agrobacterium* infiltration protocol can be effectively used to study the intracellular localization of oxidases that localize to the peroxisome.

Key words Copper-containing amine oxidase, Diamine oxidase, Yellow fluorescent protein, Infiltration, mCherry, *Nicotiana benthamiana*, Nicotine biosynthesis, *N*-methylputrescine oxidase, Peroxisome, Polyamine catabolism, Tobacco, Transient expression

1 Introduction

Copper-containing amine oxidases (CuAOs) are involved in the deamination of amine-containing molecules. For example, they mediate the deamination of polyamines that results in the formation of aminoaldehydes and H_2O_2. The copper ion is required for the formation of topaquinone, a characteristic cofactor of enzymes that belong to this family, which is derived from a conserved tyrosine residue at the catalytic site [1]. Apoplastic activities of CuAOs contribute to cell wall lignification through the formation of H_2O_2 [2]. In addition to the common extracellular CuAOs,

Rubén Alcázar and Antonio F. Tiburcio (eds.), *Polyamines: Methods and Protocols*, Methods in Molecular Biology, vol. 1694, DOI 10.1007/978-1-4939-7398-9_20, © Springer Science+Business Media LLC 2018

some CuAOs localize to peroxisomes, where they participate in the catabolism of polyamines, such as spermine, spermidine and putrescine, to γ-butyric acid with other amine oxidases and aldehyde dehydrogenase (Fig. 1) [3].

Some CuAOs also function in specialized metabolism, accepting substrates structurally related to polyamines. In tobacco, *N*-metylputrescine oxidase (MPO), which belongs to the CuAO

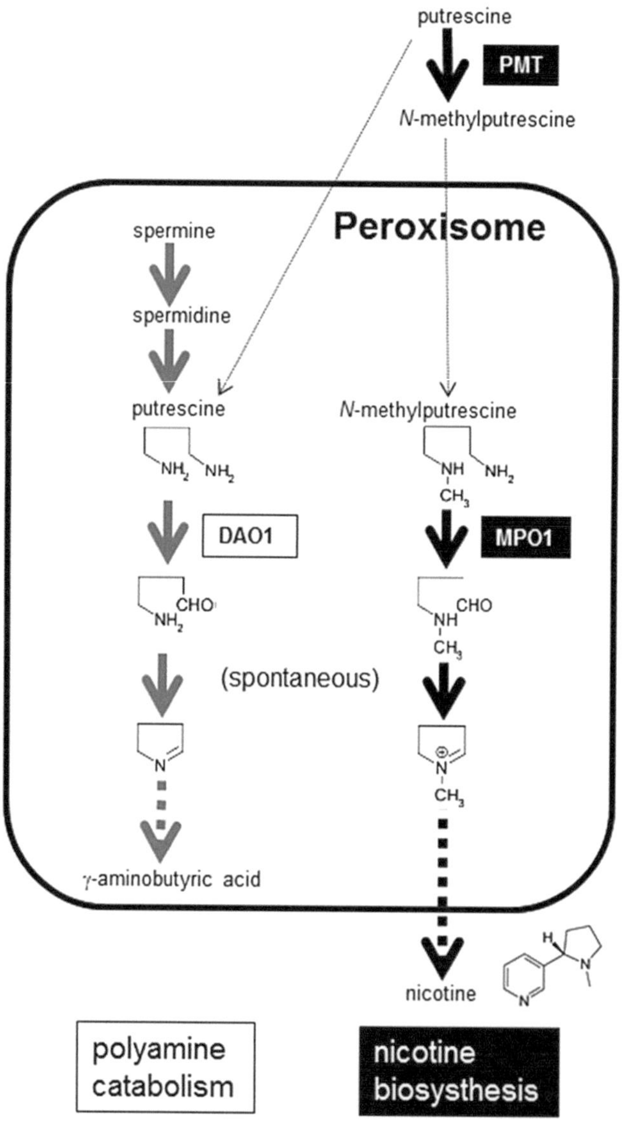

Fig. 1 Schematic diagram of the metabolic functions of MPO and DAO in peroxisomes. In the nicotine biosynthesis pathway, MPO catalyzes the conversion of *N*-methylputrscine, to 4-methylaminobutanal, which is spontaneously cyclized to *N*-methyl Δ^1-pyrrolinium cation. *N*-methylputrescine is formed from putrescine through a reaction catalyzed by putrescine *N*-methyltransferase (PMT), which is proposed to be cytoplasmic. Accepting putrescine as a substrate, DAO functions in polyamine catabolism leading to γ-aminobutyric acid

enzyme family, plays a key role in nicotine biosynthesis (Fig. 1). The nicotine biosynthetic pathway starts with the formation of *N*-methylputrescine from putrescine, which is catalyzed by putrescine *N*-methyltransferase (PMT) [4]. *N*-methylpurescine is oxidatively deaminated by MPO to 4-methylaminobutanal [5], which is spontaneously cyclized to *N*-methyl-Δ^1-pyrrolinium cation and subsequently incorporated into nicotine [6].

Two homologous cDNAs for CuAOs, *MPO1* and *diamine oxidase1* (*DAO1*), have been cloned from tobacco [7, 8]. The enzymes encoded by these genes have distinct substrate preferences: DAO1 accepts non-methylated putrescine as a substrate, converting the diamine to 4-aminobutanal which is cyclized to Δ^1-pyrroline in polyamine catabolism, whereas MPO1 is an enzyme oxidizing *N*-methylputrescine in the nicotine biosynthesis pathway (Fig. 1). Like other genes in the pathway, expression of *MPO1* is under the control of pathway-regulating transcription factors [9, 10]. As predicted from the presence of a peroxisome targeting signal type 1 (PTS1) in the protein sequences at their C-termini, MPO1 and DAO1 fusions with yellow fluorescent protein (YFP) localize to peroxisomes in *Nicotiana benthamiana* leaves when transiently expressed using *Agrobacterium* infiltration [9]. Indeed, MPO1 and DAO1 from tobacco are more closely related to the peroxisome-resident CuAO from *Arabidopsis* [3] than to the extracellular CuAOs [11], presumably forming a distinct subgroup of peroxisomal CuAOs characterized by the presence of a PTS1 consensus sequence. Peroxisomes are single-membrane organelles that harbor a number of enzymes with diverse metabolic functions [12]. Proteins containing a PTS1 localization sequence, such as the peroxisomal CuAOs, are recognized by a specific receptor in the cytosol and subsequently targeted to the peroxisome, where they are released into the matrix of the organelle [12].

Biochemical and intracellular localization studies of the two related CuAOs from tobacco suggest that MPO may have evolved from DAO by changing substrate specificities but not its localization to peroxisomes [9]. The involvement of multiple organelles, such as peroxisomes [9], plastids [13], and vacuoles [14], in nicotine biosynthesis suggests that there must be dynamic intracellular trafficking of pathway intermediates between cellular compartments, as has been proposed for other pathways [15].

Transient expression in the leaves of *Nicotiana benthamiana* through infiltration of *Agrobacterium* suspensions is a simple and versatile way to analyze the intracellular localization of proteins that are tagged with various fluorescent proteins. The relatively soft leaf tissue of *N. benthamiana* is easily infiltrated and therefore widely used for the assay. In addition, results can be obtained relatively quickly (within 2–5 days after the infiltration), facilitating the study of a large number of fusion constructs. Co-infiltration of *Agrobacterium* suspensions containing different expression constructs allows simultaneous expression of multiple proteins and defined organelle markers fused with different fluorescent proteins. This approach allows analysis of the localization of several differently tagged fluorescent proteins within the same cells. Using this approach, we have demonstrated the co-localization of YFP-MPO1 or YFP-DAO1 with peroxisome-targeted mCherry-PTS1 [9].

2 Materials

Instruments, such as a centrifuge, spectrometer, and a bacterial shaker, that are generally available in laboratories for molecular experiments are required. It is important to follow all waste disposal regulations when disposing of waste materials.

2.1 Agrobacterium Cultures

1. Electrocompetent cells of *Agrobacterium tumefaciens* strain EHA105 (*see* **Note 1**).
2. Luria Broth (LB) medium (*see* **Note 2**): dissolve 5 g bacto-tryptone, 2.5 g yeast extract, and 5 g NaCl in 500 mL distilled water, and adjust the pH to 7.0 with NaOH. To make LB agar plates, add 7.5 g agar to 500 mL LB medium. Autoclave for 20 min. When necessary, add antibiotics and acetosyringone at the required concentrations after autoclaving.
3. Kanamycin stock solution (100 mg/mL): 1 g kanamycin in 10 mL distilled water. Filter-sterilize and store at −20 °C.

2.2 Plant Material

1. Gamborg's B5 medium [16]: dissolve 3.1 g Gamborg's B-5 Basal Salt Mixture (Sigma-Aldrich), 100 mg *myo*-inositol, 1 mg nicotinic acid, 1 mg pyridoxine hydrochloride, 10 mg thiamine hydrochloride and 30 g sucrose in 1 L distilled water, and adjust the pH to 5.6 with KOH. To make Gamborg's B5 agar plates, add 8 g agar to 1 L B5 medium. Autoclave for 20 min.
2. Three- to five-week-old plantlets of *N. benthamiana*: germinate sterilized seeds on Gamborg's B5 medium under continuous illumination at 26 °C; 3–5 days after germination, transfer the seedlings to 5-cm square plastic pots filled with soil (*see* **Note 3**) and grow in the greenhouse.

2.3 Materials Needed for the Infiltration Assay

1. Infiltration solution: 10 mM MES-KOH pH 5.6, 10 mM $MgCl_2$, 150 μM acetosyringone (*see* **Note 4**). It is important to make the infiltration solution immediately before use by mixing stock solutions of all components (listed below).
2. 1 M MES [2-(*N*-morpholino)-ethanesulfonic acid]: Weigh 21.32 g MES, dissolve in 100 mL distilled water, and adjust the pH to 5.6 with KOH. Filter-sterilize and store at room temperature.

3. 1 M $MgCl_2$: Weigh 9.52 g $MgCl_2$ and dissolve in 100 mL distilled water. Autoclave for 20 min and store at room temperature.
4. 200 mM acetosyringone (4′-hydroxy-3′5′-dimethoxyacetophenone): Dissolve 78.4 mg acetosyringone in 1 mL DMSO. Filter-sterilize and store at −20 °C.
5. 5-mL plastic syringe.

2.4 Microscopy

1. Confocal laser scanning microscopy (CLSM) equipment, such as Nikon C2 CLSM with 488-nm and 544-nm lasers and emission filters of a 510/30 nm band pass for YFP and a 585/65 nm band pass for mCherry.
2. Objective slides.
3. Cover slips.

3 Methods

3.1 Plant Growth

1. Begin with healthy 3- to 5-week-old plants of *N. benthamiana* grown in soil-filled pots in the greenhouse (*see* Subheading 2.2).
2. To allow for efficient infiltration of slightly dehydrated leaves, stop watering plants approximately 16 h before infiltration.

3.2 Construction of Plasmid Vectors and Introduction into Agrobacterium

1. Clone full-length coding sequences of *MPO1* (AB289456) and *DAO1* (AB289457, formerly called *MPO2*) from tobacco into the entry vector pDONR/Zeo and then into the destination vector pGWB42 [17] to generate N-terminal YFP fusions using the Gateway Cloning Technology (Invitrogen) (*see* **Note 5**).
2. Use the same cloning approach for sequences lacking the C-terminal PTS1 site for *MPO1* and *DAO1* to generate the binary vectors for YFP-MPO1ΔPTS1 and YFP-DAO1ΔPTS1.
3. To generate a peroxisomal marker, fuse the PTS1 tripeptide (Ser-Lys-Leu) sequence to the C-terminus of the red fluorescent protein mCherry [18]. Clone the mCherry-PTS1 sequence into pGWB2 [17] using the Gateway system.
4. Introduce the plasmid vectors into competent cells of *A. tumefaciens* strain EHA105 by electroporation (*see* **Note 6**).
5. Incubate the cells in 1 mL LB medium without selective antibiotics for 2 h at 28 °C while rotating at 220 rpm.
6. Spread the cells on a LB plate containing kanamycin at 50 μg/mL and incubate at 28 °C in the dark for 2–3 days until bacterial colonies appear.

3.3 *Agrobacterium* Infiltration Assay

1. Inoculate 5 mL LB medium with a single *Agrobacterium* colony from the LB agar plate.
2. Grow overnight at 28 °C while rotating at 220 rpm.
3. Collect the cells by centrifugation at 2800 × *g* for 15 min at 4 °C.
4. Discard the supernatant and resuspend the cells in infiltration solution by pipetting (avoid using the vortex mixer).
5. Repeat **steps 3** and **4**.
6. Measure the OD_{600} of the suspension with a spectrometer and adjust the OD_{600} to 0.5.
7. Incubate for 2–3 h at room temperature.
8. (Optional) For simultaneous expression of multiple genes, mix an equal volume of *Agrobacterium* suspension for each gene; e.g., YFP-MPO1 and mCherry-PTS1 (*see* **Note 7**).
9. Fill a syringe (without needle) with the *Agrobacterium* suspension.
10. Slowly press the piston of the syringe down on the abaxial side of a *N. benthamiana* leaf (*see* **Note 8**) while applying counter-pressure with a finger on the other side of the leaf (Fig. 2). Infiltration is confirmed when a dark-green area spreads in the leaf. Infiltrate each leaf multiple times, if necessary.
11. Leave the infiltrated leaves on the plants for 2–4 days (*see* **Note 9**).

Fig. 2 Infiltration of *Agrobacterium* suspension into a leaf of a 3-week-old plant of *N. benthamiana* with a plastic syringe without a needle. Infiltration is confirmed by the spreading of a *dark-green* leaf area

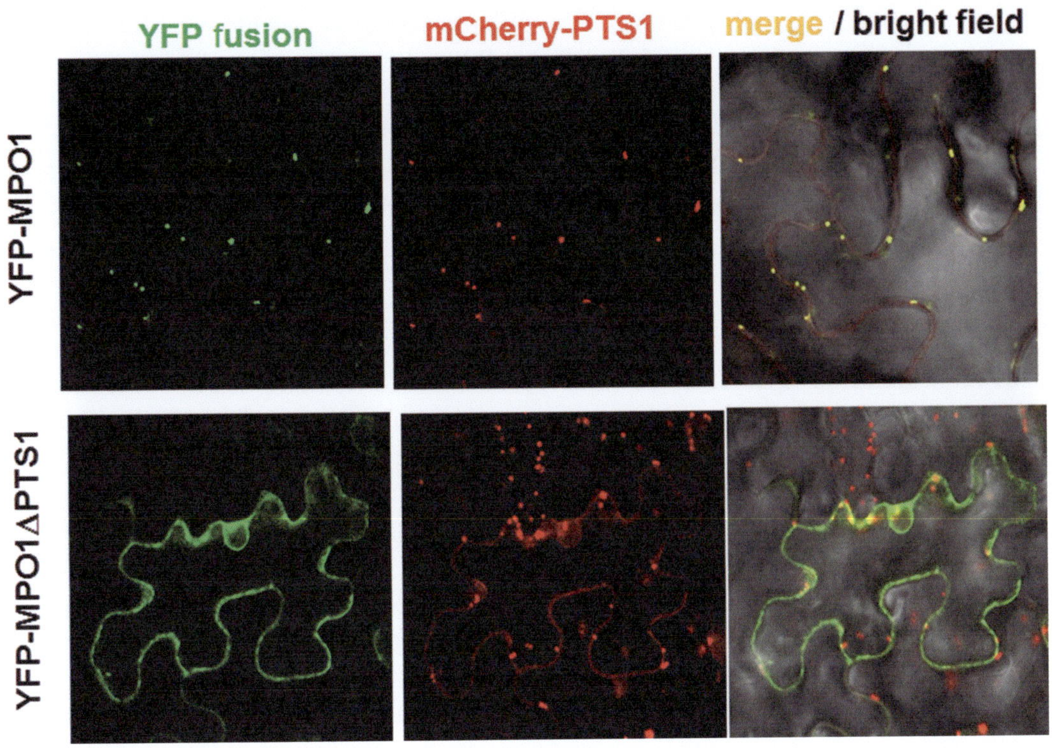

Fig. 3 Subcellular localization of YFP-MPO1 and YFP-MPO1ΔPTS1 in *N. benthamiana* leaf epidermal cells. Removal of the C-terminal three amino acid residues (Ala-Lys-Leu), corresponding to PTS1, resulted in the loss of the peroxisomal localization. mCherry-PTS1 was co-expressed as a peroxisomal marker. Although not shown here, the localization to the peroxisome and its loss was similarly observed for YFP-DAO1 and YFP-DAOΔPTS1, respectively [9]

3.4 Microscopic Observation

1. Excise a small section (3 × 3 mm) from the infiltrated regions of the leaves.
2. Mount the sections in water with the abaxial side of the leaf facing up and carefully add a cover slip to avoid including air bubbles.
3. Observe and image the fluorescence via CLSM (Fig. 3). Excite YFP and mCherry with the 488-nm and 544-nm lasers, respectively. Use a 514/30 band pass filter for YFP and a 585/65 filter for mCherry to filter and observe the emitted fluorescence.

4 Notes

1. Other *Agrobacterium* strains, such as nopaline-type GV3101 and octopine-type LBA4404, are also suitable. Agropine-type *Agrobacterium* EHA105 has resistance to rifampicin (25 μg/mL, chromosomal) and streptomycin (50 μg/mL, on Ti

plasmid). The antibiotics could be included in LB medium to ensure proper selection conditions.

2. Other *Agrobacterium* media, including YEB, YMB and AB, are also suitable for use with this protocol.
3. We use a granular horticultural soil with fertilizer (Nihon Horticulture Soil No.1, Nihon Hiryo).
4. Acetosyringone is a chemical analogous to the phenolic signal that is released from wounded plant cells. It induces the *Agrobacterium* virulence genes that play key roles in T-DNA transfer into plant cells, and thus can be used to increase the transformation efficiency when added to the culture medium [19].
5. To avoid unintended introduction of mutations, a high-fidelity DNA polymerase should be used for PCR amplification. Plasmid sequences should be verified at the necessary steps.
6. We use MicroPulser (Bio-Rad) electroporator at 1.8 kV with 0.1 cm cuvettes.
7. It has been reported that the co-infiltration of a viral protein p19 from tomato bushy stunt virus enhances the levels of transient expression by suppressing RNA silencing [20].
8. Select fully expanded young leaves. Avoid the youngest leaves and the cotyledons. The abaxial side is considered better for the infiltration because of the thinner cuticle and the higher density of stomata. Wearing disposable gloves is recommended.
9. Grow under the same conditions as before the infiltration. The fluorescence signals can be observed for the examined fusions nearly constantly during the duration.

Acknowledgment

This work was supported by the Japan Society for the Promotion of Science (grant No. 26440144).

References

1. Kumar V, Dooley DM, Freeman HC, Guss JM, Harvey I, McGuirl MA et al (1996) Crystal structure of a eukaryotic (pea seedling) copper-containing amine oxidase at 2.2 Å resolution. Structure 4:943–955
2. Cona A, Rea G, Angelini R, Federico R, Tavladoraki P (2006) Functions of amine oxidases in plant development and defense. Trends Plant Sci 11:80–88
3. Reumann S, Quan S, Aung K, Yang P, Manandhar-Shrestha K, Holbrook D et al (2009) In-depth proteome analysis of *Arabidopsis* leaf peroxisomes combined with *in vivo* subcellular targeting verification indicates novel metabolic and regulatory functions of peroxisomes. Plant Physiol 150:125–143
4. Hibi N, Higashiguchi S, Hashimoto T, Yamada Y (1994) Gene expression in tobacco low-nicotine mutants. Plant Cell 6:723–735
5. Hashimoto T, Minami A, Yamada Y (1990) Diamine oxidase from cultured roots of *Hyoscyamus niger*: its function in tropane alkaloid biosynthesis. Plant Physiol 93:216–221

6. Mizusaki S, Kisaki T, Tamaki E (1968) Phytochemical studies on the tobacco alkaloids. XII. Identification of γ-methylaminobutyraldehyde and its precursor role in nicotine biosynthesis. Plant Physiol 43:93–98
7. Katoh A, Shoji T, Hashimoto T (2007) Molecular cloning of *N*-methylputrescine oxidase from tobacco. Plant Cell Physiol 48:550–554
8. Heim WG, Skyes KA, Hildreth SB, Sun J, Lu RH, Jelesko JG (2007) Cloning and characterization of a *Nicotiana tabacum* methylputrescine oxidase transcript. Phytochemistry 68:454–463
9. Naconsie M, Kato K, Shoji T, Hashimoto T (2013) Molecular evolution of *N*-methylputrescine oxidase in tobacco. Plant Cell Physiol 55:436–444
10. Shoji T, Kajikawa M, Hashimoto T (2010) Clustered transcription factor genes regulate nicotine biosynthesis in tobacco. Plant Cell 22:3390–3409
11. Møller SG, McPherson MJ (1998) Developmental expression and biochemical analysis of the *Arabidopsis atao1* gene encoding an H_2O_2-generating diamine oxidase. Plant J 13:781–791
12. Kaur N, Reumann S, Hu J (2009) Peroxisome biogenesis and function. Arabidopsis Book 7: e0123
13. Katoh A, Uenohara K, Akita M, Hashimoto T (2006) Early steps in the biosynthesis of NAD in *Arabidopsis* start with aspartate and occur in the plastid. Plant Physiol 141:851–857
14. Kajikawa M, Shoji T, Kato A, Hashimoto T (2011) Vacuolar-localized berberine bridge enzyme-like proteins are required for a late step of nicotine biosynthesis in tobacco. Plant Physiol 155:2010–2022
15. Kutchan TM (2005) A role of intra- and intercellular translocation in natural product biosynthesis. Curr Opin Plant Biol 8:292–300
16. Gamborg OL, Miller RA, Ojima K (1968) Nutrient requirements of suspension cultures of soybean root cells. Exp Cell Res 50:151–158
17. Nakagawa T, Kurose T, Hino T, Tanaka K, Kawamukai M, Niwa Y et al (2007) Development of series of gateway binary vectors, pGBWs, for realizing efficient construction of fusion genes for plant transformation. J Biosci Bioeng 104:34–41
18. Shaner NC, Campbell RE, Steinbach PA, Giepmans BNG, Palmer AE, Tsien RY (2004) Improved monomeric red, orange and yellow fluorescent proteins derived from *Discosoma* sp. red fluorescent protein. Nat Biotechnol 22:1567–1572
19. Shelkholeslam SN, Weeks DP (1987) Acetosyringone promotes high efficiency transformation of *Arabidopsis thaliana* explants by *Agrobacterium tumefaciens*. Plant Mol Biol 8:291–298
20. Voinnet O, Rivas S, Mestre P, Baulcombe D (2003) An enhanced transient expression system in plants based on suppression of gene silencing by the p19 protein of tomato bushy stunt virus. Plant J 33:949–956

Chapter 21

Techniques Used for Functional Characterization of Polyamine Transporters

Claudio A. Pereira, Melisa Sayé, Chantal Reigada, and Mariana R. Miranda

Abstract

Transport systems are key processes in every living organism: they allow the entry of all essential nutrients into the cell and its compartments and regulate the intracellular concentrations of metabolites. The transport of cell nutrients represents the first step of many metabolic routes and may also regulate such processes. They are also responsible for reaching the effective intracellular concentration of therapeutic drugs and some mechanisms of resistance and tolerance also depend on them. However, the common techniques used to evaluate the metabolites transport in different cells types are not easy to carry out and require extensive training. In this chapter, we report detailed protocols and tips about the expression of transporters, different activity assays and transporter kinetics determination.

Key words Polyamine transport, Transporter expression, *Trypanosoma cruzi*, Transporter kinetics, Transport assays

1 Introduction

Transport systems comprise an essential feature of every living organism; they allow the entry of essential nutrients into the cell and its compartments and regulate the intracellular concentrations of metabolites. Transporters are in contact with extracellular compounds and work, not only as permeases carrying the solutes into the cell but also as environmental sensors. Transport across the plasma membrane represents the first step of many metabolic routes and may also regulate such pathways. On the other hand, reaching the effective intracellular concentration of a therapeutic drug depends exclusively on the activity of transporters and in many cases the mechanisms of resistance and tolerance also depend on them [1, 2]. However, common techniques used to evaluate metabolites transport in different cell types are not easy to carry out and require a good training since there is no "one step" transport assay kit available. The functional expression of transporter proteins

Rubén Alcázar and Antonio F. Tiburcio (eds.), *Polyamines: Methods and Protocols*, Methods in Molecular Biology, vol. 1694, DOI 10.1007/978-1-4939-7398-9_21, © Springer Science+Business Media LLC 2018

possessing several subunits or multiple transmembrane spanners is difficult and requires the appropriate model. In this chapter, a protocol is given for expression of transporters in homologous systems (*Trypanosoma cruzi*). Finally, the most relevant assays and methods for estimating the substrate specificity, kinetics properties and inhibition rates are described. Usually, it is considered that transport activities follow Michaelis Menten kinetics, therefore the estimation of their parameters as well as the mechanisms of inhibition are calculated using methods of the classic enzymology. Further approximations are also detailed to comparatively estimate the substrate specificity by competition analysis with different compounds.

2 Materials

2.1 Transporters Expression in Trypanosoma cruzi

1. Parasites: *T. cruzi* epimastigote culture in exponential growing phase was grown at 28 °C in plastic flasks (25 cm^2), containing 5 mL of BHT medium (started with 5.10^6 parasites/mL) supplemented with 10% fetal calf serum, 0.002% (m/v) hemin, 100 U/mL penicillin, and 100 μg/mL streptomycin [3]. BHT medium: BHI (Brain Heart Infusion) 33 g/L, tryptose 5 g/L, KCl 0.4 g/L, Na_2HPO_4 4 g/L, glucose 0.3 g/L.
2. Measurement of parasites density: Cells were counted using a hemocytometer and viability assays were performed using a colorimetric method for determining the number of viable cells (e.g., MTT Assay).
3. Electroporation buffers and materials: Electroporation power supply (e.g., Gene pulser II, Bio-Rad), 2 mm gap cuvettes and Geneticin (G418). Electroporation buffer: 0.5 mM $MgCl_2$, 0.1 mM $CaCl_2$ in PBS, TE buffer: 10 mM Tris–HCl pH 7.5, 1 mM EDTA [4].

2.2 Transport Assays

1. Buffers: Phosphate-buffered saline (PBS): 137 mM NaCl, 2.7 mM KCl, 10 mM Na_2HPO_4, 2 mM KH_2PO_4. Dissolve in distilled H_2O. Adjust pH to 7.4. Autoclave. Store at 4 °C up to 12 months.
2. Oil mixture: Silicone oil: dibutylphtalate/dinonylphtalate oil mixture (2:1; % v:v) [5].
3. Radiolabeled polyamines: putrescine dihydrochloride, [1,4-^{3}H (N)], (2.294 TBq/mmol). Spermidine trihydrochloride, [terminal methylenes-^{3}H(N)]. Store at −20 °C.
4. Scintillation fluid, e.g., Optiphase "Hisafe" 3 (PerkinElmer).

3 Methods

3.1 Transporters Expression in Trypanosoma cruzi

1. Clone the transporter gene in a suitable vector for the expression in *T. cruzi* (*see* **Note 1**), such as pTREX, pRIBOTEX, or pTEX [6–8]. Consider that pTREX and pRIBOTEX are integration plasmids for stable and constitutive gene expression. Alternatively, pTcINDEX [9], a similar but tetracycline inducible expression vector is suitable for *T. cruzi* transporter expression, especially for toxic proteins (*see* **Note 2**).
2. Purify the plasmid by standard plasmid miniprep and quantify by absorbance. At least 50 μg of plasmid DNA are required. Elute in TE buffer to 1 μg/mL.
3. Before electroporation, warm DNA at 80 °C 10 min to sterilize the DNA sample.
4. Harvest epimastigotes from an exponentially growing culture in BHT medium. 1–3.10^8 parasites/mL per electroporation are required.
5. Wash the parasites with 5 mL of electroporation buffer.
6. Resuspend the pellet with 350 μL of electroporation buffer and transfer to an electroporation cuvette. Add 50 μg of plasmid DNA (50 μL). Mix gently by stirring with your fingers.
7. Put the cuvette in the gene pulser. Give a single pulse of 400 V, 500 μF and immediately place the parasites in 5 mL of fresh BHT.
8. Twenty four hours after electroporation, add G418 to a final concentration of 50–500 μg/mL, depending on the expression vector used.
9. Forty eight hours after electroporation, dilute the culture 1:10 to 1:5 maintaining G418 concentration.
10. Visually follow the growth of the culture along days and replicate if necessary. In 1–2 months it should be completely selected [4].
11. Use a GFP plasmid as control for transport assays and also to follow the transfection and selection.

3.2 Short-Time Transport Assay

This rapid technique is based on the separation of cells from incubation mixture by centrifugation through a silicone oil phase (*see* Fig. 1) [5, 10].

1. Harvest cells in the exponential phase of growth by centrifugation (1500 × *g*, 5 min). Remove the supernatant and wash twice with PBS (*see* **Note 3**).

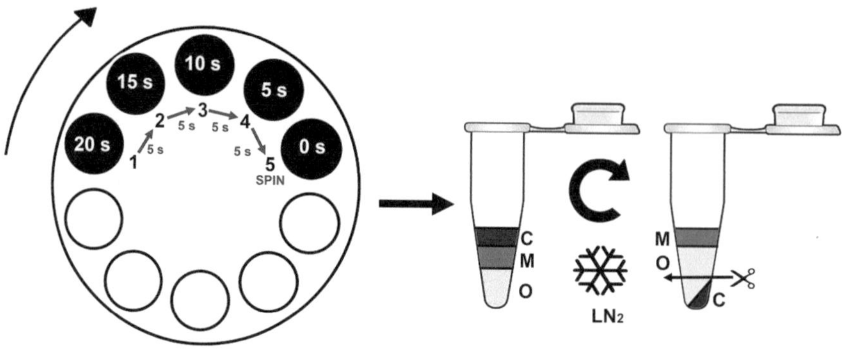

Fig. 1 Scheme of rapid transport technique. First, add silicone oil (O) to five 1.5 mL tubes. Then, add the transport mixture (M) and finally, the sample of cells (C) at times intervals of 5 s to each tube, next centrifuge. Aspirate the transport mixture and wash. Finally, aspirate the silicone oil. Another option is to cut the bottom of the tube containing the cell pellet after freeze the samples with liquid nitrogen (LN_2)

2. Resuspend the pellet in PBS to a concentration of 10×10^7 cells/mL (*see* **Note 4**). Run each assay (for one substrate concentration) at least by triplicate.
3. Pipet 100 μL of silicone oil in 1.5 mL tubes. Use five tubes for each compound concentration. Add 100 μL of PBS containing about 1 μCi of radiolabeled polyamine (*see* **Note 5**) and the inhibitor, if applicable. Centrifuge at 8000 × *g* for 30 s to ensure separation of the silicone oil and incubation medium.
4. To initiate uptake, add 100 μL of cell solution (10^7 cells) to each tube at 5 s time intervals (*see* **Note 6**). Five seconds after the addition of the last sample, stop reaction by centrifugation at 11,000 × *g* for 1 min, which separate the cells from the radiolabeled transport mixture. Store the tubes on ice until all samples are complete.
5. Repeat **step 4** for different substrate concentration.
6. Alternatives to cell pellet extraction: (A) Aspirate the medium by using a vacuum pump. To remove any residual transport mixture, wash the samples by carefully adding 800 μL PBS; then aspirate the buffer layer. Wash once more with PBS. Aspirate the silicone oil by using a Pasteur pipette connected to the vacuum pump (*see* **Note 7**). (B) Immediately after centrifugation, freeze the tubes using liquid nitrogen. Then cut the bottom of the tube containing the cell pellet and put it into a suitable volume of scintillation cocktail.
7. Alternatively (option B), resuspend the cell pellet in 200 μL of distilled water and vortex (*see* **Note 8**).
8. Add 350 μL of scintillation cocktail to the tubes and vortex to mix the contents. Transfer the tubes into vials and measure the incorporated radioactivity using a suitable liquid scintillation counter.

3.3 Long-Time Transport Assay

This technique does not use oil as described above. It is easier to perform and requires less reagents and equipment. However, the transport rates obtained by this method are less precise because of the longer measure times and the consequent underestimation of initial velocities.

1. Collect exponentially growing cells by centrifugation at $1500 \times g$ for 5 min and wash two times with PBS (*see* **Note 3**).
2. Resuspend the cell pellet in PBS at a concentration of 10×10^7 cells/mL (*see* **Note 4**). Run assays at least by triplicate.
3. Add to a 1.5 mL tube 100 μL of cell suspension (10^7 cells) and 100 μL of PBS containing the appropriate radiolabeled polyamine mixture (*see* **Note 5**) and the inhibitor, if applicable (*see* **Note 9**).
4. After incubation during the adequate time (*see* **Note 10**) and temperature, stop the reaction by adding 800 μL of ice-cold PBS. Centrifuge cells at $11{,}000 \times g$ for 1 min and wash twice with 1 mL of ice-cold PBS (*see* **Note 11**).
5. Resuspend the cell pellets in 200 μL of water (*see* **Note 8**).
6. Add 350 μL of scintillation cocktail and vortex to mix the contents. To quantify the radioactivity incorporated into the cells, use a suitable liquid scintillation counter [11].

3.4 Kinetic Parameters Determination

The expression for inward movement of substrate mediated by transporters is identical to the common form of the Michaelis Menten equation for enzyme activity: $V = V_{max}/(S + K_m)$ where V_{max} is the maximum rate of the reaction and K_m is the half-saturation Michaelis constant. First of all, the initial transport rate (V_0) has to be determined. It is common to use a high substrate concentration and evaluate the transport in a specific range of time. In order to describe the transport, its velocity must be determined during an early time interval when the amount of transported substrate is increasing in a linear manner. Next, the transport activity has to be tested using a broad range of substrate concentrations to allow the estimation of the kinetic parameters. Then, using these preliminary data, K_m and V_{max} values could be obtained by adjusting the substrate concentrations like the following: [S] = 0.1-, 0.2-, 0.5-, 1.0-, 1.5-, and 3.0-fold of estimated K_m. Finally, we can calculate the K_m and V_{max} values from the Michaelis Menten curve and the linear transformation of its equation (e.g., Lineweaver-Burke or Hanes-Woolf plots). Variations of kinetic parameters are common between assays (~30–40%) [12].

3.5 Inhibition Assays

Putative inhibitors should be tested as described under Methods 3.2 or 3.3 at least in a tenfold excess with respect to the K_m value for the substrate. First, an isotopic dilution with the unlabeled substrate should be used as positive inhibition control, and a non-inhibitor compound should be used as negative control, both in the same concentration as the putative inhibitors. Then, the kinetic parameters of the transport must be determined in absence and in presence of increasing concentrations of the inhibitor. Type of inhibition (competitive, uncompetitive or noncompetitive) can be determined by analyzing the obtained kinetic parameters [13].

3.6 IC_{50} Determination

The IC_{50} is the concentration of an inhibitor where the transport is reduced by half. IC_{50} can be determined from dose-response curves in which the data is generally plotted as transport (*y*-axis) against $\log_{10}$ of the substrate concentration (*x*-axis). The logarithmic transformation yields a sigmoidal curve and those obtained under identical conditions are helpful to compare different inhibitors efficacy and potency (*see* Fig. 2). In the example, the inhibitor 2 is more potent than inhibitor 1 (the concentration required to diminish the transport to half is lower) but inhibitor 1 has greater maximal efficacy since it reaches higher inhibition percentages [14].

In order to calculate the IC_{50} the transport has to be tested with a broad range of inhibitor concentrations (at least eight, and also without inhibitor) while the substrate remains constant around the obtained K_m value. It is desired to obtain 100% inhibition for the best curve fitting. The IC_{50} can be calculated using a curve fitting statistical software (four parameter logistic function is needed) or it can be estimated through a point-to-point calculation (this method does not fit a sigmoidal curve).

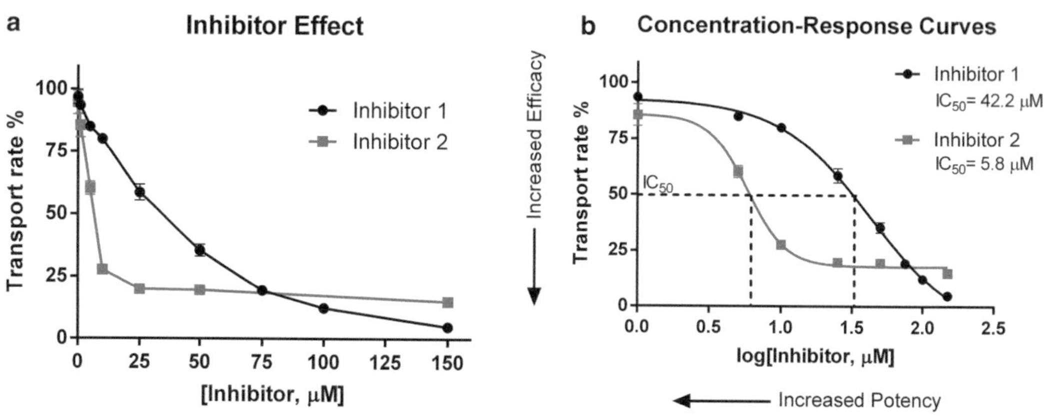

Fig. 2 IC_{50} determination. (**a**) Effect of different inhibitor concentrations on transport (linear scale). (**b**) Dose-response curves. The response is defined as the transport percentage obtained for each inhibitor concentration, and 0 μM of inhibitor is considered 100% of transport. The concentration is plot as the $\log_{10}$ inhibitor concentration. The semi-log plot is the preferred method for plotting dose-response relationships because it becomes easier to accurately determine the IC_{50} value by placing it on a linear portion of the curve

4 Notes

1. Other expression systems could be employed, such as mammalian cells, bacteria and yeast and the transport assays are very similar, especially the long-time transport assay.
2. In addition to *T. cruzi*, these protocols are suitable for expressing transporters in *Phytomonas* sp. using the same plasmids. In the case of *Leishmania* spp., *Crithida* spp., and *T. brucei*, similar constitutive and inducible plasmids are used. Transport assays are exactly the same as that of those explained in this chapter.
3. Check that PBS solution is isotonic. Alterations in the preparation of the solution can dramatically change the transport rates.
4. The cells can be subjected to nutrient starvation in order to diminish the endogenous substrate concentration. This can be achieved by 2 h pre-incubation in PBS supplemented with 2% (w/v) glucose [11].
5. Both methods (short-time and long-time assays) can be used to measure transport of other radiolabeled compounds like amino acids, nucleosides, and sugars.
6. Each substrate concentration is measured at five different times spaced at 5 s intervals (0–5–10–15–20 s), therefore each assay is a time curve where the initial velocity (V_0) is determined.
7. In this option, take care not to resuspend the pellet and aspirate cells with the silicone oil. This is a critical step.
8. The water can be replaced with detergents or other denaturing agents [10].
9. The nonspecific transport and carry over can be measured by using 100× molar excess of the corresponding substrate in transport mixtures [11].
10. The assays are measured at a single time. Initial transport rate (V_0) must be calculated (*see* Subheading 3.4) to determine the standard transport assay time.
11. In the last wash, take supernatant with pipet to minimize the measurement of radioactive residues not incorporated into the cells.

Acknowledgments

This work was supported by Consejo Nacional de Investigaciones Científicas y Técnicas (CONICET, PIP 2010-0685, 2011-0263, and 2013-0664), Agencia Nacional de Promoción Científica y Tecnológica (FONCYT PICT 2012-0559, 2013-2218 and 2015-0539). CAP, MRM are members of the career of scientific investigator; CR and MS are research fellows from CONICET.

References

1. Silber AM, Colli W, Ulrich H, Alves MJ, Pereira CA (2005) Amino acid metabolic routes in *Trypanosoma cruzi*: possible therapeutic targets against Chagas' disease. Curr Drug Targets Infect Disord 5:53–64
2. Pereira CA, Saye M, Wrenger C, Miranda MR (2014) Metabolite transporters in trypanosomatid parasites: promising therapeutic targets but... How to deal with them? Curr Med Chem 21:1707–1712. doi:CMC-EPUB-55425 [pii]
3. Camargo EP (1964) Growth and differentiation differentiation in *Trypanosoma cruzi*. I. Origin of metacyclic trypanosomes in liquid media. Rev Inst Med Trop Sao Paulo 6:93–100
4. Pereira CA, Alonso GD, Ivaldi S, Silber AM, Alves MJ, Torres HN, Flawia MM (2003) Arginine kinase overexpression improves *Trypanosoma cruzi* survival capability. FEBS Lett 554:201–205. doi:S0014579303011712 [pii]
5. Le Quesne SA, Fairlamb AH (1996) Regulation of a high-affinity diamine transport system in *Trypanosoma cruzi* epimastigotes. Biochem J 316:481–486
6. Vazquez MP, Levin MJ (1999) Functional analysis of the intergenic regions of TcP2beta gene loci allowed the construction of an improved *Trypanosoma cruzi* expression vector. Gene 239:217–225. doi: S0378111999003868 [pii]
7. Martinez-Calvillo S, Lopez I, Hernández R (1997) pRIBOTEX expression vector: a pTEX derivative for a rapid selection of *Trypanosoma cruzi* transfectants. Gene 199:71–76. doi:S0378-1119(97)00348-X [pii]
8. Kelly JM, Ward HM, Miles MA, Kendall G (1992) A shuttle vector which facilitates the expression of transfected genes in *Trypanosoma cruzi* and *Leishmania*. Nucleic Acids Res 20:3963–3969
9. Taylor MC, Kelly JM (2006) pTcINDEX: a stable tetracycline-regulated expression vector for *Trypanosoma cruzi*. BMC Biotechnol 6:32. doi:10.1186/1472-6750-6-32
10. Le Quesne SA, Fairlamb AH (1998) Measurement of polyamine transport. Cells in suspension. Methods Mol Biol 79:149–156
11. Pereira CA, Alonso GD, Paveto MC, Flawia MM, Torres HN (1999) L-arginine uptake and L-phosphoarginine synthesis in *Trypanosoma cruzi*. J Eukaryot Microbiol 46:566–570
12. Van Winkle LJ, Bussolati O, Gazzola G, McGiven J, Mackenzie B, Saier MH Jr, Taylor PM, Rennie MJ, Low SY (1999) Chapter 4 Transport kinetics: biomembrane transport. Academic Press, San Diego, pp 65–131. doi:10.1016/B978-012714510-5/50005-2
13. Kotyk A (1975) Cell membrane transport: principles and techniques. Arnost Kotyk and Karel Janacek; in collaboration with the staff of the Laboratory for Cell Membrane Transport, Czechoslovak Academy of Sciences. vol. Plenum Press, New York. Accessed from http://nla.gov.au/nla.cat-vn1835476.
14. Lees P, Cunningham FM, Elliott J (2004) Principles of pharmacodynamics and their applications in veterinary pharmacology. J Vet Pharmacol Ther 27:397–414. doi:10.1111/j.1365-2885.2004.00620.x

Chapter 22

Quantitative Trait Loci for Root Growth Response to Cadaverine in *Arabidopsis*

Nicole M. Gibbs, Laura Vaughn Rouhana, and Patrick H. Masson

Abstract

Root growth architecture is a major determinant of agricultural productivity and plant fitness in natural ecosystems. Here we describe the methods used in a Quantitative Trait Loci (QTL) study that allowed the identification of *ORGANIC CATION TRANSPORTER 1* (*OCT1*) as a determinant of root growth response to cadaverine treatment in *Arabidopsis thaliana*. This protocol screens natural accessions to characterize the variation in root growth response to the naturally occurring polyamine cadaverine, then uses recombination mapping to identify loci that are responsible for the variation existing between two accessions with contrasting phenotypes.

Key words Natural variation, Root architecture, Arabidopsis, Cadaverine, Quantitative trait loci (QTL)

1 Introduction

When growing in highly heterogeneous soil environments, plant roots have to use environmental cues to guide their growth and control their architecture, thereby allowing better anchoring as well as optimal water and nutrient acquisition. For individual plant species, distinct populations exposed to local environmental conditions have evolved different adaptive strategies allowing better use of the local resources, thereby selecting specific combinations of alleles at contributing loci. Therefore, the genetic diversity existing between plant populations can be tapped very effectively to uncover key genetic contributors to the regulation of root growth behavior and architecture. Furthermore, this approach can also be used to investigate root growth responses to environmental cues such as nutrient availability, stress, toxic chemicals (such as paraquat, a herbicide that uses polyamine transporters to move into and within plant cells [1]), or polyamines [2].

Two main genetic strategies have been developed to explore the natural variation existing between plant populations and identify contributing factors: Quantitative Trait Loci (QTL) and

Rubén Alcázar and Antonio F. Tiburcio (eds.), *Polyamines: Methods and Protocols*, Methods in Molecular Biology, vol. 1694, DOI 10.1007/978-1-4939-7398-9_22, © Springer Science+Business Media LLC 2018

Genome-Wide Association Studies (GWAS). QTL analyses have allowed the identification of contributing variant alleles within generally two populations/accessions originally chosen because they vary in the trait of interest. QTL analysis is rather powerful at uncovering rare contributing alleles provided they are represented within the two populations under investigation. GWAS, on the other hand, take advantage of linkage disequilibrium to identify genetic variants that correlate with the trait under investigation, using large numbers of distinct populations representing the diversity of the species. While inefficient at identifying rare alleles, GWAS takes fuller advantage of the overall variation existing between multiple populations to identify contributing factors, and has been successfully used to investigate the genetic architecture of root growth [3].

We have used a QTL approach to identify loci that contribute to the variation in root growth response to exogenous cadaverine existing between two *Arabidopsis thaliana* accessions, L*er* and Cvi. This analysis allowed us to identify *ORGANIC CATION TRANSPORTER 1 (OCT1)* as a QTL contributing to this variation [2]. Cadaverine is a naturally occurring diamine that is produced by plants and microbes of the rhizosphere and phyllosphere communities, and has been found to modulate Arabidopsis root architecture by affecting primary root growth, root skewing and waving on hard surfaces, and lateral root numbers (Reviewed in [4]).

Our QTL analysis took advantage of initial studies that demonstrated L*er* to be more sensitive to exogenous cadaverine than Cvi. Therefore, we investigated the genetic basis of this variation using a QTL analysis that relied on a preexisting population of 162 recombinant inbred lines (RILs) that were originally genotyped at 293 marker loci [5] to map the corresponding QTLs. This population was originally created by crossing L*er* and Cvi, self-pollinating the corresponding F1s, and recovering segregating progenies from this cross. This was followed by nine generations of self-pollinations to generate 162 RILs with mostly homozygous recombinant chromosomes resulting from meiotic crossovers between L*er* and Cvi homologs in the initial F1 heterozygotes and following generations [5].

The QTL analysis workflow described in the following protocol and illustrated in Fig. 1 starts by screening natural accessions for a trait of interest to identify lines with contrasting responses. Then, two distinct accessions are chosen as the parental lines for the RILs (L*er* and Cvi in our case). Once an RIL population has been obtained and each of its constituting lines has been mapped to determine the parental origin of each of its chromosomal segments, RILs are screened for the trait of interest. QTL-mapping software is then used to identify regions of the genome that significantly correlate with the growth response. Near isogenic lines (NILs), which have small, well-defined segments of one parental line (Cvi) overlapping with the mapped locus of interest introgressed into the

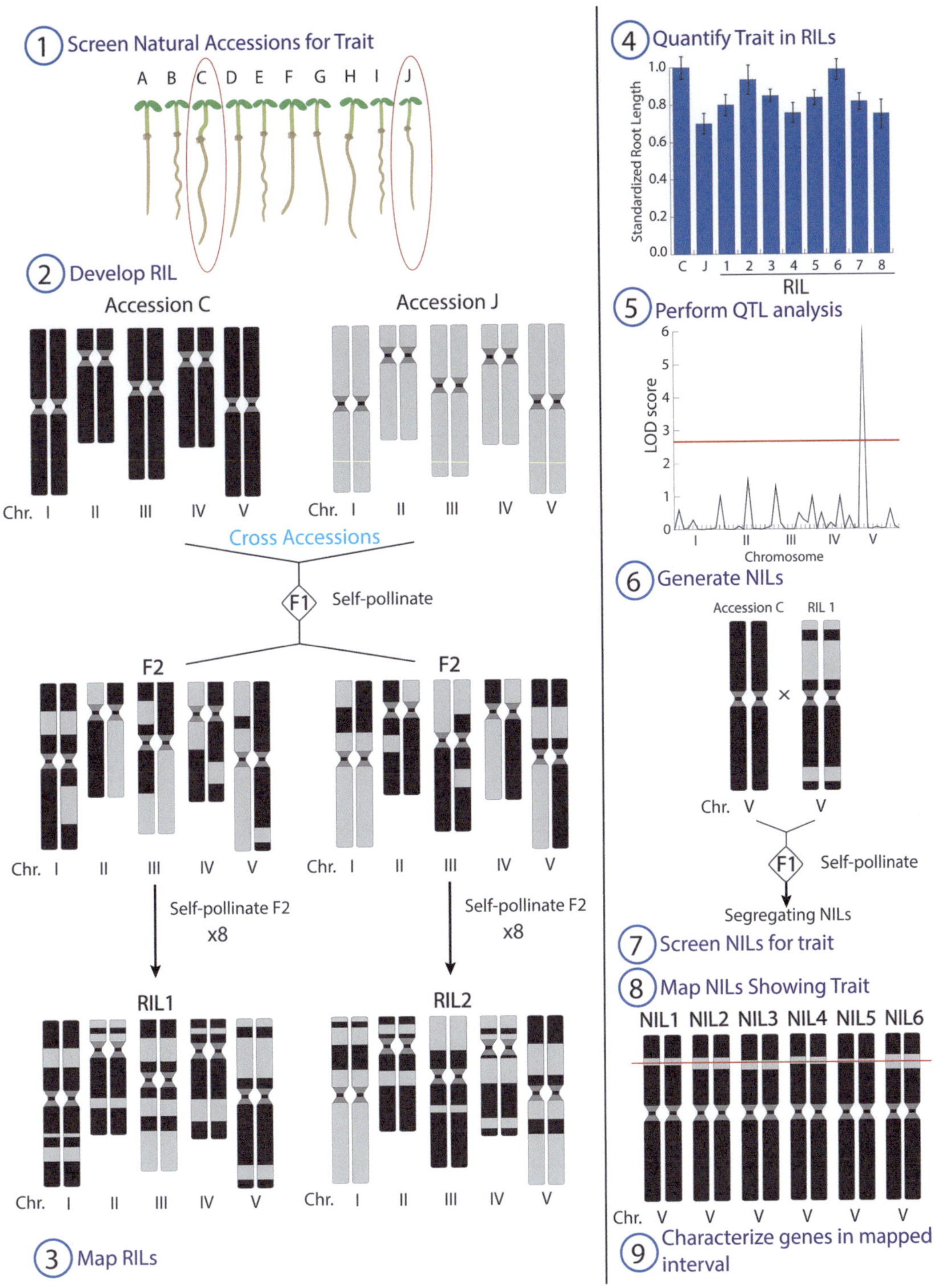

Fig. 1 Workflow for QTL analysis. (**1**) Screen natural accessions for a trait of interest. For this example, we will use root length as the desired trait. The *circled* accessions, C and J, show seedling roots with the longest and shortest root lengths, respectively. (**2**) Develop RILs. Accessions showing contrasting traits are crossed and self-pollinated for eight generations, until they are homozygous at most loci. The drawing shows the five

genome of the other parent (*Ler*), are then phenotyped which allows the mapping of the QTL of interest to a narrower region of the chromosome. The NILs carrying the smallest introgressed Cvi chromosomal fragment still containing the gene of interest are then backcrossed to the original *Ler* parent, and the corresponding F1s are self-pollinated. Segregating F2 plants are screened for retention of the trait. This last step allows identification of additional crossover events in closer proximity of the gene, leading to fine-mapping of the QTL. Finally, genes within the fine-mapped interval are characterized by looking at expression patterns, sequencing to identify associated polymorphisms, and reverse genetics to identify mutant phenotypes relevant to the trait of interest.

2 Materials

All solutions should be prepared with ultrapure water (deionized water filtered using Millipore system to 18.2 MΩcm). For all reagents, follow safety instructions, noting necessary personal protective equipment (PPE), and waste disposal procedures.

2.1 Preparing Growth Media

1. Agar, plant cell culture tested. (Sigma A1296).
2. Linsmaier and Skoog media with macronutrients, micronutrients, vitamins, 1.5% sucrose, and pH buffered to 5.7. (Caisson LSP04-1LT).

Fig. 1 (continued) Arabidopsis chromosomes, colored based on the accession they originated from (*black* for accession C and light grey for accession J). Ambiguous segments surrounding the centromeres are in darker *grey*. In the RILs, chromosome segments are also colored based on their parental origin, as determined by molecular typing in three. (**3**) Map RILs. Use molecular markers to determine the parental origin of each chromosomal segment. (**4**) Screen RILs for the trait of interest. Shown is a graph of average standardized root length for parents C and J and for RILs 1–8, with standard errors. (**5**) Perform QTL analysis. This will combine the mapping done in **step 3**, with the phenotyping shown in **step 4**, to identify chromosomal segments that contribute to the variation in the trait of interest. The graph shown is a representation of the type of plot that can be obtained using WinQTL Cartographer V2.5 (*see* **Note 22**). (**6**) Generate NILs. Using an RIL carrying a trait of interest from parent J, backcross it to parental accession C (in this case), and repeat the process until only a small segment of parent J is retained within parent-C genome. Different NILs carry distinct segments of J in an otherwise C-genome background. (**7**) Screen NILs for the trait of interest. This step will identify NILs that still carry the gene from J contributing to the trait of interest. (**8**) Map the introgressed J segments present in the NILs still showing retention of the trait of interest, using molecular markers, and use this information to identify the shortest chromosomal segment carrying this trait determinant. If this segment still contains too many genes for functional characterization, cross the NIL carrying the shortest segment of J to wild-type accession C, recover F1s and self-pollinate them to generate a segregating F2 population. Identify progeny with chromosomes that are further recombined within the introgressed segment, and identify those that still carry the trait of interest (the latter step is not shown in this figure). (**9**) Functionally characterize genes contained within the shortest introgressed chromosomal region. Use expression studies and reverse genetic analysis to identify the gene responsible for the trait of interest

3. Cadaverine dihydrochloride (>99% purity) (Sigma 33220-10G-F). Prepared as a 1 M stock in ultrapure water.
4. 10 cm × 10 cm square petri plates, nongrided. (Nunc Lab-Tek™ Petri Dishes Polystyrene 4021).
5. 50 mL conical tubes, sterile (Corning Falcon 352098).

2.2 Plating Seeds

1. Whatman filter paper grade 541, 7.0 cm (1541 070).
2. 95% ethanol.
3. Toothpicks.
4. 1.5 mL microfuge tubes.
5. 3 M Micropore surgical tape.

2.3 Root-Growth Measurements

1. High-resolution document scanner, 4800 dpi maximum resolution (Epson Perfection V33).
2. Image J software (https://imagej.nih.gov/ij/index.html).
3. Neuron J Plugin for Image J (https://imagescience.org/meijering/software/neuronj/).
4. Adobe Photoshop.
5. Microsoft Excel.

2.4 QTL Mapping

1. R software (https://www.r-project.org).
2. R/qtl add-on package (http://www.rqtl.org).

2.5 Identification of Causative Loci

RNA extraction

1. 2 mL microfuge tubes.
2. 4.5 mm steel BB (Daisy).
3. Benchtop tissue grinder (Retsch Mixer Mill MM200).
4. Ambion TRIzol reagent (ThermoFisher Scientific 15596018) *TRIzol is hazardous, read MSDS before use.
5. Direct-zol MiniPrep Kit (Zymo Research R2072).
6. Chloroform Certified A.C.S. (Fisher C298-4) *Chloroform is hazardous, read MSDS before use.
7. DNAse/RNAse-free 1.5 mL microfuge tubes (BioExpress GeneMate C-3260-1).
8. 200 μL and 1000 μL DNAse/RNAse-free filter pipette tips (VWR).
9. DNAse/RNAse-free water.
10. Nanodrop.

cDNA synthesis

1. DNase (Promega RQ1 RNase-Free DNase M6101).
2. DNase/RNase-free 8-well PCR strip tubes.
3. cDNA synthesis kit (SuperScript III First-Strand Synthesis Invitrogen 18080051).

qRT-PCR

1. Real-Time qPCR Master Mix (Bullseye Evagreen MidSci BEQPCR-S).
2. Primers.
3. DNase/RNase-free water.
4. Mineral oil.
5. DNase/RNase-free microfuge tubes.
6. DNase/RNase-free 8-well PCR strip tubes.
7. Multichannel pipette.
8. 96-well PCR plate (Framestar 480/96, for Roche Light Cycler 480 4TI-0951).
9. 96-well plate seal film (ThermalSeal RT2 Film, qPCR, Optically Clear, For Roche Light Cycler 480 TS-RT2-100).
10. Quantitative PCR instrument (Roche Lightcycler 480 II).

DNA extraction

1. 1.5 mL microfuge tubes.
2. 1.5 mL microfuge tube pestles.
3. Microscissors.
4. 95% ethanol.
5. Shorty Buffer: 0.2 M Tris–HCl, pH 9.0, 0.4 M LiCl, 25 mM EDTA, 1% SDS.
6. Molecular-grade Isopropanol (Fisher Scientific A416-500).
7. Tabletop centrifuge.
8. Paper towel.
9. TE buffer pH 8.0: 10 mM Tris–HCl pH 8.0, 1 mM EDTA.

PCR

1. DNA polymerase (Econotaq PLUS GREEN 2× Master mix Lucigen 30033).
2. Primers.
3. 1.5 mL microfuge tubes.
4. Mineral oil.
5. PCR tubes or PCR plate.

6. Thermocycler.
7. Agarose LE (GoldBio A-201-500).
8. TAE Buffer: 40 mM Tris, 20 mM acetic acid, 1 mM EDTA.
9. DNA ladder (Minnesota Molecular Hi-Lo 1010).
10. 10 mg/mL Ethidium Bromide solution.

3 Methods

3.1 Plant Growth Media Preparation

1. Prepare 0.5× strength Linsmaier and Skoog (LS) media containing macronutrients, micronutrients and vitamins with 1.5% sucrose, pH buffered to 5.7 [6] (*see* **Note 1**). If using an alternative media source and buffering and pH adjustment are required, use 2 mM 2-(N-morpholino)ethanesulfonic acid (MES) to buffer and correct the pH to 5.7 using KOH [7].
2. Prepare a 1.5% agar solution by measuring out 7.5 g agar and place into 1 L autoclave safe bottle (*see* **Notes 1** and **2**).
3. Add 500 mLs liquid LS to 7.5 g agar to generate a 1.5% agar LS solution. Mix on a magnetic stir plate.
4. Autoclave media for 20 min at 121 °C, 15 PSI. Loosen bottle caps to allow media to safely expand before autoclaving (*see* **Note 3**).
5. After autoclave cycle is complete, place into a 60 °C dry incubator until media is cooled sufficiently to handle (*see* **Note 4**).

3.2 Pouring Plates

1. Prepare cadaverine solution as a 1 M stock in ultrapure water. Cadaverine can be added at a final concentrating ranging from 50 to 500 μM to alter Arabidopsis root growth (*see* **Note 5**).
2. After media has cooled to a temperature that allows it to be handled, cadaverine can be added. Homogenize the medium by stirring with a magnetic stir bar to evenly distribute the cadaverine (*see* **Note 6**).
3. Using a sterile conical tube, measure 35 mLs of agar and pour into petri plates. Cover plate with lid and allow agar to solidify completely before moving (*see* **Notes 7** and **8**).
4. After the medium solidifies, collect the plates and wrap with plastic to prevent contamination and evaporation, and store at 4 °C until ready to use (*see* **Note 9**).

3.3 Plating Seeds

1. Start by choosing natural accessions to assess for a response. Chosen accessions should be evolutionarily diverged, as polymorphisms are required for analysis (*see* **Note 10**).
2. Transfer seeds to be sterilized to a 1.5 mL microfuge tube.
3. Add ~1 mL of 95% ethanol.

4. Vortex tube for 1 min at top speed.
5. Decant ethanol and repeat, adding 1 mL of ethanol then vortex.
6. Using a P1000 pipette, pipette to resuspend seeds in ethanol and transfer onto Whatman paper on a clean bench or within a laminar flow hood.
7. Allow seeds to dry completely on Whatman paper.
8. While seeds are drying, sterilize toothpicks using 95% ethanol, and allow to dry.
9. To plate seeds, first wet a sterile toothpick by touching condensation on the plate, and use it to pick up a dry, sterile seed from the Whatman filter paper by simple contact (*see* **Notes 11–13**).
10. Transfer this seed to the surface of the medium and deposit it on this surface by gentle contact. Seeds should be plated approximately 0.5 cm apart, in a single line.
11. For each accession, plate 10 seedlings per plate with three plates each accession. $N = 30$ seedlings.
12. Wrap plates with Micropore surgical tape and store in dark at 4 °C to stratify for a 48 to 96 h (*see* **Note 14**).

3.4 Seedling Root Growth Measurements

1. Take plates out of cold treatment and move to a growth chamber set to the following parameters: 22 °C, 16 h light/8 h dark cycle, 60 μmol m^{-2} s^{-1} cool-white light. Randomize the position of the plates within the growth chamber to avoid position-specific effects.
2. Arrange plates such that they are tilted back at a 30° angle to stimulate root waving. Indeed, some accessions of Arabidopsis have roots that wave or skew away from vertical when growing on tilted hard-agar surfaces (Reviewed in [8]); [9].
3. Allow seedlings to grow for 5 days or more after moving to the growth chamber.
4. Using a high-resolution document scanner, scan plates with agar-side down on the bed of the scanner in a dark room with the scanner lid open. Scan using at least 300 dpi image resolution and save as <TIFF> or <JPEG> file.
5. Install Image J following software developers' instructions, then download the Neuron J plug in (*see* **Note 15**) (Reviewed in [10]); [11].
6. Using Photoshop or similar image modification program, alter images to be processed with Neuron J.
   (a) Change images to grayscale.
   (b) Change images to 8-bit.
   (c) Adjust images to 400 dpi.
   (d) Save as a <TIFF> file.

7. Open the Neuron J plug-in and load the modified image. Using 'Add tracing' tool, trace the length of the root (*see* **Note 16**).
8. After all roots are traced for a particular accession, input data into Excel. Take an average of root length for each line. Other potential root growth traits that could be assessed include: root angle, Vertical Growth Index (VGI), Horizontal Growth Index (HGI), and root straightness. Vertical Growth Index (VGI) is measured by the ratio of the displacement of the root tip on the y-axis, divided by the length. Horizontal Growth Index (HGI), a measurement of skewing, is the displacement of the root tip on the x-axis, divided by the length. Straightness is the distance of the position of the root at the start of the measurement divided by the length. Root waving can be quantified by measuring the period or amplitude of the waves [12, 13].
9. Standardize each respective line to growth on the control media, and follow up with ANOVA, if comparing all accessions, or T-test, to make pairwise comparisons, to determine significance.

3.5 QTL Mapping

1. Choose a recombinant inbred line (RIL) population between parental accessions with disparate phenotypes for the trait of interest (*see* **Note 17**). The Arabidopsis Biology Resource Center (ABRC) has several sets of recombinant inbred lines (RILs) available for purchase. Many of these RILs are already well characterized with associated molecular marker maps that indicate the pattern of parental origin along the five chromosomes (*see* **Note 18**).
2. Sterilize and plate the RIL and parental seeds (*see* **Note 19**). All RILs and both parental lines should have at least three plates of ten seeds each plated in a single row about 1 cm from the top edge of the plate. This scheme allows for three biological replicates per RIL and parental line and a total of $n = 30$ seedlings. Stratify seedlings and move to the growth chamber as outlined in Subheading 3.3, taking care to randomize the position of the plates within the growth chamber. Plates should be tilted back at a 30° angle to stimulate root waving (*see* **Note 20**).
3. Once seedlings have germinated and grown for the chosen time period (6 days in our experiment), take scans of plates and quantify root growth, as outlined in **steps 4–9** of Subheading 3.4. Using Excel, calculate the mean value of each measured trait for each RIL.
4. Obtain the mean for each trait over the three separate biological trials for each RIL and parental line. Once the data are compiled, check to determine whether the data for each trait follows a normal distribution. The R program [14] http://

www.R-project.org/ has several methods to check for normality. Basic R functions for histograms, kurtosis, skewness, and qqnorm plots should first be used to look at the distribution of the data. In addition, the Shapiro-Wilk normality function ("shapiro.test") and various algorithms from the nortest package [15] can be used to evaluate if the assumption of normality is valid for each trait. If the data is not normally distributed, the data must be transformed prior to further analysis.

5. For the chosen RILs, obtain the molecular mapping data available on ABRC's website in Excel format, or use mapping data included in publications for the particular RILs (*see* **Note 21**). For each marker location, this data will delineate which parental allele the RIL has.
6. Various types of QTL analysis can be carried out in the R program using the R/qtl package [14, 16]. The R/qtl manual is available at http://www.rqtl.org.
7. Load R/qtl within R and use the "read.cross" function to import mapping (genotype) and phenotype data. R/qtl accepts this data in various formats from Excel, MapMaker, Map Manager, or QTL Cartographer. See the R/qtl resource for examples of how the data must be formatted in each case (http://www.rqtl.org/sampledata/) (*see* **Note 22**).
8. Once data is imported, it can be utilized for interval mapping using the function "scanone" and also to create two-dimensional plots using the function "scantwo". These two methods allow mapping of loci that contribute additively to a trait of interest and also areas where there might be genetic interactions such as epistasis. See the R/qtl manual for detailed instruction on how to set up data to perform these scans, or the R/qtl guide book [17].
9. Mapping methods will provide LOD scores for the probability that a QTL is located in a particular genomic region. To determine the LOD threshold for a particular data set, permutations of that data are necessary. This randomizes the given phenotypic data relative to the genotypic data to calculate a genome-wide threshold for how high a LOD score must be to consider a QTL peak to be significant. In practice, at least 1000 permutation replicates are performed, but 10,000 are preferable when possible. Permutations calculations are also available in R/qtl within functions "scanone" and "scantwo".

3.6 Identification of Causative Loci

1. To determine the gene(s) responsible for phenotypic differences between parental lines, fine mapping using Near Isogenic Lines (NILs) can be carried out to narrow down candidate regions.

2. NILs created from a variety of Arabidopsis accessions are available through ABRC. If lines are unavailable, they can be generated by back crossing RILs to one of the parental lines to create individuals with a defined segment of genetic material from one parent introgressed into the background of the other parent. NILs with an introgression in candidate regions may be screened for root growth responses as done in the QTL analysis above in Subheading 3.5.
3. Once NILs carrying the phenotype of interest have been identified, they may be backcrossed to the background parent to narrow the introgression segment. Allow the F1 seeds to self-pollinate and analyze the segregating F2 progeny, with the goal of identifying recombination breakpoints that are located closer to the QTL. Screen the F2 progeny for the trait of interest. If the trait is recessive, allow the F2 seedlings to self for another generation, identify homozygous lines, and then map the trait.
4. Ideally, after analysis of backcrossed NILs, the QTL will be mapped to a narrower segment of the chromosome (carrying few genes). Proceed by characterizing the polymorphisms present in this mapped chromosomal segment. Determine whether the polymorphisms are located within genes and result in amino acid changes, or if they are intergenic and potentially affect gene regulation.
5. To identify the causative gene, use quantitative real-time PCR (qRT-PCR) to analyze tissue-specific expression of each gene and define expression-responses to exogenous cadaverine. Gene expression analysis requires three main steps described Subheading 3.7 below: RNA extraction, cDNA synthesis, and qRT-PCR.
6. Use reverse genetics to characterize gene function. If T-DNA-insertion mutant lines are unavailable in the Arabidopsis stock centers, use CRISPR-Cas9 as an alternative method to disrupt gene function [18] (*see* **Note 23**). Using either method, generate homozygous mutant lines, and analyze root growth responses to cadaverine (*see* **Note 24**). It will be important to genotype the mutant allele to verify position of the mutation, as well as confirm the mutant is homozygous (*see* **Note 25**). DNA extraction and PCR methods are described in Subheading 3.8.
7. Similarly, overexpression lines can be generated. Transform wild-type plants (each parental accession) with a transgene that fuses the coding segment of each candidate gene with the strong CaMV 35S promoter, and determine the root-growth responses to cadaverine of each transgenic line. Correlate the response to transgene expression levels.

8. The following are examples of potential outcomes in identifying causative loci:
   (a) If gene expression correlates with response of accessions, yet few to no amino acid changes occur in the coding region, the gene can be cloned and overexpressed using a strong promoter. It is likely the level of expression that dictates the root growth response.
   (b) If little change is observed in expression, but the null mutant shows a response, likely the QTL identified an amino acid change within the coding region.
   (c) The possibility that the alleles carried by the parental accessions differ by both alterations in expression regulation and changes in amino acid coding potential should be explored using promoter swap experiments (*see* **Note 26**).
   (d) If neither changes in expression level or protein-coding potential are found, the wrong candidate gene has been chosen, and another gene in the mapped interval should be investigated (*see* **Note 27**).

3.7 Gene Expression Analysis

RNA extraction

***Steps 4–13, with exception of centrifugation, should be carried out in a fume hood. Read documentation for TRIzol (Invitrogen** MAN0001271**) and Direct-zol kit (Direct-zol RNA MiniPrep Ver. 1.1.3) before use. The described protocol is paraphrased from the manufacturer's instructions.**

1. Prepare at least ten seedlings per biological replicate, with at least three biological replicates. All samples should be grown at the same time, under the same conditions to be able to compare expression. In our case, seedlings were grown in the presence or absence of 50 μM cadverine for 7 days, then whole roots were dissected.
2. Freeze tissue in liquid nitrogen in a 2 mL microfuge tube with a 4.5 mm steel BB bead (*see* **Note 28**).
3. Using a tissue homogenizer shake tubes at 25 beats/second for 30 s. After cycle has completed, put tubes back in liquid nitrogen. Repeat once more, so tubes have been shaken twice.
4. Once all tissue is homogenized, transfer to ice and add 500 μLs TRIzol reagent and vortex.
5. Allow samples to incubate at room temperature for 10 min.
6. To perform phase separation, add 200 μLs chloroform and invert tubes. Phase separation will separate protein from nucleic acid and reduce the likelihood the RNA will degrade. Allow chloroform to sit at room temperature for 3 min.
7. Spin tubes at 4 °C at 6,000 × *g* for 15 min.

8. Remove upper, colorless phase and transfer it to a fresh tube. Avoid the in-between layer.
9. Add equal parts of 100% ethanol and mix by inverting.
10. Transfer total volume to the Direct-zol column and spin for 1 min at 13,000 × *g*.
11. Add 400 μLs Direct-zol RNA prewash and spin for 1 min.
12. Add 700 μLs RNA wash buffer and spin for 2 min.
13. Add 50 μLs of DNase/RNase-free water and spin for 1 min to elute. If higher concentration is desired, add only 30 μLs of water to the column.
14. Nanodrop samples, blanking with water, and record the amount of RNA.
15. Store samples at −80 °C until cDNA synthesis.

cDNA synthesis

***Read documentation for SuperScript III kit (Invitrogen** MAN00013460) **before use. The following protocol is paraphrased from the manufacturer's instructions, with minor modifications:**

1. Use Nanodrop readings to determine the sample with the lowest concentration of RNA.
2. Keeping RNA on ice, dilute RNA to a uniform concentration value totaling 8 μLs of RNA in DNAse/RNase-free water into DNAse/RNAse-free 8-well strip tubes. No water should be added to the sample with the lowest concentration of RNA. Return RNA stocks to −80 °C for later use (*see* **Note 29**).
3. DNase treat samples by adding 1 μL of DNase, and 1 μL of DNase Buffer.
4. Spin down in a strip-tube centrifuge, and incubate DNase-treated RNA samples at 37 °C for 30 min.
5. Return samples to ice. Add 1 μL of Stop solution to each reaction.
6. Spin samples down, and put back into thermocycler for 5 min at 65 °C.
7. Using a SuperScript III Kit, add 1 μL of oligo d(T) and 1 μL of dNTPs per reaction. Spin down, and put in thermocycler for 5 min at 65 °C.
8. Return samples to ice. Prepare a master mix of 2 μLs RT buffer, 3 μLs 25 mM $MgCl_2$, 2 μL 0.1 M DTT, 1 μL RNaseOUT and 1 μL SuperScript III RT, per reaction. Spin samples down, and put into thermocycler for 50 min at 50 °C. Directly following the 50 min incubation, increase to 85 °C for 5 min.

9. Spin samples down, and return to ice for 5 min. RNase treat samples by adding 1 μL RNase H to each reaction and incubate at 37 °C for 20 min.
10. Spin samples down, and prepare 50 μL aliquots of 1:10 diluted cDNA for qRT-PCR in DNAse/RNAse-free water (*see* **Note 30**). Samples should be stored at −20 °C until ready for use.

qRT-PCR

1. qRT-PCR primers should amplify ~200-base pair regions that span introns (*see* **Note 31**). Intron spanning regions will help determine if there is genomic DNA contamination. PCR reactions using genomic and cDNA should show distinct sizes if an intron-spanning primer set is used. Primers should also be close to the 3′ end of the gene, as reverse transcriptase may not completely transcribe long genes. BLAST primer sequences to help ensure specificity to the target gene. Reference genes for qRT-PCR are outlined in: [19] (*see* **Notes 32** and **33**).
2. Dilute primers to 10 μM in sterile TE pH 8.0. Store primers at −20 °C until use.
3. qRT-PCR should be done in a 96-well plate. All biological replicates should be in the same plate. For each sample, it is necessary to do at least two technical replicates for each gene tested.
4. Keep all reagents on ice, and protect EvaGreen from light. Vortex and then briefly spin down all reagents.
5. Aliquot 5 μLs of cDNA into 96-well plate using a multichannel pipette. For each set of samples to be tested, one water control should be used.
6. Prepare master mix. The recipe below will generate 15 μLs for a 1× reaction. Calculate the amount required based on the number of samples, and add 10% to account for pipetting errors.
   (a) 4 μLs DNase/RNase-free water.
   (b) 10 μLs EvaGreen.
   (c) 0.5 μLs Primer 1.
   (d) 0.5 μLs Primer 2.
7. After master mix is prepared, vortex and spin down.
8. Aliquot the master mix into strip tubes and use a multichannel pipette to transfer the master mix into the 96-well plate. Pipette the master mix directly into the cDNA, changing tips each time.
9. After all the reaction components are mixed, add 10 μLs mineral oil to the top of the reaction by pipetting at the top side of the well. Seal the plate with film.

Table 1
Suggested program for qRT-PCR

Cycle name	# of Cycles	Temperature (°C)	Time (min)	Temperature ramp	Acquisition
Pre-incubation	1	95 °C	5:00	4.4 °C/s	None
Amplification	45	95 °C 55 °C 72 °C	0:10 0:20 0:30	4.4 °C/s 2.2 °C/s 4.4 °C/s	None None Single
Melt curve	1	95 °C 60 °C 92 °C	0:10 1:00 N/A	4.4 °C/s 2.2 °C/s 0.57 °C/s	None None Continuous (1 acc/C)
Cooling	1	40 °C	0:10	1.5 °C/s	None

Temperature Ramp defines how quickly the temperature change occurs. Acquisition refers to how often data is quantified for each step

10. Spin the plate down in a plate centrifuge for 30 s at 650 × *g*. Make sure to have a plate balance. The plate must fit properly into the centrifuge adapter. If not, wells of the 96-well plate may be crushed.
11. Put plate into quantitative PCR instrument.
12. Start the program. A suggested program is listed in Table 1. The program may need to be modified according to primer annealing temperature, as well as manufacturer instructions for master mix.
13. After plate has cooled, output cycle number and melt curve information. Cycle number is determined at the detection threshold and is a basis for quantifying the amount of expression. The method in which cycle number is calculated is dependent on the instrument, and software used for qRT-PCR analysis. The melt curve is generated by heating the PCR reaction after the amplification cycle and measuring the fluorescence emitted as the temperature increases.
14. The melt curve should be assessed to ensure there is no genomic DNA contamination. If genomic DNA contamination is present, the melt curve should show a shouldered peak.
15. Check to make sure no amplification occurs in water samples, which would suggest contamination.
16. Gene expression can be analyzed in Excel. First, calculate the average cycle threshold (Ct, or number of cycles required for the fluorescent signal to exceed background level) of the two technical replicates for all samples (*see* **Note 34**).
17. Subtract the cycle threshold of the gene tested from the reference gene for each cDNA sample.

18. For each sample, take two raised to the difference between cycle thresholds. This can be done using the "power" function in Excel (*see* **Note 35**).
19. Take the average of three biological replicates of the control treated group. Divide all treatments by this average to standardize the control group to 1. This yields expression relative to the control group.
20. Average biological replicates for each treatment group. They will be represented as fold change in relation to the control group. Calculate standard deviation, and use a T-test to determine significance.

3.8 Genotyping Mutant Alleles

Shorty Buffer-Based DNA Extraction for Genotyping T-DNA alleles

1. Sterilize seeds carrying a candidate T-DNA allele as well as wild-type seeds, with 95% ethanol as described in Subheading 3.3.
2. With a sterile toothpick plate seeds 0.5× LS media with 1.5% agar (*see* Subheadings 3.1 and 3.2). For genotyping, ten seedlings per genotype should be plated. Wrap plates with micropore tape and store in the dark at 4 °C for 48–96 h.
3. After stratification, move plates to a growth chamber set to same parameters outlined in Subheading 3.4. For genotyping, plates can be grown vertically. After 10 days of growth, clip a cotyledon with 95% ethanol sterilized microscissors, and place the cotyledon in a microfuge tube. Label the microfuge tube, and mark the corresponding seedling on the agar-side of the plate to allow subsequent genotype assignment. Rinse and clean microscissors with with 95% ethanol, and proceed to the next seedling.
4. After all cotyledons are clipped, use a clean pestle to grind tissue (*see* **Note 36**).
5. Immediately after grinding tissue, add 400 μLs Shorty Buffer [20].
6. Centrifuge samples for 3 min at 16,000 × *g* to pellet the cell debris.
7. Transfer 300 μLs of supernatant to a fresh tube. Discard tube containing cell debris.
8. Add 300 μLs isopropanol to supernatant and mix by inverting tube 4–6 times. Do not vortex; this will cause the DNA to shear.
9. Centrifuge samples for 10 min at 16,000 × *g* to pellet DNA.
10. Decant liquid, and place tube upside down on a paper towel to dry for approximately 20 min. All liquid should evaporate before proceeding to the next step.

11. Once the tubes are dry, resuspend DNA in 300 μLs TE pH 8.0.
12. Store samples at −20 °C.

PCR

1. Design gene-specific primers to allow for 600-base pair amplification of wild type sequence. Primers should anneal around 200-base pairs on either side of the T-DNA insertion in the proposed T-DNA allele. The location of T-DNA inserts can be found by searching T-DNA Express (http://signal.salk.edu/cgi-bin/tdnaexpress). Use primer-design software, such as Primer3 (http://biotools.umassmed.edu/bioapps/primer3_www.cgi) to generate primers that are ~22-base pairs in length, 50% GC content, and a 60 °C T_m.
2. T-DNA-specific primers will be dependent on the source of the insertion mutant under investigation: For Salk and Wisc lines, see http://signal.salk.edu/tdnaprimers.2.html, or MLB1 primer [21]; for SAIL lines, see https://www.arabidopsis.org/abrc/sail.jsp; and for GABI-KAT lines, see https://www.gabi-kat.de.
3. Dilute PCR primers to 100 μM in filter-sterilized TE pH 8.0 to make a stock solution. Make a working primer solution at 10 μM in ultrapure water. Store both primer stock and working solution at −20 °C.
4. To genotype, use a PCR recipe with 20 μL reaction volumes: 5 μLs DNA, 4 μLs ultrapure water, 0.5 μLs primer 1, 0.5 μLs primer 2, 10 μLs DNA polymerase.
5. Add 25 μLs mineral oil to the top of the PCR reaction. If using a PCR plate, seal with tape.
6. Suggested cycling conditions (*see* **Notes 37** and **38**):
   (a) 94 °C 2:00 min.
   (b) 94 °C 0:20 s.
   (c) 55 °C 0:30 s.
   (d) 72 °C 0:45 s.
   (e) 72 °C 5:00 min.

   *Repeat B-D for 39 cycles.
7. Once reaction has finished, store PCR reaction at 4 °C, or immediately run on a 1% agarose gel in 1× TAE buffer containing 0.6 μg/mL Ethidium Bromide (EtBr) (*see* **Note 39**). Use DNA ladder to ensure the PCR product is of the expected size.

4 Notes

1. When preparing media, agar should be added directly into vessel it will be autoclaved in, otherwise, agar will sink and it will be unevenly distributed when aliquoted.
2. If root analysis will be done on plant species other than Arabidopsis, the concentration of agar may need to be modified to ensure the root does not penetrate the agar.
3. When autoclaving agar media it may be necessary to increase the cycle time to ensure agar is in solution.
4. After autoclaving, agar media should be poured into plates the same day. Agar media will become contaminated if left too long in incubator before pouring into plates.
5. Cadaverine stock solution is slightly yellow in color and should be stored at 4 °C. Once in solution, cadaverine should not be filtered to sterilize as cadaverine is retained in the filter and the final concentration will be altered. The stock does not appear be a source of contamination if ultrapure water is used.
6. Polyamines, while fairly stable, should be added to media after autoclaving. Other chemicals, depending on stability, may be added before autoclaving.
7. When preparing media, pour liquid agar slowly into plates to avoid bubbles. Bubbles will introduce a non-homogenous environment.
8. It is important to be fairly accurate in measuring the amount of agar media added to the plates. Variation in amount will influence the mechanical and chemical properties of the medium, and its moisture, altering root growth behavior.
9. Cadaverine-containing media are stable for up to 1 year at 4 °C. However, in our experience, putrescine-derived polyamines degrade over time in media.
10. For some plant species, sterilization with 95% ethanol will kill the embryo. If less harsh sterilization is required, there is a higher risk of fungal or bacterial contamination. If this is the case, excluding sucrose may help to deter contamination. However, it is necessary to point out that sucrose will affect plant root growth, and the concentration of sucrose should be kept consistent between trials.
11. When plating seeds, keep the distance of the seeds from the top of the plate consistent. This will ensure the amount of light received by seedlings will be uniform, and root growth is not altered by exposure to liquid condensation that occasionally occurs at the bottom of the plate after a few days of growth. The amount of light and the quality of light will affect root

growth [22, 23]. If growing Arabidopsis seeds for longer than 10 days, or using another species, first test to identify how far seedlings can be spaced without roots growing into each other. Roots of seedlings that have grown together should not be measured in subsequent steps, as root contact may have induced a touch-response.

12. Plating multiple rows of seeds on a plate will reduce the number of plates required, however, the light received by the seedlings on the lower row will be reduced, and the root growth architecture will be affected. This must be taken into account when determining significant changes in root growth. If it is necessary to plate two rows, equal number of seeds should be plated on both the top and bottom for each line, and the experimental design should be adapted to allow statistical evaluation of row effects.
13. Pushing a seed too firmly onto the agar-based medium will cause it to break the surface, an unwanted outcome because Arabidopsis roots embedded in an homogenous agar-based medium display distinct growth behaviors and architectures compared to surface-grown roots.
14. Micropore tape is used to wrap agar plates as it allows for gas exchange. Wrapping plates with Parafilm causes ethylene to build up which can influence root growth characteristics [24, 25].
15. There are many other root-tracing software systems that can be used in addition to Image J. Other software that could be used to analyze roots include: RootNav [26], ARTT [27], or the high-throughput, automated software, BRAT [28].
16. When measuring seedling roots, be sure to exclude any roots that have grown into the agar, or any seedlings that have been contaminated with bacterial or fungal growth.
17. Parental lines that show similar responses can still be used for a QTL study, as transgressive segregation often occurs, potentially leading to more extreme responses than either parental line. This is more risky than using parental lines with distinct phenotypes, but should be considered if choices in characterized RIL populations are limited.
18. If RIL lines for QTL analysis need to be generated, marker-assisted mapping will be required. Development of markers between Arabidopsis accessions Cvi and L*er* is outlined in Alonso-Blanco et al. [5]. It is expected that RILs will be mostly homozygous after eight generations of self-pollination. Alternatively, a QTL-seq method can be used in lieu of marker-based genotyping [29, 30].

19. Before characterizing seedlings for QTL analysis, it will be necessary to bulk up the seed stocks for both parental lines, RILs and NILs by growing the plants side-by-side, in the same growth chamber, under the same conditions. This will prevent artifacts related to seed sets deriving from subgroups of stressed plants exposed to altered growth conditions (when bulking is done by subgroups handled separately). Indeed, the physiological status of a plant can dramatically affect the growth characteristics of its young progeny early after germination.
20. When growing plants for QTL analysis, photoperiod, temperature, relative humidity, light intensity, and medium must be consistent for all trials, as these environmental parameters will affect root architecture.
21. If an uncharacterized RIL population is being used for these studies, MapMaker (http://archive.broadinstitute.org/ftp/distribution/software/mapmaker3/) can be used to analyze recombination frequencies and generate chromosomal maps [31].
22. QTL analysis can also be carried out using WinQTL Cartographer V2.5 (http://statgen.ncsu.edu/qtlcart/WQTLCart.htm), which has a graphical-user interface [32]. As with R/qtl, enter average trait values for each RIL along with genotypic data into WinQTL Cartographer. Map using the Kosambi method. LOD is calculated using "Composite Interval Mapping". This function was designed to reduce background from non-target QTL linked to the target of interest. Control parameters for Composite Interval Mapping used in our studies are 1 cM window size and ten control markers. The walk speed is set to 0.5, which will yield more precise measurements. Determine LOD significance thresholds by a minimum of 1000 permutations. The LOD threshold can be set manually but should be done with caution, as decreasing the threshold too far will result in false positives, and too high will cause an artifactually decreased number of significant peaks. For more information, reference WinQTL Cartagrapher manual (http://statgen.ncsu.edu/qtlcart/WQTLCart.htm).
23. CRISPR-Cas9 system induces double stranded DNA breaks at a location specified by a designed guide RNA. If repaired by non-homologous end-joining (NHEJ), mutations due to small nucleotide deletions or insertions can occur, resulting in knockout alleles in a number of plant species [18, 33]. If the homology-based repair pathway (HR) is used to fix the break, this strategy allows for gene-replacement if a donor DNA with homology to the target sequence is present, or has been introduced into the genome [34]. Interestingly, the latter strategy

allows one to swap a polymorphism in one of the parental lines for the polymorphism present in the other line, permitting determination of the impact of a particular SNP on the phenotype of interest.

24. A particular QTL peak may include multiple genes, only one of which (or a few) contributes to the trait of interest. If the loci are too closely linked to expect recombination between T-DNA lines, CRISPR-Cas9 gene knockout, or gene replacement methods could be used.
25. It is important to confirm the T-DNA insertion by genotyping. qRT-PCR will be required using primers flanking both sides of the insertion to confirm the allele is truly null.
26. Promoter-swap experiments can be done by transforming a gene under the control of another accession's promoter into a knockout allele's background. Or, ideally, CRISPR gene replacement can be used to swap out the promoters. CRISPR gene replacement is ideal as the promoter switch would be in the same location in the genome, and thus not affected by positional effects that may influence expression.
27. Polymorphisms could also have effects on non-coding RNA. Additionally, there could be epigenetic modifications that alter expression of uncharacterized genes that are linked to a polymorphism.
28. Tissue should be frozen for RNA extraction at the same time in the day-night cycle to avoid deviations in expression due to circadian rhythm.
29. Before beginning cDNA synthesis, input program into a thermocycler and pause the program between steps.
30. For all samples in an experiment, cDNA should be synthesized at the same time. Additionally, all cDNA samples should be diluted at the same time if they are going to be compared by qRT-PCR.
31. NCBI Primer-BLAST (https://www.ncbi.nlm.nih.gov/tools/primer-blast/) and Roche Universal Probe Library Assay Design Center (https://lifescience.roche.com/en us/brands/universal-probe-library.html) can be used to design primers for qRT-PCR.
32. When designing primers for qRT-PCR, it is necessary to identify conserved regions among all accessions to ensure primers bind with equal affinity.
33. The reference gene should be expressed at a similar level to the genes of interest to allow effective comparisons.
34. Technical replicates that have a cycle-value difference greater than one should not be trusted. If large differences between

technical replicates are observed, the experiment should be rerun.

35. Relative expression is quantified based on expression of a reference gene and determined relative to a calibrator using the formula: $2^{-\Delta\Delta C_T}$, where C_T is defined as the cycle threshold, or number of cycles required for the fluorescent signal to exceed background level. Applied Biosystems User Bulletin No. 2 (P/N 4303859) details calculations for qRT-PCR analysis [35].
36. After extracting DNA for genotyping, microfuge pestles should be cleaned with 10% bleach to avoid DNA contamination.
37. These cycling parameters are specific for Econotaq polymerase (*see* Subheading 2.5). If other DNA polymerases are to be used, modify the parameters according to product information.
38. If 55 °C annealing temperature does not result in amplification of wild-type DNA, adjust the temperature. If a gradient thermocycler is available, run a gradient from 50 to 60 °C.
39. Ethidium Bromide solution should be protected from light.

Acknowledgments

We thank S.-H. Su, A. Jancewicz and A. Strohm for comments, advice, and/or research contribution to the material included in this manuscript. Our work was made possible by grants from the *National Science Foundation* (IOS-1121694), the *National Aeronautics and Space Administration* (NNX14AT23G), and the University of Wisconsin–Madison College of Agricultural and Life Sciences *Hatch* program.

References

1. Li J, Mu J, Bai J et al (2013) PARAQUAT RESISTANT1, a golgi-localized putative transporter protein, is involved in intracellular transport of paraquat. Annu Rev Plant Physiol 162:470–483. doi:10.1104/pp.113.213892
2. Strohm AK, Vaughn LM, Masson PH (2015) Natural variation in the expression of ORGANIC CATION TRANSPORTER 1 affects root length responses to cadaverine in Arabidopsis. J Exp Bot 66:853–862. doi:10.1093/jxb/eru444
3. Meijón M, Satbhai SB, Tsuchimatsu T, Busch W (2013) Genome-wide association study using cellular traits identifies a new regulator of root development in Arabidopsis. Nature 46:77–81. doi:10.1038/ng.2824
4. Jancewicz AL, Gibbs NM, Masson PH (2016) Cadaverine's functional role in plant development and environmental response. Front Plant Sci 7:1237. doi:10.1248/cpb.48.1458
5. Blanco CA, Peeters A, Koornneef M (1998) Development of an AFLP based linkage map of Ler, Col and Cvi Arabidopsis thaliana ecotypes and construction of a Ler/Cvi recombinant inbred line population. Plant J 14:259–271. doi:10.1046/j.1365-313x.1998.00115.x
6. Linsmaier EM, Skoog F (1965) Organic growth factor requirements of tobacco tissue cultures. Physiol Plant 18:100–127. doi:10.1111/j.1399-3054.1965.tb06874.x
7. Clark KA, Krysan PJ (2007) Protocol: an improved high-throughput method for

generating tissue samples in 96-well format for plant genotyping (Ice-Cap 2.0). Plant Methods 3:8. doi:10.1186/1746-4811-3-8

8. Roy R, Bassham DC (2014) Root growth movements: waving and skewing. Plant Sci 221-222:42–47. doi:10.1016/j.plantsci.2014.01.007

9. Rutherford R, Gallois P, Masson PH (1998) Mutations inArabidopsis thalianagenes involved in the tryptophan biosynthesis pathway affect root waving on tilted agar surfaces. Plant J 16:145–154. doi:10.1046/j.1365-313x.1998.00279.x

10. Schneider CA, Rasband WS, Eliceiri KW (2012) NIH Image to ImageJ: 25 years of image analysis. Nat Methods 9:671–675. doi:10.1038/nmeth.2089

11. Meijering E, Jacob M, Sarria JCF et al (2004) Design and validation of a tool for neurite tracing and analysis in fluorescence microscopy images. Cytometry 58A:167–176. doi:10.1002/cyto.a.20022

12. Grabov A, Ashley MK, Rigas S et al (2004) Morphometric analysis of root shape. New Phytol 165:641–652. doi:10.1111/j.1469-8137.2004.01258.x

13. Vaughn LM, Masson PH (2011) A QTL study for regions contributing to arabidopsis thaliana root skewing on tilted surfaces. G3 1:105–115. doi:10.1534/g3.111.000331

14. R Core Team (2013). R: a language and environment for statistical computing. R Foundation for Statistical Computing, Vienna, Austria. http://www.R-project.org/.URL

15. Gross J, Ligge U ((2015)) nortest:Tests for normality. R package version 1.0–4 URL https://CRAN.R-project.org/package=nortest

16. Broman KW, Wu H, Sen S, Churchill GA (2003) R/qtl: QTL mapping in experimental crosses. Bioinformatics 19:889–890. doi:10.1093/bioinformatics/btg112

17. Broman KW, Sen S (2009) A guide to QTL mapping with R/qtl. Stat Biol Health doi: 10.1007/978-0-387-92125-9

18. Jiang W, Zhou H, Bi H et al (2013) Demonstration of CRISPR/Cas9/sgRNA-mediated targeted gene modification in Arabidopsis, tobacco, sorghum and rice. Nucleic Acids Res 41:e188–e188. doi:10.1093/nar/gkt780

19. Czechowski T, Stitt M, Altmann T et al (2005) Genome-wide identification and testing of superior reference genes for transcript normalization in Arabidopsis. Annu Rev Plant Physiol 139:5–17. doi:10.1104/pp.105.063743

20. Visscher AM, Paul A-L, Kirst M et al (2010) Growth performance and root transcriptome remodeling of arabidopsis in response to mars-like levels of magnesium sulfate. PLoS One 5:e12348–e12316. doi:10.1371/journal.pone.0012348

21. Clark KA, Krysan PJ (2010) Chromosomal translocations are a common phenomenon in Arabidopsis thaliana T-DNA insertion lines. Plant J 64:990–1001. doi:10.1111/j.1365-313x.2010.04386.x

22. Salisbury FJ, Hall A, Grierson CS, Halliday KJ (2007) Phytochrome coordinates Arabidopsis shoot and root development. Plant J 50:429–438. doi:10.1111/j.1365-313x.2007.03059.x

23. Zhang K-X, Xu H-H, Yuan T-T et al (2013) Blue-light-induced PIN3 polarization for root negative phototropic response in Arabidopsis. Plant J 76(2):308–321. doi:10.1111/tpj.12298

24. Buer CS, Wasteneys GO, Masle J (2003) Ethylene modulates root-wave responses in Arabidopsis. Annu Rev Plant Physiol 132:1085–1096. doi:10.1104/pp.102.019182

25. Buer CS (2006) Ethylene modulates flavonoid accumulation and gravitropic responses in roots of arabidopsis. Annu Rev Plant Physiol 140:1384–1396. doi:10.1104/pp.105.075671

26. Pound MP, French AP, Atkinson JA et al (2013) RootNav: navigating images of complex root architectures. Annu Rev Plant Physiol 162:1802–1814. doi:10.1104/pp.113.221531

27. Russino A (2013) A novel tracking tool for the analysis of plant-root tip movements. Bioinspir Biomim 8(2):025004. doi:10.1088/1748-3190/12/1/015001

28. Slovak R, Göschl C, Su X et al (2014) A scalable open-source pipeline for large-scale root phenotyping of arabidopsis. Plant Cell 26:2390–2403. doi:10.1105/tpc.114.124032

29. Takagi H, Abe A, Yoshida K et al (2013) QTL-seq: rapid mapping of quantitative trait loci in rice by whole genome resequencing of DNA from two bulked populations. Plant J 74:174–183. doi:10.1111/tpj.12105

30. Lim J-H, Yang H-J, Jung K-H et al (2014) Quantitative trait locus mapping and candidate gene analysis for plant architecture traits using whole genome re-sequencing in rice. Mol Cells 37:149–160. doi:10.14348/molcells.2014.2336

31. Lincoln SE, Daly MJ, Lander ES (1993) Constructing genetic linkage maps with MAPMAKER/EXP Version 3.0: a tutorial and reference manual. A Whitehead Institute for Biomedical Research Technical Report. Third Edition (Beta Distribution 3B).

http://home.cc.umanitoba.ca/~psgendb/birchhomedir/doc/mapmaker/mapmaker.tutorial.pdf

32. Wang S, Basten CJ, Zeng Z-B (2012) Windows QTL Cartographer 2.5. Department of statistics. North Carolina State University, Raleigh, NC. http://statgen.ncsu.edu/qtlcart/WQTLCart.htm
33. Ran FA, Hsu PD, Wright J et al (2013) Genome engineering using the CRISPR-Cas9 system. Nat Protoc 8:2281–2308. doi:10.1038/nprot.2013.143
34. Zhao Y, Zhang C, Liu W et al (2016) An alternative strategy for targeted gene replacement in plants using a dual-sgRNA/Cas9 design. Nature 6(23890):1–11. doi:10.1038/srep23890
35. User Bulletin #2 (1997) Relative quantitation of gene expression. Appl Biosyst http://www3.appliedbiosystems.com/cms/groups/mcb_support/documents/generaldocuments/cms_040980.pdf. Accessed 14 Mar 2017

Chapter 23

Methods Related to Polyamine Control of Cation Transport Across Plant Membranes

Isaac Zepeda-Jazo and Igor Pottosin

Abstract

Polyamines (PAs) are unique polycationic metabolites, which modulate plants' growth, development, and stress responses. As polycations, PAs interfere with cationic transport systems as ion channels and ionotropic pumps. Here, we describe the application of two techniques, MIFE to study the effects of PAs on cation fluxes in vivo and conventional patch-clamp to evaluate the PA blockage of ion currents in isolated plant vacuoles. Preparation of vacuoles for patch-clamp assays is described and solutions and voltage protocols are given, which allow separate recordings of major vacuolar channel currents and quantify their blockage by PAs.

Key words Patch-clamp, MIFE, Ion channel, H^+ pump, Calcium, Root, Vacuole, Plasma membrane

1 Introduction

Cation transport systems in plasma and vacuolar membranes of higher plants control membrane potential, regulate turgor, generation and usage of electrochemical gradients for H^+, mediate K^+ acquisition and redistribution and Na^+-K^+ exchange, as well as Ca^{2+}, ROS and electrical signaling [1–5]. Unlike their animal counterparts, constitutive plasma membrane (PM) K^+-selective and nonselective cation (NSCC) channels in plants are only weakly and, possibly, indirectly, sensitive to natural PAs [6–8]. Contrary, major vacuolar NSCCs of slow (SV) and fast (FV) vacuolar types are directly inhibited by PAs. Tandem-pore cation channel TPC1, mediating SV current conducts small mono- and divalent cations indiscriminately and is blocked by PAs spermine (Spm^{4+}) > spermidine (Spd^{3+}) > putrescine (Put^{2+}) in a voltage-dependent manner from either membrane side [9, 10]. The FV current, which conducts indiscriminately all small monovalent cations, but is inhibited by micromolar concentrations of Ca^{2+} or Mg^{2+}, could be also rapidly and reversibly inhibited by micromolar Spm^{4+} and Spd^{3+} and millimolar Put^{2+} [9, 11, 12]. Vacuolar K^+-selective (VK)

Rubén Alcázar and Antonio F. Tiburcio (eds.), *Polyamines: Methods and Protocols*, Methods in Molecular Biology, vol. 1694, DOI 10.1007/978-1-4939-7398-9_23, © Springer Science+Business Media LLC 2018

channels are relatively insensitive to PAs, so that an increase of cellular PAs, which suppress NSCCs, tends to increase the K^+ selectivity of the overall tonoplast cation conductance, with an important impact to the efficient vacuolar Na^+ sequestration [13, 14].

PAs could affect PM K^+, Na^+, and H^+ transport in a manner, which depends on species, tissue, and growing conditions [15]. It is not only relative functional expression of different ion transporters, which may underlie the diverse responses of plant tissues to PAs, but different PAs could exert adverse effects on individual ion transporters. As an example, in pea roots Put^{2+} stimulates the PM H^+ pump activity, whereas Spm^{4+} activates the H^+ pump at lower concentrations, but causes its strong inhibition at higher ones [16]. However, both Spm^{4+} and Put^{2+} similarly activate Ca^{2+} pumping across the PM in root mature zone [16, 17]. Different ROS, and most universally, hydroxyl radicals (OH•) activate nonselective cation conductance across plasma membrane, which cause a depolarization, Ca^{2+} influx to and K^+ leakage from tissues [18, 19]. This OH•-induced K^+ efflux is positively modulated by PAs, equally by diamine Put^{2+} and tetraamine Spm^{4+}, but the extent of stimulation by PAs critically depends on the overall capacity of the tissue to retain K^+, and could vary greatly between near isogenic varieties as it was shown for barley [20, 21].

Consequently, two electrophysiological techniques could be used to study cation transport across plasma and vacuolar membranes in plants. Noninvasive high-throughput MIFE technique or similar methods, utilizing self-reference ion-selective vibration probes (SIET) are optimal to study ion fluxes across the PM of intact tissues (*see* ref. 22 for a detailed MIFE and SIET comparison). Bearing in mind strong variations of ion transporters expression, which depends on tissue, zone and growing conditions, remarkable stochastic differences between individual protoplasts from the same preparation, as well as difficulties in studies of pumps and exchangers in destructive conditions, deviating from those in vivo, and, last but not the least, often indirect effects of PAs on the PM ion transporters, convenient patch-clamp is not a plausible alternative for MIFE in case of the PM cation transport. In addition, MIFE allows measurements of non-electrogenic ion transport, as for instance, the activity of Na^+/H^+ antiporters or Ca^{2+} pump with a 1 $Ca^{2+}/2$ H^+ exchange stoichiometry [16, 23], which is in principle impossible by means of voltage- or patch-clamp. On the other hand, robust expression of the two major NSCC currents, SV and FV, as well as VK in vacuoles from every plant tissue, easiness and rapidness of vacuoles isolation and patching, existence of well-established knowledge on the properties of SV, FV, and VK channels, availability of recording media and voltage protocols, which allows an easy separation of specific currents, make the patch-clamp technique the method of choice to assay PAs effects on individual vacuolar channels.

2 Materials

2.1 Materials for Plants Growth

1. Filter paper for Petri dish germination method and paper towel of medium quality for paper roll germination and growth method.
2. Stock solutions: 100 mM KCl, 50 mM $CaCl_2$, 500 mM MES-adjust pH to 6.1 with 1 N KOH, 400 mM TRIS (Trizma base).
3. Hydroponic growth solution for true hydroponics or paper rolls method: 0.5 mM KCl and 0.1 mM $CaCl_2$. Alternatively, plants can be grown in pots containing a commercially available professional potting mixture or in vertical Petri dishes for *Arabidopsis.*
4. Polystyrene Petri dishes for seed germination.
5. 1 and 3 L plastic containers for hydroponic growth.

2.2 MIFE Basic Setup and Auxiliaries

1. MIFE main amplifier and controller, with 4-channel preamplifier (supplied by UTas Research Office Commercialization Unit).
2. Inverted microscope with a long distance objective, providing a final amplification of $\times$ 40.
3. MIFE custom-assembled Narishige manipulator system (SM-17, MHW-4, MX-2), with a MIFE stepper motor drive or Eppendorf PatchMan NP2 computer-controlled micromanipulator.
4. PC, running Windows 98 or ME and having one ISA-bus slot (for the DAS08 card) with a spare slot beside it and the CIO-DAS08 card for analogue to digital conversion.
5. CHART/MIFEFLUX software (supplied by UTas Research Office Commercialization Unit).
6. Faraday cage.
7. Anti-vibration table.
8. E Serie Electrode Holder, with handle 45° Style, fits 1.2 mm capillary, Ag wire Marca Warner Instruments Cat. No. 64-1021.
9. Electrode filling station made of two general use micromanipulators and a stereomicroscope.
10. Small drying oven, to 250 °C and a custom-made metallic rack with multiple orifices to adopt bases of microelectrodes, with metallic cover (for silanization).

2.3 MIFE Media and Materials for Cation Flux Measurements

1. Capillary borosilicate glass for microelectrodes (e.g., GC150-10 capillaries with 1.5 mm O.D. and 0.86 mm ID from Harvard Apparatus Ltd).
2. Tributylchlorosilane (Fluka 90794-1 mL) to hold the LIX.
3. Ion-selective resins (*see* **Note 1**).
4. Bath solution (in mM): 0.5 KCl, 0.2 $CaCl_2$, 5 MES-KOH (pH 6.0), 2 mM TRIS. For H^+ fluxes measurements, use the same solution without a pH buffer (TRIS, MES).
5. Calibration solutions (in mM): K^+ (0.1, 0.2, 0.5 KCl), Ca^{2+} (0.1, 0.5, 1.0 $CaCl_2$), H^+ (pH of ~5.2, ~6.6, ~7.9) (*see* **Note 2**).
6. Back-filling solutions (in mM): 200 KCl (for K^+), 500 $CaCl_2$ (for Ca^{2+}), 15 NaCl plus 40 KH_2PO_4 (for H^+).
7. Reference electrode: 0.5 mm Ag/AgCl wire, filling with 3.5% agar prepared on 100 mM KCl (*see* **Note 3**).
8. For electrode back-filling: 3–5 mL plastic syringes and MICROFIL needles (MF34G-5, WPI).
9. Measuring chambers (custom-made, *see* **Note 4**).
10. For PAs treatments prepare 20 mM stock solutions of Spm^{4+}, Spd^{3+} and Put^{2+} and keep in refrigeration or at −13 °C in aliquots of 2.5 mL. In experiments, studying the effects of PAs on OH•-activated currents, copper-ascorbate mixture must be prepared from two individual solutions of 20 mM $CuCl_2$, and 20 mM sodium ascorbate ($C_6H_7NaO_6$). Prepare aliquots of 2.5 mL of each one; they must be stored at −13 °C in dark.

2.4 Materials for Preparation of Root and Leaf Protoplasts

1. Stock solutions as for MIFE plus 50 mM $MgCl_2$.
2. Enzyme solution (% in w/v, the rest in mM): 2% cellulose Onozuka RS (Yakult Honsha, Tokyo, Japan), 1.2% cellulysin (CalBiochem, Nothingham, UK), 0.1% pectolyase Y-23 (Yakult Honsha, Tokyo, Japan), 0.1% bovine serum albumin (SIGMA), 10 KCl, 10 $CaCl_2$, and 2 $MgCl_2$, 2 MES-KOH (pH 5.7 with TRIS) and final osmolality (*see* **Note 5**) is adjusted with sorbitol and verified by an osmometer (e.g., cryoscopic Osmomat 030, Germany). Mix and filter the solution through a 0.22 μm Millipore filter. Store at −13 °C in aliquots of 2.5 mL.
3. Wash solution, same as enzyme solution minus enzymes.
4. Release solution (in mM): 10 KCl for root protoplasts or 2 $CaCl_2$ (5 KCl and 1 $CaCl_2$for leaves protoplasts, plus 1 $MgCl_2$, 2 MES-KOH (pH 5.7); osmolality adjusted with D-sorbitol.
5. Bath solution (in mM): 5 KCl, 2 $CaCl_2$, 0.5 $MgCl_2$, 2 MES-KOH (pH 5.7); osmolality adjusted with D-sorbitol.
6. Rotary shaker with a temperature control.

2.5 Materials for Mechanical Isolation of Taproot and Fruit Vacuoles

1. Razor blades, small knife.
2. Preparation needles or similar.
3. Small 35 mm Petri dishes.
4. Stereomicroscope with ×20–×100 amplification (e.g., Olympus SZ series).

2.6 Patch-Clamp Basic Setup and Auxiliaries

1. Patch-clamp amplifier and headstage (e.g., Axopatch 200B, Molecular Devices).
2. Acquisition system (e.g., Digidata1550, Molecular Devices).
3. Inverted stereomicroscope, with a long-focus objective and total amplification of ×400–×600.
4. Faraday cage.
5. Anti-vibration table.
6. Good quality manual patch-clamp micromanipulator (we use piezoelectric Burleigh PCZ manipulator mounted on the Newport XYZ micropositioning platform).
7. Software (e.g., pCLAMP10, Molecular Devices).
8. Patch microelectrode puller (e.g., programmable P-97 Flaming/Brown, Sutter Instrument).
9. Microforge for patch-pipette fire-polishing (e.g., MF-900, Narishige).
10. Perfusion system for bath exchange.

2.7 Patch-Clamp Media and Materials for Vacuolar Recordings

1. To achieve maximal activity of FV and SV channels, solution at vacuolar side is set Ca^{2+}-free, Mg^{2+} is omitted and pH is set to a neutral value (pH 7.5). Alternatively, pH may be lowered to more physiological values (pH 5.5), but the free concentration of di- and multivalent cations should be virtually zero in all cases. Basic solution contains (in mM): 100 KCl, 15 HEPES-KOH (pH 7.5), 2 K_2EGTA (~2 nM free Ca^{2+}). Osmolality should be adjusted with sorbitol to the osmolality of the cell sap (measured for each sample of interest beforehand). The reference AgCl electrode contains 100 mM KCl solution and connected to the bath with agar bridge, prepared on 100 mM KCl.
2. Two types of standard cytosolic (bath in case of whole cell or outside-out patches) solutions are used: divalent cations- free (for FV assays) and with elevated cytosolic Ca^{2+} (for VK and SV channels assays). The first one is exactly the vacuolar (patch pipette) solution, described above. For VK and SV assays K_2EGTA is excluded, and 0.1 mM and 0.3 mM of $CaCl_2$ and $MgCl_2$, respectively, is added instead (*see* **Note 6**). For verification of the K^+ selectivity, K-salts in the bath are exchanged for equimolar quantities of Na^+-ones.

3. 50 mL self-standing 50 mL tubes with a screw cap for storage of bath and pipette solutions and 0.5 mL Eppendorf tubes to store stock solutions (100 mM) of PAs.
4. Patch-clamp microelectrodes are fabricated from clean glass capillaries. We use commercial 10 cm long and 1.5 mm wide (internal diameter 0.84 mm) borosilicate glass capillaries (1B150F-4, World Precision Instruments).
5. Custom-made boxes for storage of prepared microelectrodes.
6. Microelectrode holder with a suction outlet.
7. Plastic tubing.
8. For patch-pipette filling: 3 mL plastic syringes, 0.22 μm Millipore filters, long nonmetallic syringe needles (e.g., MICROFIL, World Precision Instruments).
9. Measuring chambers (custom-made).

3 Methods

3.1 Growth

1. Sterilize and germinate seeds.
2. For MIFE root measurements and protoplast isolation: hydroponic growth in darkness until roots reach several cm long. For MIFE leaves measurements and protoplast isolation: can be done by true hydroponics method or in pot method with standard potting mix. In both methods, the seedlings remain there until leaves are fully unfolded.
3. Grow seedlings under constant (25 °C) temperature conditions.
4. Barley seedlings: put 10 seeds between two layers of moistened paper towels a horizontal line at ~2 cm of the upper edge, and roll the paper into a 1 L plastic container; add 0.4 L of the growth solution.
5. Pea seedlings: germinate seeds in Petri dishes between the two layers of moistened papers filters. Once germinated, grow seedlings hydroponically on a floating mesh in plastic container above an aerated growth solution. For leaves measurements and protoplast isolation, transfer seedlings to a standard potting mix and grow them in a chamber or a glasshouse (16 h/8 h light/darkness).
6. *Arabidopsis* seedlings: spread sterilized seeds on the surface of 90 mm Petri dishes containing 0.35% phytagel, half strength Murashige and Skoog media and 1% (w/v) sucrose at pH 5.7. Seal Petri dishes with Parafilm and place them in an upright position, so roots grew down the phytagel surface without penetrating it, as described previously by Demidchik and Tester

[24]; Cuin and Shabala [25]. For *Arabidopsis* leaves, place seedling into a standard potting mix and grow them in a chamber or a glasshouse (16 h/8 h light/darkness).

3.2 Isolation of Protoplasts from Roots and Leaves

1. Root protoplasts: use hydroponically grown seedlings with a root length of few cm. Cut the roots 5 mm below the seed. For mature zone preparation discard the first 5 mm from the tip (contrary, use the part close to the apex for elongation zone preparation). Cut roots into 5–10 mm long segments and split them longitudinally under a dissecting microscope.
2. Leaf protoplasts: use completely unfolded leaves. Remove the adaxial epidermis with forceps.
3. Place split root or leaves segments into a 5 mL flask. Cover the flask openings with a Parafilm and incubate tissues with 3 mL of the enzyme solution over 15–30 min in the dark at 30 °C and agitate them at 90 rpm on a rotary shaker.
4. Transfer root or leaf segments into the measuring chamber filled with the release solution. By gently shaking, release protoplasts into the measuring chamber.

3.3 Vacuoles Isolation for Patch-Clamp Measurements

1. Vacuoles from taproots and pigmented fruits can be isolated mechanically, otherwise first the protoplasts need to be isolated from the tissue of interest (*see* **Note 6**).
2. For mechanical isolation, slice ~0.5 g of fresh tissue into segments of ~1 cm^2 area and ~1 mm thickness. Incubate them in small Petri dish for 30 min in 3 mL of solution, by 5–10% more hypertonic to the cell sap (*see* **Note 7**), but otherwise identical by its ionic composition to the bath solution for patch-clamp experiments.
3. Transmit a single slice to another Petri dish with 3 mL of clean solution, identical to the bath solution, further used for patch-clamp recording.
4. Make multiple cuts of the tissue with preparation needles and throw the rests of the slice.
5. Use a standard 20 μL automatic pipette to collect few vacuoles into a volume non-exceeding 10 μL. Transmit vacuoles to a measuring chamber. Our measuring chamber at the beginning of experiment is filled with 300 μL of bath solution. The rule of thumb for successful patch-clamping is quick, but less dirty. The fresh preparation may be used for half an hour, at longer times the gigaseal formation efficiency decreases.
6. Isolation of a vacuole from a single protoplast reduces contamination to a minimum and shortens the time to gain access to the clean tonoplast surface. For vacuole isolation we use the

same microelectrode as for patch-clamping; strong and short suction pulse normally destroys the PM, but more elastic vacuole survives (*see* **Note 8**).

3.4 Patch-Clamp Protocols and Recordings of FV, SV, and VK Currents in Plant Vacuoles

1. Pull patch electrodes in several steps and fire-polish their tips on a microforge under microscopic control. The fire-polished pipettes should be used within 2 h after their fabrication. Working patch configurations should be *outside-out* (small patches or small right-oriented vacuoles) or *inside-out* [26]. Relatively wide (2–3 μm tip opening) electrodes are used for *inside-out* and small vacuoles, higher-resistant micropipettes with a tip opening of 1–1.5 μm are optimal for single channel recordings from *outside-out* patches.
2. Select a large (few tens of μm) clean vacuole and approach it with a patch-pipette, touching but not pressing the membrane from the top. During this step keep the positive pressure within a pipette (*see* **Note 9**) and release it after touching the membrane. If a tight (GΩ) seal is not formed spontaneously, apply a light suction. If over few minutes tight seal is not formed, terminate the attempt and repeat it with the same or new vacuole, using a new microelectrode. Patch-pipettes may not be reused.
3. Tonoplast *inside-out* patches are not easy to obtain as compared to the plasma membrane ones (*see* **Note 10**). Some useful tips to improve the yield of inside-out patches are as follows. In addition to usage of relatively wide pipettes, the suction applied for giga-seal formation should be kept at a minimum. The best is when the tight seal is formed spontaneously, while touching the vacuole from the side and releasing a positive pressure. This yields vacuole-attached configuration. In case of SV currents recording conditions (high Ca^{2+} in the pipette), application of positive voltage steps evokes single channel activity. Unitary currents should be not distorted (rectangular openings and closures with a temporal resolution corresponding to a cutoff filter frequency), implying that there is no vesicle formation in the microelectrode tip (for vesicle appearance see below notes for *outside-out* patches). Pipette should be rapidly withdrawn from the vacuole. The "sidedness" of the recording configuration, *inside-* or *outside-out* may be verified by the asymmetric voltage dependence: SV channels are gated open at cytosol-positive potential and FV channels display larger but highly flickering current at large cytosol-negative potentials (<-100 mV, Fig. 1). If for a given vacuolar preparation the yield inside-out patches is too poor, to test the effects of PAs from the vacuolar side it is advisable to introduce PAs directly into the pipette solution and to work on *outside-out* patches.
4. To obtain tonoplast *outside-out* patches or right-oriented vacuolar patches (or small vesicles), first *vacuole-attached*

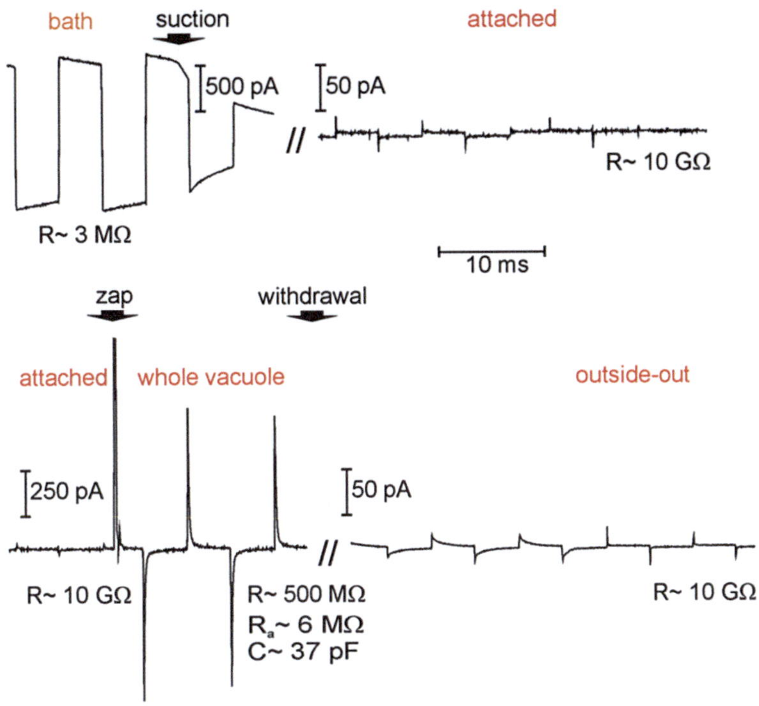

Fig. 1 Electrical events during the formation of an outside-out patch. Square-wave (5 ms, 5 mV) test voltage pulses applied all the time. Clean patch pipette inserted in the bath has low (MΩ) resistance. Touching the vacuole surface and application of a light suction causes a rapid increase of resistance, reaching a GΩ range (vacuole attached configuration). Application of a short and large voltage stimulus (zap) alone or in a combination with a short intense suction destroys the membrane patch and gains a low resistance access (*see* R_a value) to the vacuole interior. It is manifested by appearance of a large capacitance current (the area under the current transient is equivalent to the electrical charge required to polarize the whole vacuole to a new voltage level set by the test pulse). Withdrawal of the pipette causes a collapse of the capacitance current, eventually yielding a small right-oriented (outside-out) membrane patch, preserving a high-resistance (10 GΩ) seal between microelectrode glass and tonoplast

configuration has to be achieved. When high resistance seal is established, break the patch with a combination of a short large magnitude-voltage pulse (zap) and a strong suction pulse. It results in gaining low resistance access to the vacuole interior, so called *whole vacuole* configuration. If the patch-electrode is narrow and only touching the vacuole from the edge and is lifted quickly after gaining into the whole vacuole, a tiny ($C<<1$ pF) *outside- out* patch is formed at the electrode tip (*see* Fig. 1 for details). Usage of wider electrodes, superposition of a pipette tip at a larger distance from the vacuole edge and a slower withdrawal (better along Y rather than Z axis) favors the isolation of a small vacuole ($C \geq 1$ pF) from a large central one (*see* **Note 11**). The latter arrangement is optimal for FV current recordings (Fig. 2).

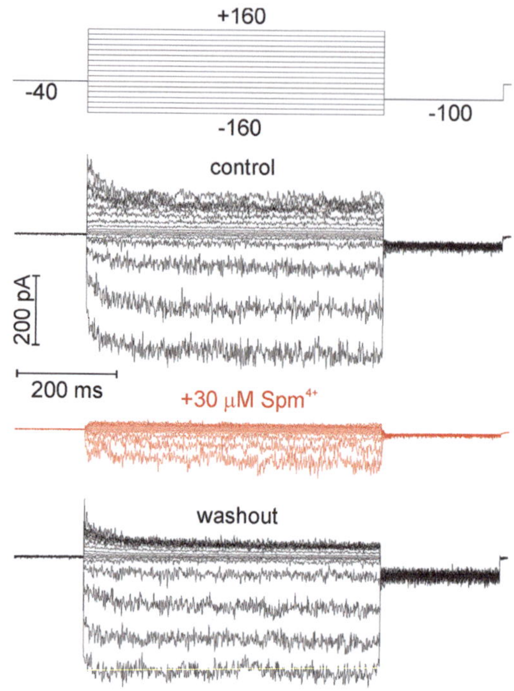

Fig. 2 Inhibition of the FV current by cytosolic spermine. Small (C ~ 1 pF) vesicle was isolated from a large sugar beet vacuole. Voltage protocol as above was applied several times, until washout of internal vacuolar solution was achieved and stable current recording resulted (control). Spermine to a final concentration of 30 μM was introduced by bath perfusion and caused a rapid inhibition of the macroscopic FV current. Washout demonstrates the reversibility of the spermine effect

5. FV current recording in small vacuoles is realized by application of a series of rectangular voltage pulses (Fig. 2). Normally, vacuole can withstand voltages as high as ±200 mV, but to be at the safe side, a narrower voltage range is recommendable, but not less than ±140 mV. For symmetrical ionic strength at both membrane sides holding potential is −40 mV (corresponds to the minimum of the FV voltage-dependent activity). Add PAs at desirable concentration by bath perfusion and evaluate their effect on the steady-state FV current at different voltages. Use compensatory circuit of the patch-clamp amplifier to define the membrane capacitance and express a specific current in pA/pF units (1 pF is approximately equivalent to a membrane surface of 100 μm^2).
6. SV and VK currents are activated at elevated Ca^{2+} concentrations at cytosolic side (bath in case of outside-out patches). SV channels greatly overnumber the VK ones, and both currents may be present in the same patch. Nevertheless, single channel currents may be separated, because VK are not voltage-

dependent and SV require cytosol-positive potentials for their activation. Use holding potentials <−100 mV to selectively record VK currents and a higher holding potential (could be variable, it should be selected in such a way that 1–2 simultaneously open SV channels could be detected, normally within ±30 mV range). Apply ±150 mV 30 ms voltage ramps to monitor single channel currents by VK or SV. Use 1–3 s pause between individual voltage ramps in case of VK channels recording, to avoid cumulative activation of SV channel activity, evoked by positive potentials (even such infrequent event takes place, the open SV channel will close during the pause at negative potentials). Recording of dominating SV channels currents could be interfered with a concomitant VK current; the latter, however, has 3-times lower unitary conductance compared to the SV one, so such events may be easily detected and discarded. More details on the separation of VK and SV unitary currents during voltage ramp protocols are described elsewhere [27].

7. During 30 ms voltage ramps it is common that SV (or VK) channels are either closed or open all the time (Fig. 3a). Select these two types of recordings. Average those containing no channel open and subtract resulting leak current from records, containing exactly one channel open. Edit resulted unitary current–voltage (I/V) relations for occurred closures or opening of extra channels. Average $n > 10$ individual I/V relations until a smooth curve results. Add desired concentration of a PA in the bath and obtain unitary I/V curve for this condition (Fig. 3b). Divide the unitary current values obtained in the presence of PA by control ones to yield the extent of PA block as a function of membrane voltage (Fig. 3c). Parameters of voltage-dependent block may be obtained by fitting the relative current like one presented in Fig. 3c by a suitable equation (*see* Eq. 3, ref. 10).
8. Warning: vesicle closure at the pipette tip (for appearance and theory *see* ref. 26). Due to its elasticity, tonoplast patch frequently forms a closed membrane vesicle. When it happens during the ramp-wave protocol, a distorted (shifted one, with a reduced conductance) unitary I/V relation results (Fig. 4). Bearing in mind that PAs also modify unitary SV channel I/V relation, formation of vesicle may affect the interpretation of PAs effects. If only current responses to voltage ramps were recorded, afterwards there is no way to separate true effect from the artifacts, caused by membrane vesicle closure. Thus, single channel currents need to be periodically inspected at fixed potentials. Vesicle formation is easily detected by distorted channel openings and closures and an apparent loss of temporal resolution (Fig. 4). If such a behavior maintains, the sample has to be discarded.

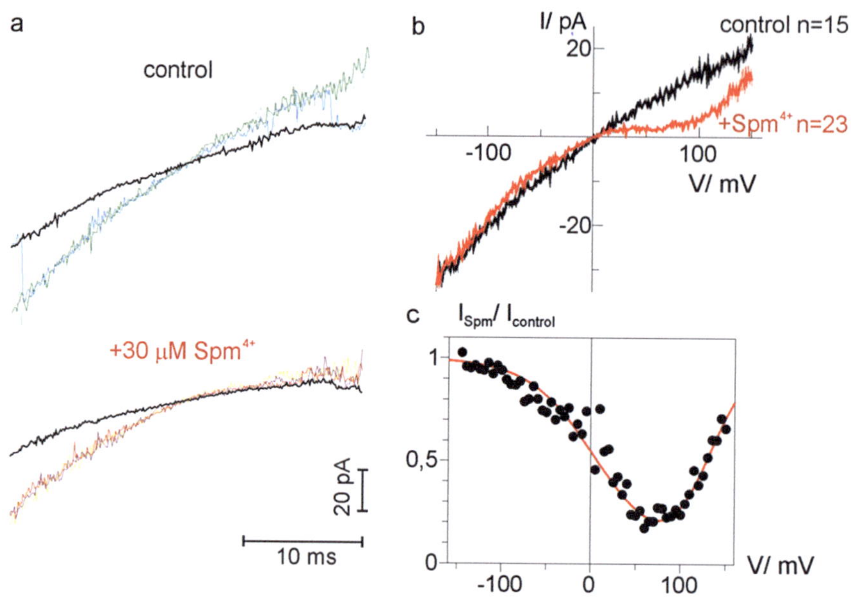

Fig. 3 Spermine block of the SV/TPC1 channel from sugar beet vacuole. (**a**) Original recordings from a small (C<<1 pF) outside-out patch of single SV current in the absence (*control*) and presence of 30 μM of Spm^{4+}·4HCl in the bath (*cytosolic side*). Records are sampled at 10 KHz and low-pass filtered at 2 KHz. Voltage ramps (30 ms) from −150 to +150 mV were applied. Leak currents (with no channel open) were averaged and drawn in *black* as a function of time (and respective voltage). Colored traces are ones recorded in the same run of 10 consequent voltage ramps as respective leak traces, contain exactly one open SV channel most of time (in *blue* trace in control an opening is seen at the beginning and a closure at the end). (**b**) Subtracting leak currents from traces with one open SV channel and substituting time points with respective voltage values one yields unitary current–voltage (I/V) relationships. The plot shows mean I/V relationships ± SE. (**c**) To obtain a voltage dependence of block, the I/V relationship in the presence of Spm^{4+} is taken relative to a control curve. For better visibility, a substitution average (substituting every five unitary current points by their mean) was performed before calculus of the relative current. Solid line is the best fit by the equation, describing permeable block (*see* Eq. 3, ref. [10])

3.5 MIFE Basics

The reader may find a recent detailed update on the noninvasive MIFE method in Newman *et al.* [22]. The principle of the method is based on the relation between flux and concentration gradient for a free diffusion within an unstirred layer. If the tissue extrudes an ion, its concentration in the medium increases, but not equally: the shorter is the distance from the surface, the higher is the concentration increase. When the ion is taken up, its concentration in the medium decreases; a decrease will be maximal at the tissue surface. If concentration of the ion of interest is measured by an ion-selective electrode in two points, one close to the surface and another at a distance, the flux can be calculated. As MIFE measures a relative change in the concentration, its sensitivity (signal-to noise-ratio) is much higher, when the ions in the external medium are diluted. When it comes to the temporal resolution, it is in the range of few seconds, which is convenient, because in most cases flux responses to

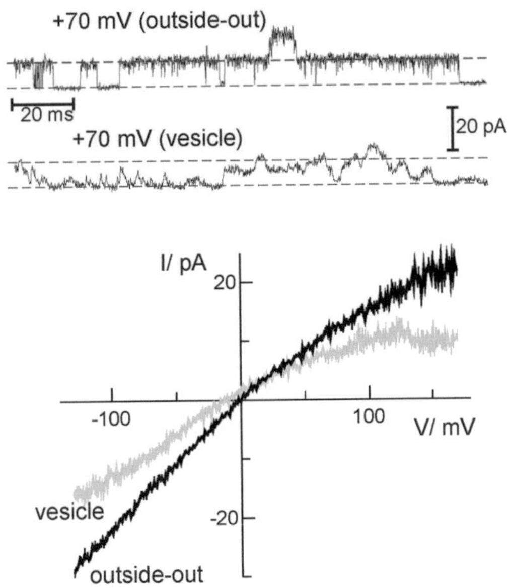

Fig. 4 Alterations of single channel currents upon a spontaneous transformation of an outside-out patch into a closed vesicle. *Dashed lines* indicate current levels for closed and open channel; at least two individual SV channels are present in the patch. NB: full closure of a vesicle may result in a total disappearance of detectable channel closures and openings

external stimuli occur within tens of minutes. An important condition for a correct flux estimate is the absence of stirring (small perturbations caused by the movement of the measuring electrode between the two positions are neglected) and that buffering capacity of the medium (important for H^+ and Ca^{2+} ions) is minimal (e.g., by the usage of low concentrations of pH buffers with a pK at least 0.5 units higher than the medium pH in case of H^+ measurements).

3.6 Microelectrodes Preparation for MIFE

1. To hold the LIX, microelectrode tips must be coated with tributylchlorosilane. Put pulled glass microelectrodes in a vertical position on a stainless-steel rack and oven-dry them at 250 °C for 1 h. 10 min before silanization cover the electrodes with steel lid. Add 55 μL of tributychlorosilane under the lid (*see* **Note 12**) at rack base and cover again by 30 min, the electrode blanks can be stored by several weeks at room temperature in a closed container.
2. Make a LIX-container by quick insertion a broken-tip microelectrode (tip diameter 50 μm) into the stock LIX. Blanks with good tip size (<3 μm-diameter), straight tip, or small debris are back-filled with corresponding back-filling solutions using a 3–5 mL syringe with a 0.22 μm Millipore filter and a MICROFIL needle avoiding air bubbles. After back-filling, under a stereo microscope put the electrode tip in touch with the front-filled electrode with LIX. The LIX must penetrate and

fill ~150 μm the tip of the electrode. Once prepared, MIFE electrodes can be stored in bath solution before use. The capillary with a LIX deposit, used for MIFE-electrode preparation, can be preserved in refrigeration and in the dark for 5 days without a loss of specific properties.

3.7 MIFE Measurements on Roots and Leaves

1. Mount the MIFE-electrode in the holder and reference AgCl electrode in the bath, connect them to the preamplifier.
2. Run the CHART software (*see* **Note 13**).
3. Calibrate each electrode against a set of three standards with a concentration range covering the concentration of the ion in question in the bath. The average response from electrodes has to have a slope of 53–54 mV for monovalent ions (e.g., H^+) and 27–28 for divalent ions (e.g., Ca^{2+}), both with a correlation greater than $R = 0.999$ (*see* **Note 14**).
4. Roots measurements: mount roots in the measuring chamber and let them for 1 h for the acclimation in the bath.
5. Leaves measurements. Mesophyll tissue; gently remove the leaf epidermis with fine forceps, cut mesophyll into segments of 5–7 mm and left them floating in a Petri dish with growth solution. Epidermal tissue; cut 5 to 7 mm leaf segments from the apical part of the leaf, avoiding major veins. In both cases, leaf segments should be immobilized in a measuring chamber and allowed for 1 h to adapt for a standard bath solution.
6. Locate the MIFE-electrode(s) onto selected tissue zone; when they are at the closest position (50 μm, position recognized by MIFE-software as M1), run a new record with ALT <S> command and switch on the MIFE motor drive (*see* **Note 15**) and start the measurements, this means the beginning of the cycle movement from position M1 (50 μm) to position M2 (100 μm) in a 10 s square-wave manner. Ensure a stable response over 10 min, which will be taken as a control.
7. After steady-state control measurements, add polyamines as chloride salts (up to 1 mM final concentration) into the measurement chamber. The effect of PAs on ion fluxes is rapid (some seconds) and it usually lasts for minutes; keep recording for at least 30 min. Special attention should be paid for PA effects on H^+ and Ca^{2+} pumping [16], i.e., concurrent measurements of H^+ and Ca^{2+} fluxes should be performed, along with a respective pharmacological analysis (Table 1). Alternatively, samples may be pre-incubated with PAs (from 10 min to 1 h) to see the modulation of the response to a certain stimulus, e.g., high NaCl (*see* **Note 16**). A special and interesting case is the PA modulation of the OH•-induced ion fluxes, K^+ efflux in particular (*see* **Note 17**). Apply PA of interest jointly with Cu/Asc OH•-generating mixture (variable combinations of concentrations, up 1 mM of PA and Cu/Asc).

Table 1
Blockers and inhibitors of plant plasma membrane ion transport

Drug	Dose	Function	Reference
TEA	10–30 mM	K^+-selective channel blocker	[28, 29]
Gd^{3+}, La^{3+}	50–200 μM	NSCC blockers	[30]
Nifedipine	0.1 mM	Cation channels blocker	[30]
Verapamil	0.1 mM	Cation channels blocker	[30]
NPPB, Niflumate, DIDS, or Ethacrynic acid	0.1 mM	Anion channels blockers (*see* **Note 22**)	[31]
Vanadate	0.5—1 mM	P-type (H^+; Ca^{2+}) ATPase inhibitor	[16, 32]
Eosin yellow, Erythrosin B	0.5 μM	Ca^{2+} – ATPase inhibitor	[16, 23]
Amiloride	1 mM	Nonspecific inhibitor of cation/H^+ exchangers	[33]

8. Noninvasive MIFE technique measures ion fluxes in a free-running manner, i.e., without any external control over membrane potential or current. Under these premises, pharmacological analysis becomes central for the identification of ion transporters, responsible for fluxes, which are monitored by MIFE. There are several pharmacological agents (*see* **Note 18**), which are proved to be useful for studies of ion transport in plants (*see* Table 1).
9. To stop the data acquisition press ALT <H> and create the . AVM file (*see* **Note 19**).
10. To estimate the magnitude of net ion fluxes, use MIFEFLUX software (*see* **Note 20**). It converts the ion's activity gradient (electrochemical potential in mV) to net ion flux (nmol m^{-2} s^{-1}) using the Nernst equation [22].
11. Once the flux estimate is done, it is easy to open the data sheet in Excel to manipulate and graph the flux kinetics (*see* **Note 21**).

4 Notes

1. Ion-selective resins (LIX) are available from FLUKA: K^+ (contains Valinomycin, Cat. No. 60031), Ca^{2+} (contains (-)-(R,R)-N,N′-(Bis(11-ethoxycarbonyl)undecyl)—N,N′-4,5-tetra-methyl −3,6- dioxaoctanediamide, Cat. No. 21048), H^+ (contains 4-Nonadecylpyridine, Cat. No. 95297).
2. Prepare all calibration solutions from the more concentrated calibration stock solution. For H^+ calibration solution prepare two 10 mM stock solutions of A- Na_2HPO_4•12 H_2O (pH 8.8)

and B- $NaH_2PO_4{\bullet}2H_2O$ (pH 4.7) and mix them to get the three pH standards: pH 5.2 add 1 mL of A plus 99 mL of B, pH 6.6 add 25 mL of A plus 75 mL of B, pH 7.9 add 90 mL of A plus 10 mL of B.

3. It is suitable to use some broken or damaged capillary glass (not appropriate for flux measurements) and seal it with Parafilm.
4. It is advisable to have at least four measurement chambers. They are manufactured from glass Petri dishes and some acrylic small pieces, glued to glass surface of Petri dish to support the plant organs with the help of a wire without damaging them.
5. Solution osmolality is a subject of variation, depending on species/tissue. For example, in case of pea mesophyll enzyme, release and bath solutions are set to 760 mOsm, 350 mOsm and 480 mOsm, respectively; these osmolality values work also for barley root protoplasts, whereas for barley leaf protoplasts 650 mOsm for all solutions is suggested.
6. Ca^{2+} and Mg^{2+} at these concentrations activate SV channels and inhibit FV ones. VK channels are fully activated at 1 μM free cytosolic Ca^{2+} [34]. So, an alternative bath for exclusive VK detection may be designed with 1 μM free Ca^{2+} and sub-millimolar Mg^{2+}, which is sufficient to inhibit the FV current, yet not sufficient to evoke a substantial SV activity at potentials below +100 mV. A higher (few mM) concentration of Ca^{2+} and Mg^{2+} in the bath causes full activation of the SV current. Nonetheless, divalent cations at these concentrations significantly block SV channel mediated currents and, because of competition with, decrease the apparent affinity for PAs.
7. Cell sap osmolality may be a subject of significant variation. For instance, for sugar beet taproots we have measured osmolality between 300 and 800 mOsm, depending on the season, age and root size.
8. Patch electrode for vacuole isolation may be the one used in a previous experiment. In case of leaf protoplasts the site of the PM disruption should be one, where chloroplasts are concentrated in the immediate proximity of the PM. This warrants that the vacuole will not be damaged.
9. Positive pressure should be applied to the interior of the patch electrode to avoid the tip contamination from the solution, especially when crossing the air-solution interface.
10. Elastic properties of the tonoplast are rather different from the plasma membrane. One of the consequences of the tonoplast higher elasticity and flexibility is a low probability of the isolation of inside-out patch from the vacuole. In most of cases, isolation is resulted in the formation of a sealed vesicle on the pipette tip (for current distortions in such vesicles *see* ref. 26).

And, for a reason, which remains not clear up to now, the reopening of such a vesicle in most of cases is resulted in right-side-oriented outside-out membrane patch. This artifact resulted, for instance, in a false identification of so called VVCA channels, which were in reality SV/TPC1 channels, recorded other-way-round (for a detailed discussion *see* ref. 35).

11. Some researchers prefer to work in the whole vacuole configuration, because of a larger ionic currents magnitude. However, the density of SV current in some preparations may be too high, which affects the quality of voltage-clamping in whole vacuole configuration. And the advantage of smaller vacuoles is that they are not adjacent to the chamber bottom but only to the pipette tip. This arrangement diminishes the probability of sample loss upon perturbations, e.g., during bath solution exchange.
12. Tributychlorosilane is very toxic. Avoid inhaling the vapor!
13. CHART: Software package aimed to control the acquisition of data by MIFE system. Runs under MS-DOS (once starting under MS-DOS press space on chart parameters page to run the software) and permits the real-time control of the amplifier configuration and the micromanipulator, while the data are being collected and written to the disk.
14. In CHART software press ALT <S> (start), to start records, the screen show the file name, press <S> (start), set the required parameters if there are some changes to do, or press <G> (go) to go immediately to CHART screen. Write a name for the file-experiment and press enter. At these point, it could be necessary to adjust the electrometer offset (with ALT <+> or <–>) to keep measurements within the 50 mV data window. Press F7 to start calibration process. Correct temperature and tape the name of the ions to measure (e.g., H^+ and Ca^{2+}), select Z for the rest of ions in the box. F7 is also used to set three concentrations of calibration solutions of each ion to measure (one at a time). When all data of ion concentrations have been recorded, press ALT-H+ <Y> (yes) to stop data acquisition. Press the next commands in order: ALT <E> (electrometer), <A> (average the data) and <C> (calibration average). CHART creates an .AVC file and displays the calibration values (ion, slope, intercept, and correlation). The H^+ electrode may require near 1 h habituation for a stable response.
15. A window opens to set the filename. It is easiest to accept the default that encodes date and time, by typing <S>. Then a new window opens to let you set the time, duration of measurements, and other values. Accept all, typing <G>. Once the records are running wait some changes from position 1 (M1) to position 2 (M2), in M1 movement, count 4 s and start the motor drive.

16. Pretreatments of 1 h with PAs have causes tissue specific effects on salt-induced K^+ leak in maize and *Arabidopsis* roots, reducing K^+ efflux in mature zone and promoting it on elongation zone in a charge-dependent manner $Spm^{4+} > Put^{2+}$ [15]. In pea leaves $Spm^{4+} \sim Put^{2+}$ strongly reduced NaCl-induced K^+ efflux [7].
17. Modulation of ROS-induced ion fluxes by PAs depends on the genotype [21] and is tissue-specific, e.g., marked differences are observed between mature or elongation root zone [36].
18. During calibration verify, whether a particular drug does not interfere with a LIX response.
19. To produce an .AVM file, press the following commands in order: ALT <E> (electrometer), <A> (average data), <M> (manipulator cycle average). Type the <Valid Time>, the <Radius> (for root measurements) and <Dist. Of Tissue> (50 μm), "Stage Time" is provided. Pressing <ENTER> moves the highlight to the next stage. Press <ENTER> to accept the order of the curve ("Kind of <F>it") to fit the data (typing <F> to cycle through the orders). An .AVM file is created. Quit CHART pressing ALT <Q>.
20. MIFEFLUX was developed to implement the flux calculations according to the published procedures. It takes output files from CHART and produces convenient ASCII text files for spreadsheet importing. In MS-DOS type MIFEFLUX to execute MIFEFLUX.exe. Type the eight characters of the .AVC (calibration) file, the software gives the opportunity to reject some wrong channel calibrations. Press <ENTER> and type the next .AVM eight characters file (flux data), chose the type of tissue <plane> for leaves or <cylinder> for roots, press <ENTER>, the .FLX file is created. You can also import it into a spreadsheet or view it with a text editor.
21. In datasheet before making a graph, remove leftover columns, leaving only the columns of time and flux. MIFE and SIET adopt opposite sign convention: in MIFE the efflux is negative.
22. Some of these blockers (NPPB, niflumate) were proved to be efficient against specific types of NSCC (OH•-induced conductance, *see* ref. [18]).

Acknowledgment

This work was supported by CONACyT grant 204910 and PRODEP grant UC-CA-4 to I. Z.-J.

References

1. Pittman JK (2012) Multiple transport pathways for mediating intracellular pH homeostasis: the contribution of H^+/ion exchangers. Front Plant Sci 3:1–8
2. Demidchik V, Straltsova D, Medvedev SS, Pozhvanov GA, Sokolik A, Yurin V (2014) Stress-induced electrolyte leakage: the role of K^+-permeable channels and involvement in programmed cell death and metabolic adjustment. J Exp Bot 65:1259–1270
3. Maathuis FJ (2014) Sodium in plants: perception, signalling, and regulation of sodium fluxes. J Exp Bot 65:849–858
4. Shabala S, Wu H, Bose J (2015) Salt stress sensing and early signalling events in plant roots: current knowledge and hypothesis. Plant Sci 241:109–119
5. Hedrich R, Salvador-Recatalà V, Dreyer I (2016) Electrical wiring and long-distance plant communication. Trends Plant Sci 21:376–387
6. Liu K, Fu H, Bei Q, Luan S (2000) Inward potassium channel in guard cells as a target for polyamine regulation of stomatal movements. Plant Phys 124:1315–1326
7. Shabala S, Cuin TA, Pottosin I (2007) Polyamines prevent NaCl-induced K^+ efflux from pea mesophyll by blocking non-selective cation channels. FEBS Lett 581:1993–1999
8. Zhao F, Song CP, He J, Zhu H (2007) Polyamines improve K^+/Na^+ homeostasis in barley seedlings by regulating root ion channel activities. Plant Phys 145:1061–1072
9. Dobrovinskaya OR, Muñiz J, Pottosin II (1999) Inhibition of vacuolar ion channels by polyamines. J Membr Biol 167:127–140
10. Dobrovinskaya OR, Muñiz J, Pottosin II (1999) Asymmetric block of the plant vacuolar Ca^{2+}-permeable channel by organic cations. Eur Biophys J 28:552–563
11. Tikhonova LI, Pottosin II, Dietz KJ, Schönknecht G (1997) Fast-activating cation channel in barley mesophyll vacuoles. Inhibition by calcium. Plant J 11:1059–1070
12. Brüggemann LI, Pottosin II, Schönknecht G (1998) Cytoplasmic polyamines block the fast-activating vacuolar cation channel. Plant J 16:101–105
13. Pottosin I, Shabala S (2014) Polyamines control of cation transport across plant membranes: implications for ion homeostasis and abiotic stress signaling. Front Plant Sci 5:154
14. Pottosin I (2015) Polyamine action on plant ion channels and pumps. In: Kusano T, Suzuki H (eds) Polyamines. Springer, Japan, pp 229–241
15. Pandolfi C, Pottosin I, Cuin T, Mancuso S, Shabala S (2010) Specificity of polyamine effects on NaCl-induced ion flux kinetics and salt stress amelioration in plants. Plant Cell Physiol 51:422–434
16. Pottosin I, Velarde-Buendía AM, Bose J, Fuglsang AT, Shabala S (2014) Polyamines cause plasma membrane depolarization, activate Ca^{2+}-, and modulate H^+-ATPase pump activity in pea roots. J Exp Bot 65:2463–2472
17. Bose J, Pottosin II, Shabala SS, Palmgren MG, Shabala S (2011) Calcium efflux systems in stress signaling and adaptation in plants. Front Plant Sci 2:85
18. Zepeda-Jazo I, Velarde-Buendía AM, Enríquez-Figueroa R, Bose J, Shabala S, Muñiz-Murguía J, Pottosin II (2011) Polyamines interact with hydroxyl radicals in activating Ca^{2+} and K^+ transport across the root epidermal plasma membranes. Plant Physiol 157:2167–2180
19. Pottosin I, Velarde-Buendía AM, Bose J, Zepeda-Jazo I, Shabala S, Dobrovinskaya O (2014) Cross-talk between ROS and polyamines in regulation of ion transport across plasma membrane: implications for plant adaptive responses. J Exp Bot 65:1271–1283
20. Chen Z, Pottosin II, Cuin TA, Fuglsang AT, Tester M, Jha D, Zepeda-Jazo I, Zhou M, Palmgren MG, Newman IA, Shabala S (2007) Root plasma membrane transporters controlling K^+/Na^+ homeostasis in salt-stressed barley. Plant Physiol 145:1714–1725
21. Velarde-Buendía AM, Shabala S, Cvikrova M, Dobrovinskaya O, Pottosin I (2012) Salt-sensitive and salt-tolerant barley varieties differ in the extent of potentiation of the ROS-induced K^+ efflux by polyamines. Plant Physiol Biochem 61:18–23
22. Newman I, Chen SL, Marshall Porterfield D, Sun J (2012) Non-invasive flux measurements using microsensors: theory, limitations and systems. In: Shabala S, Cuin TA (eds) Plant salt tolerance. Methods and protocols. Humana Press-Springer, Totowa, NJ, pp 101–118
23. Beffagna N, Buffoli B, Busi C (2005) Modulation of reactive oxygen species production during osmotic stress in *Arabidopsis thaliana* cultured cells: involvement of the plasma membrane Ca^{2+}-ATPase and H^+-ATPase. Plant Cell Physiol 46:1326–1339
24. Demidchik VV, Tester MA (2002) Sodium fluxes through non-selective cation channels

in the plant plasma membrane of protoplasts from *Arabidopsis* roots. Plant Physiol 128:379–387

25. Cuin TA, Shabala S (2007) Compatible solutes reduce ROS induced potassium efflux in Arabidopsis roots. Plant Cell Environ 30:875–885
26. Hamill OP, Marty A, Neher E, Sakmann B, Sigworth FJ (1981) Improved patch-clamp techniques for high-resolution current recording from cells and cell-free membrane patches. Pflügers Arch 391:85–100
27. Velarde-Buendía AM, Enríquez-Figueroa RA, Pottosin I (2012) Patch-clamp protocols to study cell ionic homeostasis under saline conditions. In: Shabala S, Cuin TA (eds) Plant salt tolerance. Methods and protocols. Humana Press-Springer, Totowa, NJ, pp 3–18
28. Shabala S, Demidchik V, Shabala L, Cuin TA, Smith SJ, Miller AJ, Davies JM, Newman IA (2006) Extracellular Ca^{2+} ameliorates NaCl-induced K^+ loss from Arabidopsis root and leaf cells by controlling plasma membrane K^+-permeable channels. Plant Physiol 141:1653–1665
29. Shabala S, Zhang J, Pottosin II, Bose J, Zhu M, Fuglsang AT, Velarde-Buendía A, Massart A, Hill CB, Bacic A, Wu H, Azzarello E, Pandolfi C, Zhou M, Poschenrieder C, Mancuso S, Shabala S (2016) Cell-type specific H^+-ATPase activity enables root K^+ retention and mediates acclimation to salinity. Plant Physiol 172:2445–2458
30. Demidchik V, Maathuis F (2007) Physiological roles of nonselective cation channels in plants: from salt stress to signalling and development. New Phytol 175:387–404
31. Roberts SK (2006) Plasma membrane anion channels in higher plants and their putative functions in roots. New Phytol 169:647–666
32. Percey WJ, Shabala L, Breadmore MC, Guijt RM, Bose J, Shabala S (2014) Ion transport in broad bean leaf mesophyll under saline conditions. Planta 240:729–743
33. Guo KM, Babourina O, Rengel Z (2009) Na^+/H^+ antiporter activity of the *SOS1* gene: lifetime imaging analysis and electrophysiological studies on *Arabidopsis* seedlings. Physiol Plant 137:155–165
34. Pottosin II, Martinez-Estevez M, Dobrovinskaya OR, Muñiz J (2003) Potassium-selective channel in the red beet vacuolar membrane. J Exp Bot 54:663–667
35. Pottosin II, Schönknecht G (2007) Vacuolar calcium channels. J Exp Bot 58:1559–1569
36. Pottosin I, Velarde-Buendía AM, Zepeda-Jazo I, Dobrovinskaya O, Shabala S (2012) Synergism between polyamines and ROS in the induction of Ca^{2+} and K^+ fluxes in roots. Plant Signal Behav 7:1084–1087

Chapter 24

Analysis of DNA Methylation Content and Patterns in Plants

Andreas Finke, Wilfried Rozhon, and Ales Pecinka

Abstract

DNA methylation is an epigenetic modification, which contributes to the regulation of gene expression and chromatin organization, and thus plays a role in many aspects of plant life. Here we present three methods for the detection of DNA methylation in plant tissues: high performance liquid chromatography, methylation-sensitive restriction digest followed by quantitative PCR and bisulfite conversion followed by single read sequencing. These methods are complementary and allow analysis of DNA methylation in samples from both model and non-model plant species.

Key words DNA methylation, High precision liquid chromatography, Bisulfite sequencing, Methylation-sensitive restriction endonucleases, Plants

1 Introduction

Epigenetic pathways control a plethora of biological processes including development, genome stability and stress responses [1–3]. Permissive or repressive epigenetic states are defined by specific chromatin marks and render chromatin either open or closed, respectively [4]. The 5-methyl-2′-deoxycytosine (5-mdC; DNA methylation) is a prominent epigenetic modification, which is widespread in plants and mammals [5, 6]. Three classes of DNA methylation: CG, CHG and CHH (where H is A, T or C) are found in plants. Accumulation of DNA methylation in all sequence contexts has a strong repressive effect and leads to local suppression of transcription and heterochromatinization [7].

In order to further facilitate analysis of plant DNA methylation, we present three methods developed to detect the presence of 5-mdC at different resolution levels: Reversed phase and cation exchange high performance liquid chromatography (RP and CEX HPLC); Methylation-sensitive quantitative PCR (MS-qPCR) and Bisulfite conversion (BisCo) followed by single read (Sanger) sequencing (Table 1). Each method may be used depending on the specific research question, availability of the reference sequences

Rubén Alcázar and Antonio F. Tiburcio (eds.), *Polyamines: Methods and Protocols*, Methods in Molecular Biology, vol. 1694, DOI 10.1007/978-1-4939-7398-9_24, © Springer Science+Business Media LLC 2018

Table 1
Comparison of DNA methylation analysis methods presented here

Method	Sequence information	Context specificity	Resolution	Sample throughput (per week)	Costs
RP and CEX HPLC	Not required	Not	Global	240 (RP-HPLC)[a] 800 (CEX-HPLC)[b]	Low
MS-qPCR	Required	Yes	Local	48[b]	Moderate
BisCo	Up to whole genome	Yes	Single base	12[b]	High

[a]Per HPLC system
[b]If two ROI per DNA sample are analyzed

for the region of interest (ROI) and number of samples intended for analysis.

HPLC allows estimation of the global 5-mdC content without the need for a reference genome sequence. Initially, the DNA is broken into the nucleobases or the nucleosides by hydrolysis with strong acids or enzymatic digest, respectively. However, hydrolysis has undesired side effects [8, 9], making the enzymatic digest the method of choice. The method presented here uses the enzymatic digest by nuclease P1 and DNase I, and the de-phosphorylation to nucleosides with alkaline phosphatase. Traditionally, nucleosides are analyzed by RP HPLC with UV detection [10]. However, RP chromatography has several disadvantages for the analysis of 2′-deoxycytidine (dC) and 5-mdC: (1) as the most hydrophilic nucleosides dC and 5-mdC appear as the first in the chromatogram and may overlap with only partially digested side products, and (2) the analysis time is long because all nucleosides must be eluted from the column before the next sample can be injected. These disadvantages are alleviated in the CEX chromatography, where dC and 5-mdC are the last peaks allowing very short analysis time without interference with the partially digested side products [11]. This method is simple and highly reproducible (the relative SD $< 2\%$), but requires a relatively high amount (approximately 5 μg) of DNA. Combination of RP HPLC and detection by tandem mass spectrometry offers high sensitivity and thus the amount of required DNA can be reduced to $\leq$100 ng [12]. However, this approach requires very expensive equipment and specialized expertise, and therefore is not commonly used. Recently, an alternative method was presented [13], where the nucleosides are derivatized with 2-bromoacetophenone, leading to the formation of highly fluorescent products with dC, 5-mdC but not with other nucleotides (Fig. 1). In addition, potential cytidine contamination from RNA does not interfere since its peak is well-separated. We improved the original protocol by: (1) omitting primary and secondary amines in

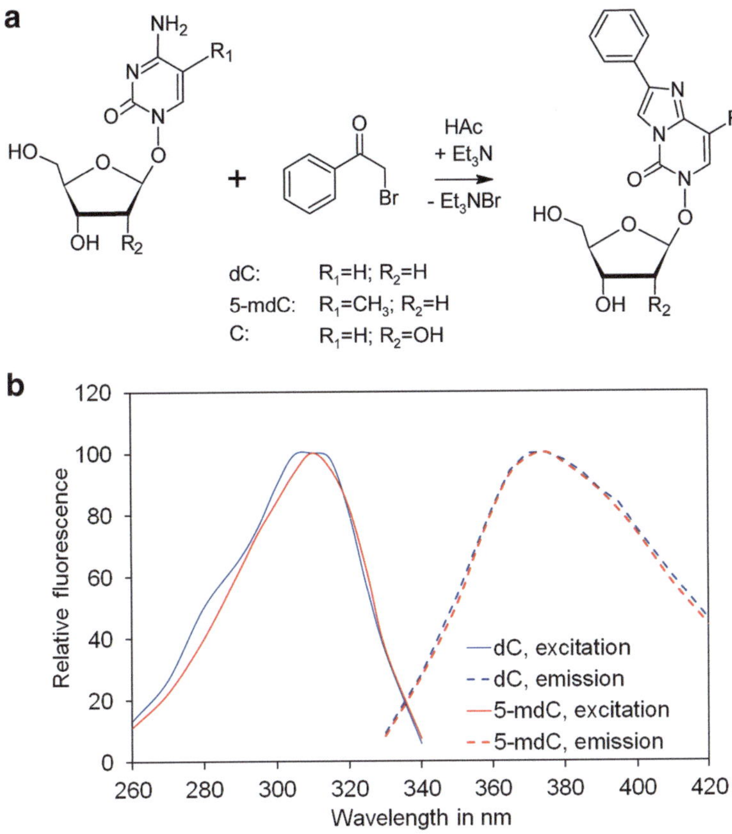

Fig. 1 Derivatization of dC and 5-mdC with 2-bromoacetophenone. (**a**) Structures of the reactants and the fluorescent products. (**b**) Fluorescence spectra of the dC and 5-mdC derivative

the digestion buffer, which form undesired side products with the reagent; (2) adding triethylamine during derivatization, which captures hydrobromic acid and thereby increases the yield and reaction velocity; (3) replacing trifluoroacetic acid with sulphuric acid, which makes the HPLC eluent system more environmentally friendly. The presented method is simple, highly reproducible (relative standard deviation of ca. 4%), and its sensitivity is similar to HPLC-tandem mass spectrometry-based techniques.

MS-qPCR allows determining methylation levels at specific target sequences. Its resolution is lower compared to BisCo, due to dependence on the presence of suitable restriction sites and enzymes, but is compensated by its speed, low costs and possibility to process simultaneously many samples. MS-qPCR is based on the inhibition of the catalytic activity of restriction endonucleases (REs) by 5-mdC present in genomic DNA. The DNA molecules non-methylated at the recognition site are digested, while the methylated ones remain intact and the region of interest (ROI) can be PCR-amplified. One of the most important steps in this analysis is the choice of ROI, a suitable amplicon size, appropriate restriction enzymes and robust controls. Typical ROIs include simple or tandem repetitive or transposon regions. Tandem repeats can be searched online using e.g. Tandem Repeat Finder (http://tandem.bu.edu/trf/trf.html) software [14]. For some species e.g. *Arabidopsis thaliana* there are also available whole genome DNA methylation browsers (e.g. [7]; http://genomes.mcdb.ucla.edu/AthBSseq/), which may greatly help in ROI design. Selection of REs is another critical parameter towards successful MS-qPCR. There are data concerning sensitivity of more than 500 commercially available REs to 5-mdC (http://rebase.neb.com). However, the sensitivity information is incomplete or even contradictory in many cases. Here, we provide a list of the enzymes, for which the methylation sensitivity was tested extensively (Table 2). The following criteria should be considered when selecting a MS RE: (1) it has a single recognition site in the ROI (*see* **Note 1**); (2) it is active at 37 °C (Table 2); (3) all enzymes used should have 100% activity in the same buffer and (4) choose one restriction enzyme that cuts outside of the amplified sequence (*see* **Note 2**). We strongly recommend including robust positive and negative controls. Positive control tests for the activity of the used enzyme as well as the presence of the restriction site by digesting a 1:100 diluted PCR product of the ROI. PCR products are free of DNA methylation and therefore should be fully digested. Successful digestion should be assessed by separation of the digestion products on an agarose gel. Incubation of the genomic DNA with an enzyme lacking a recognition site within the ROI will serve as negative non-RE control (NEC) and should yield full amplification in qPCR.

Sodium bisulfite treatment of genomic DNA leads to the deamination of dCs and their conversion to uracils, while the 5-mdCs remain unchanged. Upon PCR amplification and sequencing, the positions of dCs will appear as thymines, while the 5-mdCs will remain as cytosines. Hence, this method allows the analysis of individual DNA molecules in the strand-specific and single base pair resolution manner. The principle of bisulfite conversion of DNA was described more than 20 years ago [15, 16], but the initial protocol was improved and simplified since then [17, 18]. In addition, several companies developed kits and optimized enzymes, which greatly simplified the whole procedure and increased the success of bisulfite conversion. In our BisCo protocol, we focus on the critical aspects of bisulfite conversion such as the quality controls or design of PCR primers for amplification from bisulfite-treated DNA.

Table 2
Recommended MS REs

Enzyme	^{m}CG	^{m}CHG	^{m}CHH
AatII	++	+	+
AccI	++	+	+
AciI	++	+	+
AclI	++	–	–
AluI	–	+	+
AsiSI	++	–	–
AvaII	+	+	+
BamHI	–	+	+
BanII	–	–	++
BbvI	–	++	–
BfuCI	+	+	+
BsaAI	++	–	–
BsaHI	++	–	–
BsmAI*	+	+	+
BsmBI*	++	–	–
BspDI	++	–	–
BsrBI	++	+	+
BsrFI	++	–	–
ClaI	++	–	–
DdeI	–	++	++
EaeI	++	++	++
EegI	++	++	–
EarI	+	+	++
EciI	++	–	–
FatI	–	–	++
FseI	++	++	–
HaeIII	+	+	+
HhaI	++	+	+
HpaII	++	++	–
HphI	–	+	++
HpyCH4IV	++	–	–

(continued)

Table 2
(continued)

Enzyme	^{m}CG	^{m}CHG	^{m}CHH
KasI	++	–	–
MluI	++	–	–
MnlI	–	–	++
MspI	–	++	–
NaeI	++	++	+
NarI	++	+	+
NciI	++	–	–
NcoI	–	–	++
NheI	+	+	++
NlaIII	–	–	++
NruI	++	–	–
NspI	–	–	++
PleI	+	+	++
PmlI	++	–	–
PstI	–	++	–
PvuI	++	–	–
SalI	+	+	+
Sau96I	+	+	+
SpeI	–	–	++
SphI	–	–	++
SrfI	++	++	++
XbaI	–	–	++

(++) indicates cytosine in sequence context strictly defined by the RE recognition sequence. (+) shows C positions which are on the 3′ end of the recognition sequence and their context is defined by the bases outside of the RE recognition site. Use of such REs needs to be considered ROI to ROI. (−) Not affected by DNAmethylation in this context. REs with incubation temperature 55 °C are marked with *asterisk*. All other REs have optimal reaction temperature at 37 °C

2 Materials

2.1 DNA Extraction

1. Plant DNA extraction kit (tested with GE Healthcare and Qiagen kits) or equivalent.
2. RNase A DNase- and protease-free (10 mg/ml), store at −20 °C.

2.2 Quantification of the Global 5-mdC Content by HPLC

2.2.1 Cation Exchange Chromatography

1. Nuclease mix: nuclease P1, 2.5 U/ml and DNase I, 500 U/ml. For preparation, transfer 2.5 U nuclease P1 and 500 U DNase I (*see* **Note 3**) into a tube and add 500 μl water. Shake at 300 rpm and 4 °C for 30 min. Centrifuge at >15,000 × *g* and 4 °C for 10 min and transfer the clear supernatant to a fresh tube. Centrifuge again as described above. Transfer the clear supernatant to a fresh tube, add an equal volume of glycerol and mix well. The reagent can be used for at least 2 years if kept at −20 °C.
2. Nuclease buffer I (10×): 200 mM acetic acid, 200 mM glycine, 50 mM $MgCl_2$, 5 mM $ZnCl_2$, 2 mM $CaCl_2$, pH 5.3. For preparation dissolve 1.11 ml glacial acetic acid, 1.50 g glycine, 1.02 g magnesium chloride hexahydrate, 68 mg anhydrous zinc chloride and 29 mg calcium chloride dihydrate in approximately 80 ml water and set the pH to 5.3 by adding 4 M NaOH. Finally, add water to a total volume of 100 ml. Keep at −20 °C.
3. NaOH, 4 M: dissolve 16 g NaOH in water to a total volume of 100 ml.
4. NaOH, 100 mM: mix 25 μl NaOH, 4 M with 975 μl water. The solution must be kept tightly closed.
5. Calf intestinal alkaline phosphatase (CIAP) 1 U/μl (*see* **Note 4**).
6. Sulphuric acid, 10 mM: dilute 1 ml of 1 M sulphuric acid diluted with water to a final volume of 100 ml.
7. dC stock, 2 mM: weigh 45.4 mg 2′-deoxycytidine or 52.7 mg 2′-deoxycytidine hydrochloride (*see* **Note 5**) to the nearest 0.1 mg and transfer it quantitatively into a 100 ml volumetric flask, add approximately 80 ml water and shake until the solid has completely dissolved, which may take some time. Finally, add water to the mark. The solution is stable for many years if stored at −20 °C.
8. 5-mdC stock, 1 mM: weigh 24.1 mg 5-methyl-2′-deoxycytidine (*see* **Note 5**) to the nearest 0.1 mg and transfer it quantitatively into a 100 ml volumetric flask, add approximately 80 ml water and shake until the solid has completely dissolved, which may take some time. Finally, add water to the mark. The solution is stable for many years if stored at −20 °C.
9. Eluent: 40 mM acetic acid/sodium acetate pH 4.8 in 15% acetonitrile (ACN). For preparation add 2.38 ml glacial acetic acid and 150 ml ACN to approximately 650 ml HPLC grade water and adjust the pH to 4.8 using 4 M NaOH. Transfer the solution to a 1000 ml volumetric flask and add water to the mark. The solution can be kept at room temperature for at least 1 year.
10. HPLC column: Nucleosil SA 100-10 250 × 4 mm, Macherey-Nagel equipped with a Nucleosil SA 100-5 3 × 4 mm guard column (*see* **Note 6**).
11. HPLC system equipped with an UV detector.

2.2.2 RP-HPLC

1. Nuclease buffer II (10×): 200 mM acetic acid, 50 mM $MgCl_2$, 5 mM $ZnCl_2$ and 2 mM $CaCl_2$, pH 5.3. For preparation, add 1.11 ml glacial acetic acid, 1.02 g magnesium chloride hexahydrate, 68 mg anhydrous zinc chloride and 29 mg calcium chloride dihydrate in approximately 80 ml water and set the pH to 5.3 by adding *N*-methylmorpholine. Finally, add water to 100 ml. Keep at −20 °C.
2. Digestion premix: mix 78 μl water with 20 μl nuclease buffer II and 2 μl nuclease mix (*see* Subheading 2.2.1). This solution is sufficient for ten digestions and must be prepared immediately before use.
3. Triethylamine, 150 mM: dilute 207 μl triethylamine in water to a final volume of 10 ml. Keep at −20 °C.
4. CIAP, 0.1 U/μl: mix 2 μl CIAP, 1 U/μl (*see* **Note 4**) with 18 μl water. This solution must be prepared immediately before use.
5. 2-Bromoacetophenone, 60 mM in DMF: dissolve 11.9 mg 2-bromoacetophenone in 1 ml *N*,*N*-dimethylformamide. The solution can be kept at −20 °C for up to 1 week.
6. Triethylamine, 1 M in acetic acid: mix 139 μl triethylamine with 861 μl glacial acetic acid. Keep at −20 °C.
7. Derivatization reagent: mix 1000 μl 60 mM 2-bromoacetophenone in DMF with 20 μl 1 M triethylamine in acetic acid. This solution must be prepared immediately prior to use.
8. Eluent A: 50 mM sulphuric acid. For preparation, transfer 50 ml 1 M sulphuric acid into a 1000 ml volumetric flask and fill with distilled water to the mark. Filter through a 0.2 μm nylon or PTFE membrane filter.
9. Eluent B: acetonitrile, HPLC grade.
10. HPLC column: Nucleodor C18ec 100-5 125 × 4 mm, Macherey-Nagel (*see* **Note 7**).
11. Vacuum concentrator.
12. HPLC system equipped with a binary pump and a fluorescence detector.

2.3 Analysis of Locus-Specific DNA Methylation by MS-qPCR

2.3.1 DNA-Digest Using Methylation-Sensitive Restriction Endonucleases (MS REs)

1. Restriction endonucleases (*see* Table 2) and appropriate buffers (*see* **Note 2**).

2.3.2 Quantitative PCR with Digested DNA

1. 2× SybrGreen PCR master mix containing the passive reference dye appropriate for you qPCR device.
2. qPCR plates and adhesive film.
3. qPCR cycler.
4. Primer pairs for ROIs (*see* **Note 1**).

2.4 Analysis of Locus-Specific DNA Methylation by Bisulfite Conversion (BisCo)

2.4.1 Pre-Conversion Procedure

1. Restriction endonuclease cutting outside of your sequence of interest and fitting reaction buffer (*see* **Note 2**).
2. 3 M Sodium acetate.
3. Isopropanol.
4. 70% Ethanol.
5. Buffer AE (Qiagen).

2.4.2 Bisulfite Conversion

1. Bisulfite conversion Kit. The protocol presented here was tested using EpiTect bisulfite conversion kit (Qiagen) and Epi-JET bisulfite conversion kit (Thermo Scientific).

2.4.3 ROI Amplification, Cloning and Sequencing

1. Conversion control primers (*see* **Note 8**).
2. Target sequence-specific primers (*see* **Note 9**).
3. Taq DNA Polymerase (multiple suppliers) or MethylTaq DNA polymerase (Diagenode).
4. 10 mM dNTP mix.
5. PCR purification Kit.
6. T/A or blunt-end cloning system.
7. Competent *E. coli* DH5-alpha.
8. Liquid broth (LB) solution.
9. LB-Agar plates containing antibiotics corresponding to the used vector.

3 Methods

3.1 DNA Isolation and Quantification

1. Extract genomic DNA with your method of choice (*see* **Note 10**) and resuspend the DNA in RNAse (10 mg/ml) containing water or TE-buffer and determine the concentration by fluorimetry as described [19] (*see* **Note 11**). Isolated DNA can be stored for up to 3 months at −20 °C. For HPLC-based methods the DNA may be stored at −20 °C infinitely.
2. Optional: Check DNA integrity by gel electrophoresis of approximately 100 ng aliquot. There should be a single high molecular weight band.

3. Go to Subheading 3.2 for HPLC protocols, to Subheading 3.3 for MS-qPCR protocol or Subheading 3.4 for BisCo protocol.

3.2 Quantification of the Global 5-mdC Content by HPLC

3.2.1 Quantification of 5-mdC by Cation Exchange Chromatography

1. Transfer 4–7 μg DNA (*see* **Notes 12** and **13**) into a 1.5 ml reaction tube and add water to a final volume of 44 μl.
2. Add 5 μl 10× nuclease buffer I and 1 μl nuclease mix. Pipette up and down several times to mix properly.
3. Incubate at 37 °C overnight.
4. Add 5 μl 100 mM NaOH and 1 μl CIAP 1 U/μl. Mix by pipetting up and down several times.
5. Incubate at 37 °C for 6–24 h.
6. Add 30 μl 10 mM sulphuric acid (*see* **Note 14**), mix thoroughly and centrifuge for 5 min at 15,000 × *g*.
7. Transfer the clear supernatant into an autosampler vial and seal with a cap.
8. Prepare standard stock solutions according to Table 3 (*see* **Note 15**). The stock solutions can be used for at least 2 years if stored at −20 °C.
9. Transfer 25 μl of the standard stock solutions into autosampler vials, add 65 μl water and seal with caps.
10. Run the standards and the samples on a HPLC system with the following settings: column: Nucleosil SA 100-10 250 × 4 mm

Table 3
Preparation of standard stock solutions

Standard no.	dC in pmol/μl	5-mdC in pmol/μl	5-mdC/dC in pmol/pmol	dC stock 2 mM in μl	5-mdC stock 1 mM in μl	H_2O in μl
St1	200	0	0.00	100	0	900
St2	200	4	0.02	100	4	896
St3	200	8	0.04	100	8	892
St4	200	12	0.06	100	12	888
St5	200	16	0.08	100	16	884
St6	200	20	0.10	100	20	880
St7	200	30	0.15	100	30	870
St8	200	40	0.20	100	40	860
St9	200	60	0.30	100	60	840
St10	200	80	0.40	100	80	820
St11	200	100	0.50	100	100	800
St12	200	140	0.70	100	140	760

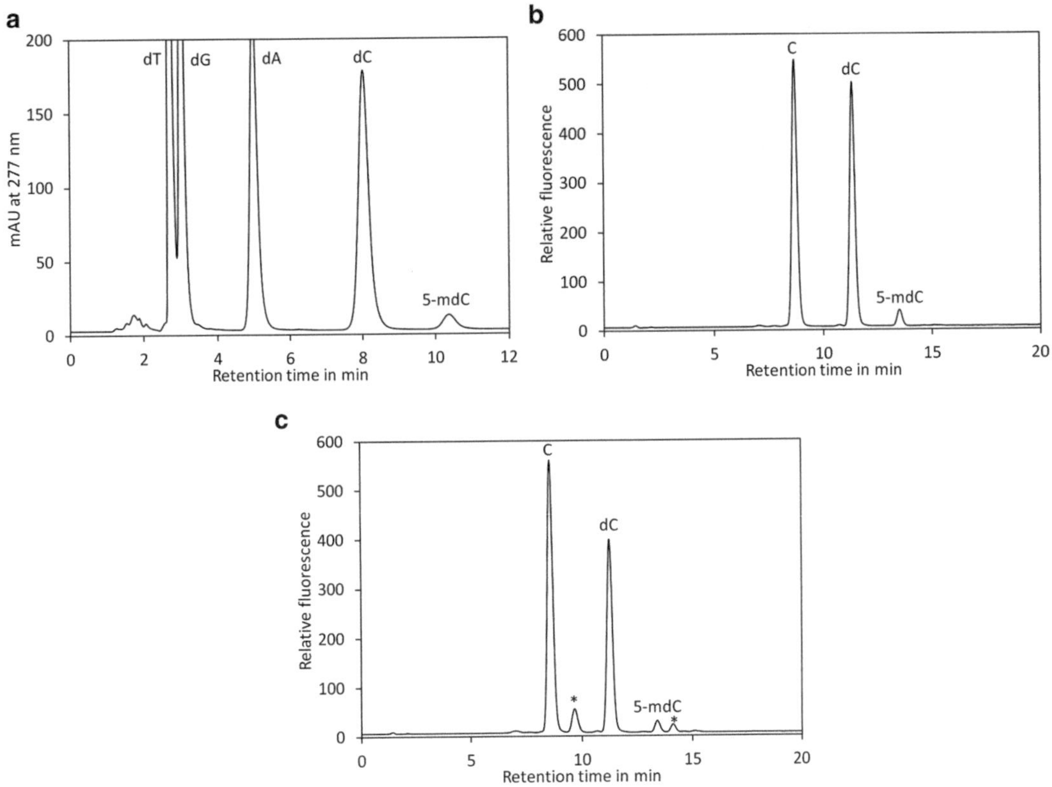

Fig. 2 Chromatograms of 5-mdC quantification. (**a**) Cation exchange chromatography of 5 μg enzymatically digested *Arabidopsis thaliana* DNA. (**b**) RP-HPLC chromatogram of 200 ng well digested *A. thaliana* DNA derivatized with 2-bromoacetophenone. (**c**) Same sample as in (**b**) but insufficiently digested (the digestion buffer had too high DNA). Insufficient digestion leads to additional peaks (indicated by *stars*) and unreliable results. The cytosine peak (C) in (**b**) and C is visible in most samples and originates from RNA contaminations

(*see* **Note 6**); elution: isocratic with 40 mM acetic acid/sodium acetate pH 4.8 in 15% acetonitrile at a flow rate of 1 ml/min; column oven temperature: 25 °C; injection volume: 50 μl; detection: UV, 277 nm; run time for one sample: 12 min. A typical chromatogram is shown in Fig. 2a.

11. A calibration curve is established by plotting a diagram with the molar ratio of 5-mdC and dC (pmol 5-mdC/pmol dC) on the *x*-axis and the ratio of the areas (area 5-mdC/area dC) on the *y*-axis. The molar ratio of 5-mdC/dC (abbreviated as $r_{5\text{-mdC}}$ in the formula shown below) of the samples is calculated using the calibration curve and converted to % 5-mdC ($\%_{5\text{-mdC}}$) using the formula $\%_{5\text{-mdC}} = 100 * r_{5\text{-mdC}}/(1 + r_{5\text{-mdC}})$.

3.2.2 Quantification of 5-mdC by RP-HPLC

1. Transfer 150–300 ng DNA (*see* **Notes 13** and **16**) into a 1.5 ml reaction tube and add water to a total volume of 10 μl.
2. Add 10 μl digestion premix and mix by pipetting up and down.

Table 4
Gradient for reversed phase HPLC

Time in min	Eluent A in %	Eluent B in %
0	88	12
16	80	20
17	20	80
22	20	80
23	88	12
30	88	12

3. Incubate at 37 °C overnight.
4. Add 2 μl 150 mM triethylamine and 2 μl CIAP, 0.1 U/μl. Mix by pipetting up and down and incubate at 37 °C overnight.
5. For preparation of standards transfer 1 μl of each standard stock solution (*see* Table 3) into 1.5 μl reaction tubes and add 2 μl nuclease buffer II and 2 μl 150 mM triethylamine to each tube.
6. Evaporate the samples and standards in a vacuum concentrator with 1 mbar final pressure for at least 1 h (*see* **Notes 17** and **18**).
7. Add 50 μl derivatization reagent to each tube and incubate at 60 °C for 1 h.
8. Centrifuge for 5 min at 16 000 × *g*.
9. Transfer the clear supernatant into an autosampler vial and seal with a cap.
10. Run the samples on a HPLC system equipped with a fluorescence detector with the following settings: column: Nucleodur C18ec 100-5 125 × 4.6 mm; gradient elution with 50 mM sulphuric acid (eluent A, *see* **Note 19**) and acetonitrile (eluent B) at a flow rate of 1 ml/min according to Table 4 shown below; column oven temperature: 25 °C; injection volume: 20 μl; detection: fluorescence with an excitation wavelength of 305 nm and an emission wavelength of 370 nm (*see* **Note 20**); run time for one sample: 30 min. A typical chromatogram is shown in Fig. 2b. An example of an incompletely digested sample is shown in Fig. 2c.
11. A calibration curve is plotted and the 5-mdC level is calculated as described in Subheading 3.2.1, **step 11**.

3.3 Analysis of Locus-Specific DNA Methylation by MS-qPCR

3.3.1 DNA Digestion Using MS RE

1. For every experimental point (e.g. condition or mutant) prepare 270 μl of a genomic DNA solution with concentration 1.5 ng/μl. This amount is sufficient for one NEC sample and treatment with nine REs
2. Add 30 μl of 10× restriction buffer and 5 μl of the RE cutting outside of the planned ROIs (*see* **Note 2**). Mix gently.
3. Divide the sample into ten 30 μl aliquots. Add 2 μl (typically 20 U) of RE per aliquot, i.e. up to nine different digests can be made. Keep one aliquot as NEC by omitting RE and adding 2 μl water.
4. Incubate all reactions at the temperature optimal for restriction digestion of your selected enzymes (37 °C for most enzymes) for at least 12 h to ensure a complete digest of the template DNA and terminate the reaction (if possible) by incubation at 65 °C for 20 min.

3.3.2 Quantitative PCR with Digested DNA

1. Add 120 μl sterile water to each tube to reach a total volume of 150 μl (*see* **Note 21**), which is sufficient to test ten different ROI if the analysis is performed in three technical replicates of 12 μl each.
2. To obtain a standard curve and assess the efficiency of the amplification prepare a dilution series (e.g. 1:2; 1:5; 1:10; 1:50; 1:100; 1:1000) of the NEC sample in water (*see* **Note 22**).
3. Prepare a MasterMix sufficient to perform three technical replicates for each digested sample, the NC and the samples of the dilution series (Table 5).
4. Pipette the master mix and template DNA into qPCR plate and seal it.
5. Spin the PCR plate for 2 min at maximum speed to remove potential air bubbles.
6. Run qPCR with the following temperature regime (Table 6).

Table 5
Recommended constitution of the qPCR reaction

Component	Volume
SybrGreen MasterMix (2×)	6 μl
Forward primer, 10 μM	0.5 μl
Reverse primer, 10 μM	0.5 μl
Sterile water	1 μl
Digested DNA	4 μl
Total	12 μl

Table 6
qPCR temperature regime

Step	Temperature	Duration
1	95 °C	10 min
2	95 °C	15 s
3	56–62 °C[a]	15 s
4	72 °C	45 s
5	72–80 °C (acquisition)	
6	Return to **step 2**	39–44 times repetition
Melting curve	62–95 °C	0.5 °C increment

[a]Depending on primer melting temperature

3.3.3 Data Analysis

1. Inspect $-dF/dT$—melting curve. A single local maximum (peak) for every analyzed ROI should be visible (*see* **Note 23**).
2. Determine whether the amplification efficiency in your standard curve dilution series falls into the accepted range of 85–115% (*see* **Note 24**).
3. Calculate the mean "threshold cycle" (C_t) of your technical replicates for each sample and digestion and for the NECs. All C_ts of the technical replicates should be within one cycle.
4. Perform the following calculations using a spreadsheet program:
   (a) $\Delta C_t = \text{mean}(C_t)_{\text{NEC}} - \text{mean}(C_t)_{\text{cut}}$
   (b) $\text{Ratio} = e^{\Delta C_t}$
   (c) $\% = \text{Ratio}^{*}100$

3.4 Analysis of Locus-Specific DNA Methylation by Bisulfite Conversion (BisCo)

3.4.1 Pre-Conversion Procedure

1. Perform an overnight digestion of approximately 500 ng DNA with DNA methylation insensitive restriction enzyme not cutting in the analyzed region (*see* **Note 25**).
2. Add 0.1 volume of 3 M sodium acetate and 0.7 volumes of room temperature isopropanol, mix and incubate at room temperature for 30 min.
3. Pellet the DNA by centrifugation (15 min at 12,000 × *g*).
4. Remove supernatant and wash the pellet once with 250 μl of 70% ethanol and dry at room temperature.
5. Resuspend the pellet in 30 μl of sterile distilled water or TE-buffer. Digested DNA can be stored at 4 °C for up to 1 month or at −20 °C for longer storage.

3.4.2 Bisulfite Conversion

1. To ensure proper and reproducible conversion as well as easy handling, we recommend use of the commercially available bisulfite conversion kits (*see* **Note 26**).
2. Perform the conversion reaction in a PCR instrument using the following program: 5 min 95 °C, 25 min 60 °C, 5 min 95 °C, 85 min 60 °C, 5 min 95 °C, 175 min 60 °C, hold 20 °C. After the cleanup procedure recommended by the manufacturer, the bisulfite converted DNA can be stored for 2 months at −20 °C.
3. Validate the conversion reaction by PCR (*see* **Notes 8** and **27**).
4. Optional: If a more comprehensive conversion rate assessment is desired, PCR product of the "converted" reaction should be cloned and approximately five clones per sample should be sequenced. Samples with the cytosine conversion rate $\geq$ 95% can be used).
5. If control PCRs and/or cloning of the PCR fragments indicate virtually full conversion, proceed to the amplification of the ROI with bisulfite primers.

3.4.3 ROI Amplification, Cloning and Sequencing

1. Shorter ROIs (<500 bp) should be PCR amplified using Hot-Start Taq Polymerase, while for longer ROIs (500 and 800 bp) we recommend a polymerase optimized for use in bisulfite sequencing (e.g. Diagenode MethylTaq).
2. Three PCR reactions containing 1, 2 and 3 μl of the bisulfite converted DNA should be prepared per ROI and experimental point (Table 7).
3. Run PCR using the following temperature regime (Table 8).
4. Validate amplicon presence and size by electrophoresis on 1% agarose gel using 5 μl of the PCR reaction.

Table 7
Recommended composition for the PCR amplification of ROI from BisCo DNA

Component	Final concentration
5× MethylTaq buffer	1×
$MgCl_2$	2 mM
dNTPs	200 μM
Forward primer	0.2 μM
Reverse primer	0.2 μM
MethylTaq DNA polymerase	1.5 U
Sodium bisulfite converted DNA	1–3 μl
Sterile water	Variable
Total volume	50 μl

Table 8
PCR temperature regime

No. of cycles	Temperature	Duration
1	95 °C	10 min
2	95 °C	30 s
3	56 –62°C[a]	1 min
4	72 °C	1 min
5	Return to **step 2**	39–44 repetitions
6	72 °C	5 min

[a]Depending on the primer annealing temperature

5. In the case of additional unspecific amplification products load the whole reaction on 1% agarose gel, cut the desired band and extract DNA by gel purification.
6. If the amplification results in a single band of expected size, perform direct cleanup of the PCR reaction using silica membrane spin columns.
7. Clone the PCR product into a sequencing vector of choice.
8. Transform *E. coli* DH5 alpha cells with 5 μl of the ligation reaction and incubate at 37 °C for 1 h. Spread transformed cells onto plates containing antibiotics corresponding to your vector and incubate at 37 °C overnight.
9. Pre-select positive colonies by colony-PCR.
10. Separate colony-PCR products on 1% agarose gel and identify those with the expected size (*see* **Note 28**).
11. Inoculate 3 ml liquid cultures with the positive colonies and incubate them overnight at 37 °C. Extract the plasmid with a standard protocol yielding sequencing quality DNA.
12. Use the isolated plasmid for single read Sanger sequencing. At least 15 clones per target should be analyzed to obtain representative and reliable results (*see* **Note 29**).

3.4.4 Analysis of Converted Sequences

1. Generate reverse complement sequences for PCR amplicons cloned and sequenced in opposite direction and trim vector and bisulfite primer-binding sequences (*see* **Note 30**).
2. Perform a pairwise alignment of the respective forward and reverse sequencing read of each plasmid to identify sequencing errors.
3. Use one read of each read pair and perform a multiple alignment with the reference sequence.
4. Remove sequence duplicates (*see* **Note 31**).

5. Analyze the data using publicly available web-based software tools CyMate (www.cymate.org) or KisMeth (http://katahdin.mssm.edu/kismeth/revpage.pl) [20, 21].

4 Notes

1. Typical candidate regions include repetitive or transposon regions flanking protein-coding genes. For some species e.g. *Arabidopsis thaliana* there are available DNA methylation browsers [7], which may greatly help in defining the ROI.
2. Secondary structures in high molecular weight DNA might influence the amplification efficiency in the qPCR and the conversion during bisulfite conversion. Restriction with an outside cutting enzyme prevents this.
3. Unit definition of nuclease P1: one unit will liberate 1.0 μmole of acid soluble nucleotides from RNA per min at pH 5.3 at 37 °C. Unit definition of DNase I: one Kunitz unit will produce a change in the absorbance of 0.001 at 260 nm/min and per ml at pH 5.0 at 25 °C using DNA as a substrate.
4. CIAP offered by many companies for dephosphorylation of vectors is suitable. Unit definition of the used alkaline phosphatase: one unit will hydrolyze 1 μmole of 4-nitrophenyl phosphate per minute at pH 9.8 (1 M diethanolamine/HCl buffer containing 0.25 mM $MgCl_2$ and 10 mM substrate) at 37 °C.
5. Use only highly pure (>99%) 2′-desoxycytidine or its hydrochloride and 5-methyl-2′-desoxycytidine. The products of some vendors contain crystal water. Use such products only if the water content is certified and adapt the amount according to the molecular weight of the hydrated from.
6. It is possible to reduce the run time by using shorter columns. However, this reduces also the resolution. For details *see* [11]. It may also be possible to use analytical and guard columns packed with silica modified with sulphonated propylphenyl residues from other manufacturers. However, the suitability should be tested in advance.
7. It may be possible to use analytical and guard columns packed with C18-modified silica from other vendors. However, only columns stable at a pH of 1 should be used and the suitability should be tested in advance.
8. Different methods were proposed to address the conversion efficiency. In species with available DNA methylation data, amplification of sequences known to be unmethylated in wild-type plants using primers that either show high affinity to fully converted or to unconverted sequences in the same

Table 9
Conversion control primers

		Sequence (5′ → 3′)
Actin 7	Act7contF (unbiased) Act7contR1 (converted) Act7contR2 (non-converted)	AATGTAAAGTGGAAATGAGAAG GTTAGATTATTTTTTAATTTTTATAGA GCTAGACCACTTTCCAACTTTTATAGA
DDM1	BScontrol1F (non-converted) BScontrol2F (converted) BScontrolR (unbiased)	CGTCTGGTGATTCACCCACTTCTGTTCTCAACG TGTTTGGTGATTTATTTATTTTTGTTTTTAATG CTCTCACTTTCTATCCCATTCTA

cytosine-rich region might be used. Conversion can be considered efficient if the PCR reactions using the "converted" primer result in a PCR product while the reactions containing the "unconverted" primer does not. We amplify promoter regions of the *A. thaliana* genes *DECREASED IN DNA METHYLATION 1* (*DDM1*; AT5G66750) and *ACTIN 7* (*ACT7*; AT5G09810) Primer sequences for BisCo efficiency control PCRs (Table 9).

PCRs to perform	(a)	Act7contF + Act7contR1
	(b)	Act7contF + Act7contR2
	(c)	BScontrol1F + BScontrolR
	(d)	BScontrol2F + BScontrolR

As PCR products are unmethylated efficient bisulfite treatment leads to full conversion of its cytosines. As with the amplification of unmethylated endogenous sequences, amplification of the heterologous sequence using primers matching the converted sequences but not those matching the unconverted should result in a product.

9. Numerous software tools for designing bisulfite-PCR primers in mammals are publicly available. These software tools should be avoided in plant experiments as they often neglect the presence of non-CG methylation and assume full conversion of the cytosine residues in CHG and CHH context, which may lead to biased primer binding.

   The following criteria should be applied during primer design: (1) ROI should be ≤500 bp for a standard Taq polymerase or ≤800 bp for an optimized bisulfite experiment polymerase; (2) primer length 25–32 nt with salt-adjusted melting temperature between 50 °C – 62 °C; (3) primers should contain maximally three degenerated residues to guarantee unbiased amplification. The bisulfite conversion reaction leads to sequence differences in originally complementary stands, which means that each of the strands needs to be analyzed using strand-specific primer pair. In the forward primer, Cs should be synthesized as Y (C or T) and in the reverse primer Gs should be synthesized as R (G or A). (4) Avoid stretches (>4) of the same base and of dinucleotides (e.g. ATATATAT). (5) If possible, perform BLAST analysis of the designed primers and reject primers with possible off-target amplification. We recommend using the Primer-BLAST algorithm and replacing the ambiguous residues R and Y in the primers by N.

10. We recommend using kit-based methods to ensure high quality of the DNA.
11. Most plant DNA preparations contain significant amounts of RNA. Treatment with RNase A, as it is included in most protocols and kits, degrades the RNA only to small fragments rather than removes it. Consequently, DNA quantification by UV spectroscopy overestimates the DNA concentration. In contrast, fluorescence-based methods discriminate RNA and degradation products and gives more reliable results. In addition, it is more sensitive than UV spectroscopy. Alternatively, the amount of DNA may be estimated by gel electrophoresis.
12. An amount of 1–10 μg DNA may be used. However, for all samples similar amounts of DNA should be used and also the concentration of the standards must be adapted.
13. The DNA must not contain more than 0.1 mM EDTA because this would interfere with digestion and de-phosphorylation by complexation of Mg^{2+} and Zn^{2+} ions.
14. Since only protonated nucleosides can interact with the stationary phase it is important to add acid to the samples in order to obtain sharp peaks.
15. The standards listed in Table 3 cover the whole range of 5-mdC contents observed in the plant kingdom. If the 5-mdC content of the investigated species is known it is possible to reduce the number of standards. For instance, for *A. thaliana*, which has a 5-mdC content of approximately 7%, standards St1 to St7 are sufficient.
16. It is also possible to use less DNA, for instance 50 ng. However, for all samples similar amounts of DNA should be used and also the concentration of the standards must be adapted. Lowering the amount of DNA may increase the relative SD of the results.
17. Evaporation over night is possible. Due to the glycerol and salts present in the enzymes and buffers usually a small drop remains even after prolonged evaporation, which does not interfere with subsequent derivatization.

18. If required, the dried samples may be stored at −20 °C for several weeks prior to derivatization.
19. Since only the protonated forms of the derivatives are highly fluorescent an eluent containing a relatively high concentration of a strong acid is required.
20. The sensitivity of the detector should be set to a level that the dC peak of the standards reaches 20–60% of the total range.
21. This step is ought to prevent high buffer salt concentrations, which can influence the efficiency of qPCR reaction.
22. In case the measure of the methylation in several DNA samples in parallel is desired, we recommend to mix aliquots of all NEC samples and to prepare the dilution series from this mix.
23. Gradual increase in temperature causes gradual dissociation of the DNA double strand and thus the release and quenching of the intercalated fluorescent dye. During melting curve acquisition the temperature (T)-dependent reduction of fluorescence (F) is determined as a negative sigmoid function. The temperature at which fluorescence is decreased by 50% is considered as dissociation (melting) temperature. Plotting of the negative first derivative ($-\mathrm{d}F/\mathrm{d}T$) of the function against the temperature leads to appearance of (ideally) a single maximum (peak) at this temperature. If more than one peak is visible, this hints to the formation of primer dimers. You can counteract this by decreasing primer concentration or design of new primers.
24. Advanced qPCR cycler software tools provide this value automatically.
25. Genomic DNA frequently forms secondary structures, which can compromise its full denaturation and hence decrease bisulfite conversion efficiency. The digestion prior to conversion aids the resolution of these secondary structures. Chosen enzymes should not cut within the ROI and should be insensitive to DNA methylation. Suitable enzymes are for example BamHI, EcoRV and ApoI.
26. To control the conversion efficiency in species without a reference genome and/or DNA methylation data a PCR product obtained from a heterologous sequence can be spiked in the DNA sample before conversion. As PCR products are unmethylated efficient bisulfite treatment should lead to full conversion of its cytosines.
27. If a heterologous PCR product was spiked in before the conversion, amplification of the heterologous sequence using primers matching the converted sequences but not those matching the unconverted should result in a new PCR product.

28. As the colony PCR primers are located in the vectors backbone the vector sequences flanking the insertion site have to be taken in consideration.
29. Plasmid inserts derived from unmethylated DNA can contain lengthy stretches of A/T, which can lead to a premature termination of the sequencing reaction. Therefore, we recommend two sequencing reactions with the forward and the reverse sequencing primer for every insert. Comparing their pairwise alignment will help identifying sequencing errors.
30. In our hand the BioEdit and MEGA program (www.megasoftware.net) proved to be suitable.
31. In significantly methylated ROIs, the methylation pattern between individual cells are very diverse. Thus, clones with identical sequences indicate PCR-generated artifacts rather than identical genomic templates and should be excluded.

References

1. Alabert C, Groth A (2012) Chromatin replication and epigenome maintenance. Nat Rev Mol Cell Biol 13:153–167
2. Li B, Carey M, Workman JL (2007) The role of chromatin during transcription. Cell 128:707–719
3. Zhang X (2012) Chromatin modifications in plants. In: Wendel FJ, Greilhuber J, Dolezel J, Leitch JI (eds) Plant genome diversity. Vol 1: Plant genomes, their residents, and their evolutionary dynamics. Springer Vienna, Vienna, pp 237–255. doi:10.1007/978-3-7091-1130-7_15
4. Kouzarides T (2007) Chromatin modifications and their function. Cell 128:693–705. doi:10.1016/j.cell.2007.02.005
5. Feng S, Jacobsen SE, Reik W (2010) Epigenetic reprogramming in plant and animal development. Science 330:622–627. doi:10.1126/science.1190614
6. Law JA, Jacobsen SE (2010) Establishing, maintaining and modifying DNA methylation patterns in plants and animals. Nat Rev Genet 11:204–220
7. Stroud H, Greenberg Maxim VC, Feng S, Bernatavichute Yana V, Jacobsen Steven E (2013) Comprehensive analysis of silencing mutants reveals complex regulation of the Arabidopsis methylome. Cell 152:352–364. doi:10.1016/j.cell.2012.10.054
8. Loring HS, Ploeser JM (1949) The deamination of cytidine in acid solution and the preparation of uridine and cytidine by acid hydrolysis of yeast nucleic acid. J Biol Chem 178:439–449
9. Shapiro R, Klein RS (1966) The deamination of cytidine and cytosine by acidic buffer solutions. Mutagenic implications. Biochemistry 5:2358–2362
10. Kuo KC, McCune RA, Gehrke CW, Midgett R, Ehrlich M (1980) Quantitative reversed-phase high performance liquid chromatographic determination of major and modified deoxyribonucleosides in DNA. Nucleic Acids Res 8:4763–4776
11. Rozhon W, Baubec T, Mayerhofer J, Scheid OM, Jonak C (2008) Rapid quantification of global DNA methylation by isocratic cation exchange high-performance liquid chromatography. Anal Biochem 375:354–360. doi:10.1016/j.ab.2008.01.001
12. Kurdyukov S, Bullock M (2016) DNA methylation analysis: choosing the right method. Biology 5:3. doi:10.3390/biology5010003
13. Lopez Torres A, Yanez Barrientos E, Wrobel K, Wrobel K (2011) Selective derivatization of cytosine and methylcytosine moieties with 2-bromoacetophenone for submicrogram DNA methylation analysis by reversed phase HPLC with spectrofluorimetric detection. Anal Chem 83:7999–8005. doi:10.1021/ac2020799
14. Benson G (1999) Tandem repeats finder: a program to analyze DNA sequences. Nucleic Acids Res 27:573–580
15. Clark SJ, Harrison J, Paul CL, Frommer M (1994) High sensitivity mapping of methylated cytosines. Nucleic Acids Res 22:2990–2997

16. Frommer M, McDonald LE, Millar DS, Collis CM, Watt F, Grigg GW, Molloy PL, Paul CL (1992) A genomic sequencing protocol that yields a positive display of 5-methylcytosine residues in individual DNA strands. Proc Natl Acad Sci U S A 89:1827–1831

17. Clark SJ, Statham A, Stirzaker C, Molloy PL, Frommer M (2006) DNA methylation: bisulphite modification and analysis. Nat Protoc 1:2353–2364

18. Patterson K, Molloy L, Qu W, Clark S (2011) DNA methylation: bisulphite modification and analysis. J Vis Exp 56:3170. doi:10.3791/3170

19. Unterholzner S, Rozhon W, Poppenberger B (2017) Analysis of in vitro DNA-interactions of brassinosteroid-controlled transcription factors using electrophoretic mobility shift assay. In: Russinova E, Caño-Delgado A (eds) Brassinosteroids. Humana Press, New York

20. Hetzl J, Foerster AM, Raidl G, Mittelsten Scheid O (2007) CyMATE: a new tool for methylation analysis of plant genomic DNA after bisulphite sequencing. Plant J 51. doi:10.1111/j.1365-313X.2007.03152.x

21. Gruntman E, Qi Y, Slotkin RK, Roeder T, Martienssen RA, Sachidanandam R (2008) Kismeth: analyzer of plant methylation states through bisulfite sequencing. BMC Bioinformatics 9:371–371. doi:10.1186/1471-2105-9-371

Chapter 25

Investigating Ornithine Decarboxylase Posttranscriptional Regulation Via a Pulldown Assay Using Biotinylated Transcripts

Anh Mai and Shannon L. Nowotarski

Abstract

Ornithine decarboxylase (ODC) is the first rate-limiting enzyme in the polyamine biosynthetic pathway. It has been well documented that ODC is tightly regulated at the levels of transcription, posttranscriptional changes in RNA, and protein degradation during normal conditions and that these processes are dysregulated during tumorigenesis. Moreover, it has been recently shown that ODC is posttranscriptionally regulated by RNA binding proteins (RBPs) which can bind to the ODC mRNA transcript and alter its stability and translation. Using a mouse skin cancer model, we show that the RBP human antigen R (HuR) is able to bind to synthetic mRNA transcripts through a pulldown assay which utilizes a biotin-labeled ODC 3′-untranslated region (UTR). The details of this method are described here. A better understanding of the mechanism(s) which regulates ODC is critical for targeting ODC in chemoprevention.

Key words Ornithine decarboxylase, Polyamines, Posttranscriptional regulation, RNA stability, Human antigen R (HuR)

1 Introduction

Ornithine decarboxylase (ODC) is the first rate-limiting enzyme in the polyamine biosynthetic pathway. It converts the amino acid ornithine to the diamine putrescine which is subsequently converted to the higher polyamines spermidine and spermine [1]. The ODC enzyme is tightly regulated at the levels of transcription, posttranscription, translation, and degradation [1–6]. However, when ODC enzyme activity is upregulated tumorigenesis can occur. In fact, the induction of ODC enzyme activity has been well characterized in numerous epithelial tumors [7, 8]. Thus, understanding the regulation of ODC under normal physiological conditions as well as tumorigenic conditions is critical.

Posttranscriptional regulation is a rapid mechanism of controlling gene expression. This level of regulation includes splicing, transport, editing, turnover, and translation of the mRNA

Rubén Alcázar and Antonio F. Tiburcio (eds.), *Polyamines: Methods and Protocols*, Methods in Molecular Biology, vol. 1694, DOI 10.1007/978-1-4939-7398-9_25, © Springer Science+Business Media LLC 2018

transcript [9]. The best characterized of these processes involves regulation of mRNA stability and translation. Through posttranscriptional regulation, a cell is able to ensure that proteins are expressed in the correct amount and location [9]. Both *cis*- and *trans*-regulatory elements influence mRNAs posttranscriptionally. These *cis*-elements include sequence and secondary structures within an mRNA which allow *trans*-factors such as microRNAs (miRs) and/or RNA binding proteins (RBPs) to bind to the mRNA and regulate its stability, translation efficiency, or both [10]. RBPs have classically been said to bind to adenosine- and uracil-rich elements within the 3′-untranslated region (UTR) of a given mRNA transcript. Numerous RBPs have been documented with some of the best studied being the stabilizing RBP human antigen R (HuR) and the destabilizing RBP tristetraprolin (TTP) [11, 12].

The posttranscriptional regulation of ODC by RBPs has been the focus of our research [3, 4]. We have shown that the stabilizing RBP, HuR, is able to bind to and stabilize the ODC mRNA transcript in a mouse model of skin carcinogenesis [3]. More recently, we have demonstrated that a destabilizing RBP, TTP, is able to bind to the ODC mRNA transcript and cause a reduction in the stability of the ODC transcript [4]. To examine ODC regulation by RBPs, we have conducted experiments that investigate whether a given RBP is able to bind to the ODC 3′UTR. To examine this question, we use a biotin pulldown assay in which a synthetic biotinylated ODC 3′UTR is incubated with the cytoplasmic fraction of our cell line of interest (Fig. 1). Streptavidin beads are used to bind to the biotinylated ODC 3′UTR and any proteins bound to the synthetic transcript are eluted from the pulldown material.

2 Materials

2.1 Cell Culture and Cell Extract Preparation (See Note 1)

1. 1× phosphate buffer: 14 mM NaCl, 2.7 mM KCl, 10 mM Na_2HPO_4, 1.8 mM KH_2PO_4, pH buffer to 7.4, and sterilize by autoclaving. Store at 4 °C.
2. Ne-PER Nuclear and Cytoplasmic Extraction kit (ThermoFisher Scientific, Carlsbad, CA).
3. 100 mm × 20 mm tissue culture-treated cell dishes.
4. Tissue culture media (*see* **Note 2**).

2.2 Preparation of Biotin-Labeled Probes

1. TriZol reagent.
2. SuperScript IV cDNA synthesis kit (Invitrogen, Carlsbad, CA).
3. Glycoblue.
4. Primers for the 3′UTR of mouse ODC (*see* **Note 3**).

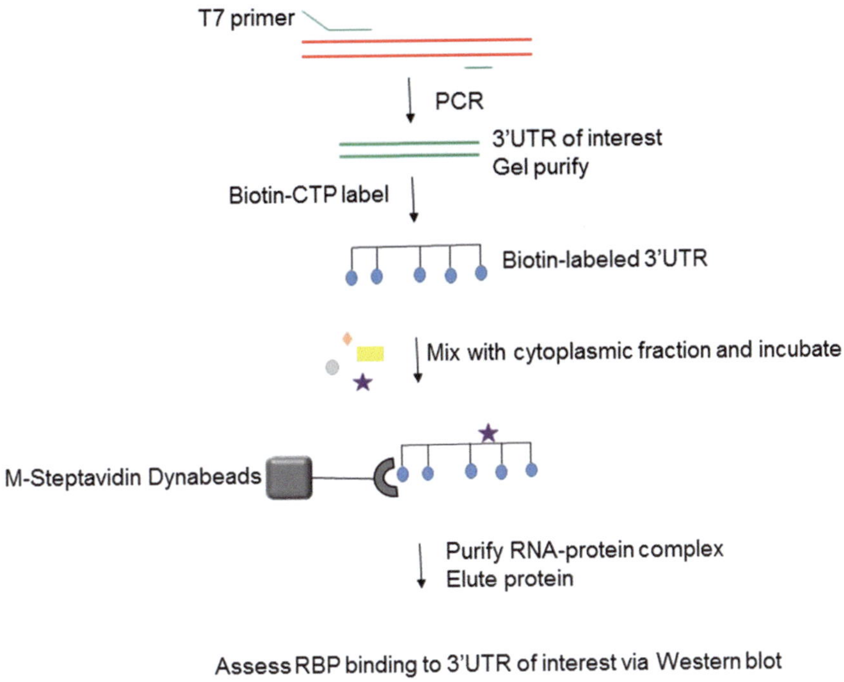

Fig. 1 The synthetic biotin-labeled RNA-protein binding assay. PCR products were made for the 3′UTR of interest. These PCR products contained the T7 promoter sequence on the 5′ end. The T7 polymerase was used to transcribe the biotin-labeled 3′UTR synthetic transcripts. These synthetic mRNAs were incubated with cytoplasmic lysate in order to allow proteins to bind to the 3′UTR sequence. Streptavidin Dynabeads were used to bind to the biotin of the labeled synthetic 3′UTRs, thus causing the 3′UTRs to pulldown any bound protein(s). The proteins were eluted and HuR binding to the synthetic 3′UTR was assessed via Western blotting

5. Taq DNA polymerase kit.
6. Agarose.
7. Ethidium bromide.
8. 100 mM dNTPs.
9. 10 mM biotin-14-CTP (Invitrogen).
10. RNAse-free water.
11. 5× transcription buffer (Invitrogen).
12. 100 mM DTT.
13. RNasin (40 U/μl) (Invitrogen).
14. T7 RNA polymerase (15 U/μl) (Invitrogen).
15. QIAquick gel extraction kit (Qiagen, Valencia, CA).
16. Biotin 14-CTP mix: 1 μl of 100 mM ATP, 1 μl of 100 mM GTP, 1 μl of 100 mM UTP, 0.9 μl of 100 mM CTP, 1 μl of 10 mM biotin-14-CTP, and 5.1 μl of RNAse-free water.
17. In vitro transcription reaction mix: 1 μg of 3′UTR PCR product, 4 μl of 5× transcription buffer, 2 μl of 100 mM DTT,

0.5 μl of RNasin (40 U/μl), 1 μl of Biotin 14-CTP mix (*see* **item 16**), 1.5 μl of T7 RNA polymerase (15 U/μl), and RNAse-free water (up to a total volume of 20 μl).

18. DNAse, 10× DNAse reaction buffer (Ambion).
19. T.E. buffer pH 8.0: 10 mM Tris–HCl, 0.1 mM EDTA in RNAse-free water.
20. G-50 column (GE Healthcare Lifescience, Marlborough, MA).
21. 100% ethanol.
22. 3 M Sodium acetate made with RNAse-free water, adjusted to pH 5.2 with glacial acetic acid.
23. 2× TENT binding buffer: 20 mM Tris–HCl, pH 8.0, 2 mM EDTA, 500 mM NaCl, 1% v/v Triton X-100, and filter sterilize.

2.3 Pulldown

1. Dynabeads M-280 Streptavidin (Invitrogen).
2. Solution A: 0.1 M NaOH, 0.05 M NaCl, and filter sterilize.
3. Solution B: 0.1 M NaCl and filter sterilize.
4. Microcentrifuge tube magnetic rack (Invitrogen).
5. 1× SDS sample loading buffer: 50 mM Tris–HCl, pH 6.8, 2% SDS, 10% glycerol, 1% β-mercaptoethanol, 12.5 mM EDTA, and 0.02% bromophenol blue.
6. HuR antibody (anti-mouse) (Santa Cruz Biotechnology, Inc., Santa Cruz, CA).
7. GAPDH antibody (anti-mouse) (Cell Signaling, Danvers, MA).

3 Methods (*See* Note 1)

3.1 Preparation of Cytoplasmic Fraction from Cell Culture

1. Grow cells to confluence on 10 cm plates and harvest by washing two times in ice-cold 1× PBS followed by trypsinizing cells according to the standard protocol for the cell line used. Once the cells detach from the plates, add ice-cold 1× PBS to each plate in order to dilute the trypsin. Transfer the cells to a 50 ml conical tube and pellet by centrifugation at 956 × *g* for 5 min at 4°C.
2. Wash the pelleted cells in 1 ml ice-cold 1× PBS by resuspension.
3. Transfer cells to a clean, RNAse-free microcentrifuge tube and pellet by centrifugation at 956 × *g* for 3 min at 4°C.
4. Remove the supernatant.
5. Several commercial kits as well as home-made buffers are available for cell fractionation. We use the NE-PER Nuclear and

Cytoplasmic kit, following the manufacturer's instructions exactly. The nucleus is not needed for these experiments; therefore they can be discarded or stored for later use. The cytoplasmic fractions are necessary for the pulldown assays (*see* **Note 4**).

6. Measure the concentration of the cytoplasmic protein collected. We typically use the Bradford method.
7. Freeze the cytoplasmic fraction and store in −80°C until use.

3.2 Preparation of Biotin-Labeled 3′ UTR Probes

1. Grow cells to confluence on 10 cm plates and harvest the total RNA in TriZol reagent following the manufacturer's instructions exactly. We use glycoblue in the precipitation step to better visualize the RNA pellet.
2. Create cDNA from the RNA from **step 1** via reverse transcription. cDNA can be made by using any commercially available cDNA synthesis kit. We use the Superscript IV cDNA synthesis kit and follow the manufacturer's instructions exactly using oligo dT as the primer.
3. Set up a PCR reaction using cDNA from **step 2** and 3′UTR primers for mouse ODC. The forward primer should contain the T7 promoter (*see* **Note 3** for details concerning the creation of the primers). Any commercially available PCR kit can be used for this step.
4. Check the PCR product quality and size by running a 2% agarose gel stained with ethidium bromide (*see* **Note 5**).
5. Purify the PCR product from **step 4** using the Qiaquick gel extraction kit. Several kits are commercially available for gel extraction of PCR products. We use the Qiaquick gel extraction kit to purify the PCR products from **step 4** by following the manufacturer's instructions exactly.
6. Measure the amount of PCR product you have by measuring the absorbance of the PCR product at 260 nm in a spectrophotometer (*see* **Note 6**).
7. Prepare the biotinylated probes by using the in vitro transcription reaction mix (from Subheading 2.2, **item 17**) and preparing this reaction mix in a clean, RNAse-free microcentrifuge tube. The 3′UTR PCR product is from **step 6**.
8. Mix and incubate the reaction mix from **step 7** at 37 °C for 2 h.
9. Treat the reaction mix with 2 μl 10× DNAse buffer and 1 μl DNAse (RNAse free).
10. Incubate the reaction sample from **step 9** for 15 min at 37 °C.
11. Deactivate the reaction by adding 2 μl DNAse inactivation buffer to the mixture (*see* **Note 7**).
12. Add 25 μl T.E. buffer to each reaction sample.

13. Pass each reaction sample through its own G-50 column following the manufacturer's instructions exactly.
14. Precipitate the reaction sample (from **step 13**) by adding 0.1 volume of 3 M sodium acetate pH 5.2, 1 μl of glycoblue, and 2.5 volumes of 100% ethanol.
15. Mix the contents of the tube well and place at −70 °C for at least 1 h.
16. Centrifuge the reaction samples (from **step 15**) at 15,294 × *g* for 20 min at 4 °C.
17. Discard the supernatant.
18. Wash the pellet with 70% ice-cold ethanol.
19. Centrifuge the reaction samples (from **step 18**) at 15,294 × *g* for 10 min at 4 °C.
20. Discard the supernatant.
21. Centrifuge the reaction samples (from **step 20**) at 15,294 × *g* for 1 min at 4 °C.
22. Remove any residual ethanol from the pellet.
23. Allow the pellet to air dry for 5 min at room temperature.
24. Resuspend the pellet in 20 μl RNAse-free 2× TENT buffer (*see* **Note 8**).

3.3 Pulldown of Cytoplasmic Material Using the Biotin-Labeled Synthetic Transcripts

1. Mix the Dynabeads M-Streptavidin by vortexing for 30 s.
2. Place 10 μl of the Dynabeads M-Streptavidin (from **step 1**) into a clean, RNAse-free microcentrifuge tube (*see* **Note 9**).
3. Wash the beads two times using 200 μl of Solution A and the magnetic rack (*see* **Note 10**).
4. Wash the beads one time using 200 μl of Solution B and the magnetic rack.
5. Remove Solution B from the beads.
6. Resuspend the beads in 10 μl RNAse-free 1× TENT buffer and set aside.
7. Place 40 μg of cytoplasmic protein (from Subheading 3.1, **step** 7) into a clean, RNAse-free microcentrifuge tube.
8. Repeat **step** 7 for two more microcentrifuge tubes. This will yield a total of 120 μg of cytoplasmic protein in three tubes.
9. Add 5 μl of the biotin-labeled synthetic mRNA transcripts (from Subheading 3.2, **step 24**) to each of the three tubes.
10. Incubate the tubes from **step 9** at room temperature for 30 min on an end-over-end rotator.
11. Collect the reaction mix from **step 10** on the bottom of the microcentrifuge tube by a short spin using a table top centrifuge.

12. Add the reaction mix from **step 11** to the 10 μl of pre-washed beads (from **step 6**). Each reaction mix tube is combined with its own tube of pre-washed beads (*see* **Note 11**).
13. Incubate the reaction mix from **step 12** at room temperature for 30 min on an end-over-end rotator.
14. Collect the reaction mix from **step 13** on the bottom of the microcentrifuge tube by a short spin using a table-top centrifuge.
15. Place the tubes on the magnetic microcentrifuge rack and settle the beads on the magnetic base.
16. Remove the supernatant and discard.
17. Using 1 ml of ice-cold 1× PBS, collect the bead samples from the three microcentrifuge tubes and combine into one microcentrifuge tube (*see* **Note 12**).
18. Settle the beads on the magnetic base and remove and discard the 1× PBS.
19. Wash the beads in 1 ml of ice-cold 1× PBS.
20. 20 Settle the beads on the magnetic base and remove and discard the 1× PBS.
21. Add 45 μl of 1× SDS sample loading buffer to each tube and resuspend the beads.
22. Boil the sample(s) for 10 min.
23. Briefly centrifuge the sample(s) so that the beads settle on the bottom of the microcentrifuge tube(s).
24. The supernatant is your pulldown material. A Western blot can be run for analysis to determine whether a given RNA binding protein is able to bind to the synthetic mRNA transcript of interest (*see* **Note 13**). Typical results are shown in Fig. 2.

4 Notes

1. For all procedures it is important to use RNAse-free tips, RNAse-free tubes, and RNAse-free water. It is also important to wear gloves.
2. In order to perform the tissue culture experiments, tissue culture reagents such as media and trypsin will need to be utilized. These reagents are specific to the cell line(s) being used.
3. In order to make the PCR product in Subheading 2.2 with a T7 promoter sequence, you need to include the T7 promoter sequence to the sense PCR primer. Also include five bases before the T7 sequence (which is underlined in the proceeding example) to aid in transcription. For example the following

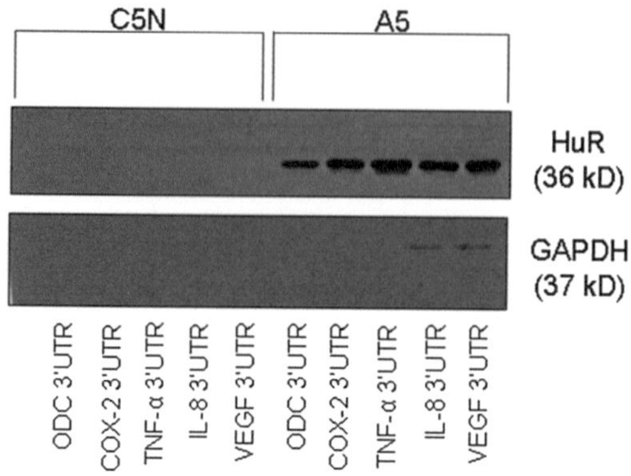

Fig. 2 ODC, COX-2, TNF-α, IL-8, and VEGF biotin-labeled 3′UTRs associate with the HuR cytoplasmic protein in transformed A5 mouse keratinocyte cells but not in the C5N nontransformed mouse keratinocyte cells. To determine whether HuR was able to bind to biotin-labeled transcripts of genes that are typically associated with neoplastic transformation in transformed A5 cells and not in nontransformed C5N cells, HuR was used in a pulldown of A5 and C5N cytoplasmic extracts as described in the Methods. A5 and C5N cytoplasmic fractions were incubated with the full length 3′UTR biotin-labeled synthetic mRNA probes for COX-2, TNF-α, IL-8, and VEGF. The ODC 3′UTR was used as a positive control. Each blot was stripped for GAPDH to assess nonspecific binding. The synthetic mRNA probes were tested for binding to HuR by Western blot analysis

sequence may be used upstream of your 3′-UTR for your gene of interest and includes the T7 promoter sequence in the sense primer: 5′-GCTTCTAATACGACTCACTATAGGGAGA-3′. For studying the mouse ODC 3′UTR we created the following sense primer (in which the T7 promoter is underlined): 5′-GCTTCTAATACGACTCACTATAGGGAGAATGC-CATTCTTGTAGCTCTTGC-3′. We also used the following antisense primer for studying the mouse ODC 3′UTR 5′-TTGCTGTTGTTGAATTTATTAC-3′. Of course, these primers will need to be modified for other species and other mRNA targets.

4. We save the cytoplasmic fraction for the pulldown because translation occurs in the cytoplasm and RBPs reside in the cytoplasm in order to function properly by stabilizing or destabilizing an mRNA target.
5. While ethidium bromide does not have to be used as the DNA intercalating agent due to its status as a mutagen, in our experience we are able to see our PCR product bands the best using this reagent.

6. Remember the probe at this point is single-stranded so you take abs 260*25* dilution factor (1:25 works well to save PCR product).
7. Try to avoid the white grains when adding the DNAse inactivation buffer.
8. The biotinylated probes may be used directly or stored at −80 °C.
9. Three tubes of pre-washed beads are necessary for each cell line or condition tested. For example, if you are looking at whether an RBP binds to ODC in a nontumorigenic versus a tumorigenic cell line then you will need six tubes of pre-washed beads (three for the tumorigenic cell line and three for the nontumorigenic cell line).
10. To use the magnetic rack place the microcentrifuge tube with the magnetic beads into the slots on the rack. You will note that the beads settle to the side of the rack with the magnet. Rotating the tube in a clockwise fashion will wash the beads. To remove excess wash or change washes simply leave the microcentrifuge tube in the rack, open the cap, and remove any buffer by holding the pipette tip away from the beads. To resuspend the beads, remove the microcentrifuge tube from the magnet, add the resuspending agent, and pipette up and down.
11. There should now be three RNAse-free microcentrifuge tubes with 10 μl of pre-washed beads, 40 μg of cytoplasmic lysate and 5 μl of biotin-labeled synthetic mRNA.
12. At the end of this step you should have one microcentrifuge tube per cell line or cellular condition with 1 ml of ice-cold 1× PBS and approximately 30 μl of beads and 120 μg of protein.
13. When conducting the Western blot, load all of the 45 μl of the pulldown material. Avoid loading any of the beads. We use a 10-well 10% SDS-PAGE gel to perform our Western blot analysis, although these parameters can be altered depending on which RBP you choose to analyze.

References

1. Pegg AE (2006) Regulation of ornithine decarboxylase. J Biol Chem 281:14529–14532
2. Zhao B, Butler AP (2001) Core promoter involvement in the induction of rat ornithine decarboxylase by phorbol esters. Mol Carcinog 32:92–99
3. Nowotarski SL, Shantz LM (2010) Cytoplasmic accumulation of the RNA-binding protein HuR stabilizes the ornithine decarboxylase transcript in a murine nonmelanoma skin cancer model. J Biol Chem 285:31885–31894
4. Nowotarski SL, Origanti S, Sass-Kuhn S, Shantz LM (2016) Destabilization of the ornithine decarboxylase mRNA transcript by the RNA-binding protein tristetraprolin. Amino Acids 48:2303–2311
5. Shantz LM (2004) Transcriptional and translational control of ornithine decarboxylase

during Ras transformation. Biochem J 377:257–264

6. Wallon UM, Persson L, Heby O (1995) Regulation of ornithine decarboxylase during cell growth. Changes in the stability and translatability of the mRNA, and in the turnover of the protein. Mol Cell Biochem 146:39–44
7. Nowotarski SL, Woster PM, Casero RA Jr (2013) Polyamines and cancer: implications for chemotherapy and chemoprevention. Expert Rev Mol Med 15:e3
8. Casero RA Jr, Marton LJ (2007) Targeting polyamine metabolism and function in cancer and other hyperproliferative diseases. Nat Rev Drug Discov 6:373–390
9. Pullmann R Jr, Rabb H (2014) HuR and other turnover- and translation-regulatory RNA-binding proteins: implications for the kidney. Am J Physiol Renal Physiol 306:569–576
10. Shyu AB, Wilkinson MF, van Hoof A (2008) Messenger RNA regulation: to translate or to degrade. EMBO J 27:471–481
11. Bakheet T, Williams BR, Khabar KS (2006) ARED 3.0: the large and diverse AU-rich transcriptome. Nucleic Acids Res 34:111–114
12. Brennan CM, Steitz JA (2001) HuR and mRNA stability. Cell Mol Life Sci 58:266–277

Chapter 26

Analysis of Cotranslational Polyamine Sensing During Decoding of ODC Antizyme mRNA

R. Palanimurugan, Daniela Gödderz, Leo Kurian, and R. J. Dohmen

Abstract

Polyamines are essential poly-cations with vital functions in all cellular systems. Their levels are controlled by intricate regulatory feedback mechanisms. Abnormally high levels of polyamines have been linked to cancer. A rate-limiting enzyme in the biosynthesis of polyamines in fungi and higher eukaryotes is ornithine-decarboxylase (ODC). Its levels are largely controlled posttranslationally via ubiquitin-independent degradation mediated by ODC antizyme (OAZ). The latter is a critical polyamine sensor in a feedback control mechanism that adjusts cellular polyamine levels. Here, we describe an approach employing quantitative western blot analyses that provides in vivo evidence for cotranslational polyamine-sensing by nascent OAZ in yeast. In addition, we describe an in vitro method to detect polyamine binding by antizyme.

Key words Polyamines, Spermidine, Spermine, ODC antizyme, Polyamine binding assay, Quantitative western blotting, Ribosomal frameshifting (RFS)

1 Introduction

Polyamines are organic poly-cations that are derived intracellularly from amino acid (arginine and methionine) precursors [1, 2]. They are essential for viability of eukaryotic cells and have a variety of functions in nucleic acids packaging, DNA replication, transcription, translation, modulation of membrane stability, and enzymatic functions [2–5]. In bacteria, they have been implicated in the control of biofilm formation [1]. The main cellular polyamines are the tri-amine spermidine and tetra-amine spermine. Spermidine is derived from the diamine putrescine and deoxy-S-adenosyl-methionine by the enzyme spermidine synthase. In higher eukaryotes and fungi, putrescine is generated solely by decarboxylation of ornithine mediated by ornithine decarboxylase (ODC) as a rate-limiting enzyme in polyamine synthesis [1, 2]. Elevated ODC levels

The original version of this chapter was revised. A correction to this chapter can be found at https://doi.org/10.1007/978-1-4939-7398-9_41

Rubén Alcázar and Antonio F. Tiburcio (eds.), *Polyamines: Methods and Protocols*, Methods in Molecular Biology, vol. 1694, DOI 10.1007/978-1-4939-7398-9_26, © Springer Science+Business Media LLC 2018

are found in many types of cancer cells to satisfy an apparently increased demand on polyamines of rapidly proliferating cells [6, 7]. Plants, in addition, employ arginine decarboxylase to generate putrescine via agmatine [1, 2]. Polyamines have an important role in plant stress tolerance [8, 9]. A key step in the regulation of polyamine biosynthesis in metazoa and fungi is the posttranslational control of ODC by ODC antizyme (OAZ), a mechanism that has not been found in plants thus far [10–12].

The active ODC enzyme is a dimer, which is in equilibrium with the monomeric form. OAZ binds to ODC monomers and thus prevents formation of active ODC dimer [10]. Binding of OAZ to ODC, in addition, leads to degradation of ODC by the proteasome, a process that does not require its ubiquitylation [13]. The fact that polyamines control the levels of OAZ by multiple mechanisms provides a homeostatic feedback regulation of polyamine levels. Apart from a posttranslational mechanism, in which polyamines inhibit the ubiquitin-dependent degradation of OAZ as shown in yeast [14], polyamines primarily regulate OAZ levels cotranslationally in fungi and metazoa. In these organisms, OAZ mRNAs contain an ORF interrupted by a STOP codon. Synthesis of full-length OAZ requires a ribosomal frameshifting (RFS) event that results in translation beyond the interrupting STOP codon, a mechanism that is conserved from yeast to humans [14–16]. Synthesis of full-length OAZ is stimulated by high polyamine concentrations. Mutational analysis of the elements of *S. cerevisiae* OAZ mRNA that influence the efficiency of its decoding revealed that critical features reside within the encoded nascent polypeptide, which negatively influences the synthesis of full-length OAZ at relatively low polyamine concentrations [17]. The polyamine sensing property of OAZ encompasses a large fraction of the polypeptide extending to its C-terminal end. At low polyamine concentrations, this element causes a ribosome pile-up on the mRNA resulting in inefficient synthesis of full-length OAZ. Deletion of more than three codons from the 3′ end of the ORF results in polyamine-insensitive constitutive synthesis of full length OAZ [17].

Here, we describe our experimental approach to analyze the translational regulation of OAZ1 in *S. cerevisiae* cells. Analogous approaches should be applicable to other organisms. In this approach, we compare the production of full-length wild-type OAZ1 or its mutant derivatives to that observed with otherwise identical constructs lacking the RFS site. To detect the effect of polyamines on the decoding efficiency, the cells are grown either in the absence or the presence of spermidine added to the culture medium. OAZ1 levels are determined by quantitative western blotting. Under these conditions, only the construct bearing the RFS site was influenced by the polyamine concentration in the media (Fig. 1). The observed polyamine sensitivity of the decoding process mediated by the nascent OAZ1 polypeptide corresponds to a

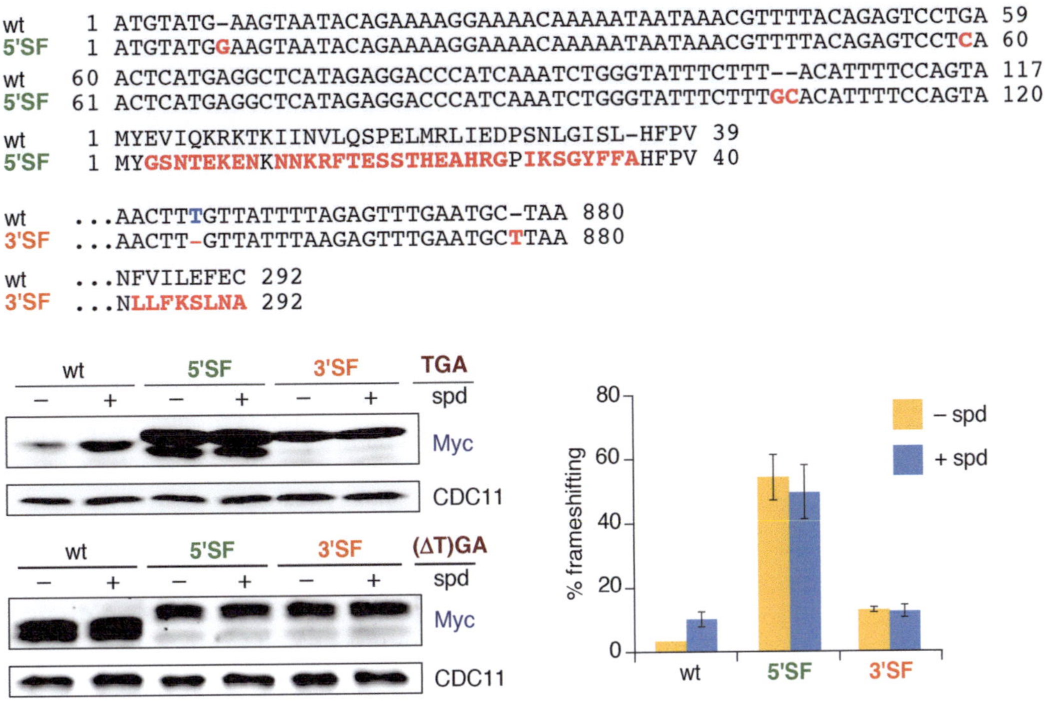

Fig. 1 Using quantitative western blotting to detect cotranslational polyamine sensing by Oaz1. Two elements in OAZ1 polypeptide affect decoding of OAZ1 mRNA. Depicted are alignments of wild-type OAZ1 with mutant variants carrying nucleotide insertions or deletions (marked in *red*) to switch reading frames, the corresponding encoded polypeptide sequences, a western blot analysis of Myc–OAZ1 levels derived from these constructs, and quantification of data. *SF* shift of frames variant. This research was originally published in [17]

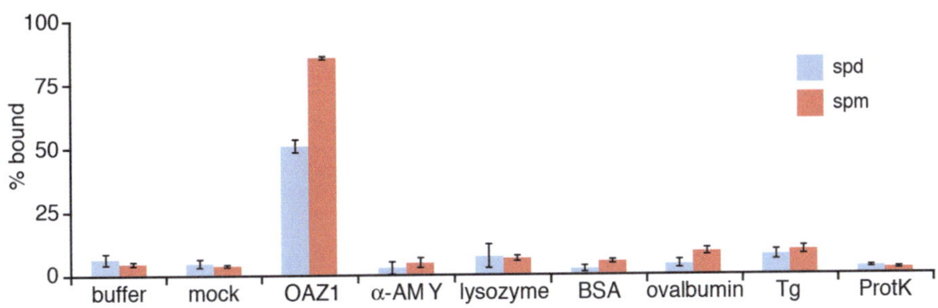

Fig. 2 Detecting in vitro binding of polyamines to OAZ1 protein. Shown are the results of measurements that determined the retention of radiolabeled polyamines during ultrafiltration. Retention of [3H]-spermidine or [14C]-spermine by 6His–OAZ1 from three independent preparations was compared to that observed with buffer only, with material from Ni-NTA mock preparations from an *E. coli* strain not expressing 6His–OAZ1, as well as with *Bacillus subtilis* α-amylase (α-AMY), chicken egg lysozyme, bovine serum albumin (BSA), ovalbumin, thyroglobulin (Tg), and proteinase K (ProtK) (each at 10 mM). Error bars, s.d.; $n = 3$. This research was originally published in [17]

polyamine binding property that can be detected with purified OAZ from different origins using an in vitro assay (Fig. 2) [17]. This assay employs radioactive polyamines and an ultrafiltration method, the details of which are described here as well.

2 Materials

Prepare buffers using deionized ultrapure water. Media used for growing cells (LB, LB-agar, YPD, YPD-agar, SD, and SD-Agar) and the respective media components are prepared using deionized water and are sterilized by autoclaving unless indicated otherwise (amino acid and nucleotide stocks are sterilized by filtration). Ampicillin and chloramphenicol stock solutions are prepared in 70% ethanol. Prepare $CuSO_4$ stock solution using ultrapure water and sterilize by filtration. Used agar plates, liquid cultures and other materials should be sterilized by autoclaving prior to disposal.

2.1 Yeast Growth and Transformation

1. *Saccharomyces cerevisiae* strain impaired in proteasome function due to a *pre1-1* mutation (YHI29/1 [*MATα, pre1-1 his3-11,15 leu2-3112 ura3*]) [18].
2. Plasmids pPM318 and pPM323 encoding either 2xMYC-OAZ1(wt) (with internal STOP codon) or 2xMYC-OAZ1(if) ("in frame", i.e. without internal STOP codon) are derivatives of the *CEN/ARS/URA3* vector YCplac33 [19].
3. 50% (w/v) glucose solution, autoclaved.
4. YPD medium: Prepare 1% (w/v) yeast extract, 2% (w/v) peptone in water, autoclave, then supplement the mix with 2% (v/v) glucose from a 50% (w/v) stock solution.
5. YPD agar plates: Prepare 1% (w/v) yeast extract, 2% (w/v) peptone, and 2% (w/v) agar in water, autoclave the mix, and then supplement it with 2% (v/v) glucose from 50% (w/v) stock solution.
6. SD medium lacking uracil: Prepare the medium in a bottle with an appropriate volume of sterile water by adding 0.67% (w/v) yeast nitrogen base without amino acids and with ammonium sulfate from an autoclaved 20× stock solution. Furthermore, add amino acids from 100× stock solutions as follows: 0.002% (w/v) arginine, 0.001% (w/v) histidine, 0.006% (w/v) isoleucine, 0.006% (w/v) leucine, 0.004% (w/v) lysine, 0.001% (w/v) methionine, 0.006% (w/v) phenylalanine, 0.005% (w/v) threonine, 0.004% (w/v) tryptophan. Finally, add 0.002% (w/v) adenine (500× stock solution prepared in 0.1 N NaOH), and supplement the mix with 2% (v/v) glucose from a 50% (w/v) stock solution.
7. SD agar plates lacking uracil: Autoclave 2% (w/v) agar in water, then supplement with the ingredients as for SD medium lacking uracil (*see* above).
8. 0.1 M and 1.0 M lithium acetate solution (autoclaved) stored at 25 °C.

9. 50% (w/v) polyethylene glycol (PEG-3350) solution (autoclaved), stored at 25 °C.
10. 10 mg/ml calf-thymus (CT) DNA prepared in water, stored at −20 °C.
11. 10 mM spermidine stock solution prepared with ultrapure water, filter-sterilized, and stored at 4 °C.
12. 100 mM $CuSO_4$ solution prepared with ultrapure water, filter-sterilized, and stored at 25 °C.
13. Autoclaved ultrapure water.
14. 0.2 μm filters and filtration units.
15. Toothpicks (autoclaved).
16. Bunsen burner.
17. Sterilized Erlenmeyer flasks and glass tubes.
18. Incubator shaker.
19. Thermo mixer.
20. Spectrophotometer and cuvettes.

2.2 SDS-Polyacrylamide Gel Electrophoresis (PAGE) and Western Blotting

1. Stacking gel buffer: 1 M Tris–HCl pH 6.8, stored at 4 °C.
2. Resolving gel buffer: 1.5 M Tris–HCl pH 8.8, stored at 4 °C.
3. 30% acrylamide/bis-acrylamide mix (e.g. Rotiphorese, Carl Roth) stored at 4 °C.
4. 10% (w/v) sodium dodecyl sulfate (SDS) (w/v) in water stored at 25 °C.
5. 10% (w/v) ammonium persulfate (APS) in water stored at 4 °C.
6. *N,N,N′,N′*-Tetramethyl-ethylenediamine (TEMED) stored at 4 °C.
7. Laemmli Running Buffer (LRB): 25 mM Tris pH 8.3, 192 mM glycine, 0.1% SDS stored at room temperature.
8. Laemmli Loading Buffer (LLB): 62.5 mM Tris pH 6.8, 2% SDS, 10% Glycerol, 0.002% bromophenol blue, add fresh 1% β-mercaptoethanol or 0.1 M 1,4 dithiothreitol (DTT) directly before usage.
9. Western Transfer Buffer (WTB): 25 mM Tris pH 8.3, 192 mM glycine, 0.1% SDS, 20% methanol.
10. Phosphate Buffered Saline (PBS): 137 mM NaCl, 2.7 mM KCl, 8.1 mM Na_2HPO_4, 1.5 mM KH_2PO_4, pH adjusted to 7.4.
11. Pre-stained protein ladder.
12. Blocking solution: 3% (w/v) dry milk powder dissolved in PBS.
13. Sodium azide 2% (w/v) in water.
14. Whatman filter paper.

15. Nitrocellulose membrane.
16. Polyacrylamide Gel Electrophoresis (PAGE) apparatus.
17. Power supply.
18. Glass plates with spacers and thin plates without spacers.
19. Multiwell combs.
20. Gel casting accessories (Gel stand and clamps).
21. Semi-dry western transfer chamber.
22. Thermo mixer.
23. Primary antibodies (anti-Myc (9B11) from Cell Signaling Technology and anti-Cdc11 from Santa Cruz Biotechnology).
24. Secondary antibodies (anti-mouse 680, anti-rabbit 800 from Rockland Immunochemicals).
25. Rocking platform.
26. Infrared western blot scanner (Li-Cor Biosciences).

2.3 Expression of OAZ1 in E. coli

1. Competent cells of *Escherichia coli* expression strain (e.g. Rosetta, Merck).
2. LB medium: 1% Tryptone, 0.5% Yeast Extract and 1% NaCl.
3. LB agar plates: 1% Tryptone, 0.5% Yeast Extract, 1% NaCl and 2% Agar.
4. Ampicillin, 100 mg/ml stock solution prepared in 70%(v/v) ethanol.
5. Chloramphenicol, 30 mg/ml stock solution prepared in 70% (v/v) ethanol.
6. 1.0 M Isopropyl-β-D-thiogalactopyranoside (IPTG) in water, filter sterilized.
7. Erlenmeyer flasks and tubes sterilized.
8. Bunsen burner.
9. Thermo mixer.
10. Incubator shaker with adjustable temperature and shaking (30 and 37 °C, 180 rpm).
11. Spectrophotometer and cuvettes.

2.4 Purification of OAZ1

1. Lysis Buffer: 50 mM Tris pH 7.8, with protease inhibitors.
2. Wash buffer: 50 mM Tris pH 7.8, 20 mM imidazole, with protease inhibitors.
3. Elution buffer: 50 mM Tris pH 7.8, 250 mM imidazole with protease inhibitors.
4. Complete EDTA-free protease inhibitor cocktail tablet (Roche).
5. 1.0 M imidazole prepared in lysis buffer.

6. Ni-NTA sepharose resin (e.g. GE healthcare).
7. Lysozyme from chicken egg (e.g. Sigma-Aldrich).
8. Dnase I (e.g. Roche).
9. Refrigerated centrifuge.
10. Rotator.
11. Spectrophotometer and cuvettes.

2.5 Polyamine Binding Assay

1. Binding buffer: 50 mM Tris pH 7.8.
2. Isotope-labeled polyamines: radioactive [^{14}C] or [^{3}H] spermine or spermidine (*see* **Note 1**).
3. Cut-off filters: 10 kDa cut-off filters with modified polyethersulfone (PES) membrane (VWR, Pall).
4. Scintillation liquid and counter.
5. Control protein solutions (e.g. BSA, Lysosyme, Ovalbumine etc.).

3 Methods

3.1 Quantitative Western Blot Analysis of Polyamine-Regulated OAZ1 Expression

1. Preparation of OAZ expression constructs: For assaying OAZ1 synthesis at various cellular polyamine levels in *S. cerevisiae*, use plasmids pPM318 and pPM323 encoding either 2xMYC-OAZ1(wt) (with internal STOP codon) or 2xMYC-OAZ1(if) ("in frame", i.e. without internal STOP codon), which were constructed in pPM90, a modified YCplac33 vector, as described previously [17]. Expression of OAZ1 from pPM318 and pPM323 is driven by the copper-regulatable P_{CUP1} promoter. To analyze the relevance of the coding sequence as compared to mRNA secondary structures, one can modify the coding capacity of the mRNA by inserting and/or deleting nucleotides thereby generating shift of frame (SF) constructs (Fig. 1). Similar constructs can be generated for the expression of antizyme genes from other organisms in yeast or other cell systems.
2. Preparation of transformation-competent yeast cells: Streak out yeast strain YHI29/1 (*pre1-1*) from glycerol stock under sterile condition (sterile hood and in close proximity to a bunsen burner flame, use autoclaved toothpicks for inoculation) on YPD plates and grow at 30 °C. Take cells from the streakout to inoculate a pre-culture of 3 ml sterile YPD medium, and incubate overnight at 30 °C with shaking at 180 rpm. Determine the optical density (OD_{600}) of the preculture using a spectrophotometer. Dilute cells into fresh YPD medium to an OD_{600} of 0.2 and incubate at 30 °C with shaking at 180 rpm. When the OD_{600} of the culture reaches 0.6–0.8,

harvest the cells by spinning the culture at 4000 × *g* for 5 min at room temperature in 15 ml sterile plastic tubes. Cells corresponding to 5 ml of an exponentially growing culture (OD_{600} 0.6–0.8) are required per transformation.

3. Yeast transformation: Prewarm all solutions to be used for yeast transformation at 30 °C. Thaw out CT-DNA on ice. After pelleting the cells (*see* above), discard supernatant. Resuspend the pellet in 1 ml sterile water (either by gently tapping the tube or by mildly pipetting the cells up and down) and transfer the suspension to a sterile 1.5 ml tube. The cells are then pelleted again by centrifugation at 4000 × *g* for 10 s. Carefully remove the supernatant using a pipette, then add 240 μl 50% PEG-3350, 36 μl 1 M lithium acetate, 5 μl CT-DNA, 5 μl plasmid DNA (total amount ~ 1 μg) and 84 μl sterile water to the cell pellet. Suspend the cells (*see* **Note 2**) first using a pipette (4–5 times) and then vortex them at maximum speed until the cells are completely suspended in the transformation mix. Incubate the suspension at 30 °C for 10 min (use thermo mixer or heat block), then shift the suspension to 42 °C for 15 min to apply a heat shock. After that, spin cells at 4000 × *g* for 5 min in a micro-centrifuge. Carefully remove supernatant. Resuspend cells in 250 μl sterile water by gentle tapping and pipetting, then spread the suspension on SD agar plates lacking uracil. Let excess liquid dry off and incubate the plates at 30 °C for 3 days to obtain uracil prototrophic transformants. Streak out cells from multiple colonies on SD plates lacking uracil as well as YPD agar plates using sterile toothpicks and incubated at 30 °C for 2 days. Cells from the individual transformants (*see* **Note 3**) can then be used to start cultures for the expression assay (*see* below). The YPD plates can be stored at 4 °C for several weeks.

4. Preculturing of yeast transformants: For assaying the expression levels of OAZ1 at various cellular polyamine concentrations, the *pre1–1* transformants (*see* above) carrying pPM318 (WT-OAZ1) or pPM323 (IF-OAZ1) on SD plates lacking uracil are used to inoculate precultures of 3 ml SD medium lacking uracil in sterile glass tubes and incubated overnight at 30 °C with shaking. In the morning, the OD_{600} of the precultures is measured using 1:10 dilutions and a spectrophotometer.

5. Expression of OAZ in media with or without polyamine supplementation: Appropriate amounts of each primary culture (calculated based on the OD_{600} values) are diluted with fresh SD medium lacking uracil to yield a final volume of 20 ml with an OD_{600} of 0.2. Add 20 μl 100 mM $CuSO_4$ to obtain a final concentration of 100 μM. Divide the culture into two flasks (labeled as "– spd" and "+ spd") each with a volume of 10 ml. Add 10 μl of 10 mM spermidine to the flask labeled as "+ spd"

to obtain 10 μM final concentration in the media. Incubate each flask at 30 °C with shaking at 180 rpm until the OD_{600} of the culture has reached 0.8–1.0 in the exponential growth phase. Harvest the cells by transferring the cultures to appropriately labeled 15 ml plastic tubes and centrifuging them at 4000 × *g* for 5 min at 4 °C. Carefully remove supernatants from all the tubes and resuspend the cell pellets in each tube with 1 ml sterile water. Transfer the cell suspensions to appropriately labeled 1.5 ml microfuge tubes and centrifuged at 4000 × *g* for 1 min. Carefully remove supernatants in each tube and freeze cell pellet in liquid nitrogen and store at −80 °C until further use.

6. Preparation of whole cell protein extracts: To analyze the samples by SDS-PAGE, thaw the frozen cell pellets on ice. Then resuspend the cells in an appropriate volume of 1xLLB with β-mercaptoethanol or DTT (35 μl for 1.0 OD_{600} cells, i.e. an amount of cells corresponding to those present in 1 ml of a culture with an OD_{600} of 1), and boil at 100 °C for 5 min to prepare the cell extracts. Afterwards, keep the tubes at room temperature for a few min and centrifuge for 1 min at maximum speed in a microfuge to pellet cell debris before loading onto a 12% polyacrylamide gel (*see* below).
7. SDS-PAGE: To set up the SDS-PAGE, first clean the glass plates with water then with 70% ethanol. Assemble the plates with spacers in a gel casting stand. Fill the 12% acrylamide resolving gel mix (5.1 ml ultrapure water, 3.75 ml 1.5 M Tris pH 8.8, 6.0 ml 30% acrylamide mix, 150 μl 10% SDS, 50 μl 10% APS and 10 μl TEMED) between the glass plates until ¾ of the volume is filled. Add 250 μl of isopropanol to the top of the resolving gel mix and allow it to polymerize at room temperature. Afterwards, remove the isopropanol and wash the gel surface with water. Fill the remaining space between the glass plates with stacking gel mix (7.32 ml ultrapure water, 1.25 ml 1 M Tris pH 6.8, 1.3 ml 30% acrylamide mix, 100 μl 10% SDS, 50 μl 10% APS and 10 μl TEMED) and carefully insert a multiwell comb without trapping air-bubbles (Always use gloves and protective eyeglasses when handling acrylamide solutions.). Wait until the acrylamide has polymerized before slowly removing the comb from the stacking gel. Wash the wells with running buffer (LRB). Place the gel onto the inner part of the electrophoresis device and fill it with LRB. Load 35 μl of cell extract for pPM318 samples and 17.5 μl for pPM323 samples. Load 2.5 μl of pre-stained protein marker in one of the remaining wells. Now fill the outer electrophoresis chamber with running buffer and connect it to a power supply. Run electrophoresis at 80 V until the dye front reaches

the bottom of the gel. Then stop the power supply, disassemble the gel plate and carefully remove the gel from the plates.

8. Protein transfer onto nitrocellulose membrane: Measure the size of the gel, then cut filter paper (6×) and nitrocellulose membrane (1×) to the same size. Clean the electrodes of the semidry western transfer chamber with ethanol and let them dry. Then place three filter papers soaked in WTB on the top of the anode plate and remove excess buffer and trapped air bubbles by rolling them out with a glass pipette or glass tube applying gentle pressure. Next, place the nitrocellulose membrane on top of the filter paper, followed by the gel. Place three more filter papers soaked in WTB on top of the gel and remove excess buffer as well as air bubble as described above. Then fix the cathode plate and close the transfer chamber with the lid. Connect the chamber to the power pack and apply constant current (0.8 mA/cm^2) for 1 h to transfer proteins from the gel to the nitrocellulose membrane. After that, the device is disassembled and the nitrocellulose membrane rinsed with water. Observe the transfer of the prestained marker to check for correct blotting. Optionally, at this step, the quality of the transfer can be additionally checked by staining the proteins on the membrane with Ponseau S.
9. Western blot antibody detection: Incubate nitrocellulose membrane in 10 ml blocking solution for 1 h at room temperature. Prepare 10 ml of primary antibody solution containing anti-Myc (1:5000 dilution) and anti-Cdc11 (1:10,000 dilution) by diluting the antibodies in blocking solution. Add sodium azide to a final concentration of 0.02% to prevent microbial growth in the antibody solution. Incubate membrane in primary antibody solution at 4 °C overnight on a rocking or shaking platform. The next morning, remove antibody solution. (The antibody solution can be stored at 4 °C for reuse). Wash nitrocellulose membrane four times with 10 ml PBS for 5 min at room temperature. Prepare 10 ml solution with secondary antibodies coupled to fluorophores (anti-mouse 680 and anti-rabbit 800, both at 1:5000 dilution) in blocking solution and with 0.02% sodium azide (*see* **Note 4**). Incubate washed membrane in secondary antibody solution in aluminum foil-covered container for 45 min at room temperature on a rocking or shaking platform. Afterwards, wash nitrocellulose membrane four times with 10 ml PBS for 5 min at room temperature (*see* **Note 5**).
10. Quantitative analysis of Western blot signals: Place either dried or wet nitrocellulose membrane probed with antibodies on the Li-cor infrared scanner (the surface onto which proteins were blotted should be placed onto the glass plate of the device). Close the lid of the scanner and open the odyssey software on

the computer connected to the scanner. Perform a two-color scan of fluorescence signals on the membrane to detect the specific signals corresponding to OAZ1 and CDC11 proteins. Quantify the signal intensities of each band using the odyssey software. In the next step, normalize the OAZ1 values to the loading control CDC11. A value for the "relative frameshifting efficiency (RFE)" of OAZ1 can be calculated using the values for the in frame control construct and for the corresponding construct with the RFS site by applying the following formula:

$$\mathrm{RFE} = 100 \times C_{(\mathrm{RFS})}/C_{(\mathrm{IF\text{ - }spd})}$$

RFE is the amount of OAZ1 produced by the construct with the RFS site (in %) relative to the in frame construct. $C_{(\mathrm{RFS})}$ is the OAZ1 signal in individual lanes obtained with the construct bearing the RFS site, and $C_{(\mathrm{IF\text{-}spd})}$ is the OAZ1 signal obtained with the in frame construct without additional polyamine. Note that under the experimental conditions described here, there was no differences observed for the in frame construct whether the cultures were supplemented with spermidine or not.

3.2 In Vitro Assay Monitoring Polyamine Binding

1. Preparation of OAZ1-encoding *E. coli* transformants: For assaying polyamine-binding, the protein of interest, such as *S. cerevisiae* OAZ1, is expressed in *E. coli* and affinity-purified (*see* **Note 6**). Thaw two 50 μl aliquots of chemical competent *E. coli* Rosetta strain on ice and transform separately with the respective empty vector (pET11a; "mock") or the expression construct (e.g. pDG240 codon-optimized *6HIS-OAZ1* in pET11a). In the next step, transformants are selected on LB agar plates containing ampicillin (100 μg/ml) and chloramphenicol (30 μg/ml) by incubating overnight at 37 °C. Single colonies from each plate are selected and used to inoculate overnight cultures (5 ml LB medium with the same concentrations of antibiotics mentioned above). These cultures are diluted with 20 ml fresh LB medium with antibiotics and grown for 6 h at 37 °C with shaking at 180 rpm.
2. Induction of OAZ1 expression in *E. coli*: Determine the OD_{600} of the cultures described in **step 1**, and dilute the cells into 250 ml fresh LB medium containing antibiotics to achieve a starting OD_{600} of 0.2. Grow cells at 30 °C until the OD_{600} has reached 0.6, then induce expression by adding IPTG to a final concentration of 1 mM and incubate for 4 h at 30 °C with shaking at 180 rpm. At this point, determine again the OD_{600} of the cultures and calculate corresponding volumes of mock and pDG240 cultures containing 550 OD_{600} cells (corresponding to an amount of cells in 550 ml culture with OD_{600} of 1). Collect cells from these volumes of cultures by centrifugation at 5,000 x *g* for 10 min at 4 °C. Discard

supernatants, resuspend cell pellets in 50 ml water, and pellet again at 10,000 rpm for 10 min at 4 °C. Carefully remove supernatants and resuspend pellets in 10 ml of lysis buffer with protease inhibitor. Freeze cell suspension in liquid nitrogen and store at −80 °C until further use.

3. Protein extraction and purification: Prior to purification, thaw the cell suspensions derived from mock and OAZ1 expression cultures on ice. Then, add 20 mg of lysozyme (dissolved in 1 ml lysis buffer) and 2 mg of Dnase I (added as powder), mix and keep on ice for 5 min. The cell lysis is initiated by vortexing the samples at maximum speed for six times for 10 s at room temperature with 1 min incubation on ice after each round of vortexing. Cell suspensions are incubated further on ice for 45 min to complete the lysis process. For analytical purpose, 100 μl of lysates are removed and added to 1.5 ml microfuge tubes. Boil samples after addition of 25 μl of 5xLLB at 100 °C for 10 min. Another 100 μl aliquot of lysate is taken and added to another sets of 1.5 ml tubes and centrifuged at 25,000 × *g* for 10 min at 4 °C. Then the supernatants are transferred to fresh 1.5 ml tubes and mixed with 25 μl of 5XLLB, the pellets are resuspended in 125 μl of 1xLLB, and all tubes were incubated at 100 °C for 10 min. Centrifuge the reminders of the lysates at 25,000 × *g* for 10 min and transfer supernatants to fresh 15 ml tubes. Adjust imidazole to a final concentration of 20 mM in the lysates and then add of 250 μl of Ni-NTA Sepharose (pre-equilibrated in lysis buffer). To allow binding of proteins to the resin, incubate the slurry in the cold room for 2 h with slow rotation. After this binding step, centrifuge the suspension at 200 × *g* for 3 min at 4 °C and carefully remove the supernatant with a pipette. 100 μl of this supernatant with unbound proteins are collected and processed for analytical purpose as described for the other samples. Wash the beads four times with 10 ml wash buffer (lysis buffer with 20 mM imidazole) for 5 min each with rotation in the cold room followed by 200 × *g* centrifugation steps. Collect 100 μl of the wash fractions and process as mentioned above for the other samples for analytical purpose. After washing, the beads are transferred to fresh 1.5 ml tubes (cut the ends of 1000 μl pipette tips to create a wider opening) and bound protein is eluted with 350 μl elution buffer containing 250 mM imidazole for 1 h in the cold room with mild rotation. The supernatant containing the eluted proteins is collected by centrifugation of the suspension at 200 × *g* for 3 min at 4 °C. Supernatants are transferred to fresh 1.5 ml microfuge tubes and stored on ice. 20 μl of each sample are collected and processed for analytical purpose as describe above. Determine the concentration of the eluted protein using bradford assay

Table 1
Setup of polyamine binding assay

Contents	Buffer	Mock	OAZ1	Control proteins
Mock eluate	–	*X* μl	–	–
OAZ1 (10 μM)	–	–	*X* μl	–
Control protein	–	–	–	*X* μl
H-3 or C-14 Spermidine or Spermine (10 μM)	20 μl	20 μl	20 μl	20 μl
Elution buffer	80 μl	80 – *X* μl	80 – *X* μl	80 – *X* μl
Total	100 μl	100 μl	100 μl	100 μl

See main text for the description of OAZ1 and mock eluate preparation, and Fig. 2 for examples of control proteins

(Bio-Rad). The purity of the eluted protein is assayed by SDS-PAGE by loading all the collected fractions from the different steps of the purification (lysis, washes, and elution) on the gel and visualization of the bands by Coomassie staining.

4. Pre-cool all the reagents required for the polyamine-binding assay on ice. The experiment involving work with isotope-labeled compounds should be carried out in appropriate isotope lab facility following all safety standards and the isotope-labeled waste disposed according to the isotope safety requirement. The assays are carried out in appropriately labeled pre-cooled 1.5 ml tubes using buffer, eluate from the mock preparation, and control proteins (the concentration and purity of the control proteins should be determined independently). The polyamine binding reactions were setup as shown in Table 1.

The content of all the tubes is mixed gently, after which the binding reactions are carried out by incubating them on ice for 60 min. In the next step, 100 μl of the polyamine binding mix are transferred to labeled cut-off filter units, which are then centrifuged at 2500 × *g* for 5 min at 4 °C. 10 μl of the "filtrate" containing only free polyamines in the collection tube as well as 10 μl of "retentate" containing Oaz1-bound polyamines and free polyamines inside the filter unit are pipetted separately into labeled 1.5 ml tubes containing 1 ml scintillation liquid. This step is repeated for all samples, and the tubes are vortexed for 1 min at room temperature. The amounts of radioactivity (counts per minute [CPM]) in each tube are measured using a scintillation counter. Percentages of polyamines bound to protein are calculated by using the following formula (*see* **Note 7**):

Percent Polyamines bound $= 100 \times (\mathrm{CPM}_{\mathrm{retentate}} - \mathrm{CPM}_{\mathrm{filtrate}})/\mathrm{CPM}_{\mathrm{retentate}}$

4 Notes

1. Providers of radioactive polyamines are American Radiolabeled Chemicals and Perkin Elmer. GE healthcare has discontinued the supply of isotope-labeled polyamines.
2. Due to high viscosity of the transformation mix, it is difficult to resuspend the cells. Therefore try to take up slowly the cell pellet along with the transformation mix with the pipette tip and release it again into the tube.
3. Keep at least four individual transformants per construct to enable a statistical analysis with independent biological replicates. Plasmid copy number variations may lead to some experimental fluctuations. Rarely, uracil-prototrophic transformants occur that do not show any expression of MYC-tagged OAZ, probably due to recombination events. Such transformants should not be considered further for the analysis.
4. The secondary antibody mix is prepared in a plastic tube covered with aluminium foil to prevent photo-bleaching of the fluorophors upon exposure to visible light.
5. After incubation with the antibody, the blot is washed and stored in a container covered with aluminium foil. The dried blots can be stored for several years when stored appropriately under light protective conditions. Avoid using detergents in PBS as they increase the background signal.
6. Expression of the authentic *S. cerevisiae OAZ1* gene in *E. coli* led to only poor yields, which could be significantly increased by using a synthetic version of the gene with a codon bias optimized for expression in *E. coli*. Similarly, we could achieve a reasonable yield of human antizyme 1 (AZ1) using a codon-adapted synthetic version of the coding sequence [17]. If genes encoding antizymes or different polyamine binding proteins from other species are employed, a similar strategy is recommended.
7. If no binding is detected, check whether the protein precipitated during the assay. To avoid precipitation of protein, perform the polyamine binding assay with freshly prepared protein immediately after its purification. Avoid freezing and thawing of purified OAZ1.

References

1. Michael AJ (2016) Biosynthesis of polyamines and polyamine-containing molecules. Biochem J 473(15):2315–2329
2. Tabor CW, Tabor H (1984) Polyamines. Annu Rev Biochem 53:749–790
3. Thomas T, Thomas TJ (2001) Polyamines in cell growth and cell death: molecular mechanisms and therapeutic applications. Cell Mol Life Sci 58:244–258

4. Childs AC, Mehta DJ, Gerner EW (2003) Polyamine-dependent gene expression. Cell Mol Life Sci 60:1394–1406
5. Palanimurugan R, Kurian L, Hegde V, Hofmann K, Dohmen RJ (2014) Co-translational polyamine sensing by nascent ODC Antizyme. In: Ito K (ed) Regulatory nascent polypeptides. Springer, Tokyo, pp 203–222
6. Nakanishi S, Cleveland JL (2016) Targeting the polyamine-hypusine circuit for the prevention and treatment of cancer. Amino Acids 48:2353–2362
7. Murray-Stewart TR, Woster PM, Casero RA Jr (2016) Targeting polyamine metabolism for cancer therapy and prevention. Biochem J 473:2937–2953
8. Liu JH, Wang W, Wu H, Gong X, Moriguchi T (2015) Polyamines function in stress tolerance: from synthesis to regulation. Front Plant Sci 6:827
9. Tiburcio AF, Altabella T, Bitrian M, Alcazar R (2014) The roles of polyamines during the lifespan of plants: from development to stress. Planta 240:1–18
10. Coffino P (2001) Regulation of cellular polyamines by antizyme. Nat Rev Mol Cell Biol 2:188–194
11. Illingworth C, Michael AJ (2012) Plant ornithine decarboxylase is not post-transcriptionally feedback regulated by polyamines but can interact with a cytosolic ribosomal protein S15 polypeptide. Amino Acids 42:519–527
12. Kahana C (2016) Protein degradation, the main hub in the regulation of cellular polyamines. Biochem J 473:4551–4558
13. Murakami Y, Matsufuji S, Kameji T, Hayashi S, Igarashi K, Tamura T, Tanaka K, Ichihara A (1992) Ornithine decarboxylase is degraded by the 26S proteasome without ubiquitination. Nature 360:597–599
14. Palanimurugan R, Scheel H, Hofmann K, Dohmen RJ (2004) Polyamines regulate their synthesis by inducing expression and blocking degradation of ODC antizyme. EMBO J 23:4857–4867
15. Matsufuji S, Matsufuji T, Miyazaki Y, Murakami Y, Atkins JF, Gesteland RF, Hayashi S (1995) Autoregulatory frameshifting in decoding mammalian ornithine decarboxylase antizyme. Cell 80:51–60
16. Ivanov IP, Atkins JF (2007) Ribosomal frameshifting in decoding antizyme mRNAs from yeast and protists to humans: close to 300 cases reveal remarkable diversity despite underlying conservation. Nucleic Acids Res 35:1842–1858
17. Kurian L, Palanimurugan R, Gödderz D, Dohmen RJ (2011) Polyamine sensing by nascent ornithine decarboxylase antizyme stimulates decoding of its mRNA. Nature 477:490–494
18. Heinemeyer W, Kleinschmidt JA, Saidowsky J, Escher C, Wolf DH (1991) Proteinase yscE, the yeast proteasome/multicatalytic-multifunctional proteinase: mutants unravel its function in stress induced proteolysis and uncover its necessity for cell survival. EMBO J 10:555–562
19. Gietz RD, Sugino A (1988) New yeast-Escherichia coli shuttle vectors constructed with in vitro mutagenized yeast genes lacking six-base pair restriction sites. Gene 74:527–534

Chapter 27

Modulation of Protein Synthesis by Polyamines in Mammalian Cells

Keiko Kashiwagi, Yusuke Terui, and Kazuei Igarashi

Abstract

Polyamines exist mainly as RNA–polyamine complexes in cells. Thus, we looked for proteins whose synthesis is enhanced by polyamines at the level of translation in mammalian cells. Here, we describe how synthesis of Cct2 (T-complex protein 1, β-subunit, a chaperonin assisting in the folding actin, tubulin, and several other proteins) and eEF1A (one of the elongation factors of protein synthesis) is stimulated by polyamines at the level of translation. Polyamines stimulated Cct2 synthesis through the stimulation of ribosome shunting during 5′-processive scanning of 40S ribosomal subunits from the m^7G-cap to the initiation codon AUG, and eEF1A synthesis through the structural change of the unusual position of a complementary sequence to 18S rRNA in eEF1A mRNA.

Key words Protein synthesis, Polyamine modulon, 5′-UTR of mRNA, Ribosome shunting, Complementary sequence to 18S rRNA (CR sequence) in mRNA

1 Introduction

Polyamines are present at mM concentrations in both prokaryotes and eukaryotes and essential for cell growth and viability [1, 2]. In mammalian cells, major polyamines are spermidine (SPD) and spermine (SPM), and about 80% of SPD and 85% of SPM existed as RNA–polyamine complexes in rat liver [3, 4]. Thus, the effects of polyamines on cell growth and viability were studied at the level of translation, and it was found that several kinds of protein synthesis were enhanced by polyamines. We proposed that a group of genes whose expression is enhanced by polyamines at the level of translation be referred to as a "polyamine modulon".

In this chapter, the procedure to identify proteins whose synthesis is enhanced by polyamines at the level of translation in mammalian cells is described. Furthermore, procedures to study how polyamines stimulate protein synthesis at the molecular level are described.

Rubén Alcázar and Antonio F. Tiburcio (eds.), *Polyamines: Methods and Protocols*, Methods in Molecular Biology, vol. 1694, DOI 10.1007/978-1-4939-7398-9_27,

2 Materials

2.1 Mammalian Cells, Plasmid, and Culture Medium

1. Mouse mammary carcinoma FM3A cells (Japan Health Science Foundation) and NIH3T3 cells (American type Culture Collection) [5] (*see* **Note 1**).
2. Plasmid pEGFP-N1 (Clontech) [5] (*see* **Notes 2** and **3**).
3. Dulbecco's modified Eagle's medium (D-MEM) and ES medium.

2.2 Inhibitors of Polyamine and eIF5A Synthesis

1. α-Difluoromethylornithine (DFMO), an inhibitor of ornithine decarboxylase [6] (*see* **Note 1**).
2. *N*′-(3-Aminopropyl)-cyclohexylamine (APCHA), an inhibitor of spermine synthase [7] (*see* **Note 3**).
3. *N*′-Guanyl-1,7-diaminoheptane (GC_7), an inhibitor of deoxyhypusine synthase [8] (*see* **Note 3**).

2.3 Preparation of Cell Lysate

1. Buffer 1: 25 mM Hepes-KOH, pH 7.8, 0.1 mM EDTA, 6 mM 2-mercaptoethanol, 5% (v/v) glycerol and 20 μM FUT-175 (6-amino-2-naphthyl-4-guanidinobenzoate dihydrochrolide), an inhibitor of serine protease.

2.4 Two-Dimensional Gel Electrophoresis of Proteins and Identification of Proteins by Edman Degradation and Mass Spectrometry

1. Ready-Prep™ Sequential Extraction Kit (Bio-Rad Laboratories).
2. PROTEIN®IEF Cell using 17 cm (pH range 3–10) Ready Strip™ IPG Strip gel (Bio-Rad Laboratories).
3. Achromobacter lysine-specific protease I [9].
4. Columns: DEAE-5PW (1 mm × 100 mm or 0.5 mm × 5 mm; Tosoh) and Inertsil ODS-3 (1 mm × 100 mm or 0.5 mm × 150 mm; GL Science Inc.).
5. Model 1100 series liquid chromatography system (Agilent Technologies).
6. Solvent A: 0.09% (v/v) aqueous trifluoroacetic acid.
7. Solvent B: 0.075% (v/v) trifluoroacetic acid in 80% (v/v) acetonitrile.
8. Procise cLC protein sequencing system (Applied Biosystems).
9. Matrix-assisted laser desorption ionization time of flight mass spectrometry (MALDI-TOF MS) on a Reflex MALDI-TOF (Bruker-Franzen Analytik) in a reflector mode using α-cyano-4-hydrocinnamic acid as a matrix.

10. Prematrix-coated MALDI target plate (Anchor Chip™ 600/384, Bruker Daltonics) using AccuSpot (Shimazu).
11. Ultraflex mass spectrometer (Bruker Daltonics) using Fuzzy control system.

2.5 Western and Northern Blot Analyses

1. Immobilon transfer membrane (Millipore).
2. ECL™ Western blotting detection reagents (GE Healthcare).
3. Quick Prep total RNA Extraction kit (GE Healthcare).
4. ECL direct nucleic acid labeling and detection system (GE Healthcare).
5. SuperScript™ II RNase H^{-} Reverse Transcriptase (Life Technologies).
6. LAS-1000 plus luminescent analyzer (Fuji Film) (*see* **Notes 2** and **3**).

2.6 Measurement of Polyamines

1. TSK gel IEX column (4 × 80 mm) (TOSOH).
2. Elution buffer: 0.35 M citrate buffer, pH 5.35, 2 M NaCl, and 20% methanol.
3. Polyamine detection buffer: 0.06% *o*-phthalaldehyde, 0.4 M boric buffer, pH 10.4, 0.1% Brij-35, and 37 mM 2-mercaptoethanol.

2.7 Transfection of Plasmids into NIH3T3 Cells

1. Lipofectamine™ Reagents (Invitrogen) (*see* **Notes 2** and **3**).

3 Methods

To study genes belonging to the polyamine modulon in mammalian cells, FM3A, and NIH3T3 cells were cultured in the absence and presence of DFMO or DFMO plus APCHA for 3 days to reduce cellular polyamine content. Proteins whose level was low in polyamine-depleted cells were isolated by two-dimensional electrophoresis and identified by Edman degradation and mass spectrometry. Then, the mechanism of polyamine stimulation of protein synthesis was examined.

3.1 Culture of FM3A Cells and Preparation of Cell Lysate

FM3A cells were cultured in D-MEM or ES medium, supplemented with 2% heat-inactivated fetal bovine serum (FBS) at 37 °C in an atmosphere of 5% CO_2 in air. To make polyamine-depleted FM3A cells, 50 μM DFMO was added to the medium. In some cases, FM3A cells were treated with GC_7, or with DFMO plus APCHA. FM3A cells (2 × 10^6 cells) were suspended in buffer 1, and lysed by repeated (three times) freezing and thawing with intermitted mechanical mixing. The supernatant was obtained by centrifugation at 17,000 × *g* for 15 min and used as cell lysate. Protein content of the cell lysate was determined by the method of Bradford [10].

3.2 Two-Dimensional Gel Electrophoresis of Proteins

Cell lysate (200 μg of protein) for two-dimensional gel electrophoresis was prepared using Ready-Prep™ Sequential Extraction Kit according to the manufacturer's protocol. The first-dimensional isoelectric focusing was performed in a PROTEIN®IEF Cell using 17 cm (pH range 3–10) Ready Strip™ IPG Strip gel. Proteins on the first-dimensional Strip gel were further separated in the second dimension by SDS-polyacrylamide gel electrophoresis (SDS-PAGE) on a 10.5% polyacrylamide gel, and stained with Coomassie Brilliant Blue R-250.

3.3 Determination of Amino Acid Sequence by Edman Degradation and Mass Spectrometry

Coomassie-stained spots were excised and treated with 0.2 μg of lysine-specific protease I [9] at 37 °C for 12 h in 0.1 M Tris–HCl (pH 9.0) containing 0.1% SDS. Peptides generated were extracted from the gel and separated on columns of DEAE-5PW and Inertsil ODS-3 connected in series with a model 1100 series liquid chromatography system. Peptides were eluted at a flow rate of 20 μL/min using a linear gradient of 0–60% solvent B, where solvents A and B were 0.09% (v/v) aqueous trifluoroacetic acid and 0.075% (v/v) trifluoroacetic acid in 80% (v/v) acetonitrile, respectively. Selected peptides were subjected to Edman degradation using a Procise cLC protein sequencing system and to MALDI-TOF MS (*see* **Note 1**). Peptides generated were separated on columns of DEAE-5PW and Inertsil ODS-3 connected in series with a model 1100 series liquid chromatography system. Peptide eluted at a flow rate 3 μL/min was directly spotted onto a pre-matrix-coated MALDI target plate using AccuSpot. MS or MS/MS analysis was performed automatically on an Ultraflex spectrometer using Fuzzy control system. α-Cyanohydroxyl cinnamic acid was used as a matrix (*see* **Note 1**).

3.4 Western Blot Analysis

Cell lysate (20 μg of protein) was separated by SDS-PAGE, transferred on the Immobilon transfer membrane, and each protein was detected by using specific antibody, followed by ECL™ Western blotting detection reagents. Concentration of the polyacrylamide gel used was 10.5% for Hnrpl and Cct2, 12% for Pgam 1. The level of each protein was quantified by a LAS-1000 luminescent image analyzer (*see* **Notes 2** and **3**).

3.5 Northern Blot Analysis

Total RNA was isolated from 2×10^7 cells using the Quick Prep Total RNA Extraction Kit. Northern blot analysis was performed using the ECL direct nucleic acid labeling and detection system with 10 μg of total RNA [11]. The first-strand cDNA used for template DNA was prepared using SuperScript™ II RNase H^- Reverse Transcriptase according to the manufacturer's protocol. PCR products used for probes were prepared as described previously [12] (*see* **Notes 2** and **3**).

3.6 Measurement of Polyamines

Polyamines were extracted from cell lysate with 5% trichloroacetic acid (TCA), and centrifuged at 27,000 × *g* for 15 min at 4 °C. The polyamines in 10 μL of the supernatant were applied to TSK gel IEX column (4 × 80 mm) heated to 50 °C. Flow rate of the elution buffer was 0.35 mL/min. Detection of polyamines was by fluorescence intensity after reaction of the effluent at 50 °C with the polyamine detection buffer. The flow rate of the polyamine detection buffer was 0.8 mL/min, the fluorescence was measured at an excitation wave length of 388 nm and an emission wavelength of 410 nm. Retention times for putrescine (PUT), spermidine (SPD), and spermine (SPM) were 9, 15, and 27 min, respectively (*see* **Note 1**).

3.7 Transfection of Plasmids into NIH3T3 Cells

NIH3T3 cells were cultured in D-MEM supplemented with 50 units/mL streptomycin, 100 units/mL penicillin G, and 10% FBS at 37 °C in an atmosphere of 5% CO_2 in air for 36 h. Then, cells were cultured in the presence and absence of 500 μM DFMO for 12 h. After changing the medium with a fresh one without FBS, cells were transfected with 4 μg of various plasmids by Lipofectamine™ Reagents according to the manufacturer's instructions and cultured for 3 h. After changing the medium with a fresh one containing FBS, cells were cultured in the presence and absence of 500 μM DFMO for further 24 h. NIH3T3 cells attached to the culture dish were washed twice with 5 mL of phosphate buffered saline (PBS), incubated with 0.4 mL of 0.25% trypsine-0.02% EDTA-4Na solution at 37 °C for 3 min, and 5 mL of D-MEM containing 10% FBS was added to the culture dish. Dispersed cells were collected by centrifugation at 300 × *g* for 5 min, washed twice with PBS, and used for Western and Northern blotting (*see* **Notes 2** and **3**).

4 Notes

1. Members of polyamine modulon in mammalian cells were searched for using FM3A cells cultured with or without 50 μM DFMO for 72 h. The cell number of DFMO-treated cells decreased to approximately 20% of the cells cultured without DFMO. Under these conditions, the content of PUT and SPD decreased greatly, and that of SPM decreased to 80% of control cells [12].

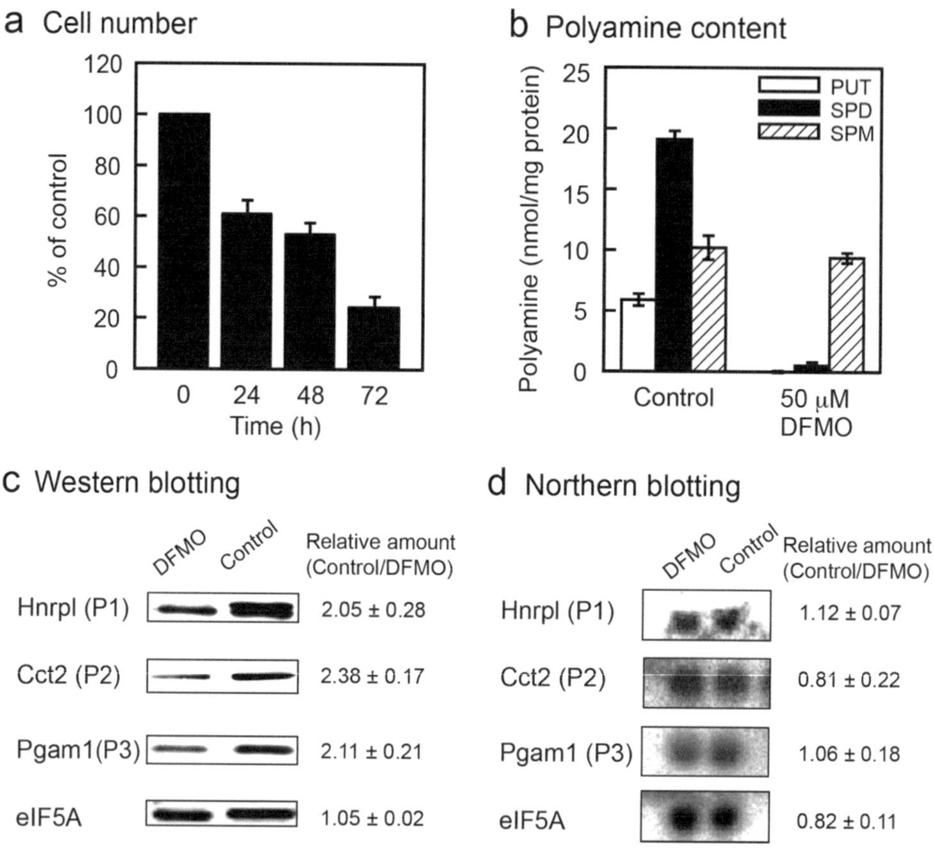

Fig. 1 Identification of members of the polyamine modulon in mammalian cells. Effect of DFMO on cell growth of FM3A cells (**a**) and polyamine content in cells cultured for 72 h (**b**). Levels of proteins (**c**) and mRNAs (**d**) of Hnrpl (P1), Cct2 (P2), Pgam 1 (P3) and eIF5A. Experiments were performed as described in Subheadings 3.1–3.6. Figures are taken from Nishimura et al. [12]

We looked for proteins whose level is low in DFMO-treated cells by two-dimensional gel electrophoresis (Fig. 1). It was found that the level of three kinds of proteins (P1, P2, and P3) decreased to less than 50% of control. These three proteins were identified by Edman degradation and mass spectrometry. Protein P1 was identified as Hnrpl (heterogeneous nuclear riboprotein L), which is a ribonucleoprotein affecting translation, mRNA stability and splicing [13]. Protein P2 was identified as Cct2 (T-complex protein 1, β-subunit), which is a chaperonin located in cytoplasm and assisting in the folding of actin, tubulin and several other proteins [14]. Protein P3 was identified as Pgam 1 (phosphoglycerate mutase 1), which is a glycolytic enzyme catalyzing the conversion of 3-phosphoglycerate to 2-phophoglycerate and modulating cellular life span [15].

2. Mechanism of polyamine stimulation of Cct2 synthesis.

   Initiation of protein synthesis in mammalian cells usually occurs by 5′-processive scanning of 40S ribosomal subunits from the m^7G-cap to the initiation codon AUG of mRNA. In the 5′-UTR (5′-untranslated region) of Cct2 mRNA, there are two hairpin structures (Fig. 2). It was tested whether polyamines can stimulate the ribosome shunting in this region. For ribosome shunting, take-off and landing sites, which are complementary in sequence to 18S rRNA, are required in the 5′-UTR of mRNA [16]. These nucleotide sequences are found in the 5′-UTR of Cct2 mRNA (Fig. 2a and b), and polyamines stimulated Cct2 synthesis through the stimulation of ribosome shunting (Fig. 2c).

3. Polyamine stimulation of eEF1A synthesis based on the structural change of the unusual position of a complementary sequence to 18S rRNA in eEF1A mRNA.

   Among 19 translation factors, only eEF1A synthesis was decreased in polyamine reduced cells. When the level of SPD increased and that of SPM decreased by treatment of cells with DFMO plus APCHA, the reduction of eEF1A was more evident, whereas the level of hypusinated eIF5A did not change compared to control. In addition, when the level of hypusinated eIF5A was greatly reduced by treatment with GC_7, the level of eEF1A did not change compared to control. The results indicate that polyamines themselves enhanced the synthesis of eEF1A (Fig. 3).

   In *E. coli*, polyamine stimulation of protein synthesis can occur when a Shine-Dalgarno (SD) sequence in the mRNA (a complementary sequence to the 3′-end of 16S rRNA) is distant from the initiation codon AUG of the mRNA [17]. So, we looked for the complementary sequence to the 3′-end of 18S rRNA, i.e. a CR sequence, on mRNAs and found it at −17 to −32 upstream from the initiation codon AUG in 18 mRNAs involved in protein synthesis except eEF1A mRNA (Fig. 4). In eEF1A mRNA, the CR sequence was located at −33 to −39 upstream from the initiation codon AUG, and polyamine stimulated eEF1A synthesis about threefold. By removing the CR sequence from the 5′-UTR of eEF1A mRNA, efficiency of translation reduced to 60% regardless the presence of polyamines. When the CR sequence was shifted to −22 to −28 upstream from AUG, eEF1A synthesis increased in polyamine-reduced cells and the degree of polyamine stimulation decreased greatly. The results indicate that a CR sequence

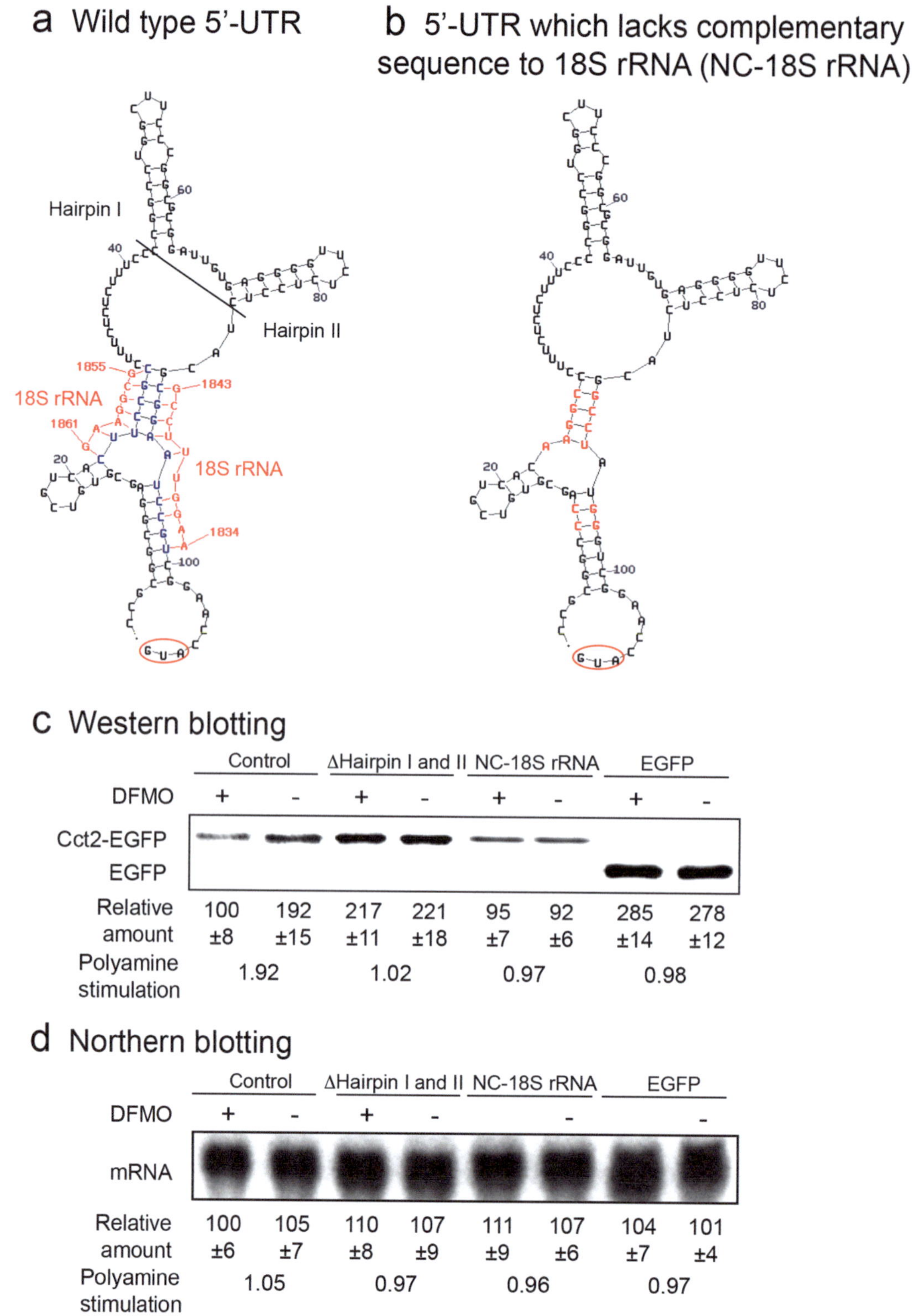

Fig. 2 Mechanism of polyamine stimulation of Cct2 synthesis. Possible secondary structure of wild-type (**a**) and mutated (**b**) 5′-UTR of Cct2 mRNA. Effect of polyamines on the synthesis of Cct2-EGFP fusion protein (**c**) and its mRNA (**d**) derived from wild-type and mutated Cct2-EGFP mRNA in NIH3T3 cells. Experiments were performed as described in Subheadings 3.4, 3.5 and 3.7. Figures are taken from Nishimura et al. [12]

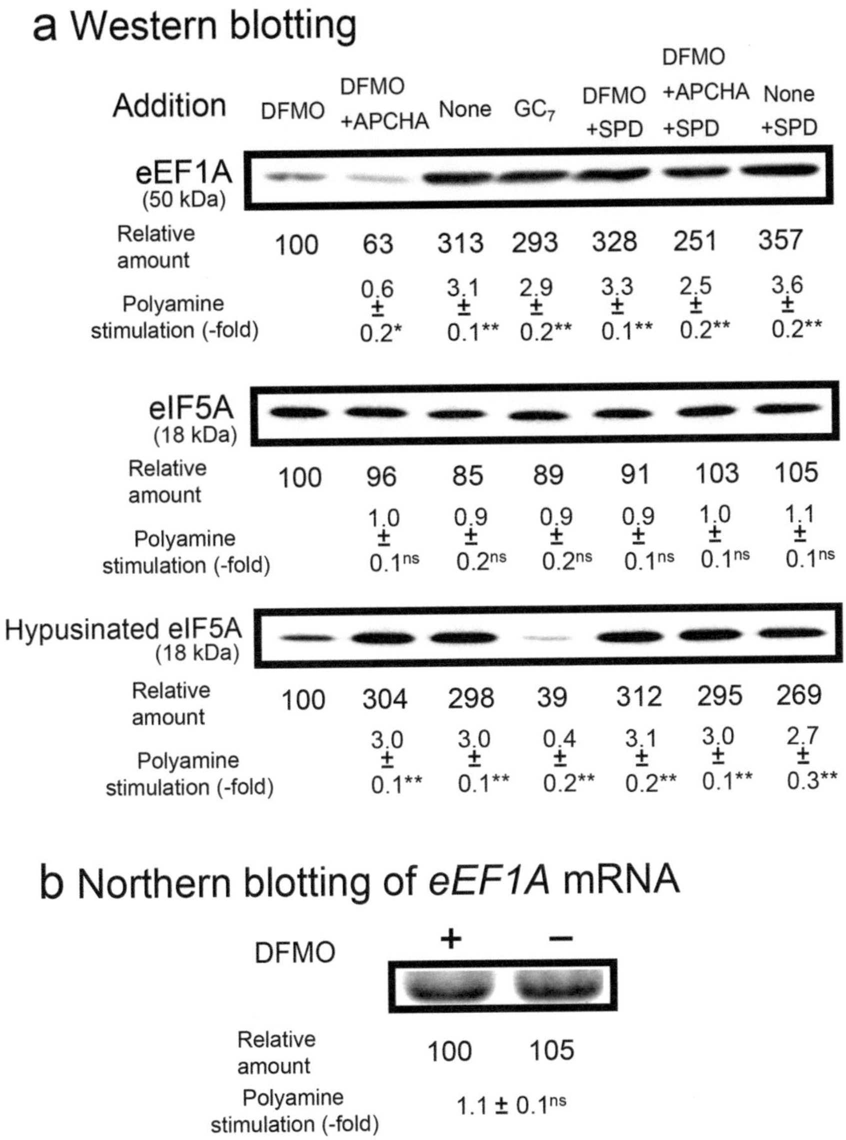

Fig. 3 Polyamine stimulation of the synthesis of eEF1A at the level of translation. The levels of eEF1A protein and its mRNA were measured by Western and Northern blotting as described in Subheadings 3.4 and 3.5. Where indicated, 50 μM DFMO, 150 μM APCHA, 2 μM GC_7, and/or 25 μM spermidine (SPD) was added to the culture medium. As a control, the levels of eIF5A and hypusinated eIF5A were examined. Values are means ± S.E. of triplicate determinations. ns, $p \geq 0.05$; *, $p < 0.05$; **, $p < 0.01$. Experiments were performed as described in Subheadings 3.4 and 3.5. Figures are taken from Terui et al. [5]

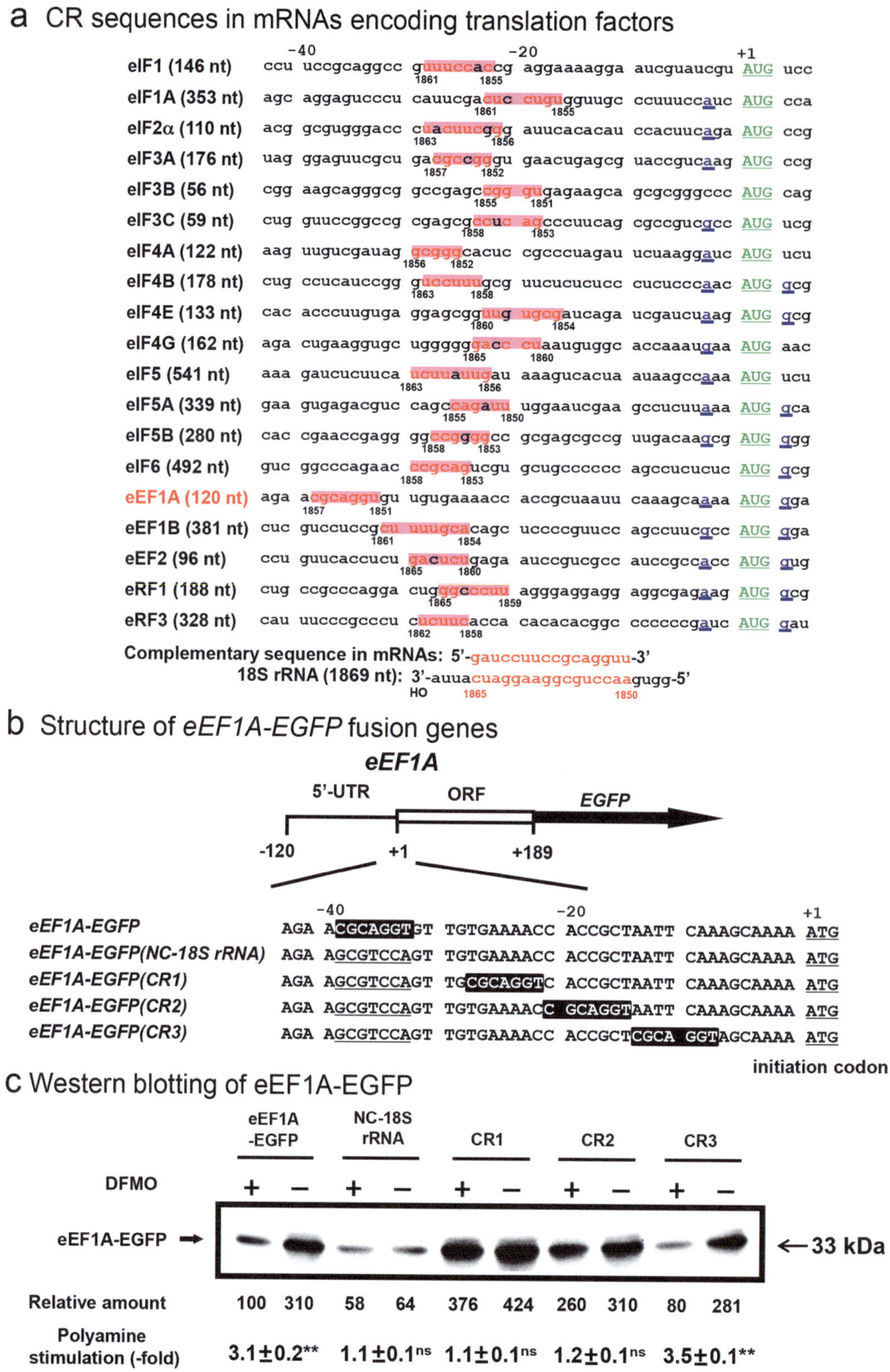

Fig. 4 Polyamine stimulation of eEF1A synthesis based on the unusual position of a complementary sequence to 18S rRNA (CR sequence) in eEF1A mRNA. (**a**) CR sequences in mRNAs encoding translation factors.

exists in many eukaryotic mRNAs, enhances translational efficiency, and the location of the CR sequence in the mRNA influences polyamine stimulation of protein synthesis.

4. We have recently found that synthesis of EXT2 (extension of polysaccharide chain by glycosyl transferase 2) is also stimulated by polyamines at the level of translation [18]. This was due to the release of microRNA (let-7b) from EXT2 mRNA by polyamines. This is a new mechanism of polyamine stimulation of protein synthesis.
5. In summary, it is thought that polyamine effects on stimulation of cell growth in both prokaryotes and eukaryotes may be similar, i.e. polyamines function at the level of RNA. Since the structure of RNA is flexible compared to DNA, polyamines stabilize RNA structure more effectively than Mg^{2+} because the distance between two positive charges of polyamines is longer than Mg^{2+}.

Acknowledgments

We are grateful to Dr. A.J. Michael for critical reading of the manuscript prior to submission.

Fig. 4 (continued) (**b**) Nucleotide sequences of various mutants at the region of CR sequence of eEF1A-EGFP gene are shown. The eEF1A mRNAs in which the position of CR sequence shifted were termed as CR1, CR2 and CR3, respectively. NC-18S rRNA, noncomplementary sequence to 18S rRNA. (**c**) The levels of eEF1A-EGFP synthesized from various eEF1A-EGFP genes are shown. Values are means ± S.E. of triplicate determinations. ns, $p \geq 0.05$; **, $p < 0.01$. Experiments were performed as described in Subheadings 3.4 and 3.7. Figures are taken from Terui et al. [5]

References

1. Igarashi K, Kashiwagi K (2000) Polyamines: mysterious modulators of cellular functions. Biochem Biophys Res Commun 271:559–564
2. Igarashi K, Kashiwagi K (2015) Modulation of protein synthesis by polyamines. IUBMB Life 67:160–169
3. Watanabe S, Kusama-Eguchi K, Kobayashi H, Igarashi K (1991) Estimation of polyamine binding to macromolecules and ATP in bovine lymphocytes and rat liver. J Biol Chem 266:20803–20809
4. Igarashi K, Kashiwagi K (2010) Modulation of cellular function by polyamines. Int J Biochem Cell Biol 42:39–51
5. Terui Y, Sakamoto A, Yoshida T, Kasahara T, Tomitori H, Higashi K, Igarashi K, Kashiwagi K (2015) Polyamine stimulation of eEF1A synthesis based on the unusual position of a complementary sequence to 18S rRNA in eEF1A mRNA. Amino Acids 47:345–356
6. Mamont PS, Duchesne MC, Grove J, Bey P (1978) Anti-proliferative properties of DL-α-difluoromethyl ornithine in cultured cells. A consequence of the irreversible inhibition of ornithine decarboxylase. Biochem Biophys Res Commun 81:58–66
7. Nishimura K, Murozumi K, Shirahata A, Park MH, Kashiwagi K, Igarashi K (2005) Independent roles of eIF5A and polyamines in cell proliferation. Biochem J 385:779–785
8. Jakus J, Wolff EC, Park MH, Folk JE (1993) Features of the spermidine-binding site of deoxyhypusine synthase as derived from inhibition studies. Effective inhibition by bis- and mono-guanylated diamines and polyamines. J Biol Chem 268:13151–13159
9. Masaki T, Tanabe M, Nakamura K, Soejima M (1981) Studies on a new proteolytic enzyme from a chromobacter lyticus M497-1: I. Purification and some enzymatic properties. Biochim Biophys Acta 660:44–50
10. Bradford MM (1976) A rapid and sensitive method for the quantitation of microgram quantities of protein utilizing the principle of protein-dye binding. Anal Biochem 72:248–254
11. Sambrook J, Fritsch EF, Maniatis T (2001) Extraction, purification, and analysis of mRNA from eukaryotic cells. In: Sambrook J, Russell DW (eds) Molecular cloning: a laboratory manual, 3rd edn. Cold Spring Harbor Laboratory, Cold Spring Harbor, NY, pp 1.32–1.34
12. Nishimura K, Okudaira H, Ochiai E, Higashi K, Kaneko M, Ishii I, Nishimura T, Dohmae N, Kashiwagi K, Igarashi K (2009) Identification of proteins whose synthesis is preferentially enhanced by polyamines at the level of translation in mammalian cells. Int J Biochem Cell Biol 41:2251–2261
13. Krecic AM, Swanson MS (1999) hnRNP complexes: composition, structure, and function. Curr Opin Cell Biol 11:363–371
14. Yokota S, Yanagi H, Yura T, Kubota H (1999) Cytosolic chaperonin is up-regulated during cell growth. Preferential expression and binding to tubulin at G_1/S transition through early S phase. J Biol Chem 274:37070–37078
15. Kondoh H, Lleonart ME, Gil J, Wang J, Degan P, Peters G, Martinez D, Carnero A, Beach D (2005) Glycolytic enzymes can modulate cellular life span. Cancer Res 65:177–185
16. Yueh A, Schneider RJ (1996) Selective translation initiation by ribosome jumping in adenovirus-infected and heat-shocked cells. Genes Dev 10:1557–1567
17. Igarashi K, Kashiwagi K (2006) Polyamine Modulon in *Escherichia coli*: genes involved in the stimulation of cell growth by polyamines. J Biochem 139:11–16
18. Imamura M, Higashi K, Yamaguchi K, Asakura K, Furihata T, Terui Y, Satake T, Maegawa J, Yasumura K, Ibuki A, Akase T, Nishimura K, Kashiwagi K, Linhardt RJ, Igarashi K, Toida T (2016) Polyamines release the let-7b-mediated suppression of initiation codon recognition during the protein synthesis of EXT2. Sci Rep 6:33549

Chapter 28

Determination of Posttranslational Modifications by 2D PAGE: Applications to Polyamines

Marta Bitrián, Antonio F. Tiburcio, and Rubén Alcázar

Abstract

Polyamines not only affect transcription and translation but also may induce a number of posttranslational modifications. The identification of polyamine-induced posttranslational modifications can be performed by 2D PAGE analyses. Here, we provide a protocol for 2D-gel electrophoresis that has been optimized for plants. The combined use of this protocol with epitope-tagged proteins expressed in plants enables the detailed analysis of posttranslational modifications induced by different polyamines in vivo.

Key words 2D gel, Protein phosphorylation, Posttranslational modifications, Kinases, Phosphatases

1 Introduction

2D-gel electrophoresis is a fundamental technique in proteomics. It enables the separation of proteins by isoelectric point (p*I*) and molecular weight, thus enhancing the resolution of proteins. In addition, some posttranslational modifications, such as protein phosphorylations and glycosilations, produce a shift in the p*I* that can be detected in 2D-gel electrophoresis.

The effects of proteins on the stimulation of translation are well known and proteomic analyses have been performed in this regard. Polyamines are part of RNA complexes and a polyamine modulon—transcripts whose translation is promoted by polyamines—has been reported [1, 2]. Evidence also indicate that PAs interact with charges at the interface of different proteins, thus affecting protein interactions [3]. Application of exogenous polyamines induces changes in the proteome, especially in proteins associated with stress protection [4–6]. Therefore, the study of proteomic changes by PAs is an active field of study. Here, we provide a protocol optimized for 2D-gel electrophoresis from plant extracts, which has been extensively tested in *Arabidopsis thaliana*. However, the protocol might be used with other plant materials.

Rubén Alcázar and Antonio F. Tiburcio (eds.), *Polyamines: Methods and Protocols*, Methods in Molecular Biology, vol. 1694, DOI 10.1007/978-1-4939-7398-9_28, © Springer Science+Business Media LLC 2018

2 Materials

2.1 Protein Extraction

1. 10% trichloroacetic acid (TCA) in acetone.
2. 0.1 M Ammonium acetate in methanol.
3. SDS buffer: 30% sucrose, 2% SDS, 0.1 M Tris–HCl pH 8.0, 0.5% β-mercaptoethanol (added prior to use) in water.
4. Phenol solution equilibrated with 10 mM Tris–HCl pH 8.0.
5. Acetone and 80% acetone in water.
6. 0.1 M ammonium acetate in methanol.
7. Urea sample buffer: 7 M Urea, 2 M thiourea, 4% CHAPS (3-[(3-Cholamidopropyl)dimethylammonio]-1-propanesulfonate), 1% DTT, 1% (pH 3–10) ampholites, 35 mM Tris-base in water (*see* **Note 1**).

2.2 Isoelectric Focusing

1. Rehydration buffer: 8 M Urea, 2% CHAPS, 1% DTT, 2% IPG buffer (pH 3–10), 0.05% bromophenol blue in water (*see* **Note 2**).
2. Mineral oil.

2.3 Strip Equilibration

1. SDS equilibration buffer: 50 mM Tris–HCl pH 8.8, 6 M Urea, 30% glycerol, 2% SDS, 0.05% bromophenol blue in water. Store at −20 °C.
2. DTT.
3. Iodoacetamide.

2.4 Sealing

1. Agarose sealing solution: 0.5% agarose, 0.05% bromophenol blue in SDS electrophoresis running buffer.
2. Electrophoresis running buffer: 25 mM Tris pH 8.3, 250 mM Glycine, 0.1% SDS.

3 Methods

3.1 Protein Extraction

Carefully read safety instructions for the manipulation of chemicals. Protein extraction must be performed in the fume hood, with appropriate protection devices.

1. Freeze the plant tissue in liquid nitrogen. Grind with mortar and pestle or tissue disruption device.
2. In the fume hood, Add 1 ml of cold acetone and mix gently.
3. Transfer to 2 ml tubes and centrifuge at 12,000 × g for 5 min at 4 °C.
4. Remove the supernatant. Add 1 ml of cold acetone to the pellet. Mix by pipetting.

5. Centrifuge at 12,000 × *g* for 5 min at 4 °C.
6. Remove the supernatant. Add 1 ml 10% TCA in acetone to the pellet. Mix by pipetting.
7. Centrifuge at 12,000 × *g* for 5 min at 4 ° C.
8. Repeat **steps 6** and 7 three more times.
9. Remove the supernatant and dry the pellet at room temperature (*see* **Note 3**).
10. Resuspend the pellet in 700 μl of SDS buffer.
11. Add 700 μl of phenol solution. Vortex 30 s.
12. Incubate at −20 °C for 2 h.
13. Centrifuge at 12,000 × *g* for 5 min at 4 °C.
14. Collect the upper phenolic phase. Transfer to a new 2 ml tube.
15. Fill the 2 ml tube with cold 0.1 M ammonium acetate in methanol.
16. Centrifuge 12,000 × *g* for 5 min at 4 °C.
17. Remove the supernatant and wash again with 1 ml cold 0.1 M ammonium acetate in methanol.
18. Centrifuge 12,000 × *g* for 5 min at 4 °C.
19. Wash the pellet two times with 1 ml of cold 80% acetone in water.
20. Dry the pellet and resuspend it in 25–200 μl of Urea sample buffer (*see* **Note 4**).
21. Quantify the amount of protein with standard methods.

3.2 Isoelectric Focusing

1. The amount of protein to load requires to be tested empirically. A good starting point is to use 50 μg of protein for loading standard 7 cm strips.
2. Bring the appropriate amount of protein to 125 μl final volume with rehydration buffer (*see* **Note 5**).
3. Pipette up and down without introducing air bubbles.
4. Load the sample along one of the lanes of the rehydration tray. Do not introduce air bubbles.
5. Using forceps, remove the plastic cover of the strip and place it on top of the sample lane.
6. After 10 min, cover with 1 ml mineral oil to avoid that the strip dries.
7. Incubate at room temperature overnight for passive rehydration (*see* **Note 6**).
8. Transfer your strip to the isoelectric focusing tray. Check that the strip is properly placed on the tray.
9. Cover with 1 ml mineral oil.

10. Place the isoelectric focusing tray on the device. Apply the voltage, timing, and settings according to the manufacturer's instructions.
11. After the isoelectric focusing, the strips can be kept frozen at −80 °C.

3.3 Strip Equilibration

1. Prepare 10 ml of 2% (w/v) DTT in SDS equilibration buffer in 50 ml plastic tube. Introduce the strip. Gently shake in orbital shaker for 10 min.
2. Prepare 10 ml of 2.5% (w/v) iodoacetamide in SDS equilibration buffer. Transfer the strip to this solution and shake for 10 min as before.

3.4 SDS-PAGE Separation

1. Prepare SDS-PAGE gels according to standard protocols.
2. Prepare the stacking gel and load on top. Insert a flat comb (without well separations but one left for loading the protein molecular weight marker). Remove the comb.
3. With the help of forceps, place the strip on top of the stacking gel.
4. Cover the strip with agarose sealing solution, previously melted in the microwave. Do not fill the marker position.
5. Follow SDS-PAGE running, blotting, and detection standard methods.

4 Notes

1. Store in aliquots at −20 °C. Filter through 0.2 μm. Add DTT prior to use. Do not heat.
2. Add IPG and DTT before use. Store in aliquots at −20 °C.
3. Do not overdry the sample, otherwise the resuspension of the pellet will be more difficult.
4. Resuspension in large volumes may lead to a weak signal for low abundant proteins.
5. The volume of 125 μl is used for strips of 7 cm. For other sizes, please check manufacturer's specifications.
6. Rehydration can also be performed actively by applying an initial step in the isoelectric focusing device.

References

1. Pegg AE, Casero RA (2011) Current status of the polyamine research field. Methods Mol Biol 720:3–35
2. Tiburcio AF, Altabella T, Bitrián M, Alcázar R (2014) The roles of polyamines during the life-span of plants: from development to stress. Planta 240(1):1–18
3. Berwanger A, Eyrisch S, Schuster I, Helms V, Bernhardt R (2010) Polyamines: naturally occurring small molecule modulators of electrostatic protein-protein interactions. J Inorg Biochem 104:118–125
4. Shi H, Ye T, Chan Z (2013) Comparative proteomic and physiological analyses reveal the protective effect of exogenous polyamines in the bermudagrass (*Cynodon dactylon*) response to salt and drought stresses. J Proteome Res 12:4951–4964
5. Li B, He L, Guo S, Li J, Yang Y, Yan B, Sun J, Li J (2013) Proteomics reveal cucumber Spd-responses under normal condition and salt stress. Plant Physiol Biochem 67:7–14
6. Sang Q, Shan X, An Y, Shu S, Sun J, Guo S (2017) Proteomic analysis reveals the positive effect of exogenous spermidine in tomato seedlings' response to high-temperature stress. Front Plant Sci 8:120

Chapter 29

Generation of EMS-Mutagenized Populations of *Arabidopsis thaliana* for Polyamine Genetics

Kostadin E. Atanasov, Changxin Liu, Antonio F. Tiburcio, and Rubén Alcázar

Abstract

In the recent years, genetic engineering of polyamine biosynthetic genes has provided evidence for their involvement in plant stress responses and different aspects of plant development. Such approaches are being complemented with the use of reverse genetics, in which mutants affected on a particular trait, tightly associated with polyamines, are isolated and the causal genes mapped. Reverse genetics enables the identification of novel genes in the polyamine pathway, which may be involved in downstream signaling, transport, homeostasis, or perception. Here, we describe a basic protocol for the generation of ethyl methanesulfonate (EMS) mutagenized populations of *Arabidopsis thaliana* for its use in reverse genetics applied to polyamines.

Key words Reverse genetics, Ethyl methanesulfonate, *Arabidopsis*, Mutagenesis, Polyamines

1 Introduction

Polyamines (PAs) are small aliphatic polycationic compounds present in all living organisms. These compounds are positively charged at physiological pH and can bind to negatively charged macromolecules such as nucleic acids, phospholipids, and some proteins [1]. PAs play important physiological roles in growth and development such as seed germination, embryogenesis, cell proliferation, and fruit ripening [2, 3]. The diamines putrescine and cadaverine, triamine spermidine, and the tetramines spermine and thermospermine are often implicated in stress responses [4–7]. Hence, genetic engineering of genes involved in their biosynthesis or degradation has been used for the development of stress-resistant plants.

The use of model species often speeds-up the identification of novel genes or pathways contributing to stress protection. One of the first researchers who proposed *Arabidopsis* as plant model was Prof. Friedrich Laibach in 1943 [8]. *Arabidopsis thaliana*, a herbaceous small-sized annual plant, has different features making it a

Rubén Alcázar and Antonio F. Tiburcio (eds.), *Polyamines: Methods and Protocols*, Methods in Molecular Biology, vol. 1694, DOI 10.1007/978-1-4939-7398-9_29, © Springer Science+Business Media LLC 2018

perfect species for plant genetics and physiological studies. It is a diploid with a small genome, it has a rapid life cycle, self-pollinates, and it contains few genomic repeats compared to other more complex plant species.

Forward and reverse genetics have been used as a powerful tool to identify novel gene functions and biological processes in plants [9]. *Arabidopsis* mutagenesis can be performed by the use of chemical agents (e.g. ethyl methanesulfonate), physical agents (X-rays, fast neutron bombardment, or accelerated ions) or biological agents such as T-DNA insertion via *Agrobacterium tumefaciens* transformation. Here, we report the chemical mutagenesis of *Arabidopsis* seeds by EMS treatment for the development of mutant populations.

2 Materials

Careful handling should be considered due to the high toxicity and mutagenic capacity of ethyl methanesulfonate (EMS). All solutions should be prepared in distilled water and kept at room temperature.

2.1 Materials Required for Mutagenesis

1. Lab coat, gloves, mask, and protective glasses.
2. Aluminum foil, orbital shaker, parafilm, micropipette, filter tips, glass bottles, 50 ml plastic tubes, and waste containers.
3. Dormancy rescue solution: 1 mg/ml KCl in water (*see* **Note 1**).
4. Ethyl methanesulfonate solution (EMS) (*see* **Note 2**).
5. EMS inactivation solution: 100 mM $Na_2S_2O_3{\cdot}5H_2O$. Dissolve 24.82 g sodium thiosulfate pentahydrate in 1000 ml of water.
6. Agarose sowing solution: 0.1% agarose in water. Autoclave and store at room temperature.

2.2 Greenhouse Materials

1. Greenhouse or plant growth cabinets suitable for *Arabidopsis thaliana*.
2. Plant trays and soil substrate.
3. Half strength Hoagland nutrient solution pH 5.7–5.8 [10, 11]. A 50× stock solution can be prepared as two individual stock solutions (A and B). To obtain solution A, dissolve 236.4 g $Ca(NO_3)_2{\cdot}4H_2O$ in water. Solution B is prepared by dissolving 101 g KNO_3, 68.1 g KH_2PO_4, 61.4 g $MgSO_4{\cdot}7H_2O$, 40 g NO_3NH_4, and 5 g of commercial mix of micronutrients including iron chelate. Dilute in water and adjust the pH with 1 M KOH.

3 Methods

3.1 Seeds Mutagenesis

1. Weigh 1 g of seeds (about 50,000 seeds) and transfer into a 50 ml tube, wrapped with aluminum foil. Add dormancy rescue solution to 50 ml and soak the seeds overnight at 4 °C.

2. Prepare an extra seed batch (about 100 seeds) as control treatment. Follow the same steps, except the EMS treatment (**step 6**).
3. Organize the fume hood with all reagents needed inside. This will reduce the risk of accidental contamination outside the working area (*see* **Note 4**).
4. After seed imbibition, discard the solution by inverting the tube or pipetting.
5. Add 135 μl of fresh EMS solution into a disposable flask containing 50 ml of water. Using a magnetic stirrer, mix the EMS reagent for 5–10 min.
6. Add the EMS solution to the seeds and incubate overnight using an orbital shaker. Protect from light by wrapping with aluminum foil.
7. Discard the supernatant.
8. Inactivate the EMS solution by adding 50 ml of 100 mM $Na_2S_2O_3$. Incubate 15 min in an orbital shaker.
9. Repeat the inactivation **steps** 7 and **8**.
10. Rinse the seeds with water. Repeat ten times.
11. At this step, mutagenized seeds are obtained.

3.2 Seed Sowing and Pooling of the M1 Generation

1. Immediately after the EMS inactivation and seed wash, add 0.1% agarose solution.
2. Mix gently for homogeneous seed dispersal.
3. Sow seeds on soil and distribute uniformly.
4. Let the mutagenized population (M_1) grow and the collect the seeds in pools of 100 plants, which will represent your M_2 population (*see* **Note 5**).
5. Screen your M_2 population for a trait of interest to identify mutant candidates using the pooled seeds.

4 Notes

1. Imbibition in 0.1% (w/v) KCl breaks dormancy of some *Arabidopsis* accessions.
2. EMS is very toxic and it is a mutagenic reagent. Because quickly oxidizes, it is recommended to use a fresh solution. Manipulation should be done in the fume hood.
3. It is recommended to verify your seed stock before the mutagenesis [12, 13].
4. EMS can be inactivated with sodium thiosulfate ($Na_2S_2O_3$) or thioglycolic acid. Prepare two waste tanks: one for solid waste with 2 g of $Na_2S_2O_3$ and another one for liquid waste

containing a solution of 100 mM $Na_2S_2O_3$. Check waste disposal regulations in your institution.

5. Not all mutations in the seed embryo are inherited in the next generation. In *Arabidopsis*, only two cells in the embryo shoot meristem contribute to seed production. Mutagenized seeds will result in chimeric M_1 plants with sectors of flowers with or without mutations [12]. Hence, a segregation ratio of 7:1 can be found in the M_2 due to the mutation of one of the two sectors (4:0 in the nonmutated sector and 3:1 in the mutated one) [13].

References

1. Shen Y, Ruan Q, Chai H et al (2016) The *Arabidopsis* polyamine transporter LHR1/PUT3 modulates heat responsive gene expression by enhancing mRNA stability. Plant J 88:1006–1021. doi:10.1111/tpj.13310
2. Walden R, Cordeiro A, Tiburcio AF (1997) Polyamines: small molecules triggering pathways in plant growth and development. Plant Physiol 113:1009–1013
3. Groppa MD, Benavides MP (2008) Polyamines and abiotic stress: recent advances. Amino Acids 34:35–45. doi:10.1007/s00726-007-0501-8
4. Yamaguchi K, Takahashi Y, Berberich T et al (2006) The polyamine spermine protects against high salt stress in *Arabidopsis thaliana*. FEBS Lett 580:6783–6788. doi:10.1016/j.febslet.2006.10.078
5. Alcázar R, Altabella T, Marco F et al (2010) Polyamines: molecules with regulatory functions in plant abiotic stress tolerance. Planta 231:1237–1249. doi:10.1007/s00425-010-1130-0
6. Tiburcio AF, Altabella T, Bitrián M, Alcázar R (2014) The roles of polyamines during the lifespan of plants: from development to stress. Planta 240:1–18. doi:10.1007/s00425-014-2055-9
7. Zarza X, Atanasov KE, Marco F et al (2016) Polyamine oxidase 5 loss-of-function mutations in *Arabidopsis thaliana* trigger metabolic and transcriptional reprogramming and promote salt stress tolerance. Plant Cell Environ. doi:10.1111/pce.12714
8. Koornneef M, Meinke D (2010) The development of Arabidopsis as a model plant. Plant J 61:909–921. doi:10.1111/j.1365-313X.2009.04086.x
9. Alonso JM, Ecker JR (2006) Moving forward in reverse: genetic technologies to enable genome-wide phenomic screens in Arabidopsis. Nat Rev Genet 7:524–536. doi:10.1038/nrg1893
10. Hoagland DR, Arnon DI (1938) The water-culture method for growing plants without soil. Water Cult method Grow plants without soil. California Agricultural Experiment Station. Circulation 347:32
11. Johnson CM, Stout PR, Broyer TC, Carlton AB (1957) Comparative chlorine requirements of different plant species. Plant Soil 8:337–353. doi:10.1007/BF01666323
12. Henikoff S, Comai L (2003) Single-nucleotide mutations for plant functional genomics. Annu Rev Plant Biol 54:375–401. doi:10.1146/annurev.arplant.54.031902.135009
13. Page DR, Grossniklaus U (2002) The art and design of genetic screens: *Arabidopsis thaliana*. Nat Rev Genet 3:124–136. doi:10.1038/nrg730

Chapter 30

Transcriptome Analysis of PA Gain and Loss of Function Mutants

Francisco Marco and Pedro Carrasco

Abstract

Functional genomics has become a forefront methodology for plant science thanks to the widespread development of microarray technology. While technical difficulties associated with the process of obtaining raw expression data have been diminishing, allowing the appearance of tremendous amounts of transcriptome data in different databases, a common problem using "omic" technologies remains: the interpretation of these data and the inference of its biological meaning. In order to assist to this complex task, a wide variety of software tools have been developed. In this chapter we describe our current workflow of the application of some of these analyses. We have used it to compare the transcriptome of plants with differences in their polyamine levels.

Key words Microarray analysis, Differential expression, Gene functional annotation, Enrichment analysis, Gene clustering

1 Introduction: Workflow on Microarray Transcriptome Analysis

Quantification of gene expression by microarrays is based on hybridization of fluorescence-tagged RNA against a surface with thousands of DNA probes fixed on it. Once hybridized, the fluorescence of each probe is quantified, generating a data array of fluorescence intensities and positions that, after statistical processing, allows simultaneous quantification of the expression level of thousands of RNAs.

Several types of microarray technologies are available, their differences are based on the use of different probe types and combinations of them, as well as which type of hybridization of the microarray is used: a one single labeled RNA (one-color arrays) or two RNA samples labeled with dyes with distinct fluorescence range (two-color arrays).

In this chapter, we present our routine analysis workflow with our preferred microarray platform, the GeneChip® Arabidopsis ATH1 Genome Array from Affymetrix, a one-color array with

Rubén Alcázar and Antonio F. Tiburcio (eds.), *Polyamines: Methods and Protocols*, Methods in Molecular Biology, vol. 1694, DOI 10.1007/978-1-4939-7398-9_30, © Springer Science+Business Media LLC 2018

22,500 probe sets that represent approximately 24,000 gene sequences, obtained from the Arabidopsis Sequencing Project [1].

A typical workflow of microarray data analysis is showed in Fig. 1:

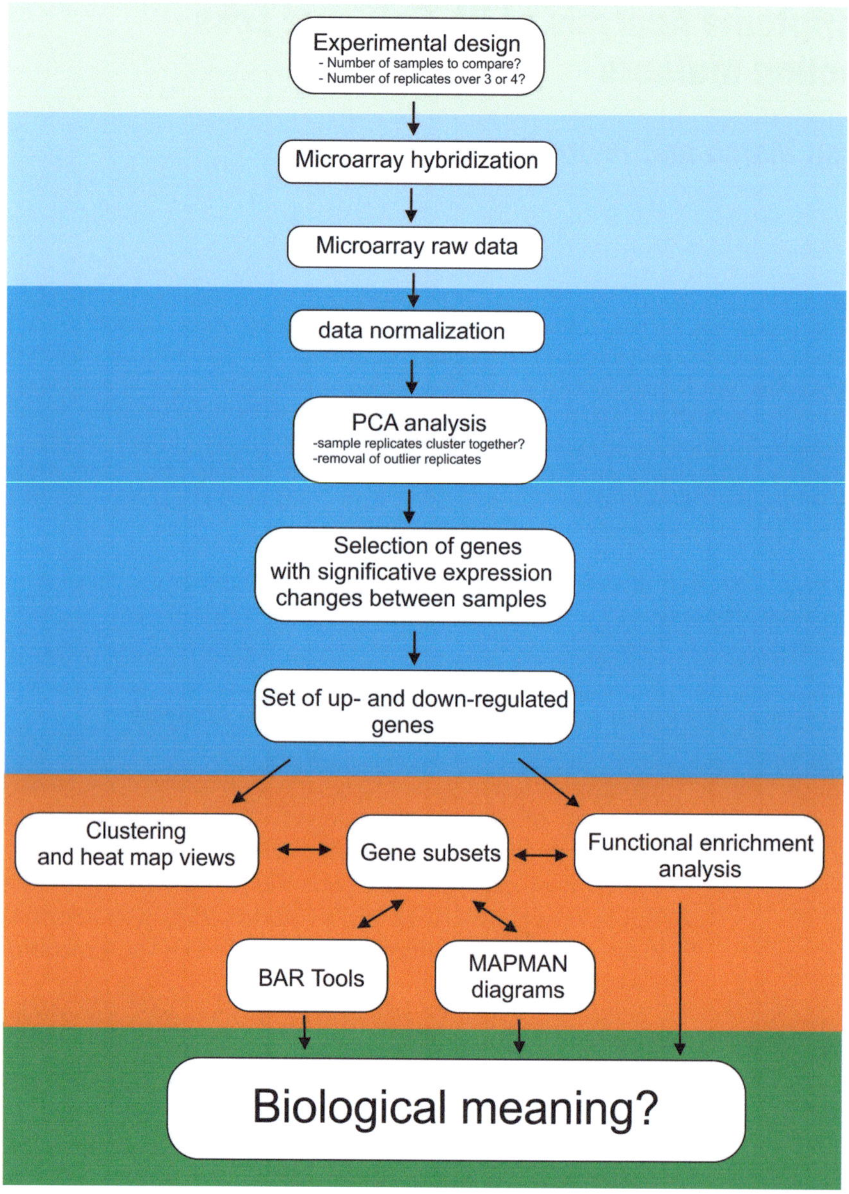

Fig. 1 Workflow of microarray data analysis. Once the experimental design is decided, RNA samples should be obtained and labeling and hybridization with microarrays should be performed. Fluorescence readings from labeled RNAs attached to microarrays probes are compiled into raw data archives that need a normalization process to make possible gene-to-gene comparisons between replicates and samples. From this comparison, a set of genes with differential expression is obtained that need a posterior analysis with several tools to try to infer a biological meaning to the gene expression changes observed

(a) Experimental design: What question we want to answer and, in consequence, which are the samples that we need to compare?

The simplest design could be a pairwise comparison of two situations, like the comparison of samples from two lines with differences in polyamine levels (mutant vs wild type). Along with the number of samples to be compared, it is important to consider a fundamental factor in the transcriptome determination: replicates are crucial to avoid "noise" in your expression data, and its number should be as high as possible, with a recommended minimum of 3–4 replicates for sample. If budget is limited, our recommendation is to favor larger number of replicas instead a larger number of samples to compare: more replicas will give you a more robust data set where statistical methods and comparisons could be more reliable.

(b) Once your samples and replicates are selected, RNA extraction, fluorescent labeling, and microarray hybridization should be done. Nowadays the most common procedure is to rely on microarray hybridization services. We recommend consulting and discussing with them the optimal RNA extraction protocol to use with your samples.

(c) After microarray hybridization, the measured intensities for each arrayed gene probe are recorded in raw data files that will need a statistical transformation to eliminate questionable or low-quality measurements. Also, all intensities have to be adjusted to a common scale in order to facilitate comparisons between samples (Normalization, Subheading 3.1).

(d) Next, a gene-by-gene comparison of the normalized data expression is necessary to identify whose genes have a differential expression between classes of samples (Subheading 3.4). Further analysis of the products of these genes may shed light on which biological processes are affected between our classes of samples (Subheadings 3.5–3.7).

2 Materials

2.1 RMAExpress Normalization of Affymetrix Microarrays

1. RMAExpress (http://rmaexpress.bmbolstad.com/).
2. CDF file specific for the model of microarray used, that can be downloaded at http://www.affymetrix.com/support/technical/libraryfilesmain.affx.
3. CEL files that contain raw data from microarray hybridization.

2.2 PCA Analysis

1. Normalized expression data file obtained on RMA normalization (Subheading 3.1).

2.3 Significance Analysis of Microarrays

1. R Software (https://www.r-project.org/).
2. R tools for windows (https://github.com/stan-dev/rstan/wiki/Install-Rtools-for-Windows).
3. samr and additional R packages (https://github.com/MikeJSeo/SAM).
4. Spreadsheet software of your choice (Microsoft Excel, Libreoffice Calc ...).
5. A file with your microarray data is required, a spreadsheet in xlsx format where:
   - (a) The first line of the file contains column names: Gene names, GENE_ IDs and microarrays identified with an integer number (replicates from each line should be tagged with the same number).
   - (b) Columns 1 and 2 should contain gene names and/or GENE_IDs.
   - (c) Remaining columns should content the expression data for each microarray in log2 format obtained from RMA normalization previously (Subheading 3.1) (Fig. 4a).

2.4 Functional Enrichment Analysis with Ontologizer

1. Oracle Java (https://www.java.com/en/download/).
2. Ontologizer (http://ontologizer.de/).
3. Graphviz (http://www.graphviz.org/).
4. Ontologizer needs in each project several input files:
   - (a) An Ontology file, which contains the GO terms, their plain definitions and their mutual relations. It can be downloaded from Gene Ontology website (http://geneontology.org/page/download-ontology). Download the most recent go.obo file at http://purl.obolibrary.org/obo/go.obo.
   - (b) Gene annotations file which should contain lines with the gene identifier (locus, gene names...) and their GO codes. *Arabidopsis thaliana* annotation file could be downloaded at http://geneontology.org/gene-associations/gene_association.tair.gz). For other plants you should try to locate it on their genome homepage or build it. More details of the format of this file are described on http://ontologizer.de/input/.
   - (c) The population set: a file that contains the set of identifiers for all the genes on the microarray. Affymetrix Arabidopsis ATH1 list of probes and locus could be downloaded from ftp://ftp.arabidopsis.org/home/tair/Microarrays/Affymetrix/. Look for the most recent "affy_ATH1_array_elements" txt file. This file could be opened with your spreadsheet software, select the column "locus" and save

it to a text file to obtain a list of TAIR_ID locus represented on the microarray. Since probes on the microarray could correspond to more than one locus, this file will need some editing to obtain a list of unique locus (*see* **Note 1**).

(d) The study sets: additional files that contain the sets of genes that are going to be tested for GO enrichment analysis (upregulated genes, or differentially expressed genes, for example). To obtain the TAIR_ID locus matching to GENE_ID probes, gene conversion tools from Subheading 3.4.3 or 3.7.1 could be used.

2.5 Functional Enrichment Analysis with GENECODIS

1. Reference list: a file that contains the set of identifiers for all the genes on the microarray. Several gene IDs are supported and could be checked in http://genecodis.cnb.csic.es/allowed/. For Arabidopsis ATH1 genome microarray could be the same list with TAIR_ID locus generated on Subheading 3.4.1.
2. The list of genes to study: a text file with the genes that are going to be tested for enrichment analysis with the same identifier format used in the reference list.

2.6 Functional Enrichment Analysis with DAVID

1. Background list: It could be loaded from preloaded microarray array list present on "Background" tab (as, for example, Arabidopsis ATH1-121501 Genome Array) or use a custom file with GENE_IDs.
2. The list of genes to study: a text file with the list the probe GENE_IDs that are going to be tested for enrichment analysis.

2.7 Generation of Gene Expression Data Files

1. Strawberry Perl for windows (http://strawberryperl.com/).
2. Spreadsheet software of your choice (Microsoft Excel, Libreoffice Calc ...).
3. Two text files:
   (a) File1.txt: A file with the gene probes of interest.
   (b) File2.txt: The file generated with RMA express normalization (Subheading 3.1) that contains the expression data for all the gene probes.

2.8 HeatMapper

1. Text file with GENE_IDs, Gene Names and expression data log2 ratios. It could be used the file generated previously on Subheading 3.5.1, and the gene name column could be added with Perl scripts (*see* Subheading 3.5, **step 2**).

2.9 The Bioanalytic Resource for Plant Biology

1. Text files with gene sets in probe GENE_IDs format or locus TAIR_IDs.

2.10 Mapman

1. Oracle Java (https://www.java.com/en/download/).
2. Mapman software (http://mapman.gabipd.org/web/guest/download).
3. Text file with GENE_IDs or TAIR_ID and expression data log2 ratios, usually the mean expression from your line replicates. First line should contain row descriptions (the first column with the gene identifiers should be named "Probesets").

3 Methods

3.1 Microarray Data Normalization

Raw data from microarray hybridization should be submitted to a first step of transformation where hybridization intensities are adjusted and balanced in an appropriated way to make possible a fair and meaningful comparison between different data samples. This process can be performed by statistical methods where the systematic effects and variations that could occur in your microarray platform are taken into account to scale the obtained data into a common specified range.

Our usual workflow is the use as hybridization platform The GeneChip® Arabidopsis ATH1 Genome Array from Affymetrix, and the normalization of raw data with Robust Multichip Average (RMA) software, RMAexpress [2, 3].

1. Start RMAExpress (Fig. 2).
2. Load microarray data by choosing "File and "Read unprocessed files".
3. Load CDF file and CEL files.
4. Once loaded, select "File", "compute RMA Measure".
5. Leave default Options (Background analysis YES, Normalization Quantile and Median Polish as Normalization Method).

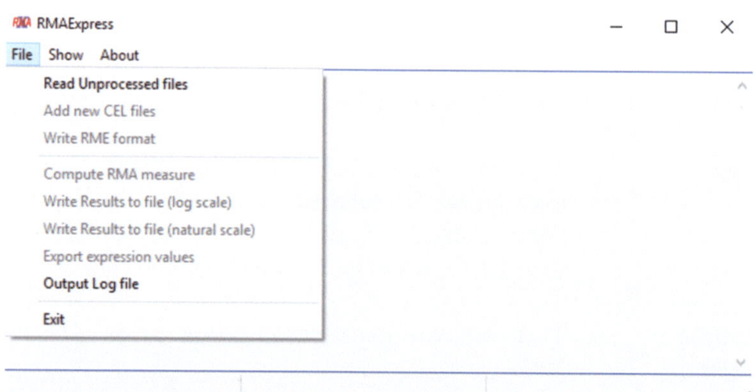

Fig. 2 RMAExpress screenshot

6. Once computed, save results to a file, log scale. A single text file will be generated where each probe is identified by a singular GENE_ID (like 244901_at) and the expression data normalized for each microarray hybridizations will appear in consecutive columns.

3.2 PCA Analysis

Once raw data obtained from microarrays hybridization are normalized, we need to validate the replicates of the samples that we are comparing. One useful statistical technique to compare microarrays expression data is Principal Component Analysis (PCA), that allow us to represent each microarray in two- or three-dimensional graphs of principal components (PC) and check if replicates from each sample have similar PC coordinates compared to the rest of replicates from other samples [4]. Routinely, we do this PCA analysis by using MEV tool from TIGR (http://mev.tm4.org) [5].

1. Enter to MEV page: http://mev.tm4.org.
2. Select "Get started".
3. Upload normalized expression data file obtained on RMA normalization.
4. Once uploaded, a link with the name of your file will appear. Click on it.
5. Select dataset/Analyses/PCA. A PCA graph will appear where each dot represents each microarray. Hover with mouse cursor to the dot to identify microarray name (Fig. 3). Good sample replicates should cluster together. Outlier replicas should be discarded to avoid tampering the posterior calculations of average expression data from each sample. PCA graph could be saved as an image.

Alternatively, PCA analysis could also be performed uploading the same file from RMA normalization into Clustvis platform (http://biit.cs.ut.ee/clustvis/) [6].

3.3 Selection of Genes with Significant Expression

When the validity of the replicates is checked, the next step is to compare expression profiles and select genes with differential expression between your samples. A variety of methods and criteria to make this selection has appeared on the literature, usually a combination of the election of a cutoff value of the expression ratio between samples (like $\geq\pm2$-fold change) combined with a statistical parameter (usually a *t*-test between replicates and samples), that consider how significant is the fold change observed.

We recommend the use of the Significance Analysis of Microarrays (SAM) [7], where this selection relays on the calculation of a "false discovery rate" (FDR), defined as "the probability that a given gene identified as differentially expressed is a false positive" [8].

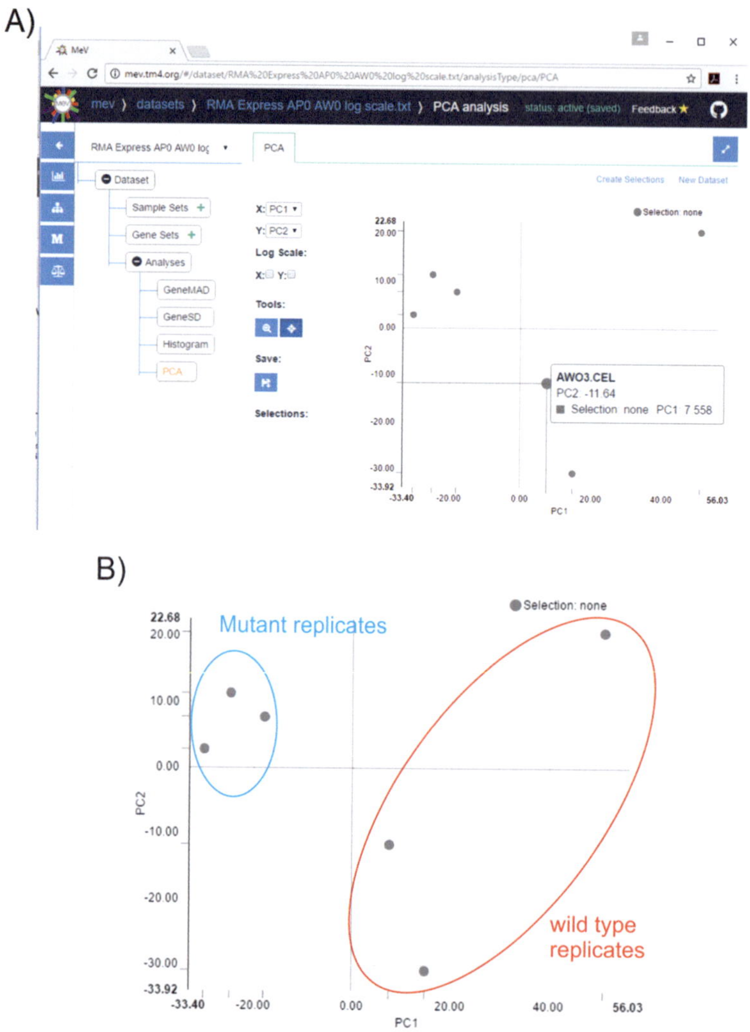

Fig. 3 MEV PCA analysis. (**a**) Example of PCA Output with MEV. Each dot corresponds to a microarray, which could be identified by mouse cursor. (**b**) PCA graph could be saved and edited with your software image of choice to represent how your replicates cluster together

1. Start an R Session.
2. With SAM installed, it could be run with typing this command sequence on your R console:

```
:
library(shiny)
    library(shinyFiles)
    runGitHub("SAM", "MikeJSeo")
```

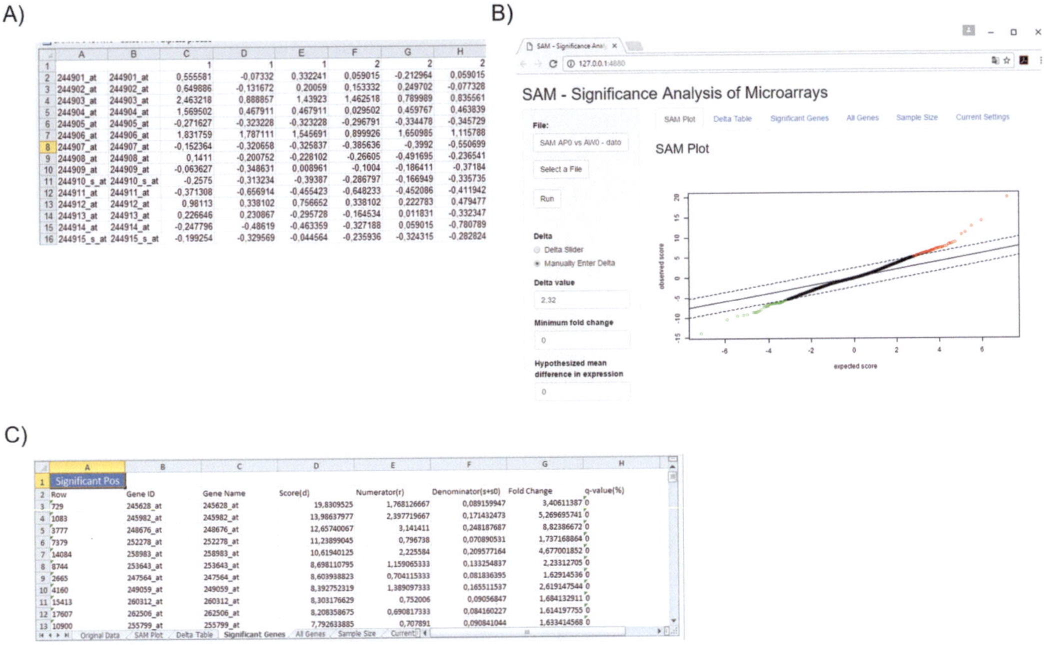

Fig. 4 SAM analysis example. (**a**) Example of input format file for SAM. The two first columns contain microarray probe GENE_IDs. The remaining columns contain log2 data from RMA normalization. Sample replicates are tagged with the same integer number (in this example 1 (wild type) and 2 (mutant)). (**b**) SAM plot. SAM select those genes outside the interval defined by Delta-value (*broken-lines*) and mark them as upregulated (*red*) and downregulated (*green*). (**c**) Results from SAM could be saved to a spreadsheet xlsx file

3. A new browser window will open with the web SAM interface.
4. Load your xlsx file.
5. Select options:
   (a) Data type: Array.
   (b) Response type: Usually we want to compare two types of samples so our method of choice is "Two-class unpaired". For other comparison methods please check SAM help.
   (c) Test Statistic: T-statistic.
   (d) Median center the arrays: no.
   (e) Are Data in log scale: yes.
   (f) Leave other options with default values.
6. Click "Run". After a period of computing, a SAM plot will appear (Fig. 4b) were the genes are represented as upregulated (red) and downregulated (green). The criterion for selection is their position outside the range defined by Delta value parameter. The higher the delta range, a more stringent selection is performed. You could check this with the Delta value slider.
7. You need to adjust the delta value to a desired value, to obtain a selection of genes with significant expression changes between

our plant lines with a low number of false positives. For this, a good starting point is the "Delta table" tab.

8. Select "Delta Table" Table A table will appear, look for the delta value (first row) with a median number of false positives (second row) close to zero.
9. Input this Delta value by selecting "Manually Enter delta" option, and click "run" again.
10. "Significant genes" tab should contain the selection of genes, grouped as upregulated (positive genes) and downregulated (negative genes) with their fold change and *q*-value on the rows located to the right.
11. Once you have obtained your desired selection, enter your desired results name and destination folder. A new xlsx file that could be opened with your spreadsheet software of choice, and where you will see the same information generated in the SAM web interface, organized in tabs in a similar order. "Significant genes" tab will contain the selection of upregulated and downregulated genes (Fig. 4c). The sets of genes obtained could be used in posterior analyses.

3.4 Functional Enrichment Analysis

After determination of genes with significant expression changes, comes the most difficult and tedious step of transcriptome analyses: to give to these expression changes a biological sense.

To answer this question is necessary to look for processes in which the products of the set of genes selected could be involved. Statistical methods are used to identify over-represented types of genes or processes. There are a lot of methods to perform this type of analysis. We recommend the use of more than one and a posterior comparison of the enriched terms obtained.

Ontologizer [9] and Genecodis [10] are our preferred starting points, as well as The Database for Annotation, Visualization and Integrated Discovery (DAVID) [11]. Aside from the biological information that could be obtained about whose processes are affected between your samples, those kind of analyses can be also a good criteria to decompose your initial set of genes with significant expression changes into smaller gene sets whose expression could be visualized by heat map diagrams (Subheading 3.6) or analyzed with other tools that could help to unravel new hints about which processes are changed between samples (Subheading 3.7).

3.4.1 Ontologizer

1. Start Ontologizer. It could be done directly in the Ontologizer website by Java webstart (http://ontologizer.de/webstart/).
2. Select "New Project", input project name an click "Next".
3. Input Ontology OBO file and Annotation file, click "Next".
4. Input the population set, click "Next".

5. Input study sets, (click next to input more than one). When all sets are introduced, click "Finish".
6. On the left panel, your project and all the input sets appear. Additional sets could be added with "new study" button, or eliminated with "remove button".
7. Select the method by which the annotated genes or gene products in the study set are analyzed for GO term overrepresentation with respect to the population set. Our recommended standard option is "Term For Term".
8. Select the statistic method for multiple testing correction (MTC). Our recommended starting test is "Westfall-Young Single-Step".
9. Click "Ontologize" to perform enrichment analysis.
10. A new window will appear with the results for each set in separated Tables GO terms enriched will appear selected and colored. The statistical threshold for p-value could be changed. After that, results for each set could be exported to several formats (ASCII text, HTML, LaTex document). Also, annotated gene products with each GO term are listed in the "Browser" window (Fig. 5).

3.4.2 Genecodis

Genecodis is a web-based service (http://genecodis.cnb.csic.es/) that offers additional enrichment analysis apart from GO Annotations, like KEGG pathways. However, it only supports *Arabidopsis thaliana* annotations.

1. Enter to Genecodis (http://genecodis.cnb.csic.es/).
2. On option 1 "Select Organism", choose "*Arabidopsis thaliana*".
3. On option 2 "Select The Annotations", choose whose annotations terms are going to be used on the enrichment analysis (GO terms, KEGG pathways, InterPro Motifs and Pubmed).
4. On option 3 "Paste of list of genes", upload the list of genes to study.
5. On option 4 "Advanced options" upload the file with the reference list.
6. On option 4 you could also adjust the statistical parameters. Default parameters (Hypergeometric statistical test and FDR p-value correction) are a good point to start.
7. Optionally, you could upload a file with your own annotations or input an e-mail and job title and a link to the results will be mailed when finished.
8. Click "Submit" and wait to the results to be showed.
9. Genecodis displays results in several ways:

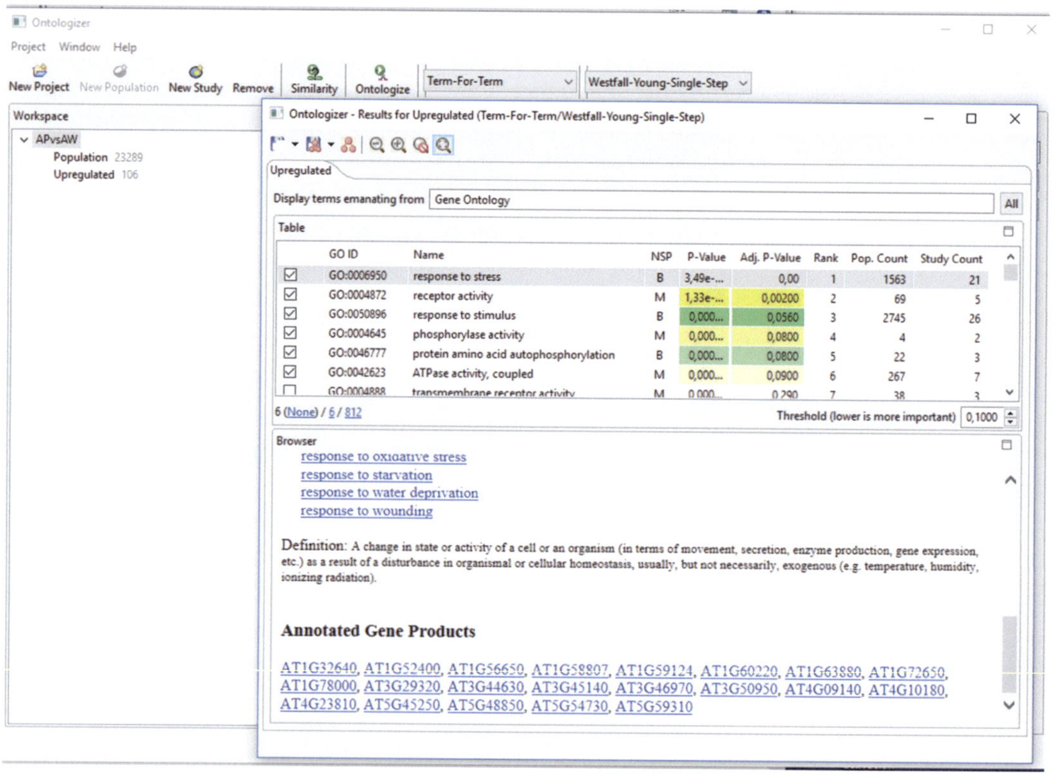

Fig. 5 Ontologizer gene enrichment analysis. Ontologizer selects enriched GO terms with an adjusted *p*-value over a specified threshold. Details from each GO term and the genes annotated with this term in the group of study could be checked in the browser window

(a) A tag cloud where the most enriched terms are displayed by letter size (Fig. 6a). The same information is available as pie chart and bar graphs if you select "other formats", where you also could filter by a minimum number of genes or a minimum *p*-value (Fig. 6b).

(b) An interactive table where the sets of genes tagged with each enriched annotation and their descriptions could be accessed (Fig. 6c).

Results are shown for two types of enriched analysis:

(a) Modular enrichment analysis, where all annotations considered are combined.

(b) Singular enrichment analysis, for each annotation type considered.

3.4.3 DAVID

DAVID (https://david.ncifcrf.gov) is another web-based platform where gene-annotation enrichment analysis could be performed by integrative annotation techniques. The current version of the tool is able to perform enrichment analysis of 40 annotation categories, including GO terms, protein–protein interactions, protein

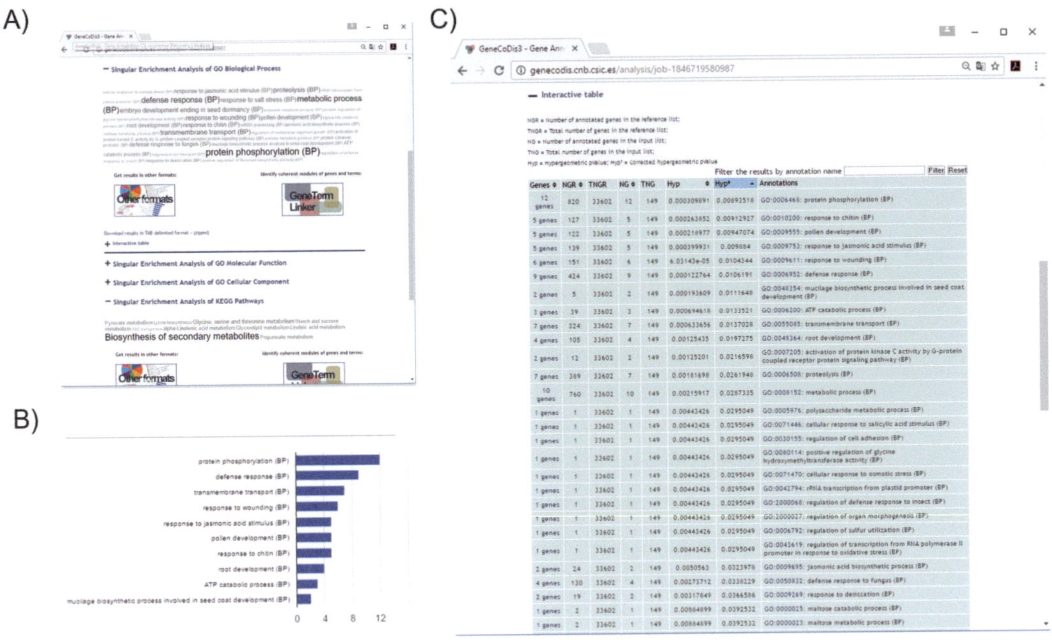

Fig. 6 Genecodis gene enrichment analysis. Genecodis make enrichment analysis of several annotation types as GO terms or KEGG pathways. Results could be showed as intuitive cloud maps (**a**) or bar graphs (**b**) among other options. Also, enriched annotation terms of interest could be seen as a more extensive Table (**c**)

functional domains, bio-pathways, sequence general features, homologies, gene functional summaries, gene tissue expressions, and literatures. Also, DAVID offers additional possibilities, like gene functional classification.

1. Enter to DAVID home page (https://david.ncifcrf.gov).
2. Click on "Start Analysis".
3. On "Upload "tab, load your data:
   (a) Paste your gene list of interest (or upload a file) in "Step 1" section.
   (b) Select gene identifier in "Step 2" (Affymetrix_3PRIME_IVT_ID).
   (c) Classify your list as "gene list" or "Background" in "Step 3".
   (d) Submit list.
   (e) Once loaded, your list will appear in "List" tab.
   (f) Select Background list from preloaded microarray array list present on "Background" tab or upload a custom file with gene ids and classify it as "Background" on "Upload" tab.
4. Once loaded, DAVID tool will point out to select the desired analysis to apply to the gene list:

(a) "Functional Annotation Tool", opens a new window where you have to select which annotation categories you want to use in the enrichment analysis. The results are showed in three possible reports:

- Functional Annotation Clustering:

  Annotation terms present on the gene list are grouped in clusters of elated annotation terms shared by similar gene members. The clustering algorithm assign Enrichment score on each cluster, based on EASE score, a modified Fisher exact *P*-Value. The higher this factor is, the more enriched and relevant is the cluster (Fig. 7a).
- Functional Annotation Chart:

  The annotation terms present in the list of genes are showed as a list ordered by their EASE score and only annotation terms below a threshold are showed.
- Functional Annotation Table: gene and annotation lists are listed with no statistics applied.

(b) "Gene Functional Classification tool":

Genes are clustered in groups based on functional similarity ordered with an enrichment score, the higher the better. Classification stringency could be modulated to generate more of fewer clusters of genes (Fig. 7b).

5. Tables generated on those analyses could be saved as text files.
6. Also, "Gene ID conversion tool" is a useful feature of this platform. Select conversion to TAIR_ID and a table will be generated with probe GENE_ID, TAIR_ID locus, and the description for each gene, that could be useful in posterior steps.

3.5 Generation of Gene Expression Data Files

As stated previously, SAM and posterior enrichment analysis of differential expressed genes are good sources to obtain set of genes of interest. To help to interpret the information obtained it could be necessary to picture their expression levels.

Expression data diagrams are usually obtained from a file with our selected genes and their expression data in each sample, which could be generated from the global expression file obtained by RMA express (Subheading 3.1), a text file with the gene IDs of the genes of interest and the use of scripts of Perl programming language.

1. Open a Perl command line and navigate to the folder where those files are located in your computer drive.

A)

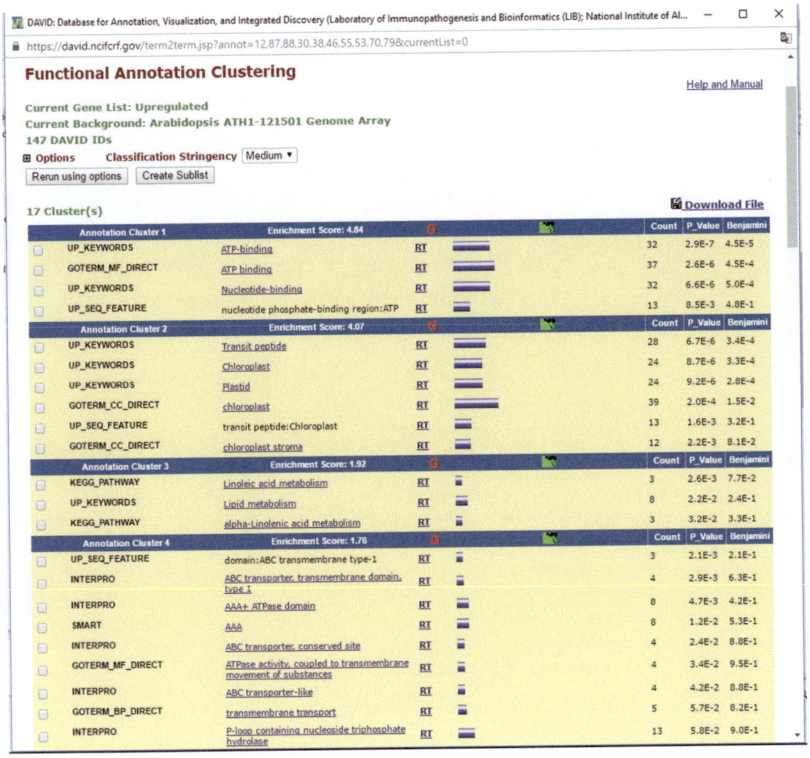

B)

Fig. 7 DAVID enrichment analysis. DAVID web portal is able to generate gene clusters that combine several annotation source by several methods, as Functional Annotation clustering (**a**) and Gene Functional classification Clustering (**b**)

```
:
cd c:\path\to\directory
```

2. This Perl script merge the lines of two files (File1.txt and File 2. txt) that share the same value in one column, whose position is defined with variables $col1 and $col2.

```
:
perl -e "$col1=0;$col2=0; ($f1,$f2)=@ARGV; open(F1,$f1);
while (<F1>) {s/\r?\n//; @F=split /\t/, $_; $line1{$F
[$col1]} .= qq~$_\n~}; warn qq~\nJoining $f1 column $col1
with $f2 column $col2\n$f1: $. lines\n~; open(F2,$f2); while
(<F2>) {s/\r?\n//; @F=split /\t/, $_; $x = $line1{$F
[$col2]}; if ($x) {$x =~ s/\n/\t$_\n/g; print $x; $merged+
+}} warn qq~$f2: $. lines\nMerged file: $merged lines\n~;"
File1.txt File2.txt > Output1.txt
```

In Perl language, the first column is numbered with 0. In our case, GENE_IDs are located in the first column in both archives and the numerical value of 0 should be given for both variable columns $col1 and $col2

3. Copy and paste this script into the Perl command window.
4. An "Output1.txt" file will be generated, where the first two columns will be the probe GENE_IDs and the other columns will contain the expression data.
5. The duplicated column of GENE_IDs could be deleted with another Perl script:

```
:
perl -e "@del_col=(0); while(<>) {s/\r?\n//; @F=split /\t/,
$_; foreach $col (sort {$b<=> $a} @del_col) {splice @F, $col,
1}; print join( qq~\t~, @F), qq~\n~;} warn qq~\nDeleted col-
umns ~, join( qq~, ~, @del_col), qq~ for $. lines\n\n~" Out-
put1.txt > Output2.txt
```

this script opens file Output1.txt and write a new file Output2.txt without the column defined by the variable @del_col. In our case @del_col could be 0 (first column) or 1 (second column).This step cloud also be omitted and the duplicated column deleted on **steps 6** and 7.

6. Copy and paste this script into the Perl command window. The "Output2.txt" file generated could be opened with your spreadsheet program of your choice (Fig. 8).
7. With spreadsheet software, expression data columns could be combined to obtain fold-change expression values (mutant or transgenic line versus reference line, for example). A quick way to do this transformation could be the generation of a reference column with the mean of the expression values found on the reference replicas, and use this value as reference. Remember that expression data is in logarithmic format and by subtraction of this reference valor, a log2 of fold change against reference value will be obtained. After that, select and export GENE_ID and log2 of fold change data in a new tab-delimited text file (as Output3.txt) (Fig. 9).
8. (Optional) Script from **step 3** could be also be used in to add additional columns with TAIR_IDs, and additional descriptions and obtain an archive that could be used to generate an informative table.

   For this, it will be necessary an additional file that shares the same common column of GENE_IDs, plus additional columns with other information. For example, a GENE_ID, TAIR_IDs and gene descriptions of a desired set of GENE_IDs that could be obtained with the "Gene ID conversion tool" of DAVID (Subheading 3.4.3), the "_at to AGI converter" tool from BAR (Subheading 3.7.1) or from the "affy_ATH1_array_elements" file (Subheading 2.4). Again, you should look for the common column present in your files (usually GENE_ID) and modify the number of @col1 and @col2 and name of files accordingly.

3.6 Expression Data Visualization with Heat Map Diagrams

Usually, a global view of gene expression levels of our desired gene set could be obtained by clustered heat map diagrams, where genes are grouped following an expression criteria and their expression levels are represented by a color code. Alternative methods could be used to observe expression patterns of gene sets (*see* Subheading 3.7). Those representations could be applied to the complete set of differentially expressed genes or smaller set of genes (up or down-regulated, tagged with an enriched GO term or KEGG pathway...).

Again, there is variety of software alternatives that could be found on the net that could be used to cluster expression data and represent it with heat map views. We recommend HeatMapper platform (http://www.heatmapper.ca/) [12] as a good place to start.

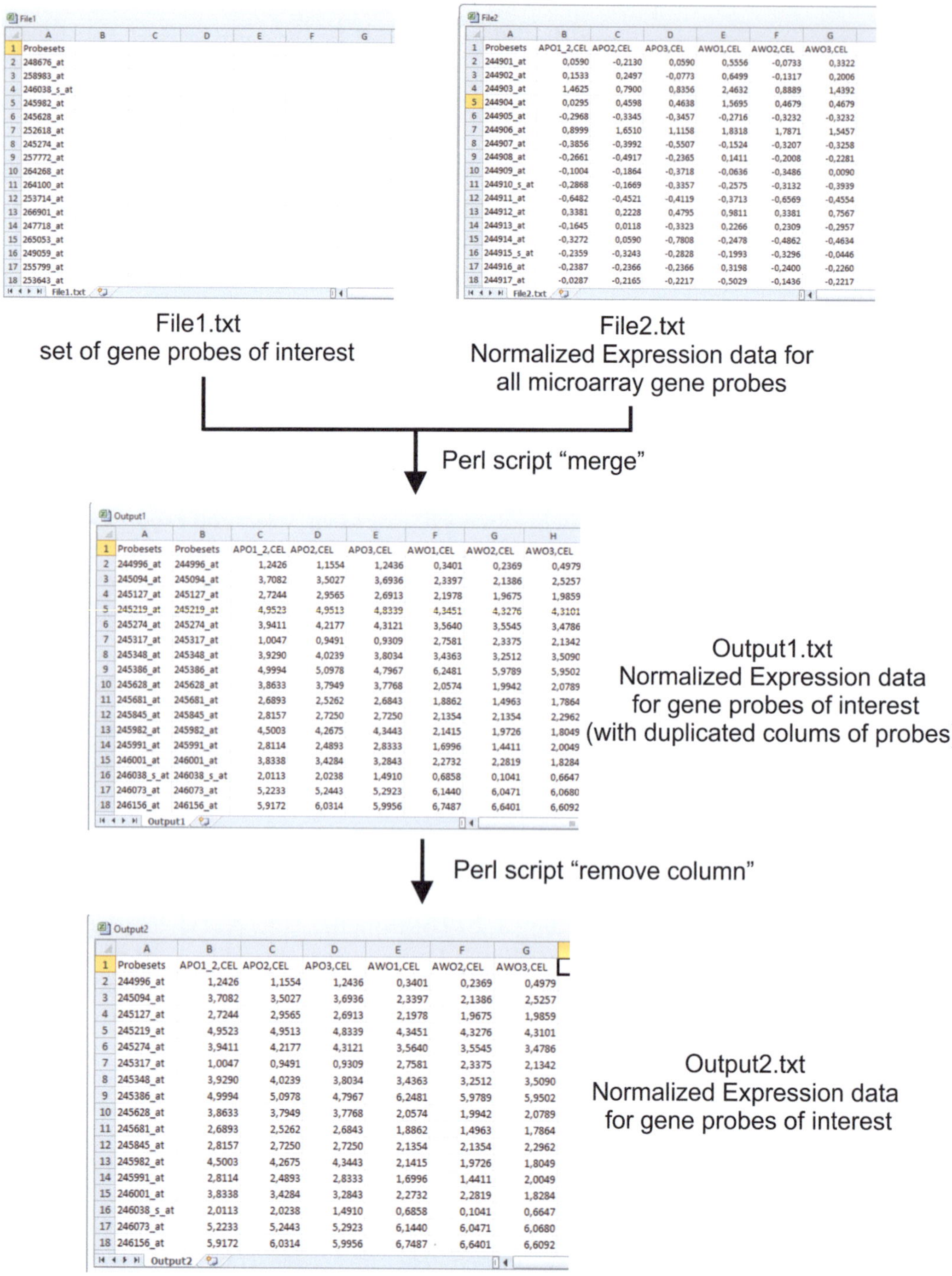

Fig. 8 Diagram of expression data file generation from the set of genes of interest with Perl scripts

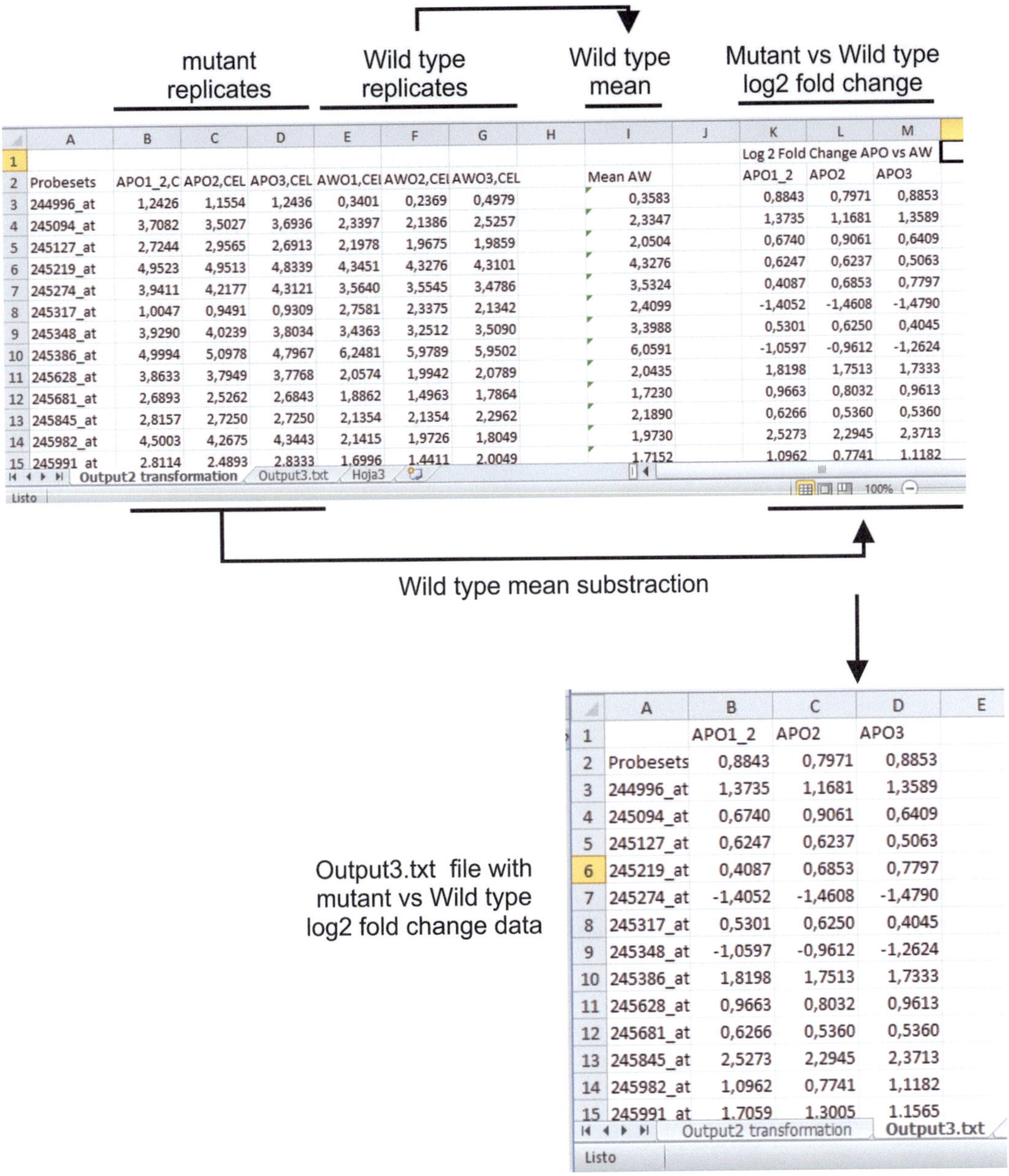

	A	B	C	D	E	F	G	H	I	J	K	L	M
1											Log 2 Fold Change APO vs AW		
2	Probesets	APO1_2,C	APO2,CEL	APO3,CEL	AWO1,CEI	AWO2,CEI	AWO3,CEL		Mean AW		APO1_2	APO2	APO3
3	244996_at	1,2426	1,1554	1,2436	0,3401	0,2369	0,4979		0,3583		0,8843	0,7971	0,8853
4	245094_at	3,7082	3,5027	3,6936	2,3397	2,1386	2,5257		2,3347		1,3735	1,1681	1,3589
5	245127_at	2,7244	2,9565	2,6913	2,1978	1,9675	1,9859		2,0504		0,6740	0,9061	0,6409
6	245219_at	4,9523	4,9513	4,8339	4,3451	4,3276	4,3101		4,3276		0,6247	0,6237	0,5063
7	245274_at	3,9411	4,2177	4,3121	3,5640	3,5545	3,4786		3,5324		0,4087	0,6853	0,7797
8	245317_at	1,0047	0,9491	0,9309	2,7581	2,3375	2,1342		2,4099		-1,4052	-1,4608	-1,4790
9	245348_at	3,9290	4,0239	3,8034	3,4363	3,2512	3,5090		3,3988		0,5301	0,6250	0,4045
10	245386_at	4,9994	5,0978	4,7967	6,2481	5,9789	5,9502		6,0591		-1,0597	-0,9612	-1,2624
11	245628_at	3,8633	3,7949	3,7768	2,0574	1,9942	2,0789		2,0435		1,8198	1,7513	1,7333
12	245681_at	2,6893	2,5262	2,6843	1,8862	1,4963	1,7864		1,7230		0,9663	0,8032	0,9613
13	245845_at	2,8157	2,7250	2,7250	2,1354	2,1354	2,2962		2,1890		0,6266	0,5360	0,5360
14	245982_at	4,5003	4,2675	4,3443	2,1415	1,9726	1,8049		1,9730		2,5273	2,2945	2,3713
15	245991_at	2.8114	2.4893	2.8333	1.6996	1.4411	2.0049		1.7152		1.0962	0.7741	1.1182

Output2 transformation / Output3.txt / Hoja3

	A	B	C	D	E
1		APO1_2	APO2	APO3	
2	Probesets	0,8843	0,7971	0,8853	
3	244996_at	1,3735	1,1681	1,3589	
4	245094_at	0,6740	0,9061	0,6409	
5	245127_at	0,6247	0,6237	0,5063	
6	245219_at	0,4087	0,6853	0,7797	
7	245274_at	-1,4052	-1,4608	-1,4790	
8	245317_at	0,5301	0,6250	0,4045	
9	245348_at	-1,0597	-0,9612	-1,2624	
10	245386_at	1,8198	1,7513	1,7333	
11	245628_at	0,9663	0,8032	0,9613	
12	245681_at	0,6266	0,5360	0,5360	
13	245845_at	2,5273	2,2945	2,3713	
14	245982_at	1,0962	0,7741	1,1182	
15	245991_at	1.7059	1.3005	1.1565	

Output2 transformation / Output3.txt

Fig. 9 Example of generation of a log2 fold change data file from normalized expression data of two lines (mutant and wild type)

1. Enter Heatmapper expression tool page: http://www.heatmapper.ca/expression/.
2. Upload text file.
3. Select Scale type: none.
4. We recommend custom color scheme: low color green, middle color black, high color red.

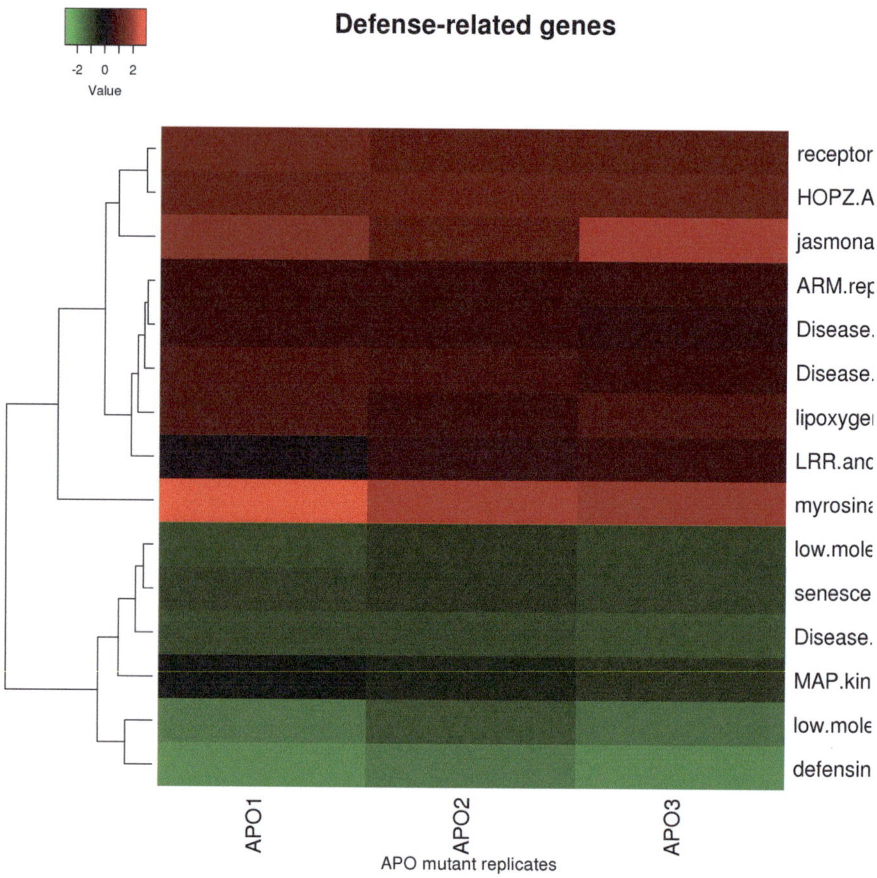

Fig. 10 Example of Heatmapper graph output. Log2 of fold change expression values of a selected set of genes (with significant expression in mutant lines and tagged with a defense-related GO term) have been clustered and represented into a heat map with the scale shown in the figure

5. Experiment with clustering and distance measurement methods to see which fits better with your data.
6. Image of the heat map obtained could be exported to several formats (Fig. 10).

3.7 Other Gene Expression Visualization Tools

The information obtained through the application of the techniques mentioned above to the set of genes with significant expression changes between our lines of study can be complemented with other tools available on the network, of which we highlight the Bioanalytic resource for Plant Biology (BAR) [13] and MAPMAN [14, 15].

3.7.1 The Bioanalytic Resource for Plant Biology (http:/bar.utoronto.ca/)

This webpage includes several tools that can be applied to the gene set of your choice, which could be uploaded or copied into the tool box present in each web page:

(a) "at to AGI converter" and "Duplicate remover" are probably the tools that you need first, to obtain a lists of TAIR_IDs from sets of GENE_IDs without duplicates, a necessary data for other tools, like Ontologizer (Subheading 3.4.1).

(b) "Venn Selector" and "Venn Superselector" are useful tools to look for shared genes between gene sets.

(c) "e-Northerns w. Expression Browser" [13] represents heat maps from public microarray data from several sources, including AtGenExpress Consortium. This analysis could be applied to a set of genes with up to 125 members, an includes a selection tool to narrow the transcriptomes to be observed, either by research area of interest (abiotic stress, development, pathogens) and/or the plant tissue or stage of plant growth. We find this tool particularly useful to check whether small gene sets selected by functional enrichment have expression patterns consistent their assigned functional category (Fig. 11a). Gene list should be uploaded in TAIR_ID format.

(d) "Classification super viewer" could be used to generate alternative graphs of GO terms present in the gene set of interest (Fig. 11b). Gene list should be uploaded in TAIR_ID format.

3.7.2 Mapman (http:/mapman.gabipd.org/web/guest/mapman)

Mapman is an alternative visualization tool of the transcriptomes generated in our hybridization experiments using Affymetrix microarrays. Gene expression levels are represented by heat map boxes placed into diagrams (pathways) where gene products are classified by several criteria. It could be used to obtain a bird-like graphic view of the transcriptome changes between your plant lines or narrow your view to a particular metabolism pathway.

We like to apply this tool to obtain an alternative view of the expression of complete set of genes with differential expression between lines and thus to be able to verify if the information obtained from the functional enrichment analysis is corroborated also by Mapman diagrams.

1. Start Mapman.
2. Choose "new pathway analysis".
3. Select desired pathway from left box (Choose Pathway).
4. Depending on your gene identifier format select Ath_AFFY_ATH1 (GENE_IDs) or Ath_AGI_TAIR9 (TAIR_IDs) mapping from central box (Choose mapping).
5. Make a new folder and upload your expression data file on "Choose experimental data" box. Select it.
6. Click "Show pathway" icon.

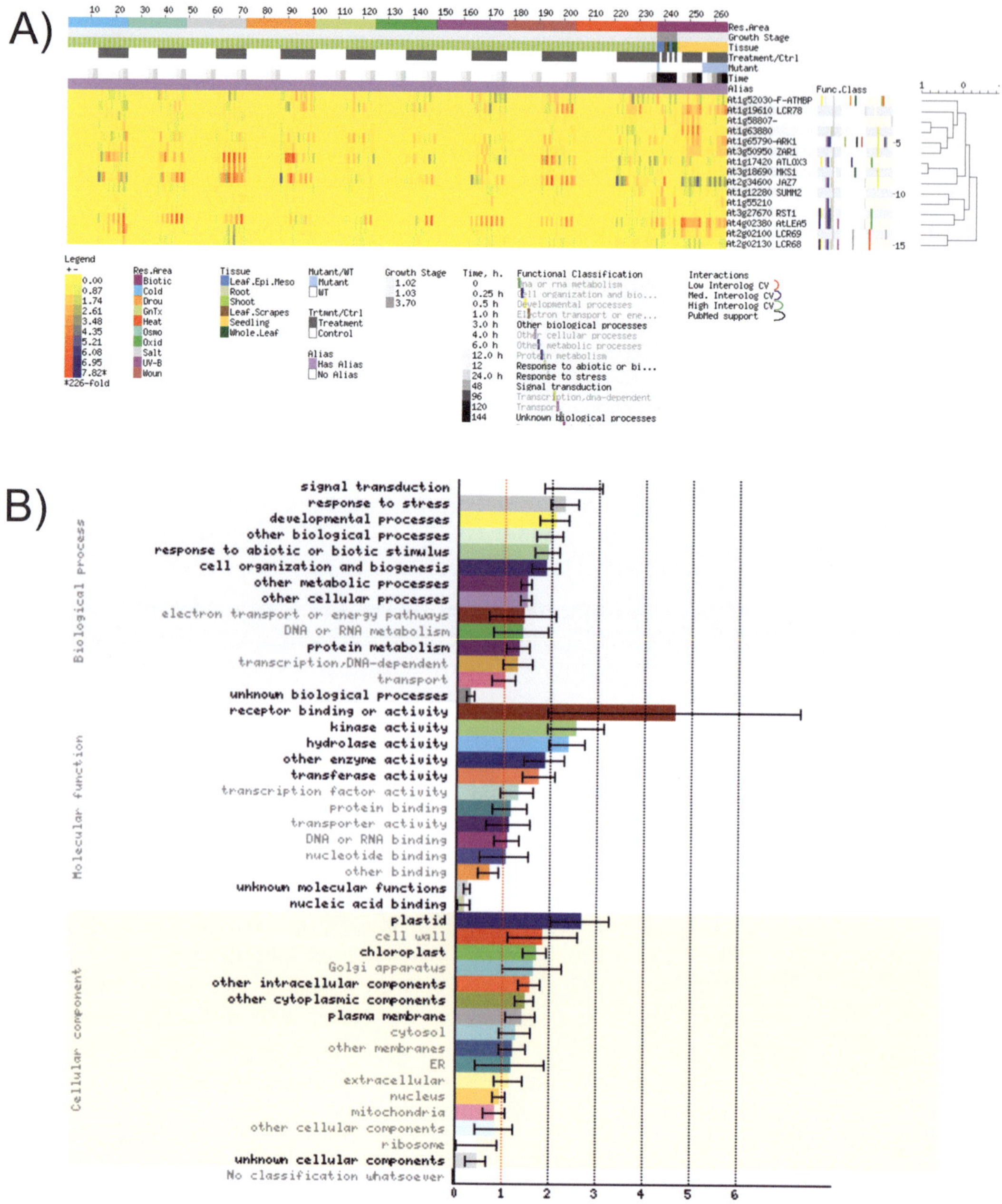

Fig. 11 Examples of BAR tools application. (**a**) E-northern output from microarray stress series of a selected set of genes (tagged with a defense-related GO term). (**b**) Classification super viewer result of a selected group of upregulated genes

7. A diagram with the desired pathway will be shown. If the pathway includes genes present in your gene set, will be shown with colored boxes where their expression levels will be represented in heat map format (Fig. 12).

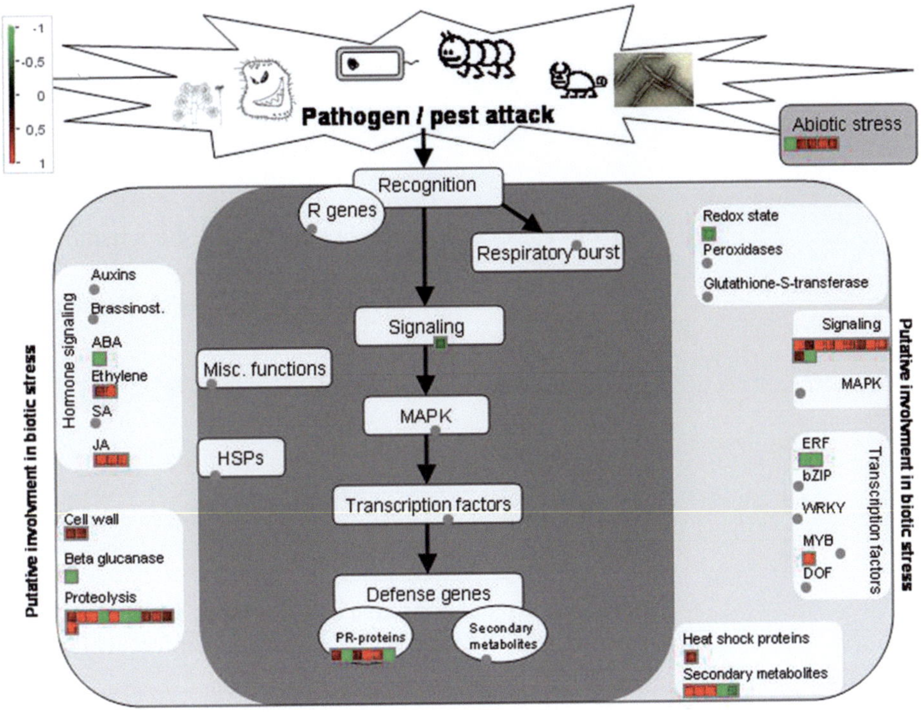

Fig. 12 MAPMAN diagram example. Stress-related genes from a set of genes with significant expression changes between mutant and wild type are represented in heat map boxes. Expression values are in log2 fold change (mutant vs wild type). Scale is shown in figure

4 Note

1. Raw "Locus" column of Affymetrix microarray elements from TAIR could have duplicates and some locus in the same line like:

```
:
AT2G45350
    AT2G45340
    AT2G45340;AT5G23600
    AT2G45350
    AT2G45310
    AT4G38280;AT2G45250
```

";" characters could be changed to non-printable new line characters "\r\n" with this Perl script, that should be copied and pasted into a Perl command line session (*see* Subheading 3.5).

```
:
perl -e "$sep= qq~;~; while(<>) {s/\Q$sep\E/\r\n/g; print
$_;} warn qq~Changed $sep to new line on $. lines\n~" locus-
file1.txt > locusfile2.txt
```

locusfile1.txt and locusfile2.txt could be changed to yout names of choice.

Also, duplicate locus lines in locusfile2.txt can be removed with this perl script:

```
:
perl -e "$unique=0; while(<>) {if (!($save{$_}++)) {print
$_; $unique++}} warn qq~\nChose $unique unique lines out of
$. total lines.\n\n~" locusfile2.txt > locusfile3.txt
```

locusfile3.txt will contain a list of unique TAIR_ID locus of the genes present on affymetrix ATH1 genome microarray:

```
:
AT2G45350
    AT2G45340
    AT5G23600
    AT2G45310
    AT4G38280
    AT2G45250
```

References

1. The Arabidopsis Genome Initiative (2000) Analysis of the genome sequence of the flowering plant Arabidopsis thaliana. Nature 408 (6814):796–815. doi:http://www.nature.com/nature/journal/v408/n6814/suppinfo/408796a0_S1.html
2. Bolstad BM, Irizarry RA, Åstrand M, Speed TP (2003) A comparison of normalization methods for high density oligonucleotide array data based on variance and bias. Bioinformatics 19 (2):185–193. doi:10.1093/bioinformatics/19.2.185
3. Irizarry RA, Bolstad BM, Collin F, Cope LM, Hobbs B, Speed TP (2003) Summaries of Affymetrix GeneChip probe level data. Nucleic Acids Res 31(4):e15. doi:10.1093/nar/gng015
4. Ringner M (2008) What is principal component analysis? Nat Biotechnol 26(3):303–304
5. Saeed AI, Bhagabati NK, Braisted JC, Liang W, Sharov V, Howe EA, Li J, Thiagarajan M, White JA, Quackenbush J (2006) TM4 microarray software suite. In: Methods in enzymology, vol 411. Academic Press, New York, pp 134–193. doi:10.1016/S0076-6879(06)11009-5
6. Metsalu T, Vilo J (2015) ClustVis: a web tool for visualizing clustering of multivariate data using principal component analysis and heatmap. Nucleic Acids Res 43(W1):W566–W570. doi:10.1093/nar/gkv468
7. Tusher VG, Tibshirani R, Chu G (2001) Significance analysis of microarrays applied to the ionizing radiation response. Proc Natl Acad Sci U S A 98(9):5116–5121. doi:10.1073/pnas.091062498
8. Slonim DK (2002) From patterns to pathways: gene expression data analysis comes of age. Nat

Genet 32(Suppl):502–508. doi:10.1038/ng1033

9. Bauer S, Grossmann S, Vingron M, Robinson PN (2008) Ontologizer 2.0—a multifunctional tool for GO term enrichment analysis and data exploration. Bioinformatics 24 (14):1650–1651. doi:10.1093/bioinformatics/btn250
10. Tabas-Madrid D, Nogales-Cadenas R, Pascual-Montano A (2012) GeneCodis3: a non-redundant and modular enrichment analysis tool for functional genomics. Nucleic Acids Res 40(Web Server issue):W478–W483. doi:10.1093/nar/gks402
11. Huang DW, Sherman BT, Lempicki RA (2008) Systematic and integrative analysis of large gene lists using DAVID bioinformatics resources. Nat Protoc 4(1):44–57. doi:http://www.nature.com/nprot/journal/v4/n1/suppinfo/nprot.2008.211_S1.html
12. Babicki S, Arndt D, Marcu A, Liang Y, Grant JR, Maciejewski A, Wishart DS (2016) Heatmapper: web-enabled heat mapping for all. Nucleic Acids Res 44(Web Server issue): W147–W153. doi:10.1093/nar/gkw419
13. Toufighi K, Brady SM, Austin R, Ly E, Provart NJ (2005) The botany array resource: e-Northerns, expression angling, and promoter analyses. Plant J 43(1):153–163
14. Thimm O, Blasing O, Gibon Y, Nagel A, Meyer S, Kruger P, Selbig J, Muller L, Rhee S, Stitt M (2004) MAPMAN: a user-driven tool to display genomics data sets onto diagrams of metabolic pathways and other biological processes. Plant J 37:914–939
15. Usadel B, Nagel A, Thimm O, Redestig H, Blaesing O, Palacios-Rojas N, Selbig J, Hannemann J, Conceicao Piques M, Steinhauser D, Scheible W, Gibon Y, Morcuende R, Weicht D, Meyer S, Stitt M (2005) Extension of the visualization tool MapMan to allow statistical analysis of arrays, display of corresponding genes, and comparison with known responses. Plant Physiol 138:1195–1204

Chapter 31

High-Throughput Phenotyping in Plant Stress Response: Methods and Potential Applications to Polyamine Field

D. Marko, N. Briglia, S. Summerer, A. Petrozza, F. Cellini, and R. Iannacone

Abstract

High-throughput phenotyping has opened whole new perspectives for crop improvement and better understanding of quantitative traits in plants. Generation of loss-of-function and gain-of-function plant mutants requires processing and imaging a large number of plants in order to determine unknown gene functions and phenotypic changes generated by genetic modifications or selection of new traits. The use of phenomics for the evaluation of transgenic lines contributed significantly to the identification of plants more tolerant to biotic/abiotic stresses and furthermore, helped in the identification of unknown gene functions. In this chapter we describe the High-throughput phenotyping (HTP) platform working in our facility, drawing the general protocol and showing some examples of data obtainable from the platform. Tomato transgenic plants over-expressing the *arginine decarboxylase 2* gene, which is involved in the polyamine biosynthetic pathway, were analyzed through our HTP facility for their tolerance to abiotic stress and significant differences in water content and ability to recover after drought stress where highlighted. This demonstrates the applicability of this methodology to the plant polyamine field.

Key words High-throughput phenotyping, Abiotic stress, Polyamines, 3D Scanalyzer Platform, Tomato

1 Introduction

Contemporary agriculture must face several new challenges imposed by environmental factors such as climate change and the increase in world population. These factors require significant advances in comprehension of the plant's response to stresses. Plant resilience to abiotic and biotic stresses will be key to fight the predicted dramatic changes in climate conditions and to guarantee the environmental sustainability. However, science must proceed at an unprecedented pace in order to acquire knowledge and provide new technologies in a timely manner. The study of phenotype by high-throughput technologies, termed plant phenomics, is an emerging field of investigation that promises to speed up

Rubén Alcázar and Antonio F. Tiburcio (eds.), *Polyamines: Methods and Protocols*, Methods in Molecular Biology, vol. 1694, DOI 10.1007/978-1-4939-7398-9_31, © Springer Science+Business Media LLC 2018

breakthrough discoveries, by closing the gap between genomics data and phenotype.

The main aim of modern phenotyping is represented by a comprehensive assessment of complex plant traits such as growth, development, tolerance, resistance to biotic and abiotic stresses, architecture, physiology, yield, and the basic measurement of individual quantitative parameters which form the basis of more complex traits. The phenomic approach has turned out to play a crucial role in better understanding and improving crop yields with the aim of building a brighter future for improving farming productivity in the whole world [1].

This new technology branch evaluates traits related to growth and crop yield with certain accuracy and precision at different scales of organization, from organs to canopies by image analysis [2]. Plant imaging is more than just 'taking pictures' as it gives information on plant architecture, photosynthetic efficiency, water content, health status and growth dynamics [3]. To achieve this goal, phenomic science enrolls experts from several disciplines such as biology, computer science, mathematics and engineering. In the last 10 years high-throughput phenotyping platforms (HTPP) have been set up in several countries (for the latest news visit the International Plant Phenotype Network IPPN available online at: www.plantphenomics.com). These platforms have the potential to investigate, with high precision, the phenotypic aspect of plants as they use automated system and precise imaging technologies (hardware and software) to analyze the complex traits of cultivated crops and model plants related to growth, yield, biotic and abiotic stresses tolerance in a controlled environment (greenhouse) or in open field.

Classic phenotypic data analysis has been largely used to unravel the interactions between genotype and environment and has been adopted for many species (plants, *Drosophila*, mice, rats, dogs) [4]. However, one of the main goal of plant phenomics is to relieve the phenotyping bottleneck by making connections between genomics, plant functions and agricultural traits [5]. Sequencing of whole plant genomes has allowed the generation of experimental models where a searchable phenotypic database, linking gene sequence to plant structure, development, composition or performance, can permit the correlation of phenotypic traits with different genomic alterations in various cultivars [6]. Rising demand for phenotyping larger quantities of plants at the same time, led to the development of high-throughput phenotyping (HTP), which enabled visualization of large amounts of plants in a non-invasive approach and to record all the data needed for phenotypical evaluation [7]. Different HTP platforms are being used, ranging from field platforms, greenhouse platforms and growth chamber platforms. In all types of platform settings, the main objective is to automate the acquisition of plant images and other physical data with which we can quantify genotype through a phenotype [8]. As an example, geometric values of plant organs were developed for leaf shape analysis and comprise terms such as projected area, center of mass, eccentricity or symmetry, and statistical moments, which result in accurate mathematical representation of plant shape without the need to store large images for comparison [8].

Despite the rapid advancement in unravelling the genome sequence and structure, the identification of molecular traits related to tolerance/resistance to abiotic stresses is far behind. This can be due to the number of genes involved and to the interactions genotype-environment [9]. Polyamines represent a class of molecules that are involved in several physiologic events in plants such as embryogenesis, organogenesis, leaf senescence, fruit development and stress responses ([10], other chapters in this book). The modification of polyamine biosynthesis obtained by overexpressing the gene *Arginine decarboxylase 2* (*ADC2*) led to the generation of *Arabidopsis thaliana* plants that have shown phenotypic differences in comparison with wild-type plants and were more tolerant to drought [11]. However, the assessment of plant phenotype during molecular breeding processes for abiotic stress tolerance is extremely important since the non-destructive analysis of the phenotype can greatly help in the selection of genotypes that better cope with the imposed stress. Consequently, the availability of HTP can speed-up the selection of stress-tolerant plants. Indeed, the ability to adapt to abiotic stress conditions was studied through the analysis of photosynthetic performances [12] or RGB and kinetic ChlF traits [13] in *Arabidopsis thaliana* using high-throughput phenotyping.

However, the characterization of the stress-response in crops via HTP is still limited. In tomato, for example, HTP was used to study the effect of genetic modifications (overexpression of transcription factors) on the plant phenotype [14] or the effect of biostimulants on the plant growth rate [15, 16]. In our facility the phenomic platform (LemnaTec Scanalyzer, www.lemnatec.com) was used to investigate the effect of several new traits introduced in tomato plants such as Transcription factors (e.g. ATHB7—[14], Heat shock factors (HsfA1), Hsp70s, and to evaluate the effect of biostimulants and fertilizers [15–17]. We obtained tomato plants overexpressing constitutively arginine decarboxylase (*ADC2* isolated from *Arabidopsis thaliana*). These plants were analyzed through the high-throughput (HTP) scanalyzer facility for their tolerance to abiotic stress and significant differences in water content and ability to recover after drought stress where highlighted [18]. In this chapter, we describe the HTP platform working in our facility, drawing the general protocol and showing some examples of data obtainable from the platform.

2 Materials

High-throughput Plant Phenomics Platform is based on a LemnaTec Scanalzer3D system (LemnaTec GmbH), equipped with:

- An automated belt conveyor system capable of accommodating 494 plants in pots, with a tracking system based on bar code and RFID for safe identification of single plants.
- Four sequential camera stations able to take images on plants NIR, UV, RGB, illumination and a special NIR for root activity.
- An automated watering system with a weighting station.
- An informatics infrastructure for data acquisition, management and elaboration.

The platform allows the quantitative, non-destructive analysis of different crops or model plants under high-throughput conditions (Fig. 1). Each plant is imaged sequentially in multiple Scanalyzer 3D camera units, employing different wavelengths that reach far beyond human vision. The result is an unprecedented number of reproducible and significant data points on any aspect of plant development.

2.1 Lights in Phenotyping Platforms

A basic phenotyping platform consists of has three imaging chambers each equipped with one or more optical sensors capable of acquiring images illuminated with:

Fig. 1 The Scanalyzer 3D (LemnaTec GmbH) phenomics imaging system at the Metapontum Agrobios Research Center. The foreground of the image contains the plant storage system with conveyor belts that carry the plants to the imaging chamber. In the background and in the direction of travel for the plants (i.e. from *right* to *left*) are; in black, the soil NIR, next the fluorescence, then the visible light, and finally the plant NIR imaging chambers

- VIS (Visible light).
- UV (Ultra Violet Light)
- NIR (Near Infrared light).

The Scanalyzer 3d System (LemnaTech GmbH) is equipped with two optical sensors per chamber which enable the investigation of plant parameters from two points of view: side view images are made from 0° to 360° (commonly for image analysis it is used 0° and 90° two side views images) and a top view image.

2.1.1 Visible Light Imaging

The VIS or RGB camera is the basic equipment in all Scanalyzer systems. Imaging obtained from visible light from 380 to 780 nm (Fig. 2) acquire information about plant colours distribution, plant morphometric parameters and plant growth morphology dynamics.

The LemnaTec image software (LemnsGrid) can easily extract many parameters per image using the VIS sensor technique, examples of which are:

- *Plant height* is the maximum distance from bottom to top of plant. It is measured in centimetres, resulting in the mean of two orthogonal side view images.

Fig. 2 Top (**a**) and side view (**b**) colour images of tomato plants from the visible light imaging chamber. The plants are illuminated by standard fluorescence light tubes (35 W/865 cool daylight) and recorded with a Scout camera (Basler)

- *Shoot area top view (SATV)* is quantified by the number of all pixels in top view vision that have been identified as a part of plant.
- *Convex Hull (HUL)* is the smallest geometric figure that contains all pixels that have been identified as part of the plant. It can be visualized as the shape formed by a rubber band stretched around the plant.
- *Compactness* is calculated based on SATV/area of a circle with the same perimeters per plant and determines if the leaves are nearer around the centroid or farther away from it, e.g. by having longer stipes.
- *Rotational Symmetry* of the complete plant is calculated based on the size-independent second moment principal axis ratio. This parameter gives us information on the whole shape of the plant and is size-independent. Rotational symmetry describes how far the leaves as a whole show a symmetry of the plant.
- *Eccentricity* as symmetry and regular growth are quite common in nature, the extent of non-centric shape can describe important effects like deviations from the regular growth scheme.
- *Surface Coverage* compares the plant area with the area of a circle covering the plant. This parameter is intended to provide a value that shows how densely the plant covers the soil within its growth area (similar to LAI = leaf area index).
- *Digital biovolume* extracted plant pixel area from all side and top view images are used to calculate a volume which is called 'digital biovolume' and corresponds to a pixel volume.

  Digital biovolume = Σpixel sideview 0° + Σpixel sideview 90° + log10 (Σpixel topview/3) [19]
  This digital trait is used as a proxy for fresh weight.
- *Colour distribution*; colour classification and quantitative assessment of shifts among colour classes is the best option to represent colour changes, especially when the colour of the plant is not completely homogeneous. The colours distribution analysis can help to detect the leaf senescence by separating the yellow and green areas of the leaf.
  Chlorosis Index can represent different health status of the plant:
  - dark green as very healthy tissue.
  - green as normal healthy tissue.
  - yellow as chlorotic tissue.
  - brown as necrotic tissue.

Fig. 3 Top (**a**) and side view (**b**) colour images of tomato plants from the fluorescence imaging chamber. The plants are illuminated with standard white fluorescence tubes (cool daylight), the light is filtered with a blue filter (wavelength below 450 nm) and the fluorescent light passes through another filter (wavelength greater than 520 nm) in front of the camera lens and recorded with a Scout camera (Basler)

2.1.2 Fluo Camera

Imaging of chlorophyll fluorescence (Fig. 3) is used as a diagnostic tool in many areas of plant physiology [20]. When the plant tissues are exposed to actinic blue light (wavelength shorter than 500 nm), this energy can follow three different fates:

(a) Phytochemistry, the productive phase of photosynthesis.
(b) Dissipated as heat.
(c) Fluorescence, re-emitted as luminous energy at lower energy [21].

Artificially irradiating the plant tissue with actinic light can be evaluated by studying the amount of re-emitted light by fluorescence related to the health status of plant photosynthetic apparatus.

The emissions of fluorescence signals in red and far-red light spectrum (with the maximum at 690 and 740 nm) are chlorophyll α molecules in the antenna and reaction center of the PSII from chloroplasts in the mesophyll cells [22]. Changes in the fluorescence emission may be used as indicators of stress and the health state of photosynthetic apparatus [23].

2.1.3 NIR Camera

In the visible spectrum reflectance of plant tissues is low. This is due to the presence of photosynthetic pigments that reflect electromagnetic radiation only in the green region (healthy tissue). Moving on

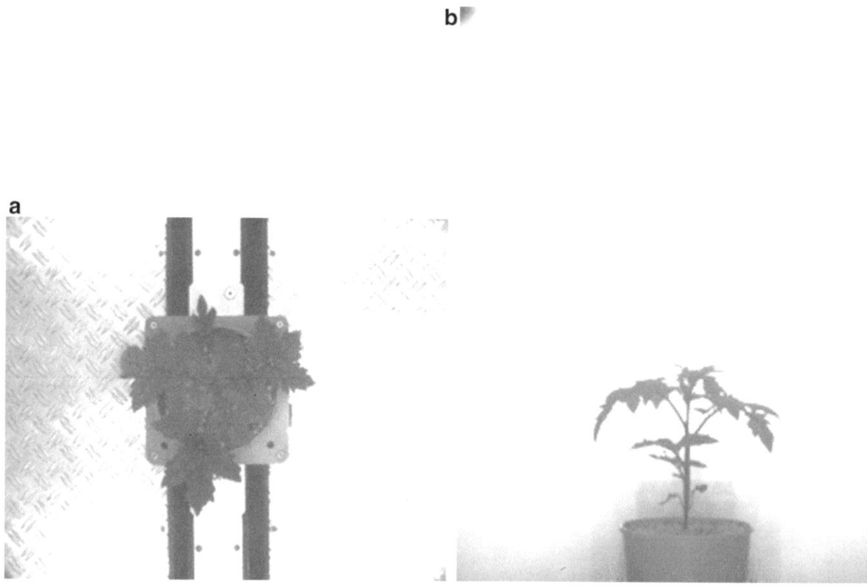

Fig. 4 Top (**a**) and side view (**b**) grey scale images of tomato plants from the near infra-red imaging chamber. The plants are illuminated by light from an array of 35 W halogenGU10 light bulbs and recorded with a NIR-3000PGE camera (Vosskueler)

to near infrared spectrum there is a considerable increase of reflected electromagnetic radiation called 'red edge' (between 800 and 1300 nm). With an increase of the wavelength (beyond 1300 nm), the reflectance decreases due to absorption of water present in the plants tissues with characteristic water absorption bands at 1450, 1930, and 2500 nm. This is the basis for using NIR imaging to study water stress of plants [24] (Fig. 4).

2.2 Scanning Whole Plant with NIR Optical Sensor for the Evaluation of the Plant Water Status

This parameter can be used as a suitable tool to help breeders selecting plant genotypes more tolerant to drought stress and supports scientists to early detection of drought stress symptoms in crops.

Every image is composed by pixels that have a digital reflectance value ranging from 0 to 255 (8 bits per pixel). To create the plant water status index each image is divided into ten homogeneous categories (by LemnaTec software *LemnaGrid*); an increase of the digital number value represents lower water content inside the plant tissue. Therefore, to determine the plant water status only the categories with lowest digital number value (shades of darker grey or highest water content) are considered.

A comparison of different imaging technique is illustrated in Table 1.

The phenotypic platform described in this paper has the following specifications:

Table 1
Comparison of different imaging techniques in plant phenotype application (modified from [3])

Imaging techniques	Sensor	Resolution	Raw data	Phenotype parameters	Examples of species	Imaging environment
Visible light imaging	Cameras sensitive in the visible spectral range	Whole organs or organ parts, time series	Gray or colour value images (RGB channels)	Projected area, growth dynamics, shoot biomass, yield traits, panicle traits, root architecture, imbibition and germination rates, early embryonic axis growth, height, size morphology, flowering time	*Arabidopsis thaliana*; barley; rice; bean	Controlled environment; field
Fluorescence imaging	Fluorescence cameras and setups	Whole shoot or leaf tissue, time series	Pixel-based map of emitted fluorescence in the red and far-red region	Photosynthetic status (variable fluorescence), leaf health status, shoot architecture	Wheat; *Arabidopsis thaliana*; natural grassland, winter wheat, corn; barley; bean; sugar beet; tomato; chicory plant	Controlled environment; field
Near infrared imaging	Near-infrared cameras	Continuous or discrete spectra for each pixel in the near infrared region	Time series or single time point analysis of shoots and canopies, single point assessment of seeds	Water content composition parameters for seeds, leaf area index	Rice; soybean; corn; barley; wheat	Controlled environment

- PostgreSQL database server
- Imaging chambers types: four imaging rooms

  First—root imaging chamber which detects soil water content (1450–1550 nm).

  Second—fluorescent light imaging—contains excitation lights reading at wavelengths shorter than 450 nm and emission readings camera that reads wavelengths longer than 520 nm. Images allow us to detect GFP or chlorophyll (excitation maximum at 450 nm, a minimum at 400 nm and a maximum wavelength of 500 nm, measurement (emission) 520 nm).

  Third—near infrared light imaging which helps us determine the water content of plant tissues (900–1550 nm).

  Fourth—colour imaging chamber where high resolution images are taken from top and side views for the phenotyping purposes (visible light 380–780 nm).
- A conveyor system.

3 Methods

Protocol For High-Throughput Phenotyping of Tomato Plants Subjected to Abiotic Stresses. The following protocol was used to study the phenotype of tomato plants overexpressing the ADC2 gene.

3.1 Organization of the Experiment

Plants are divided into groups, depending on type of experiment, in order to be labelled. Special barcode stickers are put on pots, which are later read by LemnaTec sensors.

Usually plants are divided into different genetic lines and control/stressed groups. Each group should contain no less than five replicas per line due to data statistical analysis requirements.

3.2 Experiment

Planning of the experiment is crucial and must be carefully studied. The platform allows non-invasive and dynamic imaging of plants during the experiment and readings schedule is very important, depending both on stress application and experiment length. Readings on LemnaTec platform should be always performed at the same time of the day during consecutive experimental days. Any delays in readings or application of stress factors should be avoided in order to obtain consistent data.

Plants should managed and placed on platform conveyor with caution in order to avoid any additional mechanical stress that might influence the readings. After each set of readings a check should be made on all the images have been made and stored.

There are many different applicable abiotic and biotic stresses in this type of experiments. Abiotic stress can be represented by drought, salt, heat. We can also apply biotic stresses—viruses, bacteria, fungi, parasites or insects can be applied as well.

3.3 Protocol for Image Analysis (LemnaTec Scanalyzer 3D)

The image analysis pipeline can be conducted with vendor software, such as LemnaGrid/LemnaMiner from LemnaTec, or with open-source programs and tools, such as ImageJ [25] or Python with the PlanCV/OpenCV [8] imaging resources. Both methods are valid and both capable of producing consistent results. One significant advantage of the vendor software is that has been designed for the platform and is fully integrated with the management software, while open-source software makes it possible to easily publish the complete methods of analysis on a software repository such as GitHub. In this study we describe the protocol used with the LemnaGrid and LemnaMiner software from LemnaTec GmbH.

3.3.1 Part I: Image Segmentation

The first step in image analysis for phenotyping is image segmentation. Image segmentation is the process where the image pixels related to the plant are separated from those of the background. There are many techniques for image separation. The most rapid and direct is to use the colour information of the image to create a greyscale image and subsequently threshold the image to a binary black and white image.

Colour images are generally composed of at least three different channels, most commonly these channels are Red, Green and Blue. Each channel can be represented as a greyscale image. It is also possible to represent colour images in colour spaces different than the classic RGB, some of these include HSV (Hues, Saturation, Value) or $L^* a^* b^*$ (Luminosity, *a*: red-magenta colour axis, *b*: a yellow-blue axis). The grey value from each channel can be mathematically combined, such as finding the average luminosity in the RGB colour space, $(R + G + B)/3$, or be used to increase the contrast between the plant and background pixels. The final greyscale image is then subjected to thresholding, setting a cut off value in which above the pixel becomes white and the subject and below is black, background, creating a binary image that can be used as a mask for the original colour image.

1. Images are loaded from the database (LemnaGridFunction: DB Datareader).
2. To save repetition of programming code, the two side view images which are identically processed are multiplexed (LemnaGrid Function: Batch Mux).
3. Images are converted from RAW images into normal RGB colour images (LemnaGrid Function: Demosaicing).
4. To avoid problems with high pixel colour variability the image is slightly blurred. For the segmentation process the analyses uses the original image (Lemnagrid Function: Gaussian, settings filter: kernel size 3, repeat 1). It is also possible to use the Median filter, settings, kernel size 5. (*Note*: kernel size must be an odd integer.)

5. Two grey images from the $L^*a^*b^*$ '*a*' and '*b*' channels are created. (1) LemnaGrid: LabtoGreyConverter—Green/Magenta, and (2) LemnaGrid: LabtoGreyConverter—Blue/Yellow).
6. Two greyscale images are combined mathematically by applying the following formula on corresponding pixels from the two images.
   LemnaGrid: GreyScale calculator, Formula: $b/(a + 1) * 100$
   Note: values but be integers from 0 to 255, values below 0 are 0 and values greater than 255 are truncated to 255.
7. The resulting greyscale image is subject to thresholding. (LemnaGrid: Threshold, Settings: Global threshold method, cut off 116) note values must be integers with the minimum and maximum possible values of 0 and 255 respectively.
8. Binary image is cleaned up by filling in small areas that have less than 20 pixels.
   (LemnaGrid: Fill areas, Setting: Fill areas up to 20, Fill type: foreground and background, Neighborhood: 4.)
9. Binary image is converted to contours (LemnaGrid: Universal Converter, Setting: Destination type: Image Object List).
10. The preceding steps will often include objects that are not part of the plant/subject, objects such as the plant pot or door frame. To exclude these a region where the plant is expected to be present is defined and any object outside this region can be excluded:
    (a) Create a Region of Interest (ROI) to filter out any object outside of this area (LemnaGrid: RegionofInterest Freeform maker).
    (b) Apply the filter with the ROI.

    (LemnaGrid Function: Image Object Region of Interest Filter, Settings: Method: Image Objects Center of Mass must be inside region of interest.)
11. Combine the remaining but separated objects/contours into a single object. (LemnaGrid: Image Object composition, Setting: mode: partial inside).

    In the case of Fluorescence images **steps 5** and **6** are modified in the following manner; the formula $(2 * a + \text{b})/3$ was applied to the Fluorescence image and the global two value threshold values of (255,147) side view and (170,210) were used.

    In the case of near infra-red (NIR) images, the images are already in greyscale and **steps 5** and **6** are replaced by resizing the image object contours from the Visible image to fit that of the NIR image (LemnaGrid Function: Binary image shift zoom).

3.3.2 Part II: Image Analysis

At this point a 'mask' is created which can be used to eliminate the background image pixels. The masked image can be used to analyze pixel colours and the histogram of pixel colour can serve as proxies for plant health (greater the green colour content, healthier the plant; chlorotic tissue yellow; necrotic tissue brown), water content when examining near infrared images (darker pixels have greater water content) and changes in photosynthesis by looking at fluorescence images. In addition to colour analysis values calculated from the image mask such as pixel height (plant height), compactness and pixel count (area) can be used to calculate proxies for traditional phenotyping such as plant height (height in pixels * mm/pixel), compactness and digital biovolume (a proxy for dry or wet weight).

1. Four Colour Analysis: LemnaGrid: Colour Classification, four colour classes, Dark Green (three anchor points: R95G90B55, R118G152B79,R147G160B101), Green (two anchor points; R160G212B100, R213G245B138), Yellow (two anchor points, R255G255B102, R255G255B185), and Brown (one anchor point: R146G125B95). Anchor points are represented as RGB colour values.

   Hue colour analysis. The pixel values are divided from the hue channel into 128 bin (0–255 is the pixel value range). LemnaGrid: HSI to grey converter, Settings: Hue; LemnaGrid: Histogram cc: settings: Method, Fixed Histogram 128 bins.

2. Due to the presence of filters in front of the camera that filter out wavelengths under 500 nm the bin range is limited to the first 50 values in 50 bins. LemnaGrid Function: Histogram cc: Settings: Range 0–50, levels 50.

   No conversion to hue channel greyscale was necessary for NIR images. LemnaGrid Function: Histogram cc: Settings Range 0–255, bins 128.

3. After the calculation of area (number of pixels) of the foreground (subject) in the image its value can be used to calculate a digital biomass (dry or wet weight) using a general formula [19] or it is also possible to create a linear regression from real dry/wet weight of plants if a series of plants were sacrificed and measured during the experimental phase. In addition to the area, basic information regarding the projected area of the plant from each view is automatically calculated by the LemnaTec software. Some of the values include Convex hull, compactness, caliper length, minimum enclosing circle, and minimum enclosing rectangle. These values are automatically calculated and saved with a standard Data Writer function in the LemnaTec software.

4. To create a vertical profile of plant biomass a rectangle divided into horizontal slices is created, each slice corresponding to approximately 3 cm. For each slice the plant area is calculated

in pixels. (a) LemnaGrid: Area Distribution maker: Rectangle horizontally divided, 20; (b) LemnaGrid Function: Image Object Region of Interest Filter, Method: Cut image object to region of interest).

5. To calculate the height of a plant a horizontal line is drawn along the point where the plant exits the soil. The distance (in pixels) from this line to the most distant point above and below of the image contour/object can be used to calculate the actual height in cm if the conversion factor is known. (LemnaGrid functions: Distance property, Distance to: horizontal line (y: 1360).

   At the end of each of the Image analysis pathway is a LemnaGrid function saving the results to the database. (note that side view images that were combined, termed multiplexed, must first be separated, termed demultiplexed, before being saved).

3.3.3 Part III: Image Analysis Processing

After the image snapshots (a collection of images in visible light, fluorescence and NIR each with a side view at 0°, 90° and top view) are processed by the LemnaGrid deamon software with the previously described pipeline (grid). The raw data can be saved as large csv (comma separated value) file and opened in Excel or other data analysis software such as R, where values such as the digital biovolume can be calculated. Due to the large number of replicas and consequent data statistical analysis of the results is essential.

4 Future Prospects

In this paper we described the general protocol currently used in our facility to evaluate the phenotypic features of plants subjected to abiotic stresses using the Scanalyzer 3D platform (LemnaTec). However, many details in the protocol are adjusted depending on plant species, shape, height and physiological status.

Undoubtedly, phenomics can help in identifying the key factors affecting plant growth and health, and subsequently plant productivity, since this technology allows non-destructive screening of hundreds of plants in very short time. This is useful when key metabolic pathways are altered in terms of genetic engineering or mutagenesis, implying changes in phenotypic traits.

Plant phenotyping through imaging is particularly suited to detect phenotypic changes related to abiotic stress responses, for example in connection with water use and/or photosynthesis efficiency, traits non detectable by naked eyes. Alteration of polyamine metabolic pathway (mainly suppression of genes involved in the polyamine pathway) gave often rise to the alteration in the phenotypic aspect of transgenic plants ([10, 26] other chapters in this book). Therefore, plants modified in the polyamine metabolism can benefit from plant phenomics technologies in both accurately study the phenotypic development and identify unintended effects.

Phenomics can accelerate the verification of the effect of the inserted trait or, in combination with forward genetics studies, can help in the identification of molecular traits involved in the response to environmental stimuli and lead to crop improvement.

References

1. Mishra KB, Mishra A, Klem K, Govindjee (2016) Plant phenotyping: a perspective. Indian J Plant Physiol 21(4):514–527
2. Fiorani F, Schurr U (2013) Future scenarios for plant phenotyping. Annu Rev Plant Biol 64:267–291
3. Li L, Zhang Q, Huang D (2014) A review of imaging techniques for plant phenotyping. Sensors (Basel) 14(11):20078–20111
4. Houle D, Govindaraju DR, Omholt S (2010) Phenomics: the next challenge. Nat Rev Genet 11(12):855–866
5. Montes JM, Melchinger AE, Reif JC (2007) Novel throughput phenotyping platforms in plant genetic studies. Trends Plant Sci 12 (10):433–436
6. Furbank RT, Tester M (2011) Phenomics—technologies to relieve the phenotyping bottleneck. Trends Plant Sci 16(12):635–644
7. Araus JL, Li J, Parry MA, Wang J (2014) Phenotyping and other breeding approaches for a new green revolution. J Integr Plant Biol 56 (5):422–424
8. Fahlgren N, Gehan MA, Baxter I (2015) Lights, camera, action: high-throughput plant phenotyping is ready for a close-up. Curr Opin Plant Biol 24:93–99
9. Halperin O, Gebremedhin A, Wallach R, Moshelion M (2016) High-throughput physiological phenotyping and screening system for the characterization of plant-environment interactions. Plant J 89(4):839–850
10. Tiburcio AF, Altabella T, Bitrian M, Alcazar R (2014) The roles of polyamines during the lifespan of plants: from development to stress. Planta 240(1):1–18
11. Alcazar R, Planas J, Saxena T, Zarza X, Bortolotti C, Cuevas J, Bitrian M, Tiburcio AF, Altabella T (2010) Putrescine accumulation confers drought tolerance in transgenic Arabidopsis plants over-expressing the homologous Arginine decarboxylase 2 gene. Plant Physiol Biochem 48(7):547–552
12. Rungrat T, Awlia M, Brown T, Cheng R, Sirault X, Fajkus J, Trtilek M, Furbank B, Badger M, Tester M, Pogson BJ, Borevitz JO, Wilson P (2016) Using phenomic analysis of photosynthetic function for abiotic stress response gene discovery. Arabidopsis Book 14:e0185
13. Awlia M, Nigro A, Fajkus J, Schmoeckel SM, Negrão S, Santelia D, Trtílek M, Tester M, Julkowska MM, Panzarová K (2016) High-throughput non-destructive phenotyping of traits that contribute to salinity tolerance in Arabidopsis thaliana. Front Plant Sci 7:1414
14. Mishra KB, Iannacone R, Petrozza A, Mishra A, Armentano N, La Vecchia G, Trtilek M, Cellini F, Nedbal L (2012) Engineered drought tolerance in tomato plants is reflected in chlorophyll fluorescence emission. Plant Sci 182:79–86
15. Petrozza A, Summerer S, Di Tommaso G, Di Tommaso D, Piaggesi A (2013) An evaluation of tomato plant root development and morpho-physiological response treated with VIVA® by image analysis. In: ISHS, Leuven, Belgium, pp 155–159
16. Summerer S, Petrozza A, Cellini F (2013) High throughput plant phenotyping: a new and objective method to detect and analyse the biostimulant properties of different products. In: ISHS, Leuven, Belgium, pp 143–148
17. Petrozza A, Summerer S, Di Tommaso G, Di Tommaso D, Piaggesi A (2013) Evaluation of the effect of RADIFARM® treatment on the morpho-physiological characteristics of root systems via image analysis. In: ISHS, Leuven, Belgium, pp 149–153
18. Iannacone R (2010) Translational biology approaches to improve abiotic stress tolerance in tomato. Paper presented at third meeting of FA0605: plant abiotic stress: from signalling to crop improvement, Valencia, Spain, 22nd–24th Apr 2010
19. Eberius M, Lima-Guerra J (2009) High-throughput plant phenotyping – data acquisition, transformation and analysis. Bioinformatics 7:259–278
20. Baker NR (2008) Chlorophyll fluorescence: a probe of photosynthesis in vivo. Annu Rev Plant Biol 59:89–113

21. Oxborough K (2004) Imaging of chlorophyll a fluorescence: theoretical and practical aspects of an emerging technique for the monitoring of photosynthetic performance. J Exp Bot 55 (400):1195–1205
22. Buschmann C, Lichtenthaler HK (1998) Principles and characteristics of multi-colour fluorescence imaging of plants. J Plant Physiol 152 (2–3):297–314
23. Gorbe E, Calatayud A (2012) Applications of chlorophyll fluorescence imaging technique in horticultural research: a review. Sci Hortic 138:24–35
24. Knipling EB (1970) Physical and physiological basis for the reflectance of visible and near-infrared radiation from vegetation. Remote Sens Environ 1(3):155–159
25. Schneider CA, Rasband WS, Eliceiri KW (2012) NIH image to ImageJ: 25 years of image analysis. Nat Methods 9(7):671–675
26. Kumar A, Taylor M, Altabella T, Tiburcio AF (1997) Recent advances in polyamine research. Trends Plant Sci 2(4):124–130

Sitography

http://www.plantphenomics.com

http://www.lemnatec.com

Chapter 32

Abiotic Stress Phenotyping of Polyamine Mutants

Thomas Berberich, G.H.M. Sagor, and Tomonobu Kusano

Abstract

Plant mutants in polyamine pathway genes are ideal for investigating their roles in stress responses. Here we describe easy-to-perform methods for phenotyping Arabidopsis mutants under abiotic stress. These include measurements of root growth, chlorophyll content, water loss, electrolyte leakage, and content of the reactive oxygen species hydrogen peroxide (H_2O_2) and superoxide anion (O_2-). Growth of Arabidopsis seedlings is described that enables transfer to different media for stress treatment without damaging roots.

Key words Abiotic stress treatment, Phenotyping, Root growth, Relative water content, Chlorophyll, ROS levels

1 Introduction

1.1 Abiotic Stress and Phenotyping

Plant phenotyping is the assessment of complex plant traits such as growth, development, tolerance, resistance, architecture, and physiology and the basic measurement of individual parameters that form the basis for more complex traits. Such direct measurement parameters are for example plant biomass, growth, pigment content, and other metabolite concentrations (http://www.plant-phenotyping.org/). Here we focus on easy-to-perform methods used for phenotyping *Arabidopsis thaliana* mutants under abiotic stress.

Abiotic stress, as a natural part of the environment, represents the negative impact of nonliving factors on organisms in their ecosystem beyond the normal range of variation. The abiotic factors comprise temperature (heat, cold, and frost), water availability (drought, flooding), high salinity, high- and low-light condition, starvation (e.g. phosphate), chemical contamination (e.g. heavy metals), and even mechanical stress (e.g. strong wind). Plants have evolved complex mechanisms to respond to the extreme conditions in order to survive and tolerate stress to a certain extent. The response starts from the perception of the stress followed by signal transduction, activation of stress-responsive genes, synthesis

Rubén Alcázar and Antonio F. Tiburcio (eds.), *Polyamines: Methods and Protocols*, Methods in Molecular Biology, vol. 1694, DOI 10.1007/978-1-4939-7398-9_32,

of stress-related proteins and other molecules to adjust the metabolism. Reactive oxygen species (ROS) play an important role in the stress response. In plants they are mainly generated in the redox reactions of photosynthesis and respiration and comprise superoxide radical ($O^{\cdot}_2-$), hydroxyl radical ($^{\cdot}OH$), hydrogen peroxide (H_2O_2), and singlet oxygen ($^1O_2$) which are generated mostly in chloroplasts, mitochondria, peroxisomes, and glyoxysomes, but also in the cytosol and apoplast [1, 2]. Abiotic stresses such as heat, cold, drought, salinity, or heavy metals lead to enhanced generation of ROS in plants due to disruption of cellular homeostasis [3]. They are highly reactive and toxic, causing oxidative damage to macromolecules such as lipids, proteins, and nucleic acids. On the other hand it became obvious that they are important as signaling molecules that participate in many developmental processes such as cell differentiation, seed germination, gravitropism, root hair growth, pollen tube development, programmed cell death, and senescence [4–7]. Analyses of ROS concentration in plant tissues is used to evaluate the effect of stresses.

Many details of the plant's response to abiotic stress have been unraveled, but many aspects remain unexplained including the role of polyamines in the stress responses. A complete understanding of the response and tolerance mechanisms is necessary to cope with the challenges of increasing stressful environments caused by climate change and the demand of feeding a growing world population. In this chapter we describe basic methods for investigating responses to abiotic stress of polyamine mutant plants.

1.2 Arabidopsis Mutants

Arabidopsis thaliana with a short generation time of 8 weeks from seed to seed, self-fertilization, small size, completely sequenced genome, easy transformation by Agrobacteria, and with a large collection of mutants certainly is still the most popular model organism for plant research. The earliest description of an Arabidopsis mutant dates back to 1873 [8]. Later the correct chromosome number of Arabidopsis was described and it was proved to be well suitable for genetic studies [9, 10]. The first induced Arabidopsis mutants were isolated after treating Arabidopsis with X-rays [11]. Nowadays huge collections of Arabidopsis mutant lines are available from stock centers, namely the Arabidopsis Biological Resource Center (ABRC), the Nottingham Arabidopsis Stock Center (NASC), the RIKEN Bioresource Center (BRC) and the SENDAI Arabidopsis Seed Stock Center (SASSC). Furthermore, The Arabidopsis Information Resource (TAIR) serves as a constantly updated source for data and information on Arabidopsis (https://www.arabidopsis.org/portals/mutants/). For further reading on the history of *Arabidopsis thaliana* as a model plant and the use of mutants please *see* [12–15]. Most of the methods described in this chapter on Arabidopsis mutants might be adapted to other plant species (Table 1).

Table 1
Arabidopsis polyamine pathway genes and described mutants

Gene product	Gene symbol	Ec number	Gene acc. no.	Mutants	Reference
Arginine decarboxylase	ADC1 ADC2	4.1.1.19	AT2G16500 AT4G34710	*adc1-1* *adc1-2*	[16]
Agmatine iminohydrolase	AIH	3.5.3.12	AT5G08170		
N-Carbamoylputrescine amidohydrolase	CPA	3.5.1.53	AT2G27450		
N-Carbamoylputrescine amidohydrolase	CPA	3.5.1.53	AT2G27450		
S-Adenosylmethionine decarboxylase	SAMDC1 SAMDC2 SAMDC3 SAMDC4	4.1.1.50	AT3G02470 AT5G15950 AT3G25570 AT5G18930	*samdc1-1* *bud2-1, bud2-2*	Salk_020185 Salk_007279, [17]
Spermidine synthase	SPDS1 SPDS2	2.5.1.16	AT1G23820 AT1G70310	*spds1-1* *spds2-1, spds2-2*	[18] [18], SALK_13982
Spermine synthase	SPMS	2.5.1.22	AT5G53120	*Spms-1*	[19]
(T-)spermine synthase	ACL5	2.5.1.22	AT5G19530	*acl5-1, acl5-2*	[20]
Cu-containing amine oxidase	CuAOα1 CuAOα2 CuAOα3 CuAOβ CuAOγ1 CuAOγ2 CuAOδ CuAOξ	1.4.3.6	At1g31670 At1g31690 At1g31710 At4g14940 At1g62810 At3g43670 At4g12290 At2g42490	 *cuao1-1, cuao1-2*	[21] SALK_019030.48.25, RATM12_5261-1_H [22]
Polyamine oxidase	PAO1 PAO2 PAO3 PAO4 PAO5	1.5.3.11	AT5G13700 AT2G43020 AT3G59050 AT1G65840 AT4G29720	*Atpao1-2* *Atpao2-4* *Atpao3-1* *Atpao4-1* *Atpao5-2*	SAIL_822_A11 SALK_046281 GK209F07 SALK_133599 SALK_053110

2 Materials

2.1 Plant Growth

1. Square petri dishes (120 × 120 × 15 mm).
2. Murashige and Skoog basal salt media (MS): To prepare a ½× concentrated MS solution dissolve 2.2 g MS basal salt mixture in 1 L deionized water, autoclave at 121 °C for 20 min and store at room temperature.

3. ½ MS agar plates: Weigh 2.2 g MS basal salt mixture, 10 g sucrose, 15 g bacto agar into a 1 L bottle and make up to 1 L with deionized water. Mix and autoclave at 121 °C for 20 min. After cooling to 60 °C sterile solutions of other chemicals (e.g. polyamines) can be added before pouring plates under a clean bench.
4. 70% ethanol, sterile deionized water.
5. Seed sterilization solution: 1% sodium hypochloride, 0.1% Tween-20.
6. Square plastic pots (7 cm × 7 cm × 7 cm), potting compost and vermiculite (*see* **Note 1**).

2.2 Determination of Chlorophyll Content

1. Mortar and pestle or tissue homogenizer for grinding.
2. Extraction buffer: 80% acetone in 5 mM Na-phosphate buffer, pH 7.4. First prepare 50 ml 0.2 M sodium phosphate, dibasic, dihydrate ($Na_2HPO_4{\cdot}2H_2O$, MW = 178.05) by dissolving 1.78 g in 50 ml deionized water (solution A). Also prepare 50 ml of 0.2 M sodium phosphate, monobasic, monohydrate ($NaH_2PO_4{\cdot}H_2O$, MW = 138.01) by dissolving 1.38 g in deionized water (solution B). To create 100 ml of 0.1 M Na-phosphate buffer, pH 7.4, mix 40.5 ml solution A with 9.5 ml solution B and dilute with deionized water to 100 ml. For 100 ml extraction buffer use 80 ml acetone (100%) add 5 ml 0.1 M Na-phosphate buffer, pH 7.4 and add deionized water to 100 ml.
3. Spectrophotometer for visible range (400–700 nm) and glass cuvettes.

2.3 Water Loss Assay and Relative Water Content

1. Analytical balance.
2. Blotting paper.
3. Oven set to 80 °C.
4. Fridge at 4 °C.
5. Cork borer when larger leaves are used.

2.4 Electrolyte Leakage

1. Boiling water bath or autoclave.
2. Electric conductivity meter.

2.5 Analyses of ROS Content

1. Mortar and pestle, ice bucket, heating block at 85 °C.
2. UV/vis spectrophotometer, UV-cuvettes (chamber volume 1 ml) and vis-micro cuvettes (chamber volume 70–850 μl).
3. Extraction buffer: 100 mM K- phosphate buffer (pH 7.0) containing 10% (w/v) water-insoluble polyvinylpyrrolidone

(PVP, Polyclar®). First prepare 50 ml 1 M potassium phosphate, dibasic (K_2HPO_4, MW = 174.18) by dissolving 8.71 g in 50 ml deionized water (solution A). Also prepare 50 ml of 1 M potassium phosphate, monobasic ($KH_2PO_4 \cdot H_2O$, MW = 136.09) by dissolving 6.81 g in deionized water (solution B). To create 100 ml extraction buffer weigh 10 g PVP into a graduated 100 ml cylinder add 6.15 ml solution A and 3.85 ml solution B and fill up to 100 ml with deionized water. Also prepare 0.1 M K-phosphate buffer, pH 7.0, by mixing 6.15 ml solution A with 3.85 ml solution B and dilute with deionized water to 100 ml.

4. Horseradish peroxidase (~150 U/mg) 1 mg/ml in 100 mM K-phosphate buffer (pH 7.0).
5. Hydrogen peroxide (H_2O_2) solution (30%).
6. 50 mM (w/v) ABTS [2,2′-azino-bis(3-ethylbenzthiazoline-6-sulfonic acid)] diammonium salt (MW = 548.68): Weigh 0.274 g of ABTS diammonium salt into a 10 ml graduated cylinder and add deionized water to 10 ml.
7. Incubation solution: 10 mM K-phosphate buffer (pH 7.8), 0.05% (w/v) nitro blue tetrazolium (NBT), 10 mM NaN_3 (sodium azide). Create a 100 mM K-phosphate buffer (pH 7.8) stock by mixing 9.1 ml solution A and 920 μl solution B (*see* **step 3**) and dilute the combined solutions to 100 ml with deionized water. Prepare 100 ml of a 100 mM NaN_3 (MW = 65.01) stock solution by dissolving 0.65 g in a final volume of 100 ml deionized water. For 100 ml incubation solution weigh 0.05 g NBT powder into a 100 ml graduated cylinder, add 10 ml 100 mM K-phosphate buffer (pH 7.8), 10 ml 100 mM NaN_3, and make up to 100 ml with deionized water.

3 Methods

3.1 Growth of Arabidopsis Plants

1. Surface sterilization of seeds: Place dry seeds into a 1.5 ml microcentrifuge tube and add 800 μl of 70% ethanol to wet the surface by vortexing for 15 s.
2. Collect seeds at the bottom of the tube by centrifugation (10,000 × *g*, RT, 1 min), discard the ethanol by pipetting off.
3. Add 800 μl of seed sterilization solution to the seeds, incubate for 15 min with brief vortexing every 5 min.
4. Collect seeds at the bottom of the tube as in **step 2**, discard the seed sterilization solution by pipetting off.
5. Wash the seeds at least four times with 1 ml sterile deionized water each by vortexing followed by centrifugation as in **step 2**.

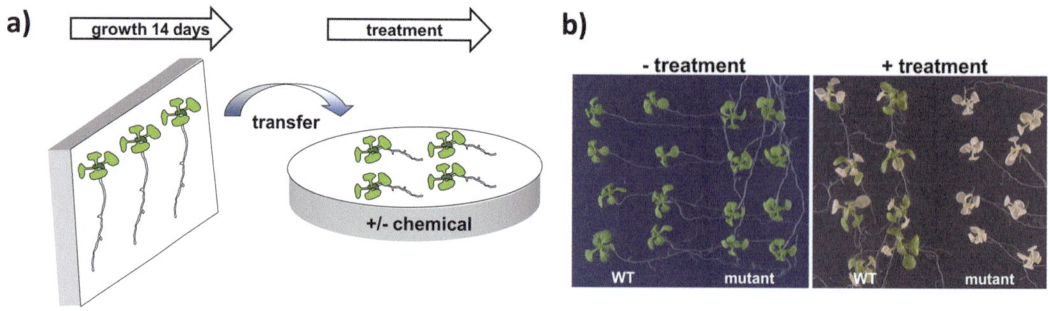

Fig. 1 (**a**) Schematic drawing of vertical growth of Arabidopsis plants on agar surface and transfer to other media. (**b**) Example of plants photographed after treatment as in (**a**)

6. Fill every pot (7 cm × 7 cm × 7 cm) with 80 g potting compost/vermiculite mixture for growing one plant per pot each. Place sterilized seeds into the center of the soil surface. For growth on agar plates seeds are transferred on top of the agar under a clean bench. Distribute the seeds in a row one by one using an automatic pipette (20 μl) and sterile pipet tips (*see* **Note 2**).
7. Keep pots or agar plates, respectively, at 4 °C for 3 days (*see* **Note 3**), then shift to normal growth condition at 22 °C under a 14 h light/10 h dark photocycle with ~150 μmol/m^2/s intensity and a humidity of 50–60% in a growth chamber. Keep soil moist by watering the pots regularly with tap water (*see* **Note 4**).
8. When plants are grown for further transfer to other media, filter paper or to examine root growth, plates are kept at vertical position with an angle of 85° to allow root growth on the agar surface (Fig. 1 *see* **Note 5**).

3.2 Stress Treatments

Use soil grown 15–20-day-old plants (*see* **Note 6**) in pots that are well watered or 14-day-old seedlings grown on agar plates for stress treatments. Always divide the plants into two groups: the first group is used as the control group and the second group is subjected to stress treatment. At least four plants/seedlings should be used for each treatment. The stress treatments can be executed in combination with chemical treatment (e.g. ±polyamines, inhibitors) when plants are grown vertically on agar plates for easy transfer on other media.

3.2.1 Temperature Stress

3.2.1.1 Heat Shock

1. For heat-shock treatment transfer the pots/plates from normal growth condition to a growth chamber set to 42 °C for 1 h.
2. Take samples for further analysis and put the pots/plates back to normal growth condition.
3. After 1 h take samples that represent the heat-shock recovery.

3.2.1.2 Cold Stress

1. For cold stress treatment transfer the pots/plates from normal growth condition to a growth chamber set to 4 °C.
2. Take samples at different time points (3, 6, 12 and 24 h) to monitor the effects over time.

3.2.1.3 Salt Stress

1. Use 14 day old seedlings grown vertically on agar plates.
2. It is not necessary to work under a clean bench for the following procedure.
3. Soak filter paper in ½ MS liquid medium (control) and ½ MS liquid medium containing 100 mM NaCl, respectively.
4. Carefully remove seedlings from the plates and transfer to the filter papers and further incubate for 12 and 24 h, respectively (*see* Fig. 1).

3.2.1.4 Drought Stress

1. For application of drought stress to soil-grown plants keep the pots under normal growth condition.
2. Keep the group of pots for treatment and control experiment, respectively, together in one tray that is exactly leveled.
3. Stop watering one group of pots while keeping the control group moist by continuing watering.
4. When plants are visibly wilted take samples for analysis including controls. Plant wilting and drying is followed daily to identify differences between wild-type plants and mutants.
5. Take photographs of the plants every day after the onset of drought treatment to demonstrate differences in wilting.
6. Test the ability to recover from drought stress (regaining full turgidity of leaves) by starting watering again with tap water poured into the tray for equal distribution to all the pots.
7. For drought stress treatment to seedlings grown on agar plates 14-day-old seedlings grown vertically are used.
8. Soak filter paper in ½ MS liquid medium (control) and use dry filter paper for stress application, respectively.
9. Carefully remove at least ten seedlings of wild type and mutants each from the plates and transfer to wet filter paper and dry filter paper, respectively.
10. After 45 min seedlings are collected and used for further analysis.

3.2.1.5 Pretreatment with Chemicals

A pretreatment with chemicals like polyamines before stress application is possible as outlined in Fig. 2. Seedlings are incubated on wet filter papers containing different concentrations of polyamines or other chemicals for certain time periods.

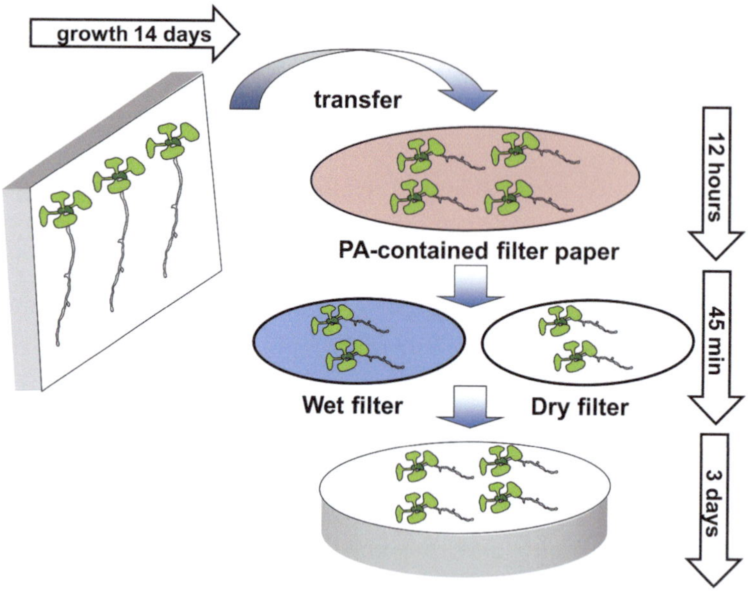

Fig. 2 Schematic drawing of drought stress application to agar-grown plants. Vertically grown seedlings are transferred onto wet or dry filter papers for different time intervals before transfer to growth media. Seedlings can be pretreated with chemicals (e.g. polyamine solution) on soaked filter paper before stress application

1. Soak filter paper in ½ MS solution containing 1 mM of selected polyamine in a petri dish. Use one filter paper soaked in ½ MS solution only as control.
2. Carefully transfer at least four 7-day-old seedlings grown vertically on ½ MS agar plates for each treatment to the respective filter paper.
3. Incubate for 12 h in closed petri dishes under normal growth condition then use seedlings for further stress treatments.

3.3 Analysis of Stress Responses

Quantitative root elongation data can be used as a rapid and quantitative assay for stress sensitivity especially for chemical treatments (e.g. salt, polyamines, heavy metals).

3.3.1 Assay of Root Elongation

1. Grow seedlings on ½ MS agar plates without addition of chemicals and containing for example 1 mM of polyamine and 100 mM NaCl in vertical position.
2. Mark the position of the root tips on the backside of the petri dish at time intervals, typically during growth period of 7–14 days.
3. At the end of the experiment, measure root elongation with a ruler or by taking a photograph of the plates followed by quantification using image analysis software (*see* **Note 7**).
4. Display results as photograph of agar plates with seedlings and root elongation in mm/day (Fig. 3).

Fig. 3 Vertically grown Arabidopsis plants that differ in root elongation

3.3.2 Quantification of Chlorophyll Content

Bleaching of leaves can be used as an indicator of plant damage by stresses. The loss of chlorophyll is visible and can be easily quantified. Chlorophyll content of leaves is determined as described by Lichtenthaler [23].

1. Accurately weigh plant leaf samples.
2. Homogenize samples with mortar and pestle or a tissue homogenizer with 5 volumes (e.g. 0.2 g in 1 ml) of 80% acetone in 5 mM Na-phosphate buffer, pH 7.5 (*see* **Note 8**).
3. Transfer the homogenate to a 1.5 ml microcentrifuge tube.
4. Clear the homogenate by centrifugation at 10,000 × g, 4 °C for 10 min and carefully transfer the supernatant into a fresh tube, keep at a dark place.
5. Measure the absorbance in a spectrophotometer with glass cuvettes at wavelengths of 663 and 645 nm for total chlorophyll content or Chlorophyll-a and Chlorophyll-b at 663 and 646 nm.
6. Equations to determine concentrations of total chlorophyll and of chlorophyll a (Chla) and chlorophyll b (Chlb) in μg/ml are as follows:

$$\text{Chl total } (\mu\text{g/ml}) = 17.32\ (A_{645}) + 7.18\ (A_{663})$$
$$\text{Chla } (\mu\text{g/ml}) = 12.21\ (A_{663}) - 2.81\ (A_{646})$$
$$\text{Chlb } (\mu\text{g/ml}) = 20.13\ (A_{647}) - 5.03\ (A_{663}).$$

7. Display the values of chlorophyll content as a bar chart with statistical analysis to rate the grade of damage to leaves.

3.3.3 Water Loss Assay and Relative Water Content (RWC)

Water loss assay under drought treatment is performed by the procedure described in Weigel and Glazebrook [24].

1. Transfer seedlings from the agar plates onto dry filter paper and determine their fresh weights immediately with a balance.
2. Repeat weighing every 10 min for 1 h after the onset of drought treatment.
3. The fresh weights at the onset of the treatment are set at 100%, and the relative water loss over time is determined.

The relative water content (RWC; or "relative turgidity") of leaves is a measurement of the hydration status relative to the maximum water holding capacity at full turgidity and indicates the "water deficit" (wilting) as a degree of stresses like drought, high or low temperature and salt. RWC is determined as described by Barr and Weatherley and by Smart and Bingham [25, 26].

1. Collect whole leaves or leaf discs (Ø 1.0–1.5 cm) punched out with a sharp cork borer from larger leaves. About 5–6 samples (replications) are taken from a single treatment or genotype.
2. Determine the fresh weights (FW) with a balance.
3. Place the samples on the surface of de-ionized water in a closed petri dish and keep at 4 °C in the dark for 6 h to overnight to hydrate to full turgidity.
4. Blot the samples on filter paper, immediately weigh to obtain fully turgid weight (TW).
5. Dry the samples in an oven at 80 °C overnight and weigh again to determine the dry weight (DW) (*see* **Note 9**).
6. RWC is calculated as: RWC (%) = [(FW − DW)/(TW − DW)] × 100

3.3.4 Electrolyte Leakage Assay

Cell membranes are one of the first targets of many plant stresses and maintenance of their integrity and stability is essential for stress tolerance. The degree of cell membrane injury can be easily estimated through measurements of electrolyte leakage and used to assess the damage caused by abiotic stress treatments [27]. The technique has been applied to quantify damages in various abiotic stress conditions such as low and high temperatures [28–30], salt stress [31] or heavy metals [32].

1. Harvest 6–8 seedlings from agar plate or five detached leaves from soil grown plants and rinse four times with deionized water, then blot on filter paper.
2. Transfer the seedlings or leaves to a 15 ml tube containing 5 ml of deionized water and incubate for 24 h at 22 °C.
3. Measure the conductivity with an electrical conductivity meter (= C1).
4. To obtain the total conductivity heat the tubes in boiling water for 20 min or autoclave at 121 °C for 15 min.
5. Let the samples cool to 22 °C and measure electric conductivity again (= C2).
6. Electrolyte leakage (EL) is expressed as percentage of total conductivity and calculated as: EL = C1/C2 × 100.

3.3.5 Analyses of Reactive Oxygen Species (ROS)

The levels of H_2O_2 are determined as described by Messner and Boll [33]. A serial dilution of H_2O_2 is freshly prepared and used for the standard curve in the reaction mixture instead of extract.

3.3.5.1 Hydrogen Peroxide Measurement

1. First create a serial dilution of H_2O_2 freshly prepared from the 30% H_2O_2 stock solution in 100 mM K-phosphate buffer (pH 7.0): Add 10 μl of 30% H_2O_2 solution to 10 ml 100 mM K-phosphate buffer (pH 7.0) in a graduated cylinder, mix and measure the absorbance at 240 nm in a UV-spectrophotometer and UV-cuvettes with 100 mM K-phosphate buffer (pH 7.0) as blank (*see* **Note 10**).
2. Calculate the H_2O_2 concentration using Lambert-Beer equation (*see* **Note 10**). The absorbance value should be about 0.4 and the H_2O_2 concentration of the dilution about 10 mM. Use the exact value for further calculations.
3. Further dilute the 10 mM solution stepwise 1:10 (100 μl + 900 μl 100 mM K-phosphate buffer (pH 7.0) to get 1 mM, 100 μM, 10 μM and 1 μM dilutions.
4. Use the dilutions of **step 3** including K-phosphate buffer only as zero H_2O_2 (*see* **Note 11**) for creating the standard curve: In 1.5 ml microcentrifuge tubes mix 200 μl of the dilutions each with 10 μl of horseradish peroxidase dissolved in 100 mM K-phosphate buffer (pH 7.0) and 10 μl of 50 mM (w/v) ABTS.
5. Incubate for 3 min at room temperature, then measure the absorbance at 415 nm with a spectrophotometer using micro-cuvettes and reaction mixture containing 200 μl 100 mM K-phosphate buffer (pH 7.0) as blank.
6. Draw the standard curve with absorbance values on the *Y*-axis and H_2O_2 amount (nmol) on *X*-axis.
7. Homogenize seedlings or detached leaves (approximately 200 mg fresh weight) in a prechilled mortar and pestle or homogenizer with 1.5 volumes of extraction buffer (i.e. 300 μl/200 mg).
8. Transfer of the slurry to a 1.5 ml microcentrifuge tube.
9. Centrifuge at 12,000 × *g*, 4 °C for 10 min.
10. Transfer 200 μl of the supernatant to a fresh tube.
11. Add 10 μl of horseradish peroxidase dissolved in 100 mM K-phosphate buffer (pH 7.0) and 10 μl of 50 mM (w/v) ABTS.
12. Incubate for 3 min at room temperature, then measure absorbance as in **step 5**.
13. Use the absorbance values and the standard curve from **step 6** to determine the H_2O_2 concentration in the samples (*see* **Note 12**).

3.3.5.2 Superoxide Measurement

The levels of O_2^- are determined based on its ability to reduce nitro blue tetrazolium (NBT) to insoluble blue formazan as described by Doke [34].

1. Cut seedlings (approximately 200 mg fresh weight) into small pieces with scissors or a sharp scalpel and transfer into a 15 ml screw cap plastic tube.
2. Add 4 ml of incubation solution, immerse the cut seedlings and incubate for 1 h at room temperature.
3. Collect the tissue at the bottom of the tube by centrifugation (8000 × *g*, 5 min, RT).
4. Transfer 2 ml of the supernatant to a 2 ml microcentrifuge tube, heat at 85 °C for 15 min, then cool rapidly on ice.
5. Read the absorbance at 580 nm with a spectrophotometer with incubation solution as blank.
6. The O_2^- content is expressed as the increase in absorbance/g fresh weight.

4 Notes

1. We use a potting compost with the following nutrient content: nitrogen (N) 150 mg/L, phosphate (P_2O_5) 150 mg/L, potassium oxide (K_2O) 400 mg/L, magnesium (Mg) 130 mg/L, sulfur (S): 110 mg/L, pH 6.0 salt (KCl) 1.5 g/L. Other composts with approximately same contents can also be used.
2. Seeds are placed one by one onto agar plates under a clean bench using an automatic pipette (20 μl) and sterile pipet tips. Seedlings can be grown in two rows (20 seeds per row) on one plate. Two parallel lines on the backside of the plate (approximately 2 cm from the top edge and one in the middle of the plate) help to distribute the seeds in rows. When plants are grown in soil equal amounts of soil are weighed into the pots which are then placed into a suitable tray that enables watering from the bottom.
3. Widely used *Arabidopsis thaliana* lines (e.g. Col-0) have moderate dormancy. Cold treatment, also called stratification, at 4 °C for 2 days will improve the rate and synchrony of germination. Cold treatment of dry seeds is not effective.
4. For all experiments, if not whole seedlings are used, it is important to have leaves at the same leaf number and stage from different plants as the stress tolerance is dependent on leaf order and developmental stage.
5. Seedlings can be grown in that way without having leaves in close contact with the agar surface. Transfer of seedlings to

other media should be carefully conducted with a suitable forceps without damaging roots.

6. Plants growing for 15–20 days are still in vegetative growth phase, i.e. pre-bolting [35].
7. An open platform for scientific image analysis is ImageJ which is available for free download from https://imagej.net/.
8. 80% acetone in water can also be used but having the solution slightly buffered omits the formation of phaeophytin by a shift to lower pH caused by the leaf extract [36].
9. In cases were the weight of seedlings or leaf discs is determined at least 200 mg of fresh material should be weighed using a balance that can accurately display 0.001 g because at the end dry weight can be less than 10% of fresh weight.
10. A H_2O_2 solution of 30% (w/w) has a density of 1.11 g/ml and therefore a molar concentration of 9.8 M. H_2O_2 solutions can be unstable, so calibration of the solution is essential for accurate results. That is why the concentration of the 1:1000 dilution of the 30% stock solution (~10 mM) is determined by absorbance at 240 nm. Using the Lambert-Beer law ($c = E/\varepsilon * d$) and the molar extinction coefficient (H_2O_2 $\varepsilon_{240} = 0.043\ \text{cm}^2/\text{mmol}$) the real concentration of H_2O_2 in the sample is determined.
11. The 0 μM H_2O_2 standard is an important negative control that shows the background absorbance coming from the reagent mixture and HRP in the absence of H_2O_2 The results are given in nmol H_2O_2 per g fresh weight.
12. The 200 μl of the 1 μM dilution contain 200 pmol H_2O_2, the 10 μM dilution 2 nmol, respectively. Calculation: X is the amount of H_2O_2 in 200 μl of the sample [nmol] (determined with the standard curve), Υ is the fresh weight used for extraction [mg]. The amount of H_2O_2 is: $X * 1.5 * 1000/\Upsilon = H_2O_2$ [nmol/gFW].

Acknowledgements

This work is financially supported by JSPS KAKENHI (No. 15K14705) to T.K.

References

1. Asada K (2006) Production and scavenging of reactive oxygen species in chloroplasts and their functions. Plant Physiol 141:391–396
2. Sandalio LM, Rodríguez-Serrano M, Romero-Puertas MC, delRio LA (2013) Role of peroxisomes as a source of reactive oxygenspecies (ROS) signalling molecules. Subcell Biochem 69:231–255
3. Gupta DK, Palma JM, Corpas FJ (eds) (2015) Reactive oxygen species and oxidative damage

in plants under stress. Springer. ISBN: 978-3-319-20421-5
4. Swanson S, Gilroy S (2010) ROS in plant development. Physiol Plant 138:384–392
5. Suzuki N, Koussevitzky S, Mittler R, Miller G (2012) ROS and redox signalling in the response of plants to abiotic stress. Plant Cell Environ 35:259–270
6. Foyer CH, Noctor G (2013) Redox signaling in plants. Antioxid Redox Signal 18:2087–2090
7. Singh R, Singh S, Parihar P, Mishra RK, Tripathi DK, Singh VP, Chauhan DK, Prasad SM (2016) Reactive oxygen species (ROS): beneficial companions of plants' developmental processes. Front Plant Sci 7:1299
8. Braun A (1873) Freunde z Berlin, p 75
9. Laibach F (1907) Zur Frage nach der Individualität der Chromosomen im Pflanzenreich. Bot Centbl Beihefte (I) 22:191–210
10. Laibach F (1943) *Arabidopsis thaliana* (L.) Henh. als Objekt für genetische und entwicklungsphysiologische Untersuchungen. Bot Archiv 44:439–455
11. Reinholz E (1947) Auslösung von Röntgenmutationen bei *Arabidopsis thaliana* (L.) Heynh. und ihre Bedeutung für die Pflanzenzüchtung und Evolutionstheorie. Field Inform Agency Tech Rep 1006:1–170
12. Meyerowitz EM (2001) Prehistory and history of Arabidopsis research. Plant Physiol 125:15–19
13. Somerville C, Koornneef M (2002) A fortunate choice: the history of Arabidopsis as a model plant. Nat Rev Genet 3:883–889
14. Koornneef M, Meinke D (2010) The development of Arabidopsis as a model plant. Plant J 61:909–921
15. Krämer U (2015) The natural history of a model organism: planting molecular functions in an ecological context with Arabidopsis thaliana. eLife 4:e06100
16. Urano K, Hobo T, Shinozaki K (2005) Arabidopsis ADC genes involved in polyamine biosynthesis are essential for seed development. FEBS Lett 579:1557–1564
17. Ge C, Cui X, Wang Y, Hu Y, Fu Z, Zhang D, Cheng Z, Li J (2006) BUD2, encoding an S-adenosylmethionine decarboxylase, is required for Arabidopsis growth and development. Cell Res 16:446–456
18. Imai A, Matsuyama T, Hanzawa Y, Akiyama T, Tamaoki M, Saji H, Shirano Y, Kato T, Hayashi H, Shibata D, Tabata S, Komeda Y, Takahashi T (2004) Spermidine synthase genes are essential for survival of Arabidopsis. Plant Physiol 135:1565–1573
19. Imai A, Akiyama T, Kato T, Sato S, Tabata S, Yamamoto KT, Takahashi T (2004) Spermine is not essential for survival of Arabidopsis. FEBS Lett 556:148–152
20. Hanzawa Y, Takahashi T, Komeda Y (1997) ACL5: an Arabidopsis gene required for internodal elongation after flowering. Plant J 12:863–874
21. Tavladoraki P, Cona A, Angelini R (2016) Copper-containing amine oxidases and FAD-dependent polyamine oxidases are key players in plant tissue differentiation and organ development. Front Plant Sci 7:824
22. Wimalasekera R, Villar C, Begum T, Scherer GFE (2011) Copper amine oxidase1 (CuAO1) of *Arabidopsis thaliana* contributes to abscisic acid- and polyamine-induced nitric oxide biosynthesis and abscisic acid signal transduction. Mol Plant 4:663–678
23. Lichtenthaler HK (1987) Chlorophylls and carotenoids: pigment of photosynthetic biomembranes. Methods Enzymol 148:350–382
24. Weigel D, Glazebrook J (2002) Arabidopsis, a laboratory manual. Cold Spring Harbor Laboratory Press, Cold Spring Harbor, NY
25. Barr HD, Weatherley PE (1962) A reexamination of the relative turgidity technique for estimating water deficit in leaves. Aust J Biol Sci 15:413–428
26. Smart RE, Bingham GE (1974) Rapid estimates of relative water content. Plant Physiol 53:258–260
27. Sukumaran NP, Weiser CJ (1972) An excised leaflet test for evaluating potato frost tolerance. Hort Sci 7:467–468
28. Vainola A, Repo T (2000) Impedance spectroscopy in frost hardiness evaluation of Rhododendron leaves. Ann Bot 86:799–805
29. Takagi T, Nakamura M, Hayashi H, Inatsugi R, Yano R, Nishida I (2003) The leaf-order-dependent enhancement of freezing tolerance in cold-acclimated Arabidopsis rosettes is not correlated with the transcript levels of the cold-inducible transcription factors of CBF/DREB1. Plant Cell Physiol 44:922–931
30. Ismail AM, Hall AE (1999) Reproductive-stage heat tolerance, leaf membrane thermostability and plant morphology in cow-pea. Crop Sci 39:1762–1768
31. Sreenivasulu N, Grimm B, Wobus U, Weshke W (2000) Differential response of antioxidant compounds to salinity stress in salt-tolerant and salt-sensitive seedlings of foxtail millet (Setaria italica). Physiol Plant 109:435–442
32. De B, Mukherjee AK (1996) Mercuric chloride induced membrane damage in tomato cultured cells. Biol Plant 38:469–473

33. Messner B, Boll M (1994) Cell suspension cultures of spruce (Picea abies): inactivation of extracellular enzymes by fungal elicitor-induced transient release of hydrogen peroxide (oxidative burst). Plant Cell Tissue Organ Cult 39:69–78
34. Doke N (1983) Generation of superoxide anion by potato tuber protoplasts during hypersensitive response to hyphal wall components of Phytophtora infestans and specific inhibition of the reaction with supressors of hypersensitivity. Physiol Plant Pathol 23:359–367
35. Boyes DC, Zayed AM, Ascenzi R, McCaskill AJ, Hoffman NE, Davis KR, Görlach J (2001) Growth stage-based phenotypic analysis of Arabidopsis: a model for high throughput functional genomics in plants. Plant Cell 13:1499–1510
36. Lorenzen CJ (1965) A note on the chlorophyll and phaeophytin content of the chlorophyll maximum. Limnol Oceanogr 10:482–483

Chapter 33

Phenotypic and Genotypic Characterization of Mutant Plants in Polyamine Metabolism Genes During Pathogenic Interactions

Franco R. Rossi, Fernando M. Romero, Oscar A. Ruíz, Maria Marina, and Andrés Gárriz

Abstract

Plants respond to pathogen attack by modifying defense gene expression and inducing the production of myriad proteins and metabolites. Among these responses, polyamine (PA) levels suffer remarkable modifications. Evidences demonstrate that plants make use of the polyamine biosynthetic pathway and the oxidative catabolism of these compounds in order to mount adequate defenses against pathogens. In *Arabidopsis thaliana*, putrescine is synthesized exclusively through the arginine decarboxylase (ADC) pathway, this enzyme exists as two isoforms named ADC1 and ADC2. Even though both isoforms participate in the response to pathogen attack, the mechanisms modulating ADC activity are not completely understood. Therefore, studies to clarify their roles are necessary. In this chapter, we describe the methods that can be applied for the study of plant–pathogen interactions using Arabidopsis *adc* mutant plants.

Key words Putrescine, Arginine decarboxylase, *adc* mutants, Plant defense, *Pseudomonas viridiflava*, *Botrytis cinerea*

1 Introduction

Polyamines (PAs) are natural amines required for all living cells. In plants, the levels of the most abundant PAs Putrescine (Put), Spermidine (Spd), and Spermine (Spm) suffer dramatic changes in response to different stressful conditions [1]. In general, the concentration of Put is increased in response to stress, whereas the changes in the level of Spd and Spm depend on the stressful factor [2].

In most plants, Put synthesis is carried out by the enzymes arginine decarboxylase (ADC, EC 4.1.1.19) and ornithine decarboxylase (ODC, EC 4.1.1.17), which use arginine and ornithine as substrates, respectively [3]. These enzymes play a fundamental role in maintaining the homeostasis of PAs, providing the precursor used by Spd and Spm synthase to generate a molecule of Spd and

Rubén Alcázar and Antonio F. Tiburcio (eds.), *Polyamines: Methods and Protocols*, Methods in Molecular Biology, vol. 1694, DOI 10.1007/978-1-4939-7398-9_33, © Springer Science+Business Media LLC 2018

Spm, respectively. Thus, plants require ADC and/or ODC activity for normal development [4].

In recent years, considerable advances have been obtained on the physiological functions of PAs during plant–pathogen interactions. Thus, it has been demonstrated in different pathosystems that the accumulation of these compounds and its subsequent oxidation by PA oxidases play an essential role in the defense against pathogenic viruses, bacteria, and fungi [5–8]. However, this subject still remains to be fully comprehended. In particular, the processes underlying the regulation of the PA metabolism and the activation of the defense mechanisms mediated by these compounds have been poorly explored.

Arabidopsis thaliana is a plant species widely used as a model to study plant–pathogen interactions. This is mainly based on the availability of a vast amount of homozygous mutant lines in almost all known genes, including those involved in the PA metabolism [9]. This species constitutes the only known superior plant bearing the ADC pathway for Put biosynthesis and lacking ODC activity. Two isoforms of this protein were described, encoded by *ADC1* and *ADC2* genes [10], which show different expression patterns during distinct biotic and abiotic stresses [7, 11, 12]. Recently, it has been shown that the inoculation of *A. thaliana* seedlings with the natural pathogen *Pseudomonas viridiflava* leads to the accumulation of Put due to the induction of *ADC1*, whereas *ADC2* expression remained unaltered. However, Put also accumulates in plant lines lacking *ADC1* due to the upregulation of *ADC2*, demonstrating that these isoforms might play partially redundant functions [13]. In turn, the expression of both isoforms is induced in adult plants in response to pathogens such as *Pseudomonas syringae* or *Botrytis cinerea*, suggesting that different regulatory mechanisms may modulate *ADC* gene expression during distinct stages of plant development ([14], Rossi et al., unpublished results). Several techniques can be applied to evaluate immunity of these lines. In this trend, most of plants respond to pathogen attack by activating distinct biochemical and physiological defense mechanisms in order to counteract pathogens propagation. These responses include the reinforcement of cell walls by the production of phenolic compounds and callose, as well as the accumulation of toxic compounds such as H_2O_2. In addition, defense hormones induce the expression of particular genes encoding pathogenesis-related (PR) proteins such as *PR1* and *PDF1.2* genes, which are regulated by salicylic and jasmonic acids, respectively. Thus, plants respond against pathogens by activating a wide arsenal of defense mechanisms [15, 16].

The aim of this chapter is to describe molecular and biochemical techniques used to characterize plant mutants in *ADC* genes. We also describe common techniques usually applied to explore plant defense responses to pathogens. The chapter is focused on the interactions between *A. thaliana* and two known pathogens as the hemibiotrophic bacteria *P. viridiflava* and the necrotrophic fungus *B. cinerea*. However, it is noteworthy to acknowledge that the majority of the techniques described below may be used for the study of other mutant lines on other genes.

2 Materials

2.1 Arabidopsis thaliana Culture

1. Murashige and Skoog basal medium: 20.625 mM NH_4NO_3, 0.099 mM H_3BO_3, 2.99 mM $CaCl_2$, 0.0001 mM $CoCl_2{\cdot}6H_2O$, 0.0001 mM $CuSO_4{\cdot}6H_2O$, 0.099 mM $FeSO_4{\cdot}7H_2O$, 1.5 mM $MgSO_4$, 0.099 mM $MnSO_4{\cdot}H_2O$, 0.001 mM $Na_2MoO_4{\cdot}2H_2O$, 0.005 mM KI, 18.81 mM KNO_3, 1.25 mM KH_2PO_4, 0.029 mM $ZnSO_4{\cdot}7H_2O$, 3% sucrose and 0.8% phytagar. Adjust pH with 1 N KOH.
2. Substrate mixture, white peat:perlite:sand (1:1:1).
3. Hoagland 1× nutrient solution: 6 mM KNO_3, 4 mM $Ca(NO_3)_2{\cdot}4H_2O$, 2 mM $MgSO_4{\cdot}7H_2O$, 1 mM $NH_4H_2PO_4$, 0.09 mM EDTA $NaFeO_8{\cdot}2H_2O$, 0.009 mM $MnCl_2{\cdot}4H_2O$, 0.046 mM H_3BO_3, 0.0003 mM $CuSO_4{\cdot}5H_2O$, 0.0008 mM $ZnSO_4{\cdot}7H_2O$, 0.0001 mM $Na_2MoO_4{\cdot}2H_2O$.

2.2 ADC Activity Determination (Fig. 1)

1. Extraction buffer: 15 mM KH_2PO_4, 20 mM sodium ascorbate, 0.5 mM EDTA, 1 mM phenylmethylsulfonyl fluoride (PMSF), 10 mM DTT, 1 mM Pyridoxal phosphate (PLP), 85 mM Na_2HPO_4, pH 7.5. Prepare just before use.
2. Reaction mixture: 20 mM L-arginine, 2 mM urea and 50 nCi [U-^{14}C]-L-arginine.
3. 5% percloric acid.
4. Aqueous/organic scintillation cocktail.
5. Scintillation counter.
6. Bradford's reagents and 1 mg/mL Bovine Seric Albumin (BSA) solution.

2.3 Botrytis Cinerea Culture and Determination of Necrotic Area in Infected Leaves

1. Potato Dextrose Agar (PDA): Potato Starch (from infusion) 4 g/L, Dextrose 20 g/L, Agar 15 g/L and Potato Dextrose Broth (PDB): Potato Starch (from infusion) 4 g/L, Dextrose 20 g/L. Alternatively, PDB might be prepared by adding 20 g dextrose to 1 L potato infusion (also add 20 g agar to make PDA). Potato infusion is prepared by boiling 300 g of sliced unpeeled potatoes in 1 L of distilled water for 1 h, and straining broth through cheesecloth to recover the potato juice. Then, distilled water is added to complete 1 L (*see* **Note 1**).
2. Freeze-dried tomato leaves.

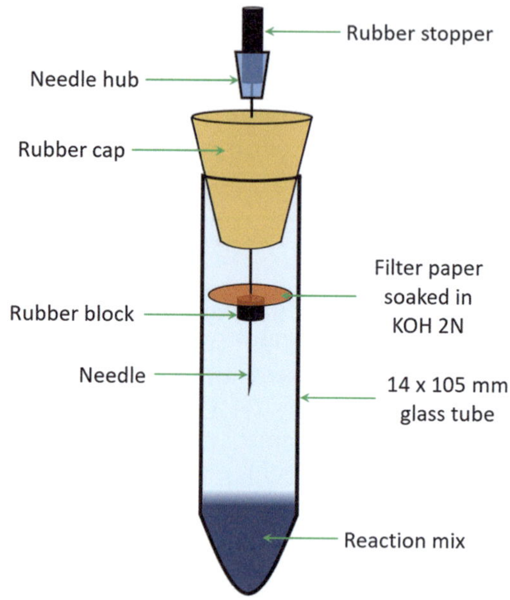

Fig. 1 Schematic representation of reaction tubes. The system consists of a needle piercing though a rubber cap and sealed with a small rubber stopper. This rubber cap together with the rubber stopper avoids $^{14}CO_2$ leakage from the reaction tube. Affix a piece of filter paper soaked in KOH to the needle. A rubber block prevents the filter paper from falling inside the reaction mix

3. Hematocytometer.
4. Stereoscopic microscope with camera.
5. Image processing software like Image-Pro Plus® (MediaCybernetics) or ImageJ.

2.4 Pseudomonas Viridiflava Culture, Plant Inoculation and in Planta Growth Assay

1. King's B medium: 10 g proteose peptone, 1.5 g anhydrous K_2HPO_4, 15 mL glycerol, 5 mL 1 M $MgSO_4$ and 20 g agar. Complete to 1 L with distilled water and sterilize by autoclaving.
2. 10 mM $MgCl_2$, pH 7.0.
3. Blue polypropylene pellet pestles (*see* **Note 2**).
4. Stereoscopic microscope with digital camera.
5. Image processing software like Image-Pro Plus® (MediaCybernetics) or ImageJ.

2.5 Assessment of Plant Cell Death

1. Staining solution: 3.75 mL distilled water, 3.75 mL lactic acid, 3.75 mL glycerol, 3.75 mL equilibrated phenol, 0.15 mL Trypan Blue (25 mg/mL) in water.
2. 2.5 g/mL Chloral hydrate solution.
3. 50% glycerol.

4. Stereoscopic microscope with digital camera.
5. Image processing software like Image-Pro Plus® (MediaCybernetics) or ImageJ.

2.6 Assessment of the Accumulation of Phenolic Compounds

1. Fixing solution: 10% formaldehyde, 5% acetic acid, 45% ethanol in H_2O.
2. Epifluorescence microscope (excitation filter, 330–380 nm; emission filter, 435–485 nm).

2.7 Determination of Callose Deposition

1. Fixing and distaining solution: ethanol/lactophenol 2:1 (vol/vol).
2. Staining solution: 0.01% aniline blue powder in 150 mM K_2HPO_4, pH 9.5.
3. 50% glycerol.

2.8 Assessment of ROS Accumulation

1. 15 μM 2′,7′-dichlorofluorescein diacetate (DCFH-DA).

3 Methods

3.1 Seedling Culture

1. Prepare 1× Murashige and Skoog (MS) medium by dissolving 3.52 g MS powder (and 24 g sucrose in 600 mL of distilled water, adjust pH to 6.5–6.7 with KOH 1 N and complete the volume to 800 with water.
2. Add phytagar to 0.8% (w/v).
3. Sterilize by autoclaving by 20 min at 121 °C.
4. Allow the medium to cool at 55–60 °C and pour 20 mL into each Petri dish (90 mm × 15 mm). Work under sterile conditions in a laminar flow hood. Place the lids on the plates and allow them to cool for 30–60 min (until solidified).
5. Disinfect seeds by dipping them in 75% ethanol for 30 s, followed by soaking in 5% commercial bleach for 15 min. Wash them at least three times with sterile distilled water.
6. Sow 25–30 seeds per plate and seal with parafilm. Cold stratify seeds for 48 h at 4 °C. Then transfer plates to the growth chamber.
7. Grow seedlings for 15 days under the appropriated photoperiod at 22 ± 1 °C.

3.2 Adult Plants Culture

1. Prepare substrate mixture and autoclave for 2 h. Repeat autoclaving after 1–2 days to assure the sterility of the substrate.
2. Sow seeds directly on the substrate in plastic pots. Cold stratify seeds for 48 h at 4 °C and transfer to growth chamber.

3. Grow plants for 4-weeks under 16/8 h (day/night) photoperiod. Set the temperature at 24 ± 0.5 and 20 ± 0.5 °C for day and night, respectively. Water with 0.5× Hoagland's nutrient solution (*see* **Note 3**).

3.3 Characterization of A. thaliana ADC Mutant Lines

3.3.1 Confirming Homozygous Insertion Lines by PCR

1. Identify ADC mutant lines from the Arabidopsis Biological Resource Center (ABRC, https://abrc.osu.edu/) or Nottingham Arabidopsis Stock Center (NASC, http://arabidopsis.info/) databases.
2. Look for the insertion site and flanking sequences using the Arabidopsis Gene Mapping Tool, T-DNA express (http://signal.salk.edu/cgi-bin/tdnaexpress).
3. Check homozygous mutants by PCR using genomic DNA as template and different sets of primers made by the combination of the T-DNA left border a1 (Lba1)-specific primer and ADC1/ADC2 sequence-specific primers. A throughout description of the steps to follow are available at The Arabidopsis Information Resource's webpage (https://www.arabidopsis.org/portals/mutants/identification.jsp). Some *ADC1* and *ADC2* mutant lines had been already checked and are available (*see* **Note 4**).

3.3.2 Determination of ADC Activity

1. Immediately after harvesting the tissue, homogenize Arabidopsis leaves or seedlings with a mortar and pestle in liquid N_2 (*see* **Note 5**).
2. Transfer 200–250 mg of tissue powder to a 1.5 mL tube and resuspend in 0.25 mL extraction buffer. Vortex vigourously.
3. Centrifuge for 20 min at 12,000 × *g* at 4 °C.
4. Recover the supernatant into a new tube, and keep at 4 °C until further use.
5. Transfer 190 μL of the supernatant to a reaction tube. Add 10 μL reaction mixture and close with rubber cap. Ensure that reaction tubes are completely sealed to avoid $^{14}CO_2$ leaking out.
6. Incubate under gentle agitation (75 rpm) by 45 min at 37 °C. Avoid contact between reaction mixture and paper affixed to needle.
7. Stop the reaction by adding 0.2 mL 5% (v/v) perchloric acid (PCA) through needle using a syringe. To this, take out the rubber stopper from needle hub and inject PCA inside the reaction mix using a syringe. The needle hub must be sealed again later with the rubber stopper as quickly as possible.
8. Incubate under gentle agitation by 40 min at 37 °C. During this period, the $^{14}CO_2$ released is trapped in the filter paper.

9. Uncap tubes and take the filter paper away avoiding the contact with the solution and tube walls.
10. Transfer the filter paper to a scintillation vial and add 800 μL of scintillation cocktail.
11. Measure the radioactive emission using a scintillation counter.
12. Determine protein content in the supernatants using the Bradford's method [17] and bovine serum albumin (BSA) as the reference standard.
13. The enzyme activity is expressed as pmol CO_2 released per hour per μg total protein.

3.4 Inoculation of ADC Mutant Plants with Botrytis Cinerea

1. Plate out conidia of *B. cinerea* strain B05.10 on PDA supplemented with 40 mg/mL of tomato leaf powder (*see* **Note 6**).
2. Keep plates from 7 to 10 days in darkness at room temperature to allow sporulation.
3. Harvest conidia by adding 1–2 mL of sterile 0.02% Tween-20 solution over the mycelia, and carefully scraping and removing the mycelia with a spatula. Transfer the solution containing mycelia to a 15 mL tube and vortex thoroughly for 1–2 min.
4. Filter the suspension by using a syringe (without needle) with cotton on the top and recover conidia in a new tube.
5. Centrifuge at 10,000 × *g* for 5 min. Discard supernatant and resuspend the pellet in sterile water. Repeat this step once more again.
6. Quantify conidia with a hemacytometer. Adjust the inoculum concentration to 5×10^4 conidia per mL in PDB medium supplemented with 10 mM sucrose and 10 mM KH_2PO_4.
7. Incubate the conidia for 2–3 h at room temperature without shaking (*see* **Note 7**).
8. Select two leaves from 4-week-old plants (*see* **Note 8**).
9. Perform the inoculation by placing two droplets of 5 μL of the conidial suspension on the adaxial surface of leaves, one on each side of the main vein. Mock plants must be treated similarly, but using PDB instead of conidial suspension.
10. Incubate inoculated plants in trays covered with PVC film to obtain high humidity conditions. Follow-up the infection by several days, the first necrotic symptoms should appear at 24 hpi.

3.5 Determination of the Necrotic Area in Botrytis Cinerea Infected Leaves

1. Inoculate leaves following the protocol before mentioned.
2. After 24 and 48 hpi, cut leaves by their petioles using a razor-blade (*see* **Note 9**).

3. Expand the leaves as much as possible between two 20 cm × 20 cm × 3 mm transparent glass pieces (*see* **Note 10**).
4. Take photos of infected leaves using stereoscopic microscope. Include in each photo a ruler to be used as reference.
5. Determine necrotic areas using an image processing software.

3.6 Inoculation of Seedlings with Pseudomonas viridiflava

1. Prepare inoculums of *P. viridiflava* by streaking bacteria from −80 °C stocks onto King's B agar-medium plates. Incubate plates at 28 °C for 24 h.
2. Make a sub-culture onto King's B agar-medium plates and incubate plates at 28 °C for 24 h.
3. Scrape bacterial cells off plates by adding 5 mL of 10 mM $MgCl_2$, pH 7.0.
4. Centrifuge at 10,000 × *g* for 5 min, discard supernatant and suspend the pellet in 10 mM $MgCl_2$, pH 7.0, repeat this step once more.
5. Measure optical density (OD) at 600 nm using a spectrophotometer.
6. Adjust the bacterial suspension to yield 5×10^8 colony forming units (CFUs) per milliliter ($OD_{600nm} = 0.6$).
7. Inoculate seedlings by applying 5 μL drops of bacterial suspension in the center of the rosette (for gene expression analysis) or in the most developed leaves of each seedling (for incidence, susceptibility, and biochemical assays) (*see* **Note 11**). Control plants must be treated similarly, but using 10 mM $MgCl_2$, pH 7.0 instead of bacterial suspension.
8. Seal plates with plastic wrap and return them to the same growing conditions.

3.7 Assessment of Bacterial Multiplication in Infected Tissues

1. Avoiding any damage, take out 5–6 seedlings from each plate (biological replicates). Disinfect seedlings with a solution of 2% commercial bleach by 2 min and wash them thoroughly twice with sterile water. Transfer the seedlings to a new tube.
2. Add 400 μL of 10 mM $MgCl_2$, pH 7. Grind the tissue with blue polypropylene pellet pestles or with an automated sample disruptor until complete homogenization. Typically, 24, 48, and 72 hpi, are time points selected to follow the multiplication of the bacteria.
3. Make tenfold serial dilutions in 10 mM $MgCl_2$, pH 7.0 until reaching a dilution of 10^{-5} of the initial homogenates.
4. Plate out 5 μL drops of each dilution in a King's B plate. Use a grid as a pattern guide for placing each drop.
5. Incubate plates at 28 °C for 24 h.
6. Count CFUs under stereoscopic microscope (*see* **Note 12**).

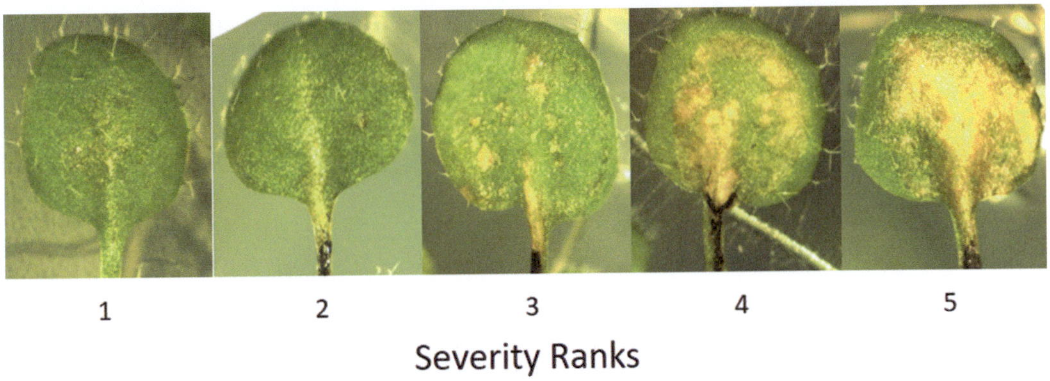

Fig. 2 Ranking of severity in infected leaves. Five severity ranks between 1 (lowest severity) and 5 (highest severity) are established on the basis of the percentage of the total leaf surface exhibiting symptoms as follows: 1, less than 10%; 2, 11–25%; 3, 26–50%; 4, 51–75%; and 5, 76% to completely necrotic leaf

3.8 Determination of Disease Incidence and Severity

1. Inoculate a single leaf per seedling with a 5 μL drop following the procedure detailed in Subheading 3.4. Identify each inoculated leaf by marking with a permanent marker avoiding any damage.
2. Disease incidence, defined as the percentage of diseased plants, is evaluated at different time points after inoculation by determining the number of diseased plants among a total of 25–30 inoculated seedlings.
3. Disease severity, defined as the level of damage in diseased plants, is evaluated by estimating the percentage of diseased tissues in inoculated leaves. Assign disease symptoms in leaves to different severity ranks as shown in Fig. 2.
4. Monitor incidence as well as severity by taking photos at different time points as far as 144 hpi.

3.9 Biochemical and Physiological Defense Responses

3.9.1 Cell Death

1. Submerge leaves or seedlings in staining solution and incubate them for 6 h at 37 °C.
2. Rinse twice with distillated water and clear with chloral hydrate. Rinse again twice with distillated water, equilibrate the samples in 50% glycerol by 1 h at room temperature.
3. Mount the samples in slides and visualize under a binocular microscope.

3.9.2 Autofluorescence Analysis (Accumulation of Phenolic Compounds)

1. Submerge leaves or seedlings in fixing solution for 15 min.
2. Clear samples in 50% ethanol for 20 min, following by an overnight incubation in 95% ethanol.
3. Remove ethanol, rinse twice with distillated water and equilibrate samples in 50% glycerol by 1 h at room temperature.

4. Observe the autofluorescence of phenolic compounds using an epifluorescence microscope (excitation filter, 330–380 nm; emission filter, 435–485 nm).

3.9.3 Callose Deposition

1. Infiltrate leaves with fixing and distaining solution, and incubate at 60 °C for 30 min.
2. Rinse the samples thoroughly in distillated water and incubate them overnight in staining solution (0.01% aniline blue in 150 mM K_2HPO_4, pH 9.5).
3. Wash twice with distilled water.
4. Equilibrate samples in 50% glycerol for 1 h at room temperature.
5. Visualize callose-derived fluorescence using an epifluorescence microscope and the same filters set as for autofluorescence.

3.9.4 ROS Accumulation

1. Detach leaves and submerge their petioles (or whole seedlings in case they are being evaluated) in 15 μM 2′,7′-Dichlorodihydrofluorescein diacetate (DCFH-DA) for 30 min.
2. Rinse the samples thoroughly with water at least twice in order to eliminate any rests from the fluorescent dye.
3. Visualize green fluorescence by epifluorescence microscopy using 460 nm excitation filter and >515 nm emission filter.

4 Notes

1. Usually, *B. cinerea* grows faster and the conidia production is promoted on non-commercial PDB and PDA.
2. Alternatively, a TissueLyser automated sample disruptor (Quiagen) can be used to disrupt samples. Disrupt 2 min at 25 oscillations/s.
3. *adc1* and *adc2* mutant lines grow and reproduce normally without exogenous Put supply.
4. CS9657 (*adc1-3*), CS9658 (*adc1-2*), CS9659 (*adc2-3*), and CS5960 (*adc2-4*) mutant lines on Col-0 background are available at ABRC and were already used by Cuevas et al. [18] and Rossi et al. [13].
5. Samples could be stored for no more that 1–2 weeks at −80 °C since the enzymatic activity decays considerably after this period.
6. By adding tomato leaf powder it is possible to maintain high virulence after several consecutive subcultures.
7. Incubating conidia in the presence of sucrose and KH_2PO_4 promotes germination.

8. It is convenient to use third to foutth leaves pairs, under normal growth conditions these leaves are the biggest and are fully expanded.
9. Because of the high virulence of *B. cinerea*, leaves could be completely necrotic after 72 hpi.
10. Under some growing conditions, Arabidopsis leaves might show a rolling aspect. In this case, it is recommendable to make a small cut with a razorblade on the middle of leaf tip.
11. Two-weeks-old Arabidopsis seedlings normally present 3–4 true leaves pairs.
12. Bacterial number is expressed as CFUs/mg of Fresh Weight or CFUs/seedling.

Acknowledgments

This study was supported by grants from the ANPCyT (PICT2015/2956 and PICT2013/0477). Rossi F, Gárriz A, Marina M, Romero FM, and Ruiz OA are members of the Research Staff of the Consejo Nacional de Investigaciones Científicas y Técnicas (CONICET, Argentina).

References

1. Galston AW, Sawhney RK (1990) Polyamines in plant physiology. Plant Physiol 94:406–410
2. Tiburcio AF, Altabella T, Bitrián M, Alcázar R (2014) The roles of polyamines during the lifespan of plants: from development to stress. Planta 240(1):1–18
3. Fuell C, Elliott KA, Hanfrey CC, Franceschetti M, Michael AJ (2010) Polyamine biosynthetic diversity in plants and algae. Plant Physiol Biochem 48:513–520
4. Bagni N, Tassoni A (2001) Biosynthesis, oxidation and conjugation of aliphatic polyamines in higher plants. Amino Acids 20:301–317
5. Marina M, Maiale SJ, Rossi FR, Romero MF, Rivas EI, Gárriz A, Ruiz OA, Pieckenstain FL (2008) Apoplastic polyamine oxidation plays different roles in local responses of tobacco to infection by the necrotrophic fungus *Sclerotinia sclerotiorum* and the biotrophic bacterium *Pseudomonas viridiflava*. Plant Physiol 147:2164–2178
6. Gonzalez ME, Marco F, Minguet EG, Carrasco-Sorli P, Blázquez MA, Carbonell J, Ruiz OA, Pieckenstain FL (2011) Perturbation of spermine synthase gene expression and transcript profiling provide new insights on the role of the tetraamine spermine in Arabidopsis defense against *Pseudomonas viridiflava*. Plant Physiol 156:2266–2277
7. Jiménez-Bremont JF, Marina M, Guerrero-González MDLL, Rossi FR, Sánchez-Rangel D, Rodrí-guez-Kessler M, Ruiz OA, Gárriz A (2014) Physiological and molecular implications of plant polyamine metabolism during biotic interactions. Front Plant Sci 5:1–14
8. Marina M, Sirera FV, Rambla JL, Gonzalez ME, Blázquez MA, Carbonell J, Pieckenstain FL, Ruiz OA (2013) Thermospermine catabolism increases *Arabidopsis thaliana* resistance to *Pseudomonas viridiflava*. J Exp Bot 64:1393–1402
9. Alcazar R, Bitrián M, Zarza X, Tiburcio AF (2012) Polyamine metabolism and signaling in plant abiotic stress protection. In: Recent advances in pharmaceutical sciences II, Transworld Research Network pp 29–47
10. Hanfrey C, Sommer S, Mayer MJ, Burtin D, Michael AJ (2001) Arabidopsis polyamine biosynthesis: absence of ornithine decarboxylase and the mechanism of arginine decarboxylase activity. Plant J 27:551–560
11. Groppa MD, Benavides MP (2008) Polyamines and abiotic stress: recent advances. Amino Acids 34:35–45

12. Alcázar R, Altabella T, Marco F, Bortolotti C, Reymond M, Koncz C, Carrasco P, Tiburcio AF (2010) Polyamines: molecules with regulatory functions in plant abiotic stress tolerance. Planta 231:1237–1249
13. Rossi FR, Marina M, Pieckenstain FL (2015) Role of arginine decarboxylase (ADC) in *Arabidopsis thaliana* defence against the pathogenic bacterium *Pseudomonas viridiflava*. Plant Biol (Stuttg) 17:831–839
14. Kim S-H, Kim S-H, Yoo S-J, Min K-H, Nam S-H, Cho BH, Yang K-Y (2013) Putrescine regulating by stress-responsive MAPK cascade contributes to bacterial pathogen defense in Arabidopsis. Biochem Biophys Res Commun 437:502–508
15. van Verk MC, Gatz C, Linthorst HJM (2009) Transcriptional regulation of plant defense responses. Adv Bot Res 51:397–438
16. Rossi FR, Gárriz A, Marina M, Romero FM, Gonzalez ME, Collado IG, Pieckenstain FL (2011) The sesquiterpene botrydial produced by *Botrytis cinerea* induces the hypersensitive response on plant tissues and its action is modulated by salicylic acid and jasmonic acid signaling. Mol Plant-Microbe Interact 24:888–896
17. Bradford M (1976) A rapid and sensitive method for the quantitation of microgram quantities of protein utilizing the principle of protein-dye binding. Anal Biochem 72:248–254
18. Cuevas JC, López-Cobollo R, Alcázar R, Zarza X, Koncz C, Altabella T, Salinas J, Tiburcio AF, Ferrando A (2008) Putrescine is involved in Arabidopsis freezing tolerance and cold acclimation by regulating abscisic acid levels in response to low temperature. Plant Physiol 148:1094–1105

Chapter 34

Real-Time In Vivo Monitoring of Reactive Oxygen Species in Guard Cells

Ky Young Park and Kalliopi A. Roubelakis-Angelakis

Abstract

The intra-/intercellular homeostasis of reactive oxygen species (ROS), and especially of superoxides ($O_2^{\cdot -}$) and hydrogen peroxide ($O_2^{\cdot -}$) participate in signalling cascades which dictate developmental processes and reactions to biotic/abiotic stresses. Polyamine oxidases terminally oxidize/back convert polyamines generating H_2O_2. Recently, an NADPH-oxidase/Polyamine oxidase feedback loop was identified to control oxidative burst under salinity. Thus, the real-time localization/monitoring of ROS in specific cells, such as the guard cells, can be of great interest. Here we present a detailed description of the real-time in vivo monitoring of ROS in the guard cells using $H_2O_2^{-}$ and $O_2^{\cdot -}$ specific fluorescing probes, which can be used for studying ROS accumulation generated from any source, including the amine oxidases-dependent pathway, during development and stress.

Key words Reactive oxygen species localization, ROS monitoring, Hydrogen peroxide, Superoxide ion, NADPH-oxidase, Polyamines, Amine oxidases, Guard cells

1 Introduction

Reactive oxygen species (ROS) in the cells are generated by enzymatic and non-enzymatic reactions and their intra/intercellular titers is the result of the balance between generation/scavenging [1, 2]. Superoxide ions.

($O_2^{\cdot -}$) are generated mainly by the respiratory burst oxidase homologs NADPH-oxidases, which are encoded by the *Rboh* genes. Subsequent dismutation of $O_2^{\cdot -}$ by superoxide dismutase (SOD) is one of the major routes for hydrogen peroxide (H_2O_2) production [3–5]. In addition, polyamines (PAs) are highly reactive aliphatic polycations, whose homeostasis affects a vast range of dynamic developmental and metabolic processes and also responses to biotic/abiotic stresses [6–8]. PAs are catalyzed by amine oxidases (AOs), the diamine oxidases (DAOs or copper containing AOs) and the flavin-containing PA oxidases (PAOs). They localize either inter- (i.e. apoplast) or intracellularly (i.e. cytoplasm and

Rubén Alcázar and Antonio F. Tiburcio (eds.), *Polyamines: Methods and Protocols*, Methods in Molecular Biology, vol. 1694, DOI 10.1007/978-1-4939-7398-9_34, © Springer Science+Business Media LLC 2018

peroxisomes). DAOs oxidize mainly Put, but also Spd and Spm (with much lower efficiency) yielding H_2O_2 and aminoaldehydes. The apoplastic PAOs terminally oxidize Spd and Spm yielding aminoaldehydes and H_2O_2, while the intracellular ones, back-convert PAs producing H_2O_2, an aminoaldehyde and a PA with one less aminogroup (in the order tetramine → triamine → diamine) [9–12]. ROS homeostasis signals downstream cellular processes related to growth and development as well as stress responses [13–15]. Guard cells can be efficiently used for real-time in vivo monitoring of the accumulation of H_2O_2 and $O_2^{\cdot-}$ intra/intercellularly, derived from various enzymatic/non-enzymatic sources including the AOs [16], in order to assess their contribution/cross-talk during development or under stress [17]. The localization/quantification of $O_2^{\cdot-}$ and H_2O_2 is implemented by using fluorescing probes and fluorescence confocal images [18–26] (Figs. 1, 2, 3 and 4).

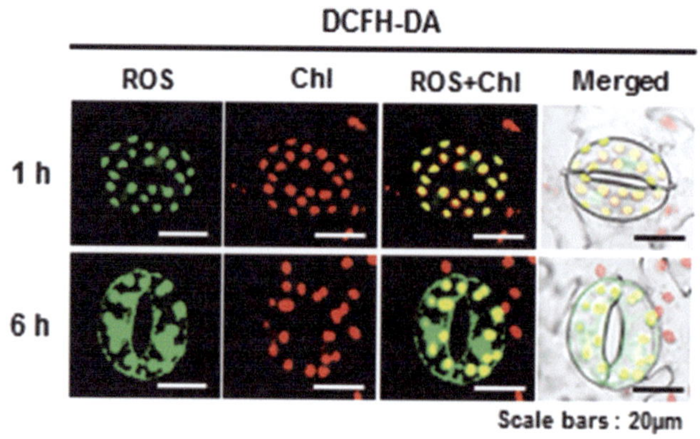

Fig. 1 ROS detection in guard cells of *Nicotiana tabacum* post-NaCl treatment. Confocal laser scanning microscopy (CLSM) images of DCFH-DA fluorescence (*green*; DCFH-DA) and chlorophyll autofluorescence (*red*) at 1 and 6 h post-NaCl treatment

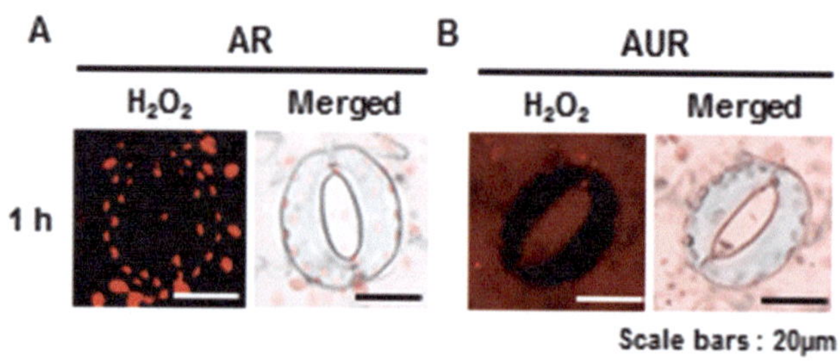

Fig. 2 Intracellular/extracellular H_2O_2 in guard cells of *Nicotiana tabacum* post-NaCl treatment. (**a**) Intracellular H_2O_2 in AR fluorescence. (**b**) Extracellular H_2O_2 in AUR fluorescence

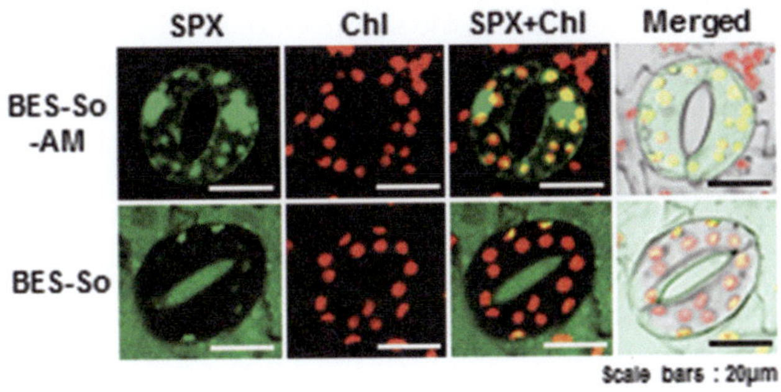

Fig. 3 Intracellular/extracellular $O_2^{\cdot -}$ in guard cells of *Nicotiana tabacum* post-NaCl treatment. Intracellular $O_2^{\cdot -}$ in BES-So-AM fluorescence. Extracellular $O_2^{\cdot -}$ in BES-So fluorescence

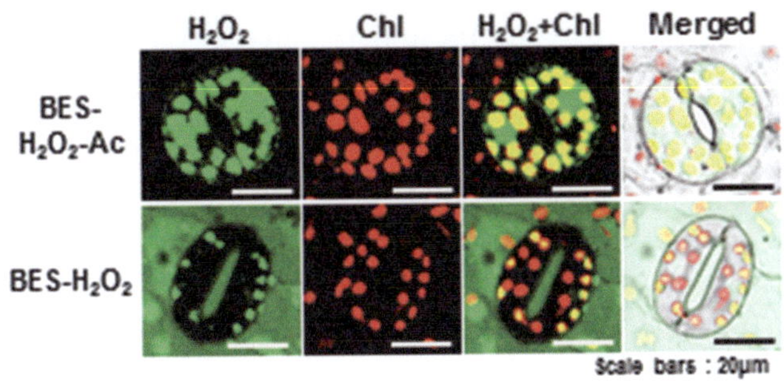

Fig. 4 Intracellular/extracellular H_2O_2 in guard cells of *Nicotiana tabacum* post-NaCl treatment. Intracellular H_2O_2 in BES-H_2O_2-Ac fluorescence. Extracellular H_2O_2 in BES-H_2O_2 fluorescence

2 Materials

All solutions are prepared with sterilized distilled water (SDW) and analytical-grade reagents. The solutions and the reagents are stored at room temperature, unless otherwise specified. All waste disposal regulations should be followed when disposing waste materials.

2.1 Intracellular Localization of ROS with 2′,7′-Dichlorofluorescin Diacetate (DCFH-DA)

1. Potassium phosphate buffer 20 mM, pH 6.0: Prepare two 1 M solutions KH_2PO_4 (monobasic) and K_2HPO_4 (dibasic). The pH of the dibasic will be alkaline (around pH 8), and the monobasic will be weakly acidic (around 6.5, maybe 6.8). Weigh 17.418 g K_2HPO_4, transfer to a glass beaker and add water to a volume 100 mL of dibasic solution. Weigh 13.609 g KH_2PO_4, transfer to a glass beaker and add water to a volume 100 mL of monobasic solution. Add about 85 mL water to a glass beaker, and then add slowly 1.32 mL of 1 M stock K_2HPO_4 and 8.68 mL of 1 M stock KH_2PO_4. Mix and adjust pH to 6.0. Make up to 100 mL with water for making a concentrated stock of 0.1 M. Autoclave at 121 °C for 20 min. Store at room temperature. Before use, dilute with water from the concentrated stock to 20 mM potassium phosphate buffer, and then adjust again pH to 6.0.

2. Reagent stock solution of DCFH-DA: 10 μM DCFH-DA reagent working solution. Prepare 10 mM DCFH-DA (Sigma Chemicals, St Louis, MO, USA) stock solution with dimethyl sulfoxide (DMSO). Weigh 2.4 mg of DCFH-DA white powder and transfer to 1.5 mL tube. Add 500 μL of DMSO for a concentrated stock of 10 mM DCFH-DA and vortex well. Dilute from 10 mM stock to 1 mM stock with DMSO for storage in separate tubes. Store at −20 °C in a black box to protect from light (*see* **Note 1**). The day of the experiment, a working solution of 10 μM DCFH-DA is prepared from the 1 mM stock solution with 20 mM potassium phosphate buffer.
3. 2′,7′-Dichlorofluorescein diacetate.
4. Potassium Phosphate, monobasic (KH_2PO_4).
5. Potassium Phosphate, dibasic (K_2HPO_4).
6. DMSO at room temperature (*see* **Note 2**).
7. Confocal laser scanning microscopy (FV300, OLYMPUS, Japan) equipped with a filter set compatible with blue argon laser (488 nm) and green helium neon (543 nm) laser.
8. Glass microscope slides.
9. Glass cover slips.

2.2 Intracellular Localization of $O_2^{\cdot-}$ with BES-So-AM

1. Reagent solution of BES-So-AM ($C_{31}H_{19}O_{13}NF_4S$, MW = 721.54): 30 μM BES-So-AM reagent working solution. For preparing 3 mM reagent stock solution of BES-So-AM (WAKO, Japan), dissolve 1 mg of BES-So-AM powder with 216 μL of DMSO in a 1.5 mL tube. Store at 4 °C in a black box for protection from light. Just prior to use, dilute 3 mM BES-So-AMstock solution to final concentration of working reagent at 30 μM with 20 mM potassium phosphate buffer.
2. DMSO.
3. Confocal laser scanning microscopy.

2.3 Extracellular Localization of $O_2^{\cdot-}$ with BES-So

1. Reagent solution of BES-So ($C_{28}H_{15}O_{11}NF_4S$, MW = 649.48): 30 μM BES-So reagent working solution. For preparing 3 mM reagent stock solution of BES-So, dissolve 1 mg of BES-So powderin 195 μL of DMSO using an 1.5 mL tube. Store at 4 °C in a black box to protect from light. Just prior to use, dilute 3 mM BES-Sostock to final concentration of working reagent at 30 μM with 20 mM potassium phosphate buffer.

2.4 Intracellular Localization of H_2O_2 with Amplex red (AR) and BES-H_2O_2-Ac

1. Reagent solution of AR (10-acetyl-3,7-dihydroxyphenoxazine, $C_{28}H_{11}F_7O_8S$, MW = 257.25): 50 μM AR reagent working solution. Keep one vial of Amplex® Red reagent (Amplex® Red Hydrogen Peroxide/Peroxidase Assay Kit) from −20 °C into ice and DMSO at room temperature. Dissolve 154 μg of AR

with 600 μL of DMSO in a vial at concentrated stock solution of 1 mM AR, which is stored at −20 °C. For preparing the reagent working solution of AR, just prior to use, dilute 1 mM AR to final concentration of 50 μM with 50 mM sodium phosphate buffer.

2. Reaction buffer for AR: 50 mM sodium phosphate buffer, pH 7.4. Prepare two 1 M solutions of NaH_2PO_4 (monobasic) and Na_2HPO_4 (dibasic). Weigh 12.0 g NaH_2PO_4, transfer to a glass beaker and add water to a volume 100 mL. Weigh 14.2 g Na_2HPO_4, transfer to a glass beaker and add water to a volume 100 mL. Add 85 mL water to a glass beaker, and then add slowly 7.74 mL of 1 M NaH_2PO_4 stock and 2.26 mL of 1 M Na_2HPO_4 stock. Mix and adjust pH to 7.4. Make up to 100 mL with water for making a concentrated stock of 0.1 M. Autoclave at 121 °C for 20 min. Store at room temperature. Before use, dilute from the concentrated stock to 50 mM potassium phosphate buffer with SDW, and then adjust again pH to 7.4.
3. Reagent solution of BES-H_2O_2-Ac (3′-O-Acetyl-6′-O-pentafluorobenzenesulfonyl-2′-7′-difluorofluorescein, MW = 640.44): 50 μM BES-H_2O_2-Ac reagent working solution. For preparing 5 mM reagent stock solution of BES-H_2O_2-Ac (WAKO, Japan), dissolve 1 mg of BES-H_2O_2-Ac powder with 320 μL of DMSO in a 1.5 mL tube, which is stored at 4 °C. Just prior to use, dilute 5 mM BES-H_2O_2-Ac to final concentration of working reagent of 50 μM with 20 mM potassium phosphate buffer.

2.5 Extracellular Localization of H_2O_2 with Amplex Ultra Red (AUR) and BES-H_2O_2

1. Reagent solution of AUR (MW = ~300): 50 μM AUR reagent working solution. Prepare one vial of 1 mg Amplex® UltraRed reagent (Invitrogen, USA) with 340 μL DMSO to make a concentrated stock solution of 10 mM AUR. Vortex well to dissolve. For storage of AUR in separate tubes, dilute from 10 mM AUR stock to 1 mM AUR stock with DMSO and then store at −20 °C. For preparing the reagent working solution of AUR, just prior to use, dilute 1 mM AUR stock to final concentration of 50 μM with 50 mM sodium phosphate buffer.
2. Reaction buffer for AUR: 50 mM sodium phosphate buffer, pH 7.4.
3. Reagent solution of BES-H_2O_2 ($C_{26}H_9O_7F_7S$, MW = 598.40): 50 μM BES-H_2O_2 reagent working solution. For preparing 5 mM reagent stock solution of BES-H_2O_2 (WAKO, Japan), dissolve 1 mg of BES-H_2O_2 powder with 300 μL of DMSO in a 1.5 mL tube, which is stored at 4 °C. Just prior to use, dilute 5 mM BES-H_2O_2 to final concentration of working reagent of 50 μM with 20 mM potassium phosphate buffer.

3 Methods

3.1 Preparation of Leaf Epidermal Tissues

1. Healthy seedlings are grown until appropriate size is reached, when they are used to study the effect of ROS homeostasis in plant growth and developmental phenomena.
2. For abiotic/biotic stress studies, plants are treated with pathogens/abiotic factors, respectively for the desirable time period. In the case of abiotic stress, such as salinity or heavy metals, detached leaves are stressed in a sealed test tube to prevent water loss.

3.2 Loading of ROS ($O_2^{\cdot -}/H_2O_2$) Detection Reagent into Samples for Staining

1. Lower epidermal strips are peeled from leaves and floated on a working reagent solution with dye (*see* **Note 3**).
2. They are incubated with ROS detection solution for indicated time at room temperature in the dark (*see* **Notes 4** and **5**). Ten min for ROS with DCFH-DA, 30–60 min for intracellular $O_2^{\cdot -}$ with BES-So-AM, 30–60 min for extracellular $O_2^{\cdot -}$ with BES-So, 30–60 min for intracellular H_2O_2 with Amplex red (AR) and BES-H_2O_2-Ac, 30–60 min for extracellular H_2O_2 with Amplex ultra red (AUR) and BES-H_2O_2.

3.3 In Situ Detection of ROS ($O_2^{\cdot -}/H_2O_2$) with Confocal Microscopy

1. Lower epidermal strips of 25 mm^2 are stained with ROS ($O_2^{\cdot -}/H_2O_2$) detection reagents, and examined with confocal microscopy equipped with a blue argon laser (488 nm excitation line) for DCFH-DA, BES-So-AM, BES-So, BES-H_2O_2-Ac, and BES-H_2O_2, and a green helium/neon (HeNe) laser (543 nm excitation line) for AR and AUR.
2. Thereafter they are washed quickly with reaction buffer, and mount onto a microscope slide (*see* **Note 6**). Immediately the tissue is covered with a cover slip and the fluorescence images are examined as soon as possible with confocal scanning laser microscope equipped with a cooled CCD camera.
3. The fluorescence confocal images are acquired at different filter sets with 10% of laser power for each detection reagent (*see* **Note 7**). Laser excitation in 480 nm and emission detection at 535 nm for ROS with DCFH-DA. Laser excitation in 505 nm and emission detection at 544 nm for intracellular $O_2^{\cdot -}$ with BES-So-AM. Laser excitation in 505 nm and emission detection at 544 nm for extracellular $O_2^{\cdot -}$ with BES-So. Laser excitation in 571 nm and emission detection at 585 nm for intracellular H_2O_2 with AR. Laser excitation in 485 nm and emission detection at 530 nm for intracellular H_2O_2 with BES-H_2O_2-Ac. Laser excitation in 568 nm and emission detection at 581 nm h for extracellular H_2O_2 with AUR. Laser excitation in 485 nm and emission detection at 530 nm for extracellular H_2O_2 with BES-H_2O_2 (Table 1).

Table 1
Filter set for each type of ROS with a specific detection reagent

Type of ROS	Detection reagent	Excitation laser (nm)	Emission laser (nm)
Intracellular ROS	DCFH-DA	480	535
Intracellular superoxide	BES-So-AM	505	544
Extracellular superoxide	BES-So	505	544
Intracellular H_2O_2	Amplex red	571	585
Intracellular H_2O_2	BES-H_2O_2-Ac	485	530
Extracellular H_2O_2	Amplex ultra red	568	581
Extracellular H_2O_2	BES-H_2O_2	485	530

4. Image analysis can be subsequently performed with Fluoview application software (OLYMPUS FLUOVIEW FV31S-SW) and exported into Adobe PhotoShop 7.0 for image presentation.

4 Notes

1. The detection reagents of both ROS, $O_2^{\cdot-}$ and H_2O_2 are susceptible to photo-oxidation. Thus, they must be protected from light and repeated freeze/thaw cycles should be avoided.
2. DMSO is hazardous. Reagents containing DMSO should be handled using equipment and practices appropriate for the hazards posed by such materials. Avoid contact with skin and eyes and do not swallow.
3. The reagents must be prepared just prior to use. Working solutions should be freshly prepared. Because diluted ROS probes in a reaction mixture oxidize more readily and the presence of moisture facilitates the decomposition of the dye, excess diluted probes should be discarded.
4. The retention time for loading of reagent onto samples has to be decided for optimal detection, because different exposure time may be required for each dye. This depends on the specificity of dye, plant tissues and organs or plant species.
5. Use of longer retention time for the reagent loading onto plant tissues may result to extracellular labeling which will appear on fluorescence microscopy image.
6. Bubbles can be formed in the slide, which may disturb viewing with the dissecting microscope. To exclude, slowly slide the cover glass with forceps until the bubble has moved away from the tissue. Do not lift the cover glass. This will damage the

tissue or result in tissue loss. If you want to make bubble-free slide, add one drop of 5% glycerol on the tissues before placing cover glass.

7. The detection dyes are susceptible to photo-oxidation. Thus, low light conditions should be used during fluorescence microscopy applications whenever possible. Therefore, in situ analysis of fluorescence image using ROS detection reagents should be performed rapidly within 2 min with the laser power as low as possible (about 10%). This reduces artifacts as much as possible.

Acknowledgement

This work was supported by EU and Greek national funds through the research funding program THALES (MIS 377281 to K.A.R.-A.). Also, by the fund (NRF-2014R1A2A2A01005145) from National Research Foundation of Korea to K.-Y. Park and implemented in the frame of COST Action FA1106.

References

1. Gilroy S, Suzuki N, Miller G, Choi WG, Toyota M, Devireddy AR, Mittler R (2014) A tidal wave of signals: calcium and ROS at the forefront of rapid systemic signalling. Trends Plant Sci 19:623–630
2. Foyer CH, Noctor G (2015) Stress-triggered redox signaling: What's in pROSpect? Plant Cell Environ. doi:10.1111/pce.12621
3. Torres MA, Dangl JL, Jones JD (2002) Arabidopsis *gp91phox* homologues *AtrbohD* and *AtrbohF* are required for accumulation of reactive oxygen intermediates in the plant defence response. Proc Natl Acad Sci U S A 99:517–522
4. Kwak JM, Mori IC, Pei ZM, Leonhardt N, Torres M, Dangl JL (2003) NADPH oxidase *Atrbohd* and *Atrbohf* genes function in ROS-dependent ABA signaling in Arabidopsis. EMBO J 22:2623–2633
5. Wang GF, Li WQ, Li WY, Wu GL, Zhou CY, Chen KM (2013) Characterization of rice NADPH oxidase genes and their expression under various environmental conditions. Int J Mol Sci 14:9440–9458
6. Paschalidis KA, Roubelakis-Angelakis KA (2005a) Sites and regulation of polyamine catabolism in the tobacco plant. Correlations with cell division/expansion, cell cycle progression, and vascular development. Plant Physiol 138:2174–2184
7. Paschalidis KA, Roubelakis-Angelakis KA (2005b) Spatial and tem- poral distribution of polyamine levels and polyamine anabolism in different organs/tissues of the tobacco plant. Correlations with age, cell division/expansion, and differentiation. Plant Physiol 138:142–152
8. Wu JY, Shang ZL, Wu J, Jiang XT, Moschou PN, Sun WD, Roubelakis-Angelakis KA, Zhang SL (2010) Spermidine oxidase-derived H_2O_2 regulates pollen plasma membrane hyperpolarization-activated Ca^{2+}-permeable channels and pollen tube growth. Plant J 63:1042–1053
9. Angelini R, Cona A, Federico R, Fincato P, Tavladoraki P, Tisi A (2010) Plant amine oxidases "on the move": an update. Plant Physiol Biochem 48:560–564
10. Moschou PN, Sanmartin M, Andriopoulou AH, Rojo E, Sanchez-Serrano JJ, Roubelakis-Angelakis KA (2008) Bridging the gap between plant and mammalian polyamine catabolism: a novel peroxisomal polyamine oxidase responsible for a full back-conversion pathway in *Arabidopsis*. Plant Physiol 147:1845–1857
11. Moschou PN, Sarris PF, Skandalis N, Andriopoulou AH, Paschalidis KA, Panopoulos NJ, Roubelakis-Angelakis KA (2009) Engineered polyamine catabolism preinduces tolerance of

tobacco to bacteria and oomycetes. Plant Physiol 149:1970–1981

12. Moschou PN, Paschalidis KA, Delis ID, Andriopoulou AH, Lagiotis GD, Yakoumakis DI, Roubelakis-Angelakis KA (2008) Spermidine exodus and oxidation in the apoplast induced by abiotic stress is responsible for H_2O_2 signatures that direct tolerance responses in tobacco. Plant Cell 20:1708–1724
13. Pal M, Szalai G, Janda T (2015) Speculation: polyamines are important in abiotic stress signaling. Plant Sci 237:16–23
14. Tiburcio AF, Altabella T, Bitrián M, Alcázar R (2014) The roles of polyamines during the lifespan of plants: from development to stress. Planta 240:1–18
15. Baxter A, Mittler R, Suzuki N (2014) ROS as key players in plant stress signaling. J Exp Bot 65:1229–1240
16. Song Y, Miao Y, Song CP (2014) Behind the scenes: the roles of reactive oxygen species in guard cells. New Phytol 201:1121–1140
17. Gémes K, Kim Y-J, Park K-Y, Moschou PN, Andronis E, Valassaki C, Roussis A, Roubelakis-Angelakis KA (2016) An NADPH-oxidase/polyamine oxidase feedback loop controls oxidative burst under salinity in *Nicotiana tabacum*. Plant Physiol. doi:10.1104/pp.16.01118
18. Si F, Liu Y, Yan K, Zhong W (2015) Mitochondrion targeting fluorescent probe for imaging of intracellular superoxide radical. Chem Commun 51:7931–7934
19. Maeda H, Yamamoto K, Nomura Y, Kohno I, Hafsi L, Ueda N, Yoshida S, Fukuda M, Fukuyasu Y, Yamauchi Y, Itoh N (2005) A design of fluorescent probes for superoxide based on a nonredox mechanism. J Am Chem Soc 127:68–69
20. Maeda H, Yamamoto K, Kohno I, Hafsi L, Itoh N, Nakagawa S, Kanagawa N, Suzuki K, Uno T (2007) Design of a practical fluorescent probe for superoxide based on protection-deprotection chemistry of fluoresceins with benzenesulfonyl protecting groups. Chemistry 13:1946–1954
21. Kristiansen KA, Jensen PE, Møller IM, Schulz A (2009) Monitoring reactive oxygen species formation and localisation in living cells by use of the fluorescent probe CM-H_2DCFDA and confocal laser microscopy. Physiol Plant 136:369–383
22. Watkins JM, Hechler PJ, Muday GK (2014) Ethylene-induced flavonol accumulation in guard cells suppresses reactive oxygen species and moderates stomatal aperture. Plant Physiol 164:1707–1717
23. Allan AC, Lapidot M, Culver JN, Robert Fluhr R (2001) An early tobacco mosaic virus-induced oxidative burst in tobacco indicates extracellular perception of the virus coat protein. Plant Physiol 126:97–108
24. Ortega-Villasante C, Burén S, Barón-Sola A, Martínez F, Hernández LE (2016) In vivo ROS and redox potential fluorescent detection in plants: present approaches and future perspectives. Methods 109:92–104
25. Lachauda C, Silvaa D, Amelota N, Béziata C, Brièrea C, Cotellea V, Grazianaa A, Grata S, Mazarsa C, Thuleaua P (2011) Dihydrosphingosine-induced programmed cell death in tobacco BY-2 cells is independent of H_2O_2 production. Mol Plant 4:310–318
26. Ashtamker C, Kiss V, Sagi M, Davydov O, Fluhr R (2007) Diverse subcellular locations of cryptogein-induced reactive oxygen species production in tobacco Bright Yellow-2 cells. Plant Physiol 143:1817–1826

Chapter 35

Genome-Wide Association Mapping Analyses Applied to Polyamines

Luis Barboza-Barquero, Paul Esker, and Rubén Alcázar

Abstract

Genome Wide Association Studies (GWAS) allow the use of natural variation to understand the genetics controlling specific traits. Efficient methods to conduct GWAS in plants have been reported. This chapter provides the main steps to conduct and analyse GWAS in *Arabidopsis thaliana* using polyamine levels as trait. This approach is suitable for the discovery of genes that modulate the levels of polyamines, and can be used in combination with different types of stress.

Key words Natural variation, GWAS, Polyamines

1 Introduction

Genome Wide Association Studies (GWAS) make use of natural variation to conduct gene mapping. To identify the genetics controlling natural variation three methods can be considered: (1) Quantitative Trait Loci (QTL) analysis using the progeny of crosses among accessions, (2) bulked sample analysis (BSA) which employs selected and pooled individuals (based on extreme phenotypes) derived from biparental populations [1], and (3) GWAS using individuals collected from different parts of the world [2]. QTL mapping approaches have a low gene mapping resolution meaning loci with many (even hundreds) potential gene candidates controlling the trait of interest are mapped. Therefore, additional experiments are required that involve fine-mapping [2]. Another method is BSA which has been successfully used to map polyamine transporters involved in paraquat tolerance in *Arabidopsis* [3]. The advantage of the BSA, compared with QTL or GWAS, is the reduction in scale and costs of the mapping experiments, mainly because only individuals with extreme phenotypes are used. Similarly to QTL analyses, BSA requires further experiments and crosses to fine map the gene(s) controlling the trait. In contrast with QTL

Rubén Alcázar and Antonio F. Tiburcio (eds.), *Polyamines: Methods and Protocols*, Methods in Molecular Biology, vol. 1694, DOI 10.1007/978-1-4939-7398-9_35, © Springer Science+Business Media LLC 2018

and BSA analyses, GWAS facilitates direct mapping of genes affecting a phenotype without the need for experimental crosses [4, 5]. In the specific case of polyamines, GWAS have been applied to map genes involved in tolerance to the polyamine oxidase inhibitor guazatine, in which loss-of-function mutants of *CHLOROPHYLLASE* genes are more tolerant to this herbicide than wild-type genotypes [6].

One relevant factor when considering the use of GWAS is the population structure, which means that some genotypes can be in linkage disequilibrium with each other, for instance, due to a common origin, thus leading to false genotype-to-phenotype associations [5]. To correct the population structure, protocols using linear mixed models have been applied, in which a kinship matrix inferred from the genotypes is considered in the analysis as a nonrandom effect [7, 8]. Furthermore, methodologies have been developed in which there is no need to correct for population structure, allowing the mapping of genomic regions, which could have not been mapped using linear mixed models [9]. Some difficulties with the use of GWAS include the presence of epistatic interactions, and the involvement of rare alleles in the traits under analysis [10].

2 Materials

2.1 Representative set of *Arabidopsis thaliana* Accessions

For conducting GWAS, naturally occurring variation is needed. In the case of *Arabidopsis thaliana*, a large number of natural accessions are available in germplasm stocks such as the Nottingham Arabidopsis Stock Centre (NASC), the Arabidopsis Biological Resource Center (ABRC), and the RIKEN Bioresource Center (BRC)/SENDAI Arabidopsis Seed Stock Center (SASSC). Also, it is possible to consider using published populations from GWAS, including those that have 107 individuals [11], 473 [12], or 1386 [13]. Currently, online platforms to conduct GWAS are available for Arabidopsis community research, e.g., GWAPP [13] and easyGWAS [14], and those continue to expand the number of accessions and mapping tools available for the research community in a user-friendly environment.

2.2 Genotype Data

Allele data from each accession is required to conduct GWAS. Most studies have genotyped accessions using SNP arrays, which after quality control have yielded between 216,130 SNPs [11] and 213,497 SNPs [12] for Arabidopsis. SNP data with minor allele frequency lower than 10% is usually filtered out of the analysis. Since most of the Arabidopsis accessions are genetically stable, once an accession is genotyped it is possible to continue to use the genotype data in additional independent studies. As mentioned above, online platforms for GWAS contain the allele data required for the analysis [13, 14]. Furthermore, next generation sequencing technologies enables the efficient genotyping of materials suitable for GWAS [15, 16]. This can be especially useful in crops in which no genotype data is typically available for specific accessions. For instance, recent GWAS in rice identified new genes associated with agronomical traits [17].

2.3 Phenotype Data

Polyamine level quantification is determined using high-performance liquid chromatography [6, 18].

3 Methods

3.1 Data Sets Preparation

1. Calculate descriptive statistics for the phenotype data (*see* **Note 1**).
2. Check for normality. For instance, perform histograms and Shapiro Wilks tests for normality. GWAS assume that the phenotype data has a normal distribution. Studies have reported that lack of normality can affect the identification of the causal polymorphisms [19] (*see* **Note 2**).
3. Prepare the phenotype file (Table 1).
4. Prepare the genotype file in a transposed ".tped" format. Table 2 shows an example of the genotype file.
5. Prepare the kinship matrix to correct for the population structure using the EMMAX-Kin program [8, 20].

Table 1
Example of the phenotype file

1	CS28636	2.97
2	CS28637	1.73
3	CS28640	6.15

Only three accessions are shown. The first column is the family ID, the second the individual ID, and the third column is the phenotype value. The file can be saved using a txt format

Table 2
Example of the genotype file matrix

1	m1	657	2	2	2	2	2	2
1	m2	3102	2	2	2	2	1	1
1	m3	4648	2	2	2	2	1	1

Only three markers are shown. The first column is the chromosome; the second the marker name; the third the position in the genome. Starting from the fourth column, the SNP data (in binary format in this case 1 and 2 representing the alleles) for each accession is displayed. In this example data are presented for six accessions

3.2 Mapping and Results Interpretation

1. Locate all data files under the same folder.
2. Run the mapping procedure. In the case of EMMAX [8, 20] run the following script in a Linux environment:

   ./emmax -v -d 10 -t tped_prefix -p phenotype_file.txt -k kinship_file -o output_prefix.txt
3. Plot and interpret the results. The output files will appear in the same folder where all the data sets are saved (*see* **Note 3**).
4. Do a quality control on the results with a QQ-plot (*see* **Note 4**).
5. Find causable genes associated with significant markers. Because the physical position of the SNP markers is known, then it is possible to know the exact gene where they are located.
6. Conduct pairwise linkage disequilibrium (LD) analysis between SNPs located in the region with the highest associations (*see* **Note 5**).
7. Validate associations with mutant analysis. The use of mutants, for instance T-DNA mutants available at NASC can be employed to validate the phenotype of individuals carrying mutations in genes carrying markers with high association scores.

4 Notes

1. This initial analysis is important to evaluate the reproducibility of each of the phenotypes. A high value of broad sense heritability (H_2) (closer to one) indicates a high reproducibility among the different replicates, which means a high genotype effect, as well as high precision for the method to quantify the polyamines. If these values are low, it is important to check the data for outliers that interfere with the reproducibility of the replicates, discard possible technical errors, and even considering modifications in the experimental design (e.g., increase the number of repetitions).
2. A lack of normal distribution in the phenotype can affect the GWAS, nonetheless, it has also been noted that the use of transformation can increase the rate of false positive associations [19].
3. The .ps file contains the *P-values* of the association tests. It is convenient to create a new matrix file where one column contains the marker name, another the position in the genome and a last column with the obtained *P-values* for each marker. It is also possible to add another column with the chromosome number, since that can be helpful in the results interpretation. Plotting the results can be conducted using R packages, such as

"qqman" [21], in which tools are available for performing Manhattan plots. Furthermore, it is possible to include a correction for multiple testing. This correction is necessary since, the more markers tested, the higher the possibilities of finding associations just by chance (for more details *see* [22]). One stringent correction is the Bonferroni method, which is calculated by dividing the significance level (α) by the number of tested hypotheses. Most of the GWAS reporting strong associations have a group of markers associated with the trait rather than single markers with strong associations. Most of those single associations may be considered as false positives.

4. To make a QQ-plot is a good practice to check for confounding effects. QQ-plots can be conducted using the qqman R package [21]. It plots the observed *P-value* for all tested associations between phenotypes and SNPs on the y axis versus the expected uniform distribution of the *P-values* under the null hypothesis of no association on the x axis [21].
5. LD analysis can be performed with the R package LD heatmap [23]. LD indicates how an SNP is inherited with another, and the R^2 values are used to measure allelic correlations, ranging from 0 to 1, being 1 a complete LD between two markers. High R^2 values between pairs of SNPs indicate the possibilities of additional markers providing similar information as the one fount for the association under study [24].

References

1. Zou C, Wang P, Xu Y (2016) Bulked sample analysis in genetics, genomics and crop improvement. Plant Biotechnol J 14:1941–1955
2. Weigel D (2011) Natural variation in Arabidopsis: from molecular genetics to ecological genomics. Plant Physiol 158:2–22
3. Fujita M, Fujita Y, Iuchi S et al (2012) Natural variation in a polyamine transporter determines paraquat tolerance in Arabidopsis. Proc Natl Acad Sci U S A 109:6343–6347
4. Baxter I, Brazelton JN, Yu D et al (2010) A coastal cline in sodium accumulation in *Arabidopsis thaliana* is driven by natural variation of the sodium transporter AtHKT1;1. PLoS Genet 6:e1001193
5. Korte A, Farlow A (2013) The advantages and limitations of trait analysis with GWAS: a review. Plant Methods 9:29
6. Atanasov KE, Barboza-Barquero L, Tiburcio AF, Alcázar R (2016) Genome wide association mapping for the tolerance to the polyamine oxidase inhibitor guazatine in *Arabidopsis thaliana*. Front Plant Sci 7:401
7. Kang HM, Zaitlen NA, Wade CM et al (2008) Efficient control of population structure in model organism association mapping. Genetics 178:1709–1723
8. Kang HM, Sul JH, Service SK et al (2010) Variance component model to account for sample structure in genome-wide association studies. Nat Genet 42:348–354
9. Klasen JR, Barbez E, Meier L et al (2016) A multi-marker association method for genome-wide association studies without the need for population structure correction. Nat Commun 7:13299
10. Ingvarsson PK, Street NR (2011) Association genetics of complex traits in plants: Tansley review. New Phytol 189:909–922
11. Atwell S, Huang YS, Vilhjálmsson BJ et al (2010) Genome-wide association study of 107 phenotypes in *Arabidopsis thaliana* inbred lines. Nature 465:627–631
12. Li Y, Huang Y, Bergelson J et al (2010) Association mapping of local climate-sensitive quantitative trait loci in *Arabidopsis thaliana*. Proc Natl Acad Sci U S A 107:21199–21204

13. Seren U, Vilhjalmsson BJ, Horton MW et al (2012) GWAPP: a web application for genome-wide association mapping in Arabidopsis. Plant Cell 24:4793–4805
14. Grimm DG, Roqueiro D, Salome P et al (2016) easyGWAS: a cloud-based platform for comparing the results of genome-wide association studies. Plant Cell 29:5–19
15. Elshire RJ, Glaubitz JC, Sun Q et al (2011) A robust, simple genotyping-by-sequencing (GBS) approach for high diversity species. PLoS One 6:e19379
16. Ott A, Liu S, Schnable JC, et al (2017) Tunable genotyping-by-sequencing (tGBS®) enables reliable genotyping of heterozygous loci. bioRxiv
17. Yano K, Yamamoto E, Aya K et al (2016) Genome-wide association study using whole-genome sequencing rapidly identifies new genes influencing agronomic traits in rice. Nat Genet 48:927–934
18. Marcé M, Brown DS, Capell T et al (1995) Rapid high-performance liquid chromatographic method for the quantitation of polyamines as their dansyl derivatives: application to plant and animal tissues. J Chromatogr B Biomed Sci Appl 666:329–335
19. Goh L, Yap VB (2009) Effects of normalization on quantitative traits in association test. BMC Bioinformatics 10:415
20. Kang HM (2010) Efficient Mixed-Model Association eXpedited (EMMAX) http://genetics.cs.ucla.edu/emmax/
21. Turner SD (2014) qqman: an R package for visualizing GWAS results using Q-Q and manhattan plots. bioRxiv
22. Noble WS (2009) How does multiple testing correction work? Nat Biotechnol 27:1135–1137
23. Shin J-H, Blay S, McNeney B, Graham J (2006) LDheatmap: an R function for graphical display of pairwise linkage disequilibria between single nucleotide polymorphisms. J Stat Soft 16: Code Snippet 3
24. Bush WS, Moore JH (2012) Chapter 11: Genome-wide association studies. PLoS Comput Biol 8:e1002822

Chapter 36

Polyamine Metabolism in Climacteric and Non-Climacteric Fruit Ripening

Ana Margarida Fortes and Patricia Agudelo-Romero

Abstract

Polyamines are small aliphatic amines that are found in both prokaryotic and eukaryotic organisms. These growth regulators have been implicated in abiotic and biotic stresses as well as plant development and morphogenesis. Several studies have also suggested a key role of polyamines during fruit set and early development. Polyamines have also been linked to fruit ripening and in the regulation of fruit quality-related traits.

Recent studies indicate that during ripening of both climacteric and non-climacteric fruits, a decline in total polyamine contents is observed together with an increased catabolism of these growth regulators.

In this review, we explore the current knowledge on polyamine biosynthesis and catabolism during fruit set and ripening. The study of the role of polyamine metabolism in fruit ripening indicates the possible application of these natural polycations to control ripening and postharvest decay as well as to improve fruit quality traits.

Key words Fruit ripening, Grape, Polyamines, Polyamine catabolism, Polyamine oxidase, Tomato, *Vitis vinifera*

1 Introduction

An appropriate fruit development, followed by a correct maturation, guarantee fruits with an excellent quality for the consumers. Ripening is the process in which fruit biochemistry and physiology are altered including changes in color, texture, aroma, flavor, and nutritional characteristics [1]. Fruit ripening is dependent on a highly coordinated network of endogenous and exogenous signals involving hormones among others. Fleshy fruits are classified into climacteric and non-climacteric, depending on the presence or absence of the climacteric increase in the respiration and ethylene production during ripening [2]. Indeed in climacteric fruits (i.e. tomato, banana, apple, pears, mangoes, papaya, and avocado), ethylene constitutes a major cue that controls most ripening features. Although ethylene does not play such a major role in

Rubén Alcázar and Antonio F. Tiburcio (eds.), *Polyamines: Methods and Protocols*, Methods in Molecular Biology, vol. 1694, DOI 10.1007/978-1-4939-7398-9_36,

non-climacteric fruit ripening (i.e. grape, citrus, strawberry, pepper, and oil palm fruit) [3] a complex interaction among different phytohormones occurs involving also ethylene [4]. In grape, abscisic acid (ABA), brassinosteroids (BRs), and ethylene have been reported to promote ripening, while auxin delays some ripening-associated processes and also interacts with other hormones such as ABA and ethylene (reviewed by [4]). Recently, a decline of polyamine contents during grape ripening along with an increased catabolism of these growth regulators was observed in three cultivars of grapevine [5]. Moreover, recent studies in tomato employing genetics approaches have provided insight about the involvement of polyamines in fruit ripening [6–10]. In addition, polyamines have also been linked to fruit ripening by controlling the quality of climacteric and non-climacteric fruits [5, 11–13].

Polyamines (PAs) are small aliphatic amines that are found in both prokaryotic and eukaryotic organisms. In plants, the most abundant polyamines are putrescine (Put), spermidine (Spd) and spermine (Spm). Nevertheless, less abundant polyamines have also been reported such as cadaverine, thermospermine, norspermidine, and norspermine. PAs can be present in free form, as well as in soluble conjugated, and insoluble bound forms. In the case of soluble conjugated, PAs are covalently conjugated to small molecules such as phenolic compounds; whereas in the insoluble bound form, PAs are covalently bound to macromolecules such as nucleic acids and proteins. Physiologically, PAs are growth regulators that have been implicated mainly in abiotic and biotic stresses [14–19]; as well as a wide range of metabolic processes such as plant development and morphogenesis [20–23], senescence [24, 25], and fruit development and ripening [6, 13, 26, 27].

The biosynthetic pathway of PAs is represented in Fig. 1. In plants, the diamine Put synthesis proceeds through arginine decarboxylase via agmatine (Agm) or/and ornithine decarboxylase, from arginine (Arg) and ornithine (Org), respectively. Conversion of diamine Put to triamine Spd and tetramine Spm requires successive addition of aminopropyl moieties by Spd synthase and Spm synthase, respectively [15]. Catabolic pathway of PAs is mainly starring by Diamine oxidases (CuAO; EC 1.4.3.6) and Polyamine oxidases (PAO; EC 1.5.3.3) (Fig. 1). Diamine oxidases are copper-containing enzymes and catabolize Put [14]. Polyamine oxidases are flavin adenine dinucleotide-dependent (FAD) enzymes and are involved in the terminal catabolism of triamine Spd and tetramine Spm. Additionally, PAs pathway has been linked to a larger metabolic network in which ethylene, γ-aminobutyric (GABA), nitric oxide (NO), Krebs cycle (TCA), and abscisic acid (ABA) are involved [15].

Polyamine catabolism is active in several developmental processes such as fruit ripening [5, 28] and leave senescing [25]. Nevertheless, PA catabolism is poorly studied during fruit ripening

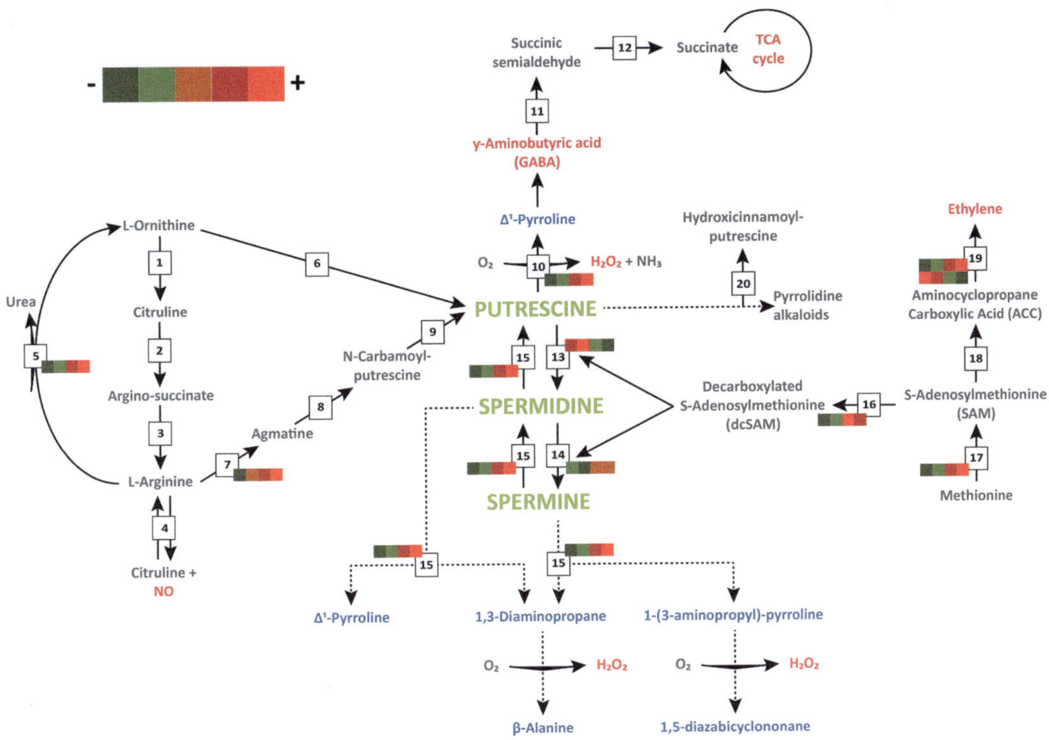

Fig. 1 Metabolism of polyamines, their interaction with other metabolic routes and heatmaps of gene expression obtained in three cultivars of grape during fruit ripening. The heatmaps present the tendency in gene expression among three Portuguese cultivars as assessed by realtime qPCR in Agudelo-Romero and coworkers [5]. The figure was adapted from [15]. Continuous lines indicate PAs biosynthetic pathway and related metabolites. *Dashed lines* display the formation of putrescine-derived alkaloids, polyamine conjugation and catabolic processes. Numbers represent the following enzymes: (*1*) ornithine-carbamoyl transferase, (*2*) arginosuccinate synthase, (*3*) arginosuccinate lyase, (*4*) nitric oxide synthase, (*5*) arginase, (*6*) ornithine decarboxylase, (*7*) arginine decarboxylase, (*8*) agmatine iminohydrolase, (*9*) N-carbamoylputrescine amidohydrolase, (*10*) diamine oxidase, (*11*) γ-aminobutyrate aminotransferase, (*12*) succinic semialdehyde dehydrogenase, (*13*) spermidine synthase, (*14*) spermine synthase, (*15*) polyamine oxidase, (*16*) SAM decarboxylase, (*17*) SAM synthetase, (*18*) ACC synthase, (*19*) ACC oxidase, (*20*) putrescine hydroxycinnamoyl transferase

though several studies support the idea that PAs promote fruit ripening and fruit quality. In fact, when exogenous PAs are supplied to fruits improvements are noticed in quality features such as fruit set, fruit size, postharvest decay [11, 29, 30] and decreased fungal infection [30, 31]. Furthermore, it has been reported that preharvest applications of a strong inhibitor of polyamine oxidase to grapevine berries (cv. Trincadeira) produce phenotypic and quality changes in the berries. Altogether, these evidences support an important role of PAs anabolism and catabolism during fruit development and ripening [13].

2 Polyamine Metabolism During Fruit Set and Development

Fruit set is one of the most important processes in the sexual reproduction of flowering plants and is regulated by endogenous and exogenous signals. This stage can be defined as the transition of a quiescent ovary to a rapidly growing young fruit [32]. During the first few weeks after fertilization fruits undergo an intensive cell division phase which then ceases and is followed by a an increase in size and weight due to cell enlargement [33]. Early studies in pea using exogenous spermine indicated an important role of this polyamine in ovary senescence and fruit set and development [34]. Ovaries that fail to set fruits are fated to senesce; changes in the spermine levels were shown to modulate ovary fate in tomato [35]. Additionally, Slocum and Galston [36] noticed that during ovary enlargement of tobacco, ODC activity increased between anthesis and fertilization stages, while polyamine levels remained constant until fertilization when they increased dramatically.

Several studies have suggested a key role of polyamines during climacteric and non-climacteric fruit set and early development namely in apple [37], avocado [38], grape [5, 39], mango [40], oil palm [41], olive [42], orange [43], pea [34], peach [44–47], pepper [48], tobacco [36], and tomato [35, 49].

In general, higher polyamine levels have been noticed at early stages of fruit development accompanied by a gradual decrease afterwards. Biasi and coworkers [37] reported that in apple free polyamines content was high during the first weeks after full bloom and then decreased gradually. The same holds true for others climacteric fruits such as avocado, peach, mango, and olive [38, 40, 42, 44].

Similar findings have been reported in non-climacteric fruits such as grape [5]. In this study, polyamine levels from three genotypes of grapevine were assessed during one ripening season and in all of them the same pattern was observed with total free and soluble polyamines higher in green fruits coinciding with cell division phase. The decline in polyamine levels during ripening stages was also observed for the three genotypes and are in agreement with the results reported by [39] in the grapevine cultivar "Muscar Bailey A" and also in pepper, another non-climacteric fruit [48]. Interestingly, a different trend was observed in non-climacteric orange and oil palm fruits [41, 43]. Tassoni and coworkers [43] reported that the amount of free Put increased, both in the peel and in the flesh during ripening and decreased in over ripe oranges, reaching maximum values in the peel and in the flesh of ripe fruits. A similar trend was also observed for Spd though with a much lower amplitude. In addition, Put in both soluble and insoluble conjugated forms reached a maximum in the peel of ripe fruits.

Polyamines have been shown to play a role in the stimulation of cell division and growth in several higher plants [50]. In grapevine Trincadeira cultivar, which presents a remarkable irregular ripening among seasons, polyamine metabolism was studied during two consecutive seasons. In one of the seasons the berries presented higher weight at fruit set stage comparing to the other season and this was correlated with higher levels of PAs [5]. The same was observed in two cultivars of olives (*Olea europaea* L.), in which differences were noticed in the fruit size and polyamine concentrations [42]. These authors suggested that changes in PA metabolism during early fruit development may be part of a mechanism controlling the transitions from cell division to the developmental acquisition of cell size, depending on PA concentrations maintained within strict limits.

Polyamines are synthetized from arginine and ornithine through ADC and ODC enzymes [15]. The role of ADC seems to be prevalently involved in stress responses [15, 51–53] which actually occur during fruit ripening [26], whereas ODC seems to play a key role during cell division [54]. In avocado, a polyamines burst occurred during the onset of fruit development, then polyamines levels declined until ripe stage. ODC and ADC activities showed a similar pattern, presenting a peak during cell proliferation but with ODC showing higher activity than ADC [38]. The same holds true in peach "Biscoe," where polyamines levels were significantly higher along with ADC and ODC activities at early stages of fruit development. Remarkably, ODC activity was also higher than ADC [44]. In two additional studies in peach, ADC and ODC presented a peak in activity at the same stages [12, 47]. These studies support the idea that polyamines are synthetized with the involvement of ADC and ODC at the beginning of fruit development eventually as a source of nitrogen for cell division and growth [50]. In addition, it was also suggested that PA levels influence the efficacy of ATP production and the regulation of sink strength in the early stages of development of fast-growing organs such as fruits [55, 56]. However, ODC pathway does not appear to contribute to the biosynthesis of PAs in non-climacteric fruits such as grapes where the major pathway of PAs biosynthesis is via arginine and agmatine [5, 26, 57].

In tomato, in situ localization of ADC and ODC transcripts conducted at early stages of fruit development (14 days after flowering) revealed the expression of these genes in ovules, endosperm and vascular bundles; suggesting the polyamine biosynthetic enzymes are involved in cellular growth [28]. In fruits analyzed 28 days after flowering, just before mature green stage, ADC and ODC transcripts presented an intense localization in the locular parenchyma. This fact was related to a putative role of PAs in cell expansion and tissue development [28]. However, it should be noted that transcript profiles of both enzymes may not correlate with their activities [12].In fact, maximum ADC expression can be associated with non-dividing expanding cells, including fruit tissues [35, 58].

Taken together, these observations support the hypotheses that polyamines might play a key role in the regulation of the transition from flower to fruit as well as during fruit set and growth. However, the specific physiological role of polyamines during fruit set and development, along with the mechanisms influencing the stimulation of cell division and growth are still largely unknown.

3 Polyamine Metabolism During Fruit Ripening

3.1 Polyamine Biosynthesis in Climacteric and Non-Climacteric Fruit Ripening

Polyamine metabolism in grape was examined using molecular and biochemical approaches in three important Portuguese cultivars (Trincadeira, Touriga Nacional, and Aragonês) during two productive seasons [5] (Fig. 1). Microarray and real-time qPCR studies revealed upregulation of a gene coding for arginine decarboxylase (ADC) during grape ripening in all the varieties [5, 26]. This suggested that biosynthesis of polyamines was being activated during ripening in this non-climacteric fruit where ethylene, that shares a common biosynthetic precursor (SAM), plays a less clear and eventually pivotal role than in climacteric fruits. In fact, genes coding for ACC synthase (ACS) and ACC oxidase (ACO) were essentially downregulated at ripe stage when compared to *veraison* suggesting a decrease in ethylene biosynthesis during ripening as previously reported for other grape varieties [59, 60]. However, the increase in the expression of *ADC* was not accompanied by an increase in free and conjugated polyamines that presented a strong decrease [5]. In all cultivars the levels of Putrescine and Spermidine presented a strong decrease during ripening whereas Spermine had a more or less constant content during ripening. Low free polyamine contents of the ripe/overripe grape pericarp have been reported for other grape cultivars [39, 57] and other non-climacteric fruits such as pepper and citrus [43, 48]. In climacteric fruits such as peach, polyamines also reach minimum levels at ripening when the growth of fruit ceases and the climacteric ethylene emission starts [11, 12]. Nevertheless, Tsaniklidis and coworkers [28] showed recently that in tomato Put levels increased progressively during fruit maturation and reached maximum at red ripe stage. In contrast, Spd and Spm levels exhibited their lowest values at this stage accounting for the low total free polyamine content observed in ripe tomato. ADC gene expression was also minimal for these samples highlighting that ADC is not mainly regulated at the transcriptional level [5]. In another report in the non-climacteric oil palm fruit, putrescine and spermidine increased during fruit ripening [41]. In peach, a climacteric fruit [47] verified that the concentrations of free polyamines, progressively decreased until harvest with the exception of Put, which showed a second peak just before the onset of ethylene production. In postharvest fruit, minor changes in concentrations of Spd and Spm were observed, whereas Put concentration peaked on the harvest day.

Anyhow, the involvement of higher polyamines in fruit ripening has been described in several reports and they seem to be mainly related to inhibition/modulation of fruit ripening [10, 45, 61]. Metabolite profiling by using NMR from transgenic tomato lines transformed with yeast SAM decarboxylase showed an accumulation in Spermidine and Spermine, and revealed that these higher polyamines influence multiple cell pathways in tomato fruit ripening [62]. Levels of Choline, Glu, Gln, Asn, Citrate, Malate, and Fumarate increased and those of Asp, Thr, Val, glucose, and sucrose decreased in the transgenics compared to the wild-type and azygous control lines [62]. In grape, the decrease in polyamines observed during ripening may contribute to the previously detected decrease in organic acids and glutamate and increased sucrose and glucose, Valine and Threonine [26, 63]. This fact suggests further relevance of higher polyamines in the regulation of TCA cycle and nitrogen metabolism [5, 64, 65]. Polyamines may also act as cellular signals in intricate crosstalk with hormonal pathways, as suggested by several studies [45, 66–68].

It should be noted however, that different metabolisms may exist depending on the fruit and even the cultivar [5, 28]. Interestingly, polyamine and ethylene metabolisms seem more dependent on the grape cultivar (genotype) than on the vintage. Differences in the expression of genes involved in polyamine metabolism were indeed noticed among cultivars suggesting a complex regulation of this pathway [5]. While genes coding for *SPDS* were mostly downregulated in all grape cultivars at ripe stage (Figs. 1 and 2), different expression patterns of genes coding for *SPMS* could be noticed: upregulation, downregulation or no modulation. In tomato, *SPDS1* and *SPDS2* presented opposite expression trends [28] while *SPMS* exhibited higher expression at the pre-climacteric stages. Furthermore, in grape *SAMDC1* presented an increase during ripening mainly in Touriga Nacional and Aragonês cultivars whereas a gene coding for SAMDC2 exhibited a strong cultivar-dependent profile [5]. *SAMDC1* was also highly expressed in ripe tomatoes [28]. Altogether, these findings indicate different roles of the diverse isoenzymes in polyamine homeostasis. The fine tuning of isoenzymes' activities and polyamine contents may be involved in the onset of particular ripening stages, tissue specific regulation and cultivar/fruit type specificities.

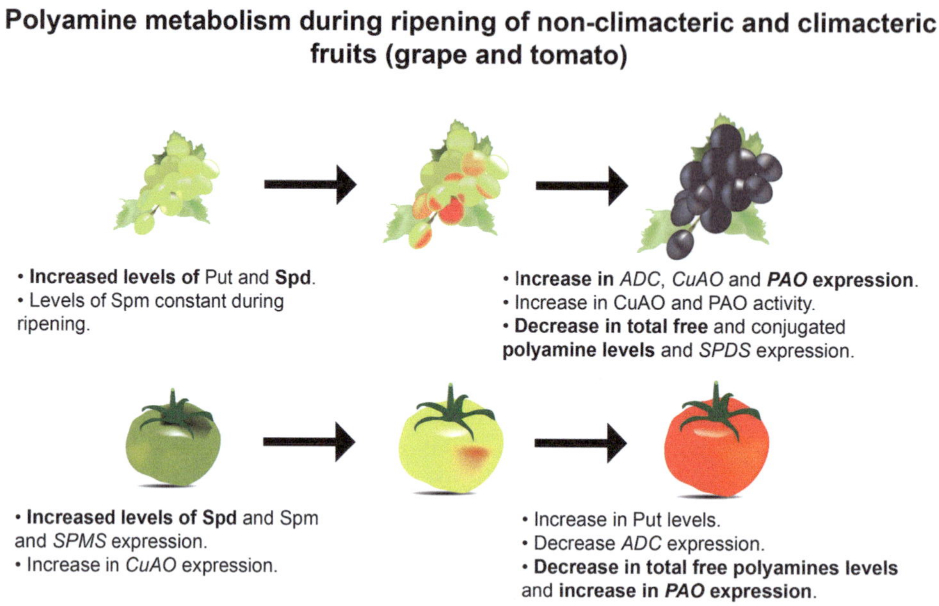

Fig. 2 Polyamine metabolism during ripening of non-climacteric and climacteric fruits (grape and tomato). Common events are highlighted in *bold* and include decrease in free polyamines levels in ripe fruits together with an increase in *PAO* expression

3.2 Polyamine Catabolism in Climacteric and Non-Climacteric Fruit Ripening

Polyamine catabolism has been poorly studied during fruit ripening. In an innovative study, Agudelo-Romero and coworkers [5] showed that polyamine catabolism may play a role in grape ripening. Previously, an increase in GABA was detected during grape ripening [26] suggesting that the oxidation of polyamines produced a decreased in the polyamine titers. In ripe grape, the decrease in polyamines' content was accompanied by upregulation of genes coding for diamine oxidase (CuAO) and polyamine oxidase (PAO), together with a very significant increase in their enzymatic activity and in the hydrogen peroxide content [5]. Interestingly, though slight differences in expression were observed, all cultivars exhibited upregulation of polyamine catabolic genes during ripening (Figs. 1 and 2). These results provided, for the first time, strong evidence of a role of polyamine catabolism in grape ripening.

In a more recent study in tomato, Tsaniklidis and coworkers [28] obtained a different expression pattern of a gene coding for CuAO comparing to grape. This gene was more expressed during fruit set and mature green stages than at ripe stages. However, a peak in *PAO* expression was also noticed for red ripe stage comparing to mature green stages which suggests that enhanced polyamine catabolism may play also a role in climacteric fruit ripening (Fig. 2). It has been suggested that enzymes producing H_2O_2 like amine oxidases along with their contribution to PA homeostasis, are implicated in senescence processes leading to fruit maturation

[69, 70]. PAO and CuAO derived H_2O_2 may serve as a signaling compound leading to the acceleration of the ripening process [65] or to signal downstream stress responses to abiotic and biotic stress factors that are enhanced during ripening along with sugar accumulation [5]. In fact, enhanced polyamine metabolism along with GABA accumulation was noticed in grapes infected with *Botrytis cinerea* [17]. Hydrogen peroxide resulting from CuAO has been shown to contribute to cell reinforcement during plant–pathogen interaction [71]; H_2O_2 is produced by amine oxidases as co-substrate for peroxidase- driven reactions [14, 72]. Recently, H_2O_2 involved in lignification of the exocarp of a pear mutant was shown to originate from the oxidation of spermidine by the enzyme PAO [73]. Several reports indicate that amine oxidases are involved in defense mechanisms in the course of both incompatible and compatible host–pathogen interactions (reviewed by [74]). Interestingly, the proteins products of genes coding for CuAO and PAO that were shown to be upregulated during grape ripening presented a putative peroxisomal target as described for other amine oxidases from Arabidopsis [5]. Plant peroxisomes have been described to play diverse roles in development as well as in abiotic and biotic stress responses [75].

In order to gather insights on the effect of the PA catabolism inhibition in vivo, Agudelo-Romero and coworkers [17] studied the transcriptional and metabolic profilings of grape berries following application of guazatine, a potent inhibitor of polyamine oxidase activity. The authors found an increase of putrescine, proline, threonine, and 1-*O*-ethyl-β-glucoside in guazatine-treated samples. These changes were accompanied by modulation of gene expression involved in amino acid and carbohydrate metabolisms.

Guazatine treatment had also profound changes in hormonal metabolism. Transcriptional profiling suggested an increase in ABA contents and alterations in ABA signaling due to inhibition of polyamine oxidase activity [13]. Several studies suggest Put and ABA reciprocally promote each other's biosynthesis under stress to increase the plants' adaptive potential [15, 76–78]. Furthermore, an integrated metabolome and transcriptome analysis revealed ABA-dependent transcriptional regulation of the biosynthesis of the branched-chain amino acids, saccharopine, proline, and polyamine [77]. These last two compounds were also found to increase in guazatine-treated samples. The possibility of cross-talk of polyamines with other growth regulators was emphasized by the upregulation of genes involved in ethylene and jasmonate signaling and downregulation of genes involved in brassinosteroid and auxin signaling following guazatine treatment [13]. Transgenic tomato generated with reduced endogenous methyl jasmonate levels [9] presented significant changes in the primary metabolism, especially the aminome (amino acids and polyamines) of ripening fruits. In peach fruit, the ripening delay induced by jasmonates is associated with increased polyamine levels [68]. Spermidine application to young developing peach fruits leads to a slowing down of ripening by impairing ripening-related ethylene and auxin metabolism and signaling [45]. On the other hand, functional interaction and synergism between ABA and ethylene during grape berry ripening and after harvest has been shown [79]. While ethylene seems to induce ABA, auxin negatively regulates ABA-induced ripening processes in grape [80]. ABA also seems to trigger ethylene biosynthesis and ripening of tomato [81, 82]. In mango, [40] suggested that the biosynthesis of ethylene and free polyamines may not be competitive during fruit ripening. Additionally, [83] concluded that tomato fruit possesses the capability to simultaneously consume SAM during ripening to ensure a high rate of ethylene and polyamine production. In olive, [67] suggested that polyamine-dependent ethylene signaling and biosynthesis pathways participate, at least partially, during mature fruit abscission. On the other hand, application of guazatine led to a decrease in polyamine catabolism, increase in free polyamine contents, and induced multipoint cross-talk with auxin, cytokinin, brassinosteroid, ABA, and ethylene pathways [13]. In climacteric fruits such as tomato, the effect of such treatments is expected to present differences comparing to non-climacteric fruits such as grapes. However, it is clear from the above that polyamine catabolism is involved in ripening of both type of fruits and probably through interaction with other growth regulators.

4 Perspectives on Modulation of Fruit Ripening by Polyamines

The study of the role of polyamine metabolism in fruit ripening has highlighted the possible application of these natural polycations to control ripening and postharvest decay as well as to improve fruit quality traits. In fact, application of exogenous polyamines in particular at pre-harvest stage has been conducted in several fruit species such as date palm [84], mango [85], pear and nectarine [86, 87], sweet orange [88], and kiwi [89].

On the other hand, studies employing reverse genetics approaches have provided direct evidence for the involvement of polyamines in fruit ripening [7, 9, 10]. Numerous reports described improved quality traits in tomato by manipulation of polyamine metabolism. Transgenic tomatoes fruits from plants ectopically expressing a yeast SAMDC/SPDS had longer shelf life and altered fatty acid profile and phytonutrient content [10, 64, 90]. Overexpression of SAMDC and SPDS significantly influenced carotenoid concentration in these fruits [91, 92]. These data suggests these enzymes are prime targets in manipulating the nutritional value of tomato fruits.

Such transgenic material constitutes a good genetic resource to analyze the biological effects of polyamines and their interaction with other signaling pathways. The active response of tomato fruit to engineered high levels of polyamines predicates that it will be possible to modulate ripening and influence nutrient levels of the fruit by rational design of genes with precision-based and ripening stage-specific promoters [93].

However, for other fruit species such as grape neither stable nor transient expression protocols have been efficiently established. The understanding of polyamine metabolism in these fruits will benefit from the development of fruit agroinjection coupled with gene editing tools [94]. The impact of the inhibition of polyamine catabolism in the regulatory networks of ripening and fruit quality which has been little studied up to now could be clarified by using these approaches. Furthermore, polyamines are able to interact with chromatin and this may lead to epigenetic modifications of DNA and histones (reviewed by [16]), this fact opens new exciting frontiers of research focusing on how polyamine metabolism may affect fruit ripening.

For fruits lacking such biotechnological resources (agroinjection, stable genetic transformation) genes coding for enzymes involved in polyamine catabolism have been studied mostly by heterologous expression [95] or by exogenous application of modulators of polyamine catabolism [13]. In this last work, application of guazatine led to inhibition of *PAO* expression and PAO activity, and to alterations in cell wall metabolism and primary (amino acids, fatty acids, sugars) and secondary metabolisms (stilbenes, flavonoids). In addition, genes involved in terpenes' metabolism were differentially expressed between guazatine and mock-treated samples, supporting a role of polyamine catabolism in aroma development during grape ripening [13]. Flavor, color, texture, sweetness are important fruit quality standards, therefore this topic deserves attention in order to increase quality and nutritional content of fruits. Other important quality standards are shelf life and resistance against biotic and abiotic stresses. Manipulation of polyamine levels either by modulating their biosynthesis or catabolism or eventually conjugation with other compounds will certainly contribute in the future for obtaining better yields and improved nutritional characteristics of fruits produced under more sustainable conditions.

Acknowledgments

Funding to A.M.F. was provided by the Portuguese Foundation for Science and Technology (SFRH/BPD/100928/2014, IFCT050, PEst-OE/BIA/UI4046/2014).

We would like to thank Dr. Caparrós-Martín JA for his invaluable help with Fig. 2.

References

1. Giovannoni JJ (2004) Genetic regulation of fruit development and ripening. Plant Cell 16: S170–S180
2. Cherian S, Figueroa CR, Nair H (2014) 'Movers and shakers' in the regulation of fruit ripening: a cross-dissection of climacteric versus non-climacteric fruit. J Exp Bot 65 (17):4705–4722
3. Liu M, Pirrello J, Chervin C, Roustan J-P, Bouzayen M (2015) Ethylene control of fruit ripening: revisiting the complex network of transcriptional regulation. Plant Physiol 169 (4):2380–2390
4. Fortes AM, Teixeira RT, Agudelo-Romero P (2015) Complex interplay of hormonal signals during grape berry ripening. Molecules 20 (5):9326–9343
5. Agudelo-Romero P, Bortolloti C, Pais MS, Tiburcio AF, Fortes AM (2013) Study of polyamines during grape ripening indicate an important role of polyamine catabolism. Plant Physiol Biochem 67:105–119
6. Mattoo AK, Chung SH, Goyal RK et al (2007) Overaccumulation of higher polyamines in ripening transgenic tomato fruit revives metabolic memory, upregulates anabolism-related genes, and positively impacts nutritional quality. J AOAC Int 90(5):1456–1464
7. Srivastava A, Sang HC, Fatima T, Datsenka T, Handa AK, Mattoo AK (2007) Polyamines as anabolic growth regulators revealed by transcriptome analysis and metabolite profiles of tomato fruits engineered to accumulate spermidine and spermine. Plant Biotechnol 24 (1):57–70
8. Handa AK, Mattoo AK (2010) Differential and functional interactions emphasize the multiple roles of polyamines in plants. Plant Physiol Biochem 48(7):540–546
9. Kausch KD et al (2012) Methyl jasmonate deficiency alters cellular metabolome, including the aminome of tomato (Solanum lycopersicum L.) fruit. Amino Acids 42(2–3):843–856
10. Nambeesan S, Datsenka T, Ferruzzi MG, Malladi A, Mattoo AK, Handa AK (2010) Overexpression of yeast spermidine synthase impacts ripening, senescence and decay symptoms in tomato. Plant J 63(5):836–847
11. Bregoli AM et al (2002) Peach (Prunus persica) fruit ripening: aminoethoxyvinylglycine (AVG) and exogenous polyamines affect ethylene emission and flesh firmness. Physiol Plant 114 (3):472–481
12. Ziosi V, Scaramagli S, Bregoli AM, Biondi S, Torrigiani P (2003) Peach (Prunus persica L.) fruit growth and ripening: transcript levels and activity of polyamine biosynthetic enzymes in the mesocarp. J Plant Physiol 160 (9):1109–1115
13. Agudelo-Romero P et al (2014) Perturbation of polyamine catabolism affects grape ripening of Vitis vinifera cv. Trincadeira. Plant Physiol Biochem 74:141–155
14. Cona A, Rea G, Angelini R, Federico R, Tavladoraki P (2006) Functions of amine oxidases in plant development and defence. Trends Plant Sci 11(2):80–88
15. Alcázar R et al (2010) Polyamines: molecules with regulatory functions in plant abiotic stress tolerance. Planta 231:1237–1124
16. Minocha R, Majumdar R, Minocha SC (2014) Polyamines and abiotic stress in plants: a complex relationship. Front Plant Sci 5:175
17. Agudelo-Romero P et al (2015) Transcriptome and metabolome reprogramming in Vitis vinifera cv. Trincadeira berries upon infection with Botrytis cinerea. J Exp Bot 66(7):1769–1785
18. Liu J-H, Wang W, Wu H, Gong X, Moriguchi T (2015) Polyamines function in stress tolerance: from synthesis to regulation. Front Plant Sci 6:827
19. Pál M, Szalai G, Janda T (2015) Speculation: polyamines are important in abiotic stress signaling. Plant Sci 237:16–23
20. Applewhite PB, Kaur-Sawhney R, Galston AW (2000) A role for spermidine in the bolting and flowering of Arabidopsis. Physiol Plant 108 (3):314–320
21. Tiburcio AF, Altabella T, Bitrián M, Alcázar R (2014) The roles of polyamines during the lifespan of plants: from development to stress. Planta 240(1):1–18
22. Jancewicz AL, Gibbs NM, Masson PH (2016) Cadaverine's functional role in plant development and environmental response. Front Plant Sci 7:1–8
23. Fortes AM et al (2011) Arginine decarboxylase expression, polyamines biosynthesis and reactive oxygen species during organogenic nodule formation in hop. Plant Signal Behav 6 (2):258–269
24. Pandey S, Ranade SA, Nagar PK, Kumar N (2000) Role of polyamines and ethylene as modulators of plant senescence. J Biosci 25:291–299

25. Sobieszczuk-Nowicka E (2016) Polyamine catabolism adds fuel to leaf senescence. Amino Acids 49:49–56
26. Fortes AM, Agudelo-Romero P, Silva MS et al (2011) Transcript and metabolite analysis in Trincadeira cultivar reveals novel information regarding the dynamics of grape ripening. BMC Plant Biol 11:1–34
27. Tavladoraki P, Cona A, Angelini R (2016) Copper-containing amine oxidases and FAD-dependent polyamine oxidases are key players in plant tissue differentiation and organ development. Front Plant Sci 7:824
28. Tsaniklidis G, Kotsiras A, Tsafouros A et al (2016) Spatial and temporal distribution of genes involved in polyamine metabolism during tomato fruit development. Plant Physiol Biochem 100:27–36
29. Valero D, Martinez-Romero D, Serrano M, Riquelme F (1998) Influence of postharvest treatment with putrescine and calcium on endogenous polyamines, firmness, and abscisic acid in lemon (Citrus lemon L. Burm cv. Verna). J Agric Food Chem 46:2102–2109
30. Mirdehghan SH, Rahimi S (2016) Pre-harvest application of polyamines enhances antioxidants and table grape (Vitis vinifera L.) quality during postharvest period. Food Chem 196:1040–1047
31. Khosroshahi MRZ, Esna-Ashari M, Ershadi A (2007) Effect of exogenous putrescine on post-harvest life of strawberry (Fragaria ananassa Duch.) fruit, cultivar Selva. Sci Hortic 114:27–32
32. Sotelo-Silveira M, Marsch-Martínez N, de Folter S (2014) Unraveling the signal scenario of fruit set. Planta 239:1147–1158
33. Palavan N, Galston AW (1982) Polyamine biosynthesis and titer during various developmental stages of Phaseolus vulgaris. Physiol Plant 55:438–444
34. Carbonell J, Navarro JL (1989) Correlation of spermine levels with ovary senescence and with fruit set and development in Pisum sativum L. Planta 178:482–487
35. Alabadí D, Agüero MS, Pérez-Amador MA, Carbonell J (1996) Arginase, arginine decarboxylase, ornithine decarboxylase, and polyamines in tomato ovaries' changes in unpollinated ovaries and parthenocarpic fruits Induced by auxin or gibberellin. Plant Physiol 112:1237–1244
36. Slocum RD, Galston AW (1985) Changes in polyamine biosynthesis associated with postfertilization growth and development in tobacco ovary tissues. Plant Physiol 79:336–343
37. Biasi R, Bagni N, Costa G (1988) Endogenous polyamines in apple and their relationship to fruitset and fruit growth. Plant Physiol 73:201–205
38. Kushad MM, Yelenosky G, Knight R (1988) Interrelationship of polyamine and ethylene biosynthesis during avocado fruit development and ripening. Plant Physiol 87:463–467
39. Shiozaki S, Ogata T, Horiuchi S (2000) Endogenous polyamines in the pericarp and seed of the grape berry during development and ripening. Sci Hortic 83:33–41
40. Malik AU, Singh Z (2004) Endogenous free polyamines of mangos in relation to development and ripening. J Am Soc Hort Sci 129:280–286
41. Teh HF, Neoh BK, Wong YC et al (2014) Hormones, polyamines, and cell wall metabolism during oil palm fruit mesocarp development and ripening. J Agric Food Chem 62:8143–8152
42. Gomez-Jimenez MC, Paredes MA, Gallardo M et al (2010) Tissue-specific expression of olive S-adenosyl methionine decarboxylase and spermidine synthase genes and polyamine metabolism during flower opening and early fruit development. Planta 232:629–647
43. Tassoni A, Germanà MA, Bagni N (2004) Free and conjugated polyamine content in Citrus sinensis Osbeck, cultivar Brasiliano N.L. 92, a Navel orange, at different maturation stages. Food Chem 87:537–541
44. Kushad MM (1998) Changes in polyamine levels in relationship to the double-sigmoidal growth curve of peaches. J Am Soc Hortic Sci 123:950–955
45. Torrigiani P, Bressanin D, Beatriz Ruiz K et al (2012) Spermidine application to young developing peach fruits leads to a slowing down of ripening by impairing ripening-related ethylene and auxin metabolism and signaling. Physiol Plant 146:86–98
46. Ziosi V, Bregoli AM, Bonghi C et al (2006) Transcription of ethylene perception and biosynthesis genes is altered by putrescine, spermidine and aminoethoxyvinylglycine (AVG) during ripening in peach fruit (Prunus persica). New Phytol 172:229–238
47. Liu J, Nada K, Pang X et al (2006) Role of polyamines in peach fruit development and storage. Tree Physiol 26:791–798
48. Serrano M, Martinez-Madrid MC, Riquelme F, Romojaro F (1995) Endogenous levels of polyamines and abscisic acid in pepper fruits during growth and ripening. Physiol Plant 95:73–76

49. Rastogi R, Davies PJ (1991) Polyamine metabolism in ripening tomato fruit. II. Polyamine metabolism and synthesis in relation to enhanced putrescine content and storage life of alc tomato fruit. Plant Physiol 95:41–45
50. Smith TA (1985) Polyamines. Annu Rev Physiol 35:117–143
51. Perez-Amador MA, Leon J, Green PJ, Carbonell J (2002) Induction of the arginine decarboxylase ADC2 gene provides evidence for the involvement of polyamines in the wound response in Arabidopsis. Plant Physiol 130:1454–1463
52. Urano K, Yoshiba Y, Nanjo T et al (2003) Characterization of Arabidopsis genes involved in biosynthesis of polyamines in abiotic stress responses and developmental stages. Plant Cell Environ 26:1917–1926
53. Urano K, Yoshiba Y, Nanjo T et al (2004) Arabidopsis stress-inducible gene for arginine decarboxylase AtADC2 is required for accumulation of putrescine in salt tolerance. Biochem Biophys Res Commun 313:369–375
54. Acosta C, Pérez-Amador MA, Carbonell J, Granell A (2005) The two ways to produce putrescine in tomato are cell-specific during normal development. Plant Sci 168:1053–1057
55. Ioannidis NE, Kotzabasis K (2014) Polyamines in chemiosmosis in vivo: a cunning mechanism for the regulation of ATP synthesis during growth and stress. Front Plant Sci 5:71
56. Paschalidis KA, Roubelakis-Angelakis KA (2005) Sites and regulation of polyamine catabolism in the tobacco plant. Correlations with cell division/expansion, cell cycle progression, and vascular development. Plant Physiol 138:2174–2184
57. Bauza T, Kelly MT, Blaise A (2007) Study of polyamines and their precursor amino acids in Grenache noir and Syrah grapes and wine of the Rhone Valley. Food Chem 105:405–413
58. Rastogi R, Dulson J, Rothstein SJ (1993) Cloning of tomato (Lycopersicon esculenfum Mill.) arginine decarboxylase gene and its expression during fruit ripening. Plant Physiol 103:829–834
59. Deluc LG, Grimplet J, Wheatley MD et al (2007) Transcriptomic and metabolite analyses of Cabernet Sauvignon grape berry development. BMC Genomics 8:429
60. Pilati S, Perazzolli M, Malossini A et al (2007) Genome-wide transcriptional analysis of grapevine berry ripening reveals a set of genes similarly modulated during three seasons and the occurrence of an oxidative burst at vèraison. BMC Genomics 8:428
61. Tassoni A, Watkins CB, Davies PJ (2006) Inhibition of the ethylene response by 1-MCP in tomato suggests that polyamines are not involved in delaying ripening, but may moderate the rate of ripening or over-ripening. J Exp Bot 57:3313–3325
62. Mattoo AK, Sobolev AP, Neelam A et al (2006) Nuclear magnetic resonance spectroscopy-based metabolite profiling of transgenic tomato fruit engineered to accumulate spermidine and spermine reveals enhanced anabolic and nitrogen-carbon interactions. Plant Physiol 142:1759–1770
63. Ali K, Maltese F, Fortes AM et al (2011) Monitoring biochemical changes during grape berry development in Portuguese cultivars by NMR spectroscopy. Food Chem 124:1760–1769
64. Neelam A, Cassol T, Mehta RA et al (2008) A field-grown transgenic tomato line expressing higher levels of polyamines reveals legume cover crop mulch-specific perturbations in fruit phenotype at the levels of metabolite profiles, gene expression, and agronomic characteristics. J Exp Bot 59:2337–2346
65. Moschou PN, Wu J, Cona A et al (2012) The polyamines and their catabolic products are significant players in the turnover of nitrogenous molecules in plants. J Exp Bot 63:5003–5015
66. Serrano M, Romojaro F, Casas JL, Acosta M (1991) Ethylene and polyamine metabolism in climacteric and nonclimacteric carnation flowers. Hortscience 26:894–896
67. Parra-Lobato MC, Gomez-Jimenez MC (2011) Polyamine-induced modulation of genes involved in ethylene biosynthesis and signalling pathways and nitric oxide production during olive mature fruit abscission. J Exp Bot 62:4447–4465
68. Ziosi V, Bregoli AM, Fregola F et al (2009) Jasmonate-induced ripening delay is associated with up-regulation of polyamine levels in peach fruit. J Plant Physiol 166:938–946
69. Delis C, Dimou M, Flemetakis E et al (2006) A root- and hypocotyl-specific gene coding for copper-containing amine oxidase is related to cell expansion in soybean seedlings. J Exp Bot 57:101–111
70. Mateos RM, Jiménez A, Román P et al (2013) Antioxidant systems from pepper (Capsicum annuum L.): involvement in the response to temperature changes in ripe fruits. Int J Mol Sci 14:9556–9580
71. Rea G, Metoui O, Infantino A et al (2002) Copper amine oxidase expression in defense responses to wounding and Ascochyta rabiei invasion. Plant Physiol 128:865–875

72. Angelini R, Tisi A, Rea G et al (2008) Involvement of polyamine oxidase in wound healing. Plant Physiol 146:162–177
73. Heng W, Wang Z, Jiang X et al (2016) The role of polyamines during exocarp formation in a russet mutant of "Dangshansuli" pear (Pyrus bretschneideri Rehd.) Plant Cell Rep 35:1841–1852
74. Angelini R, Cona A, Federico R et al (2010) Plant amine oxidases "on the move": an update. Plant Physiol Biochem 48:560–564
75. Hu J, Baker A, Bartel B et al (2012) Plant peroxisomes: biogenesis and function. Plant Cell 24:2279–2303
76. Cuevas JC, López-Cobollo R, Alcázar R et al (2008) Putrescine is involved in Arabidopsis freezing tolerance and cold acclimation by regulating abscisic acid levels in response to low temperature. Plant Physiol 148:1094–1105
77. Urano K, Maruyama K, Ogata Y et al (2009) Characterization of the ABA-regulated global responses to dehydration in Arabidopsis by metabolomics. Plant J 57:1065–1078
78. Toumi I, Moschou PN, Paschalidis KA et al (2010) Abscisic acid signals reorientation of polyamine metabolism to orchestrate stress responses via the polyamine exodus pathway in grapevine. J Plant Physiol 167:519–525
79. Sun L, Zhang M, Ren J et al (2010) Reciprocity between abscisic acid and ethylene at the onset of berry ripening and after harvest. BMC Plant Biol 10:257
80. Davies C, Boss PK, Robinson SP (1997) Treatment of grape berries, a nonclimacteric fruit with a synthetic Auxin, retards ripening and alters the expression of developmentally regulated genes. Plant Physiol 115:1155–1161
81. Zhang M, Yuan B, Leng P (2009) The role of ABA in triggering ethylene biosynthesis and ripening of tomato fruit. J Exp Bot 60:1579–1588
82. Mou W, Li D, Bu J et al (2016) Comprehensive analysis of ABA effects on ethylene biosynthesis and signaling during tomato fruit ripening. PLoS One 11(4):e0154072
83. Van de Poel B, Bulens I, Oppermann Y et al (2013) S-adenosyl-L-methionine usage during climacteric ripening of tomato in relation to ethylene and polyamine biosynthesis and transmethylation capacity. Physiol Plant 148:176–188
84. Tavakoli K, Rahemi M (2014) Effect of polyamines, 2,4-D, isopropyl ester and naphthalene acetamide on improving fruit yield and quality of date (Phoenix dactylifera L.) Int J Hortic Sci Technol 1:163–169
85. Malik AU, Singh Z (2006) Improved fruit retention, yield and fruit quality in mango with exogenous application of polyamines. Sci Hortic 110:167–174
86. Franco-Mora O, Tanabe K, Tamura F, Itai A (2005) Effects of putrescine application on fruit set in "Housui" Japanese pear (Pyrus pyrifolia Nakai). Sci Hortic 104:265–273
87. Torrigiani P, Bregoli AM, Ziosi V et al (2004) Pre-harvest polyamine and aminoethoxyvinylglycine (AVG) applications modulate fruit ripening in Stark Red Gold nectarines (Prunus persica L. Batsch). Postharvest Biol Technol 33:293–308
88. Saleem BA, Malik AU, Anwar R, Farooq M (2008) Exogenous application of polyamines improves fruit set, yield and quality of sweet oranges. Acta Hortic 774:187–194
89. Jhalegar MJ, Sharma RR, Pal RK et al (2011) Analysis of physiological and biochemical changes in kiwifruit (Actinidia deliciosa cv. Allison) after the postharvest treatment with 1-methylcyclopropene. J Plant Biochem Biotechnol 20:205–210
90. Kolotilin I, Koltai H, Bar-Or C et al (2011) Expressing yeast SAMdc gene confers broad changes in gene expression and alters fatty acid composition in tomato fruit. Physiol Plant 142:211–223
91. Mehta RA, Cassol T, Li N et al (2002) Engineered polyamine accumulation in tomato enhances phytonutrient content, juice quality, and vine life. Nat Biotechnol 20:613–618
92. Neily MH, Matsukura C, Maucourt M et al (2011) Enhanced polyamine accumulation alters carotenoid metabolism at the transcriptional level in tomato fruit over-expressing spermidine synthase. J Plant Physiol 168:242–252
93. Mattoo AK, Handa AK (2008) Higher polyamines restore and enhance metabolic memory in ripening fruit. Plant Sci 174:386–393
94. Fortes A, Gallusci P (2017) Plant stress responses and phenotypic plasticity in the epigenomics era: perspectives on the grapevine scenario, a model for perennial crop plants. Front Plant Sci 8:82
95. Wang W, Liu J-H (2016) CsPAO4 of Citrus sinensis functions in polyamine terminal catabolism and inhibits plant growth under salt stress. Sci Rep 6:31384

Chapter 37

Application of Polyamines to Maintain Functional Properties in Stored Fruits

María Serrano and Daniel Valero

Abstract

Polyamines are natural compounds involved in many growth and developmental processes with ubiquitous presence in all cells. Research in fruits has been developed to get a better understanding of the role of polyamines, both endogenous and exogenous, especially during the ripening and senescence processes. However, in recent years and given the relationship between fruit consumption and human health, the study of antioxidant compounds responsible for these beneficial effects is of increasing interest.

This chapter focuses on the role of polyamines on the content of bioactive compounds with antioxidant activity as well as in the activities of the main antioxidant enzymes in fruits.

Key words Postharvest, Putrescine, Spermidine, Spermine, Antioxidants, Phenolics, Anthocyanins, Carotenoids, Antioxidant enzymes

1 Introduction

The polyamines have a long history, and the first evidence was obtained in 1678 by Antonie van Leuwenhoek which described the presence of crystalline substances in the human semen, but was in 1878 when these crystals were identified as spermine by Laudenburg and Abel [1]. In 1924 putrescine (Put), spermidine (SPd), and spermine (Spm) were synthesized [2]. During the mid-twentieth century the experiments with polyamines were carried out in bacteria while experiments with animals appeared in the 1970s with special interest in the association between polyamines and diseases and largely focused on cancer cells leading to a synthesis of the inhibitor of polyamine synthesis DFMO (α-difluoromethylornithine). In plants, the first evidence to the occurrence of Put was dated in 1911 but was in 1971 [3] when it was concluded that Put, Spd, and Spm are ubiquitous organic cations in higher plants.

Polyamines are involved in many aspects of plant development and considered as important molecules associated with both abiotic

Rubén Alcázar and Antonio F. Tiburcio (eds.), *Polyamines: Methods and Protocols*, Methods in Molecular Biology, vol. 1694, DOI 10.1007/978-1-4939-7398-9_37, © Springer Science+Business Media LLC 2018

and biotic stresses, the most studied in fruits being chilling injury (CI). Thus, polyamines have been reported to stimulate cell division, dormancy breaking, germination, development of flower buds, fruit set and growth [4–8]. However, one of the main aspects of the relationship between polyamines and fruit are those related to fruit ripening. Generally, there is an inverse relationship between the content of endogenous polyamines and ethylene production which is attributed to the fact that both share the common precursor S-adenosylmethionine (SAM) in their biosynthesis pathway [6]. This relationship has justified the large body of knowledge about the application of polyamines, either at pre-harvest (during fruit growth and ripening on tree) or after harvest (postharvest treatments) to delay the postharvest ripening process and maintain fruit quality attributes such as colour, firmness, acidity, and total soluble solids during postharvest storage, either at ambient temperature or under cold storage [6, 9, 10]. However, the role of these treatments on the content of bioactive compounds and antioxidant activity has been poorly studied. In this chapter we present the recent literature about the effect of polyamines as pre- or postharvest treatments on the content of phytochemicals with antioxidant properties in fruit, at harvest and during storage, given the health beneficial effects attributed to these compounds.

2 Fruit and Health Beneficial Effects

Nowadays, it is widely accepted that nutrition has a great influence on health and increasing evidence suggests that diets rich in fruits and vegetables may prevent a wide range of diseases, mainly cardiovascular diseases (by reducing risk factors such as diabetes, hypertension, and hyperlipidemia, platelet aggregation, blood pressure, insulin resistance index, and obesity), several kind of cancers, such as colon, oesophagus, oral cavity, stomach, pancreas, prostatic, breast, and ovary (by regulating gene expression in cell proliferation and apoptosis), neurodegenerative diseases, brain and immune dysfunction, and even against bacterial and viral diseases [11–13]. These effects are due to phytochemical compounds with antioxidant activity, the most important being phenolics, including anthocyanins, carotenoids, vitamins (C and E), and glucosinolates [14–18]. In addition, it is important to note that whole fruit intake provides more health beneficial effects than one of their constituent, because of additive and synergistic effects [13, 14].

Dietary polyamines from plant origin are considered very important for human health. The body pool of polyamines in human body is maintained by three sources: endogenous or de novo biosynthesis, produced by the gut microbiota, and exogenous intake through the diet [19]. From these sources, the polyamines derived from the diet are the most important from the point of view of quantity, since the capacity to synthetize polyamines decreases with age. Some of the health-beneficial effects of dietary polyamines are those related to protection against oxidative stress, maintenance of gut integrity, modulation of inflammation and immune functions, among others [20]. It seems that polyamine antioxidant activity leads to reduction in both cell membranes and DNA [21].

The average intake of dietary polyamines is different depending on the country. In Europe, the highest intake of total polyamines (700 μmol per day) has been estimated in the Mediterranean regions compared with northern Europe. This difference could justify the protective role attributed to the Mediterranean diet. Among the different food groups, fruit provides the highest amount of Put (500 μmol/kg) while cheese contains over 600 μmol/kg of Spd [19–21].

3 Polyamines, Bioactive Compounds and Antioxidant Activity in Fruits

3.1 Carotenoids

Carotenoids are a group of natural-lipophilic pigments with a general structure of C_{40} tetraterpenoids responsible for yellow to red colour in fruits. The carotenoids are classified as carotenes (ß-carotene, lycopene, etc.) and xanthophylls. In addition, carotenoids can be acyclic (e.g. lycopene), monocyclic (γ-carotene), or dicyclic (α- and ß-carotene). The most studied carotenoids have been ß-carotene, lycopene, lutein and zeaxanthin which exhibit a potent antioxidant activity. The antioxidant ability of these carotenoids (carotenes and xanthophylls) follows the sequence from high to low: lycopene > ß-cryptoxanthin ≈ ß-carotene > lutein ≈ zeaxanthin > α-carotene > canthaxanthin [6].

The first evidences regarding the effects of polyamines on carotenoids content in fruits were obtained by Malik and Singh [22] who found that mango tree treatments by foliar spraying with Put, Spd, or Spm (0.01, 0.1, and 1 mM) at final fruit set stage increased total carotenoids in the pulp at harvest time as compared with fruits from control trees, the main effect being observed with Put treatments (95%) followed by Spd (33%). More recently, Mehta et al. [23] reported that ripe fruits from transgenic tomato plants having the yeast SAMDC gene had higher Spd and Spm concentrations and threefold more lycopene concentration than did the red fruits from the parental lines. Accordingly, higher levels of polyamines and lycopene were found in transgenic tomatoes overexpressing human *SAMDC* gene. The increase in lycopene concentration was attributed to enhanced levels of gene transcripts involved in lycopene biosynthesis in transgenic tomatoes [24]. Similarly, plants overexpressing mouse *ODC* gene yielded tomato fruits with higher polyamine concentration and higher concentration of lycopene as compared with fruits from control plants [25]. These effects of PAs are of special significance since carotenoids, including lycopene, are bioactive compounds having antioxidant capacity and proved beneficial effects in human health [11, 12, 26]. On the contrary, pre- and postharvest Put application to "Angelino" plum led to a linear reduction in the levels of carotenoids, which were more pronounced with increased concentrations of Put and storage periods [27]. This effect could be due to the inhibition of ethylene production and postharvest ripening process as a consequence of Put treatments, since carotenoids increase during ripening and during postharvest storage in plum fruits [6, 16].

3.2 Phenolic Compounds

Phenolic or polyphenolics are secondary plant metabolites that exhibit a wide range of physiological roles in plants such as pigmentation, growth, and resistance to pathogens, among many other functions. The main phenolics in fruits and vegetables are classified according to their basic skeleton as C_6-C_1 (phenolic acids), C_6-C_3 (hydroxycinnamic acids), C_6-C_2-C_6 (stilbenes), and C_6-C_3-C_6 (flavonoids). Phenolics as a group represent the strongest antioxidants in plant foods, although the antioxidant activity of individual phenolic compounds may vary depending on their chemical structure. The antioxidant activity of phenolics is attributable to the electron delocalization over the aromatic ring and their high redox potential, which allows them to act as reducing agents, hydrogen donors, and singlet oxygen quenchers. Generally, total phenolics increase as ripening advanced for most fruits, although concentration depends on several factors including species, cultivar, growth conditions, and environmental factors [6, 16, 28].

Different effects of polyamine treatment on phenolic content of fruits have been reported depending on the applied polyamine, concentration, moment of application, and fruit species. Thus, in table grape postharvest Spm treatment at 0.5 and 1 mM maintained total phenol concentration over the controls along 75 days of cold storage, while Spm at 1.5 mM reduced significantly the berry phenolic content [29]. On the other hand, postharvest Put treatment at 2 mM of table grapes maintained higher concentration (1.5-fold with respect to controls) of total phenolics during low-temperature (0 °C) storage [30]. These authors also measured the individual concentration of catechin (flavonoid) and observed that it was maintained at higher concentration (twofold) after 60 days of cold storage +5 days of shelf life, which was correlated with the higher total antioxidant activity obtained at the end of the experiment. In other two table grape cultivars, polyamines have been also applied as pre-harvest treatments (40 and 20 days before harvest) at 1 or 2 mM, and results revealed that treated berries showed higher antioxidant activity and total phenolics at harvest and during cold storage, although the effectiveness of the treatments was affected by cultivar, polyamine type, and concentration [31].

In mango, total phenolics and total antioxidant capacity increased during ripening at ambient temperature and during cold storage at 11 °C although these increases were higher in fruits treated with Put at 0.5, 1, 1.5, and 2 mM before storage [1]. By other hand, in kiwifruit total phenolics decreased during storage at room temperature, although the application of Spd or Spm (0.5, 1 and 1.5 mM), by dipping treatment previously to storage, led to retention of total phenolic concentration along storage, the main positive effect being found for Spm at 1.5 mM [33]. Similar results were observed in pomegranate as a consequence of Put or Spd postharvest treatments, Spd treatment being more effective than Put treatment when applied as vacuum infiltration [34]. On the contrary, Koushesh-saba et al. [35] reported lower phenolic content in Put- and Spd-treated apricots than in controls. Thus, the effects of polyamines on total phenolic content (TPC) remain elusive.

3.3 Anthocyanins

Anthocyanins are the water-soluble pigments responsible for the red, blue, and purple colour of fruits, and have been described as potent antioxidants. Anthocyanins are located in the vacuole and classified as flavonoids with glycosilated derivatives of the 3,5,7,3′-tetrahydroxyflavylium cation. The free aglycones (anthocyanidins) are highly reactive with sugars to form the glycosides and all anthocyanins are O-glycosilated. The main aglycones found in fruits are pelargonidin, cyanidin, peonidin, delphinidin, petunidin, and malvidin while the most relevant sugars are D-glucose, L-rhamnose, D-galactose, D-xylose, and arabinose [36]. Anthocyanins have shown higher antioxidant activity than other phenolic compounds, with cyanidin being the most common anthocyanidin and the 3-glucoside the most active anthocyanin with antioxidant activity [6].

Pre-harvest treatment with Put or Spd (at 1 or 2 mM) of "Olhoghi" and "Rishbaba" table grape cultivars led to berries with higher content of anthocyanins at harvest, especially for 2 mM Put [31]. These authors also found that after 55 days of cold storage, anthocyanin concentrations had decreased in control berries while remained at significant higher levels in all Put- or Spd-treated ones, probably due to the effect of polyamines on delaying the postharvest ripening process. In table grape "Flame Seedless" cultivar, it has been reported that postharvest Spm treatment (0.5, 1 and 1.5 mM) induced a progressive increase in total anthocyanins during prolonged storage, while in control samples there was an initial increase (during 45 days of storage) but a sharp decrease occurred during next 30 days (end of the experiment) of storage. Interestingly, the effect of Spm treatment on increasing berry anthocyanin concentration was dose-dependent [29].

In addition, in "Mollar de Elche" pomegranate, postharvest treatments with Put or Spd at 1 mM (either by dipping or infiltration under low pressure) maintained the concentration of aril anthocyanins during cold storage with respect to controls [34]. In other pomegranate cultivar ("Mridula") treatment by immersion with 2 mM Put retained also higher anthocyanin during postharvest storage leading to arils with higher antioxidant activity [37]. The mechanism by which Put and Spd induce these effects is still unknown, although they may be related to their antisenescent effects related to the suppression of membrane lipid peroxidation and maintenance of the integrity of membranes [6, 38].

3.4 Vitamins

Vitamins are a group of nutrients necessary for human body due to their biochemical and physiological functions. Vitamins are classified into lipid-soluble and water soluble, the vitamins A, D, E, and K being lipophilic, while C and B are hydrophilic. Tocopherols (vitamin E) are the major lipid-soluble antioxidant vitamins in fruits, while vitamin C is the major hydrophilic antioxidant vitamin, although recent evidences indicate that vitamin D could also have a role as antioxidant [39].

With respect to the relation and/or effect of polyamines in ascorbic acid content of fruits, it has been found that in transgenic tomato plants, overexpressing mouse *ODC* gene, tomato fruits had higher polyamine concentration than those of the wild type, which was correlated with an increase in ascorbic acid content as compared with fruits from control plants [25]. Similarly, genetic modification of tomato fruit ripening by overexpressing human-*SAMDC* led to higher endogenous levels of polyamines and enhancement of ascorbic acid content [24]. This increase in ascorbic acid content in transgenic tomatoes was attributed to their lower ethylene production with respect to the wild type, since ascorbic acid is used as a cofactor for 1-aminociclopropane-1-carboxilic acid oxidase, the last enzyme in the ethylene biosynthesis pathway.

By other hand, in two tomato cultivars, postharvest treatments with Put or Spd (at 1 or 2 mM) did not affect ascorbic acid concentration, although the combination of both polyamines significantly increased the level of ascorbic acid during 15 or 25 days of storage at 2 °C, especially for the combination of Put 1 mM + Spd2 mM [40].

In pomegranate, treatment with Spd alone or in combination with calcium chloride showed a net increase in ascorbic acid content in the arils during 4 months of storage at 2 °C [41]. Accordingly, postharvest treatment of pomegranates with 2 mM Put alone or in combination with carnauba wax revealed that ascorbic acid declining trend was much pronounced in control as compared to that found in treated fruit arils, either in pomegranates stored at chilling temperature (3 °C) or at safe temperature (5 °C). Among the applied treatments, the combination of Put and carnauba wax gave the best results in terms of ascorbic acid retention at 3 and 5 °C storage temperatures [37]. Postharvest treatments with Put or Spd at 1 mM (either by immersion or vacuum-infiltration) in "Mollar de Elche" pomegranate induced higher content of ascorbic acid which was detected immediately 1 day after treatments and lasted along storage, the effect being higher in those fruits treated under vacuum-infiltration [34].

These effects of PAs increasing or maintaining high ascorbic acid content during storage are of special interest, since ascorbic acid is a bioactive compound with antioxidant capacity and proved beneficial effects in human health [11, 12].

4 Polyamines and Antioxidant Enzymes

In plant cells, the reactive oxygen species (ROS), such as superoxide radical ($O_2{\bullet}^{\downarrow}$), peroxide radical ($O_2{\bullet}^{2\downarrow}$), hydrogen peroxide ($H_2O_2$), and hydroxyl radical ($OH{\bullet}^{\downarrow}$), are randomly generated as a consequence of normal metabolism, mainly in reactions catalyzed by oxidase and lipoxygenase and in ß-oxidation of fatty acids. The ROS content in plant cell is dependent on their producing systems and scavenging mechanism, both enzymatic and non-enzymatic ones [42]. As commented above, non-enzymatic antioxidant compounds are reduced forms of ascorbate and glutathione, tocopherols, phenolics, alkaloids, and carotenoids. In addition, at physiological concentrations, polyamines are potent scavengers of hydroxyl radicals while Spd and Spm are also able to quench both singlet oxygen and hydrogen peroxide [43]. By other hand, enzymatic scavenging mechanisms include mainly superoxide dismutase (SOD), catalase (CAT), peroxidase (POD), and ascorbate peroxidase (APX). SOD detoxifies $O_2{\bullet}^{\downarrow}$ free radicals by converting them to O_2 and H_2O_2, which is further converted to H_2O and O_2 by CAT, APX, and POD. CAT catalyzes the decomposition of hydrogen peroxide to water and oxygen, while APX uses ascorbate and H_2O_2 as substrates producing water and dehydroascorbate as products, the last one being converted to ascorbate by glutathione reductase enzyme. In addition, H_2O_2 can be also reduced to water by POD, by using organic molecules such as phenols as electron donor [44].

In this sense, it has been reported that treatment of cherry tomatoes with arginine, the amino acid precursor of polyamines, resulted in increased SOD, CAT, and APX activities compared with the control, which were associated with the reduction of chilling injury development in treated fruit [45]. Moreover, in tomato plants under drought stress, an increase in Put and Spd concentrations were found in tomato fruit, as well as in the activities of SOD and CAT, the latter being responsible for the reduction in the oxidative damage induced by the stress conditions [46].

In mango, the activity of SOD, POD and CAT increased during ripening at ambient temperature and during cold storage at 11 °C although these increases were higher after treatments with Put at 0.5, 1, 1.5, and 2 mM [32]. Similarly, higher activities of CAT, POD and SOD were found in two apricot cultivars

("Bagheri" and "Asgarabadi") during storage at 1 °C plus 2 days at 20 °C treated with 1 mM Put or Spd than in control fruits, which were related to a higher resistance to stress and a longer commercial life [35]. These studies reveal that postharvest polyamine treatment increases the ability of fruit tissues to eliminate ROS and contributes to alleviate the chilling injury (CI) symptoms observed in control fruits.

Thus, the effects of exogenous application of polyamines on increasing the antioxidant enzyme activities would lead to preserve membrane integrity and to reduce accumulation of ROS and in turn to alleviate fruit stress, especially chilling injury, as well as to delay the postharvest ripening process with additional benefits on maintaining fruit quality properties during postharvest storage.

5 Conclusions

In this chapter we provide the recent information regarding the effect of polyamines on maintaining functional properties in stored fruits. With respect to the non-enzymatic antioxidant systems, the exogenous application of polyamines, as pre- or postharvest treatments, generally resulted in increases on the content of antioxidant compounds such as phenolics (including anthocyanins), carotenoids, and vitamins, especially ascorbic acid. On the other hand, polyamines are also effective on increasing the activity of several antioxidant enzymes, such as SOD, CAT, POD, and APX. Both systems could contribute to cleaning the ROS generated during the postharvest ripening process, and in turn to delay the postharvest ripening and senescence processes and extending the shelf life of fruits. Nevertheless, the precise physiological and molecular mechanisms by which polyamines increase both antioxidant systems remain elusive and deserve further research.

References

1. Bachrach U (2010) The early history of polyamine research. Plant Physiol Biochem 48:490–495
2. Rosenheim O (1924) The isolation of spermine phosphate from semen and testis. Biochem J 18:1253–1263
3. Smith TA (1971) The occurrence, metabolism and functions of amines in plants. Biol Rev Camb Philos Soc 46:201–204
4. Alburquerque N, Egea J, Burgos L, Martínez-Romero D, Valero D, Serrano M (2006) The influence of polyamines on apricot ovary development and fruit set. Ann Appl Biol 149:27–33
5. Groppa MD, Benavides MP (2008) Polyamines and abiotic stress: recent advances. Amino Acids 34:35–45
6. Valero D, Serrano M (2010) Postharvest biology and technology for preserving fruit quality. CRC/Taylor & Francis, Boca Raton
7. Asadi R, Ardebili ZO, Abdossi V (2013) The modified fruit quality by application of different kinds of polyamines in apricot tree (*Prunus armeniaca*). J Appl Environ Biol Sci 3:28–31
8. Tiburcio AF, Altabella T, Bitrián M, Alcázar R (2014) The roles of polyamines during the lifespan of plants: from development to stress. Planta 240:1–18

9. Valero D, Martínez-Romero D, Serrano M (2002) The role of polyamines in the improvement of the shelf life of fruit. Trends Food Sci Technol 13:228–234
10. Serrano M, Zapata PJ, Martínez-Romero D, Díaz-Mula H, Valero D (2016) Polyamines as an eco-friendly postharvest tool to maintain fruit quality. In: Siddiqui MW (ed) Eco-friendly technology for postharvest produce quality. Elsevier Inc., London, UK
11. Martin C, Zhang Y, Tonelli C, Petroni K (2013) Plants, diet, and health. Annu Rev Plant Biol 64:19–46
12. Nile SH, Park SW (2014) Edible berries: bioactive components and their effect on human health. Nutrition 30:134–144
13. Rodriguez-Casado A (2016) The health potential of fruits and vegetables phytochemicals: notable examples. Crit Rev Food Sci Nutr 56:1097–1107
14. Grosso G, Galvano F, Mistretta A, Marventano S, Nolfo F, Calabrese G, Buscemi S, Drago F, Veronesi U, Scuderi A (2013) Red orange: experimental models and epidemiological evidence of its benefits on human health. Oxid Med Cell Longev. Article ID 157240. doi:10.1155/2013/157240
15. Sharma P (2013) Vitamin C rich fruits can prevent heart disease. Ind J Clin Biochem 28:213–214
16. Valero D, Serrano M (2013) Growth and ripening stage at harvest modulates postharvest quality and bioactive compounds with antioxidant activity. Stewart Postharvest Rev 3:5
17. Woodside JV, McGrath AJ, Lyner N, McKinley MC (2015) Carotenoids and health in older people. Maturitas 80:63–68
18. Zanotti I, Dall'Asta M, Mena P, Mele L, Bruni R, Ray S, Del Rio D (2015) Atheroprotective effects of (poly) phenols: a focus on cell cholesterol metabolism. Food Funct 6:13–31
19. Ali MA, Poortvliet E, Strömberg R, Yngve A (2011) Polyamines in foods: development of a food databse. Food Nutr Res 55:5572. doi:10.3402/fnr.v55i05572
20. Hunter DC, Burritt DJ (2012) Polyamines of plant origin – an important dietary consideration for human health. In: Rao V (ed) Phytochemical as nutraceuticals – global approaches to their role in nutrition and health. Chapter 12. InTech, Croatia. doi:10.5772/2375
21. Kalač P (2014) Health effects and occurrence of dietary polyamines: a review for the period 2005-mid 2013. Food Chem 161:27–39
22. Malik AU, Singh Z (2006) Improved fruit retention, yield and fruit quality in mango with exogenous application of polyamines. Sci Hortic 110:167–174
23. Mehta RA, Cassol T, Li N, Ali N, Handa AK, Mattoo AK (2012) Engineered polyamine accumulation in tomato enhances phytonutrient content, juice quality, and vine life. Nat Biotechnol 20:613–618
24. Madhulatha P, Gupta A, Gupta S, Kumar A, Pal RK, Rajam MV (2014) Fruit-specific overexpression of human S-adenosylmethionine decarboxylase gene results in polyamine accumulation and affects diverse aspects of tomato fruit development and quality. J Plant Biochem Biotechnol 23:151–160
25. Pandey R, Gupta A, Chowdhary A, Pal RK, Rajam MV (2015) Overexpression of mouse ornithine decarboxylase gene under the control of fruit-specific promoter enhances fruit quality in tomato. Plant Mol Biol 87:249–260
26. Friedman M (2013) Anticarcinogenic, cardioprotective, and other health benefits of tomato compounds lycopene, α-tomatine, and tomatidine in pure form and in fresh and processed tomatoes. J Agric Food Chem 61:9534–9550
27. Khan AS, Singh Z, Abbasi NA, Swinny EE (2008) Pre- or post-harvest applications of putrescine at low temperature storage affect fruit ripening and quality of 'Angelino' plum. J Sci Food Agric 88:1686–1695
28. Serrano M, Díaz-Mula HM, Valero D (2011) Antioxidant compounds in fruits and vegetables and changes during postharvest storage and processing. Stewart Postharv Rev 1:1
29. Harindra Champa WA, Gill MIS, Mahajan BVC, Bedi S (2015) Exogenous treatment of spermine to maintain quality and extend postharvest life of table grapes (*Vitis vinifera* L.) cv. Flame Seedless under low temperature storage. LWT Food Sci Technol 60:412–419
30. Shiri MA, Ghasemnezhad M, Bakshi D, Sarikhani H (2013) Effect of postharvest putrescine application and chitosan coating on maintaining quality of table grape cv. 'Shahroudi' during long-term storage. J Food Proc Preserv 37:999–1007
31. Mirdehghan SH, Rahimi S (2016) Pre-harvest application of polyamines enhances antioxidants and table grape (*Vitis vinifera* L.) quality during postharvest period. Food Chem 196:1040–1047
32. Razzaq K, Khan AS, Malik AU, Shahid M, Ullah S (2014) Role of putrescine in regulating fruit softening and antioxidative enzymes systems in 'Samar Bahisht Chaunsa' mango. Postharvest Biol Technol 96:23–32
33. Jhalegar MJ, Sharma RR, Pal RK, Rana V (2012) Effect of postharvest treatments with

polyamines on physiological and biochemical attributes of kiwifruit (*Actinidia deliciosa*) cv. Allison Fruits 67:13–22

34. Mirdehghan SH, Rahemi M, Serrano M, Guillén F, Martínez-Romero D, Valero D (2007) The application of polyamines by pressure or immersion as a tool to maintain functional properties in stored pomegranates arils. J Agric Food Chem 55:755–760
35. Koushesh sab M, Arzani K, Barzegar M (2012) Postharvest polyamine application alleviates chilling injury and affects apricot storage ability. J Agric Food Chem 60:8947–8953
36. Castañeda-Ovando A, Pacheco-Hernández ML, Páez-Hernández ME, Rodríguez JA, Galán-Vidal A (2009) Chemical studies of anthocyanins: a review. Food Chem 113:859–871
37. Barman K, Asrey R, Pal RK, Kaur C, Jha SK (2014) Influence of putrescine and carnauba wax on functional and sensory quality of pomegranate (*Punica granatum* L.) fruits during storage. J Food Sci Technol 51:111–117
38. Lester GE (2000) Polyamines and their cellular anti-senescence properties in honey dew muskmelon fruit. Plant Sci 160:105–112
39. Asensi-Fabado MA, Munné-Bosh S (2010) Vitamins in plants: occurrence, biosynthesis and antioxidant function. Trends Plant Sci 15:582–592
40. Javanmardi J, Rahemi M, Nasirzadeh M (2013) Post-storage quality and physiological responses of tomato fruits treated with polyamines. Adv Hort Sci 27:173–181
41. Ramezanian A, Rahemi M, Maftoun M, Bahman K, Eshghi S, Safizadeh MR (2010) The ameliorative effects of spermidine and calcium chloride on chilling injury in pomegranate fruits after long-term storage. Fruits 65:169–176
42. Apel K, Hirt H (2004) Reactive oxygen species: metabolism, oxidative stress, and signal transduction. Annu Rev Plant Biol 55:373–399
43. Mozdzan M, Szemra J, Rysz J, Stolarek RA, Nowak D (2006) Anti-oxidant activity of spermine and spermidine re-evaluated with oxidizing systems involving iron and copper ions. Int J Biochem Cell Biol 38:69–81
44. Tareen MJ, Abbasi NA, Hafiz IA (2012) Postharvest application of salicylic acid enhanced antioxidant enzyme activity and maintained quality of peach cv. 'Flordaking' fruit during storage. Sci Hortic 142:221–228
45. Zhang X, Shen L, Li F, Zhang Y, Meng D, Sheng J (2010) Up-regulating arginase contributes to amelioration of chilling stress and the antioxidant system in cherry tomato fruits. J Sci Food Agric 90:2195–2202
46. Sánchez-Rodríguez E, Romero L, Ruíz JM (2016) Accumulation of free polyamines enhances antioxidant responses in fruits of grafted tomato plants under water stress. J Plant Physiol 190:72–76

Chapter 38

Acrolein: An Effective Biomarker for Tissue Damage Produced from Polyamines

Kazuei Igarashi, Takeshi Uemura, and Keiko Kashiwagi

Abstract

It is thought that the major factor responsible for cell damage is reactive oxygen species (ROS), but our recent studies have shown that acrolein (CH_2=CH-CHO) produced from spermine and spermidine is more toxic than ROS. Thus, (1) the mechanism of acrolein production during brain stroke, (2) one of the mechanisms of acrolein toxicity, and (3) the role of glutathione in acrolein detoxification are described in this chapter.

Key words Acrolein, Brain infarction, Spermine, Spermine oxidase, GAPDH (glyceraldehyde-3-phosphate dehydrogenase), GSH (glutathione)

1 Introduction

Since polyamines [spermidine (SPD) and spermine (SPM)] exist in mammalian cells at mM concentrations [1], it is thought that acrolein and H_2O_2, one of the component of ROS [superoxide anion radical ($O_2\bullet^-$), hydrogen peroxide (H_2O_2) and hydroxyl radical (•OH)], are produced effectively from SPM by SPM oxidase (SMO) during brain infarction. Thus, we compared the relative importance of acrolein and H_2O_2 during tissue damage using a photochemically induced thrombosis (PIT) model mouse [2]. It was found that protein-conjugated acrolein (PC-Acro) at the locus of brain infarction increased approximately 25.6-fold, while protein-conjugated 4-hydroxynonenal (PC-HNE), a product produced from unsaturated fatty acid by ROS, increased only 3.4-fold (Fig. 1) [3]. We also found that the toxicity of acrolein is nearly equal to 4-HNE. These results are consistent with those obtained in a cell-culture system [4] using serum amine oxidase [5, 6] instead of SMO. It is known that serum amine oxidase catalyzed the degradation of SPM into acrolein and H_2O_2 [5, 6] like SMO. In a cell-culture system, acrolein was much more toxic than H_2O_2 (Fig. 2) [3].

Rubén Alcázar and Antonio F. Tiburcio (eds.), *Polyamines: Methods and Protocols*, Methods in Molecular Biology, vol. 1694, DOI 10.1007/978-1-4939-7398-9_38, © Springer Science+Business Media LLC 2018

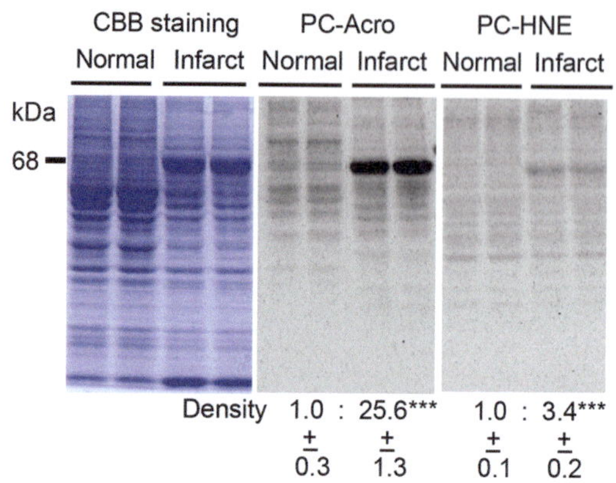

Fig. 1 Increase in markers of cell damage during brain infarction in PIT mice. Total proteins (20 μg) were stained with Coomassie Brilliant Blue R250 after SDS-PAGE. Levels of PC-Acro and PC-HNE were measured by Western blotting as described in Subheading 3.3. Figures are taken from Saiki et al. [3]

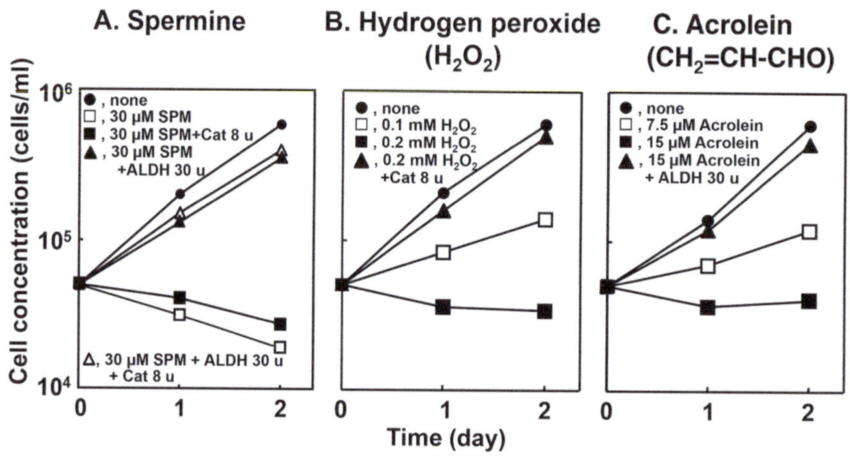

Fig. 2 Effect of spermine, H_2O_2 and acrolein on cell growth of FM3A cells cultured in the presence of 2% FBS. Cells were cultured as described in Subheading 3.2. Chemicals and enzymes shown in the figure were added to the medium together with 2% FBS. *Cat* catalase, *ALDH* aldehyde dehydrogenase. Figures are taken from Sharmin et al. [4]

We also looked at the role of ROS during brain infarction. It was found that •OH, one of the components of ROS, caused the degradation of ribosomal RNA (Fig. 3) [3]. Since it is well known that SPM mainly exists as a spermine–RNA complex [7], spermine is released from ribosomes through ribosomal RNA degradation by •OH (Fig. 3) [3]. So, ROS is a trigger of acrolein production.

In this chapter, we would like to present the results of the following: (1) how acrolein is produced during brain infarction, (2) how acrolein causes tissue damage, and (3) how acrolein is normally detoxified in tissues.

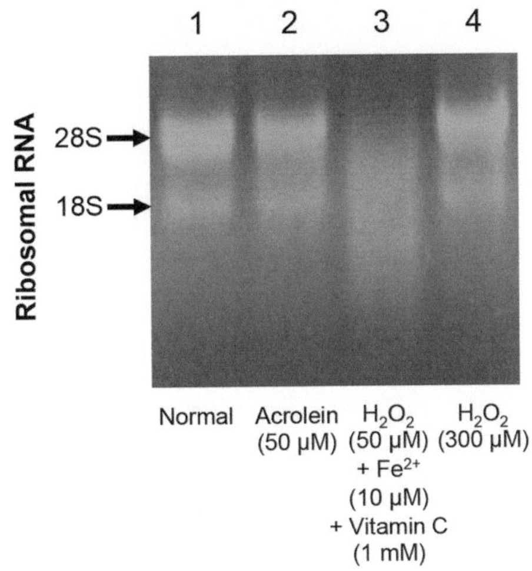

Fig. 3 Effect of acrolein, •OH and H_2O_2 on RNA degradation. Experiments were performed as described in Subheading 3.4. Figures are taken from Saiki et al. [3]

1. Low concentrations of Ca^{2+} (10–30 μM) influxed by NMDA receptors released SPM from ribosomes during brain infarction [8]. So, acrolein was produced from free SPM.
2. Inactivated GAPDH (glyceraldehyde-3-phosphate dehydrogenase) produced by acrolein translocated to the nucleus, and caused cell death (Fig. 4) [9]. This is one mechanism of acrolein toxicity.
3. When acrolein toxicity-attenuating FM3A and Neuro2a cells were isolated, the concentration of glutathione was increased, so that detoxification occurred through acrolein conjugation with GSH [10].

2 Materials

2.1 Photochemically Induced Thrombosis Model Mice

1. Male C57BL/6 mice (7-week-old) (*see* **Note 1**) were purchased from Japan SLC Inc. (Hamamatsu, Japan).
2. 3% Isoflurane.
3. Rose Bengal.
4. Xenon lamp.
5. 5% Triphenyltetrazolium chloride solution.
6. NIH image program.

2.2 Cell Culture

1. Mouse mammary carcinoma FM3A cells (Japan Health Sciences Foundation) (*see* **Note 2**).
2. Mouse neuroblastoma Neuro2a cells (DS Pharma Biomedical Co., Ltd.) (*see* **Note 3**).

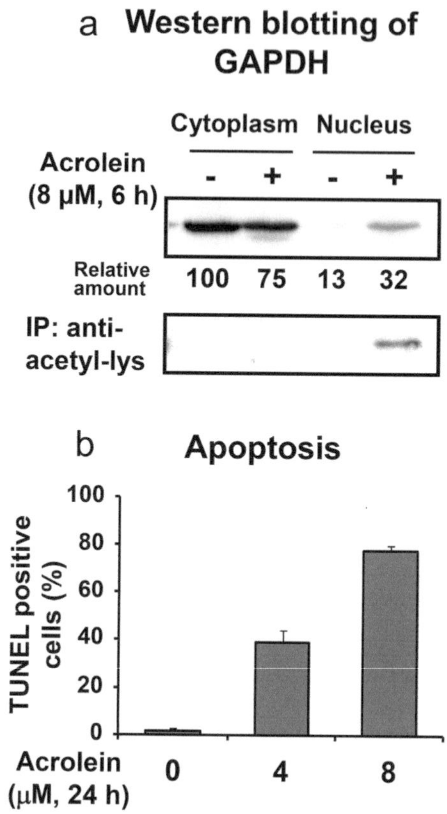

Fig. 4 Cellular distribution of GAPDH and increase in TUNEL positive cells after treatment with acrolein. (**a**) FM3A cells were treated with 0 or 8 μM acrolein for 6 h, and cytoplasm and nuclei were isolated. GAPDH in each fraction was identified by Western blotting. Immunoprecipitation by anti-acetyl-lysine was performed to estimate the level of acetyl-lysine in GAPDH. (**b**) FM3A cells were treated with 0, 4 and 8 μM acrolein for 24 h and percentage of TUNEL positive cells was shown by counting approximately 500 cells. Figures are taken from Nakamura et al. [9]

3. D-MEM (Low-glucose).
4. 0.25% Trypan blue.

2.3 Mutagenesis of Cells

1. Ethyl methanesulfonate.
2. Stepwise increase in acrolein concentration.

2.4 Measurement of PC-Acro and Polyamines

1. Ultra-Turrax homogenizer.
2. Buffer A: 10 mM Tris–HCl, pH 7.5, 1 mM dithiothreitol, 10% glycerol, 0.2 mM EDTA, and 0.02 mM FUT-175 (6-amino-2-naphthyl-4-guanidinobenzoate), a protease inhibitor.

3. ACR-LYSINE ADDUCT ELISA System (NOF Corporation, Japan) (*see* **Note 1**).
4. Microplate reader Bio-Rad Model 550 (CORONA ELECTRIC, Japan).
5. High performance liquid chromatography (HPLC).

2.5 Western Blotting

1. Antibody against N^{ε}-(3-formyl-3,4-dehydropiperidino-lysine) (FDF-lysine) (NOF Corporation) (*see* **Note 1**).
2. Antibody against 4-hydroxynonenal (4-HNE) (Abcam) (*see* **Note 1**).
3. Antibody against glyceraldehyde-3-phosphate dehydrogenase (GAPDH) (Santa Cruz) (*see* **Note 4**).
4. Antibody against acetylated lysine (Cell Signaling Technology) (*see* **Note 4**).
5. ECL Western blotting reagents (GE Healthcare Bio-Siences).
6. LAS-3000 luminescent image analyzer (Fuji Film).

2.6 Separation of RNA and Protein

1. Buffer B: 10 mM Hepes-KOH, pH 7.5, 100 mM KCl, and 2 mM magnesium acetate.
2. Gel electrophoresis with 1.2% agarose containing 2.2 M formaldehyde (*see* **Note 5**).
3. Ethidium bromide.
4. Sodium dodecyl sulfate-polyacrylamide gel electrophoresis (SDS-PAGE).
5. Coomassie Brilliant Blue R-250.

2.7 Measurement of Ca^{2+}, GAPDH Activity and GSH

1. Atomic absorption spectrometry.
2. Fura-2.
3. KDalert™ GAPDH Assay Kit (Applied Biosystems) (*see* **Note 4**).
4. Total glutathione assay kit (Northwest Life Science Specialties, LLC) (*see* **Note 3**).

2.8 Terminal Deoxynucleotidyl Transferase dUTP Nick End Labeling (TUNEL) Assay

1. In Situ Cell Death Detection kit Fluorescein (Roche) (*see* **Note 4**).
2. 0.5 μg/mL Propiodium iodide.
3. Antibody against acetyl-lysine (Cell Signaling Technology).
4. Scanning Microscope LSM 510 META Laser (Carl Zeiss).

3 Methods

The toxicity of acrolein was examined using photochemically induced thrombosis (PIT) model mice, cell culture systems, and measurements of various components involved in acrolein toxicity.

3.1 Preparation of Photochemically Induced Thrombosis (PIT) Model Mice

Seven-week-old male mice weighing 22–26 g were anesthetized with inhalation of 3% isoflurane. Anesthesia was continued with 1.5% isoflurane during the operation, and body temperature was kept at 37 ± 1 °C with a heating pad. The thrombotic occlusion of the middle cerebral artery (MCA) was induced by photochemical reaction [11]: an incision was made between the left orbit and the external auditory canal, and the temporalis muscle was detached from dura mater to expose the proximal section of the MCA. Immediately after intravenous injection of photosensitizer, Rose Bengal (20 mg/kg), through a jugular vein, green light (wave length: 540 nm) emitted from a xenon lamp illuminated the MCA for 10 min. After MCA occlusion, incised skin was restored. At 24 h after the induction of PIT stroke, the brain was removed and sectioned into 2 mm thick coronal slices. Each slice was incubated with 5% triphenyltetrazolium chloride solution at 37 °C for 30 min. The volume of infarction was analyzed on a Macintosh computer using the NIH image program [2] (*see* **Note 1**).

3.2 Cell Culture

Mouse mammary carcinoma FM3A cells ($1–2 \times 10^4$ cells/mL) were cultured in ES medium, supplemented with 50 units/mL streptomycin, 100 units/mL penicillin G and 2% heat-inactivated fetal bovine serum (FBS) at 37 °C in an atmosphere of 5% CO_2 in air as described previously [12] (*see* **Note 2**). Mouse neuroblastoma Neuro2a cells ($3–5 \times 10^4$ cells/mL) were cultured in D-MEM (Low-glucose) supplemented with 10% FBS and Non-Essential Amino Acids (Sigma) at 37 °C in an atmosphere of 5% CO_2 in air. The viable cell number was counted in the presence of 0.05% Trypan blue.

3.3 Measurements of PC-Acro and PC-HNE

The level of PC-Acro and PC-HNE was measured by Western blotting. Brain tissues at the locus of infarction of PIT model mice and at the same locus of control mice were homogenized using Ultra-Turrax homogenizer in 0.5 mL of Buffer A in Subheading 2.4. Total proteins (20 μg) were stained with Coomassie Brilliant Blue R-250 after SDS-PAGE [13], and the levels of PC-Acro and PC-HNE were measured by Western blotting [14] using 20 μg protein of tissue homogenate and monoclonal antibodies against PC-Acro and PC-HNE (*see* **Note 1**).

3.4 Assay for RNA Degradation

Rabbit reticulocyte lysates were prepared by the method of Pelham and Jackson [15]. Polysomes were obtained by centrifugation at 150,000 × *g* for 2.5 h of the reticulocyte lysate. The sedimented polysomes were mixed with Buffer B in Subheading 2.6, centrifuged as above, and dissolved in the same buffer. RNA degradation was tested by incubating Buffer B (40 μL) containing 1 A_{260} unit of rabbit reticulocyte polysomes, acrolein and ROS shown in Fig. 3 at 37 °C for 10 h. Hydroxyl radical (•OH) was produced from H_2O_2 in the presence of 10 μM Fe^{2+} and 1 mM vitamin C. After the incubation, RNA was isolated by gel electrophoresis with 1.2% agarose containing 2.2 M formaldehyde, and stained with 0.03% ethidium bromide (*see* **Note 5**).

3.5 Measurements of Ca^{2+}, GSH and GAPDH

Ca^{2+} ions in brain tissues and cells were extracted with 5% trichloacetic acid (TCA), and measured using 5% TCA supernatant in the presence of lanthanium chloride (1 mg/mL) by means of atomic absorption spectrometry [16] or Fura-2 [17] (*see* **Note 1**). The level of GSH was measured using total glutathione assay kit according to the manufacturer's instructions (*see* **Note 3**). For measurement of GAPDH activity, cells (2 × 10^6 cells) were suspended in 0.1 mL of Buffer A, and lysed by repeated (three times) freezing and thawing with intermittent mechanical mixing. The supernatant was obtained by centrifugation at 17,000 × *g* for 15 min and used as cell lysate for measurement of GAPDH activity. GAPDH activity was measured using KDalert™ GAPDH Assay Kit (Applied Biosystems). The level of GAPDH in cytoplasm or nucleus was estimated by Western blotting (*see* **Note 4**).

3.6 TUNEL Assay

TUNEL reaction was performed in TUNEL reaction mixture containing 0.5 μg/mL propidium iodide using In Situ Cell Death Detection Kit Fluorescein (Roche). TUNEL positive cells were counted under a microscope (*see* **Note 4**).

4 Notes

1. The level of PC-Acro at the locus of infarction was measured using PIT model mice (Fig. 1). The level of PC-Acro at the locus of infarction was 25.6-fold higher compared with that at the normal part of brain (Fig. 1). Infarction volume was determined by staining 2 mm thick coronal slices with triphenyltetrazolium. This stains the viable brain red, whereas infarct tissue remained unstained. The average volume of infarction at 24 h after photoinduction was 36 mm^3. We have previously shown that N^1,N^4,N^8-tribenzylspermidine (TB34) is a strong channel blocker of the NMDA receptor, which aggravates brain infarction through Ca^{2+} influx [18]. The level of Ca^{2+} at the locus of infarction increased approximately 6.5-fold compared

to that of Ca^{2+} at the corresponding locus of normal mice. Size of brain infarction decreased significantly by TB34 treatment, and Ca^{2+} accumulated at the locus of brain infarction was reduced significantly by TB34. In parallel, the level of PC-Acro was reduced by TB34. In contrast, the level of PC-HNE, which is produced from unsaturated fatty acids through oxidative stress, was very low at the locus of brain infarction (Fig. 1). The results indicate that Ca^{2+} toxicity is mainly caused by acrolein rather than ROS, and acrolein is produced from SPM and SPD.

2. SPM inhibits cell growth in the presence of FBS. The toxicity is caused by factors produced from SPM by amine oxidase: SPM is converted to acrolein and H_2O_2 by bovine serum amine oxidase or SMO. Addition of aldehyde dehydrogenase, but not catalase, to the culture medium could prevent the toxic effect of SPM on cell growth. Furthermore, the concentration of acrolein necessary for inhibition of cell growth was much lower than that of H_2O_2 in the inhibition of cell growth by SPM (Fig. 2) [4], indicating that acrolein is much more toxic than H_2O_2.
3. We then tried to isolate cells with reduced sensitivity to acrolein toxicity to clarify how acrolein is detoxified under cell culture conditions. Neuro2a cells were mutagenized by treatment with 0.1% ethyl methanesulfonate, and cultured in medium containing acrolein. Concentrations of acrolein were gradually increased in a stepwise manner from 10 to 35 μM over 6 months, and acrolein toxicity-attenuating cells were isolated [10]. The IC_{50} of acrolein in neuroblastoma Neuro2a cells was 4.2 μM, whereas it was 8.4 μM in acrolein toxicity-attenuating Neuro2a (Neuro2a-ATD1) cells. In Neuro2a-ATD1 cells, the concentration of glutathione (GSH) was increased, so that detoxification occurred through acrolein conjugation with GSH. In Neuro2a-ATD1 cells, phosphorylation of transcription factors (C-jun and NF-κB) necessary for expression of genes encoding γ-glutamylcysteine ligase catalytic unit (GCLC), and glutathione synthetase (GSHS) involved in GSH synthesis was stimulated, so that transcription of two genes increased in Neuro2a-ATD1 cells. These results support the idea that GSH plays important roles in detoxification of acrolein. Recently we have characterized another acrolein toxicity-attenuating Neuro2a (Neuro2a-ATD2) cell line. Expression of genes for polyamine oxidases (SMO and acetylpolyamine oxidase) was decreased at the level of transcription in Neuro2a-ATD2 cells, because of the reduction of transcription factors, FosB of AP-1 and C/EBPβ, which are involved in transcription of genes for polyamine oxidases. The results

indicate that acrolein is mainly produced from SPM, not from unsaturated fatty acids [19].

4. Acrolein interacts with Cys, Lys and His residues in proteins. Thus we tried to identify protein(s) conjugated with acrolein using 100,000 × *g* supernatant of FM3A cells treated with 40 μM acrolein for 9 h. The level of a protein approximately 37 kDa strongly decreased in acrolein-treated FM3A cells compared with control cells. This protein was identified as glyceraldehyde-3-phosphate dehydrogenase (GAPDH) by determining the peptide sequences by LC-MS/MS. Acrolein interacted with Cys-150 at the active site of GAPDH, and also with Cys-282. When cells were treated with 8 μM acrolein, the activity of GAPDH was greatly reduced. In addition, it was shown that acrolein-conjugated GAPDH translocated to the nucleus and was acetylated, and the number of TUNEL positive cells was increased, indicating that cell death is enhanced by acrolein-conjugated GAPDH (Fig. 4). The results indicate that inactivation of GAPDH is one mechanism that underlies cell toxicity caused by acrolein [3].
5. We also tried to clarify the role of H_2O_2 produced from SPM by spermine oxidase together with acrolein. As shown in Fig. 3, hydroxyl radical (•OH) produced from H_2O_2 in the presence of Fe^{2+} and vitamin C caused the degradation of ribosomal RNA. So, SPM bound to ribosomes was released and acrolein is produced [8]. The results indicate that •OH functions as an initiator of acrolein production.

Acknowledgments

We are grateful to Dr. A.J. Michael for critical reading of the manuscript prior to submission.

References

1. Igarashi K, Kashiwagi K (2010) Modulation of cellular function by polyamines. Int J Biochem Cell Biol 42:39–51
2. Saiki R, Nishimura K, Ishii I, Omura T, Okuyama S, Kashiwagi K, Igarashi K (2009) Intense correlation between brain infarction and protein-conjugated acrolein. Stroke 40:3356–3361
3. Saiki R, Park H, Ishii I, Yoshida M, Nishimura K, Toida T, Tatsukawa H, Kojima S, Ikeguchi Y, Pegg AE, Kashiwagi K, Igarashi K (2011) Brain infarction correlates more closely with acrolein than with reactive oxygen species. Biochem Biophys Res Commun 404:1044–1049
4. Sharmin S, Sakata K, Kashiwagi K, Ueda S, Iwasaki S, Shirahata A, Igarashi K (2001) Polyamine cytotoxicity in the presence of bovine serum amine oxidase. Biochem Biophys Res Commun 282:228–235
5. Tabor CW, Tabor H, Bachrach U (1964) Identification of the aminoaldehydes produced by the oxidation of spermine and spermidine with purified plasma amine oxidase. J Biol Chem 239:2194–2203
6. Bachrach U (1970) Oxidized polyamines. Annu N Y Acad Sci 171:939–956
7. Watanabe S, Kusama-Eguchi K, Kobayashi H, Igarashi K (1991) Estimation of polyamine

binding to macromolecules and ATP in bovine lymphocytes and rat liver. J Biol Chem 266:20803–20809

8. Nakamura M, Uemura T, Saiki R, Sakamoto A, Park H, Nishimura K, Terui Y, Toida T, Kashiwagi K, Igarashi K (2016) Toxic acrolein production due to Ca^{2+} influx by the NMDA receptor during stroke. Atherosclerosis 244:131–137
9. Nakamura M, Tomitori H, Suzuki T, Sakamoto A, Terui Y, Saiki R, Dohmae N, Igarashi K, Kashiwagi K (2013) Inactivation of GAPDH as one mechanism of acrolein toxicity. Biochem Biophys Res Commun 430:1265–1271
10. Tomitori H, Nakamura M, Sakamoto A, Terui Y, Yoshida M, Igarashi K, Kashiwagi K (2012) Augmented glutathione synthesis decreases acrolein toxicity. Biochem Biophys Res Commun 418:110–115
11. Tanaka Y, Marumo T, Omura T, Yoshida S (2007) Quantitative assessments of cerebral vascular damage with a silicon rubber casting method in photochemically-induced thrombotic stroke rat models. Life Sci 81:1381–1388
12. Ayusawa D, Iwata K, Seno T (1981) Alteration of ribonucleotide reductase in aphidicolin-resistant mutants of mouse FM3A cells with associated resistance to arabinosyladenine and arabinosylcytosine. Somatic Cell Genet 7:27–42
13. Laemmli UK (1970) Cleavage of structural proteins during the assembly of the head of bacteriophage T4. Nature 227:680–685
14. Nielsen PJ, Manchester KL, Towbin H, Gordon J, Thomas G (1982) The phosphorylation of ribosomal protein S6 in rat tissues following cycloheximide injection, in diabetes, and after denervation of diaphragm. A simple immunological determination of the extent of S6 phosphorylation on protein blots. J Biol Chem 257:12316–12321
15. Pelham HR, Jackson RJ (1976) An efficient mRNA-dependent translation system from reticulocyte lysates. Eur J Biochem 67:247–256
16. Rappaport ZH, Young W, Flamm ES (1987) Regional brain calcium changes in the rat middle cerebral artery occlusion model of ischemia. Stroke 18:760–764
17. Grynkiewicz G, Poenie M, Tsien RY (1985) A new generation of Ca^{2+} indicators with greatly improved fluorescence properties. J Biol Chem 260:3440–3450
18. Igarashi K, Shirahata A, Pahk AJ, Kashiwagi K, Williams K (1997) Benzyl-polyamines: novel, potent *N*-methyl-D-aspartate receptor antagonists. J Pharmacol Exp Ther 283:533–540
19. Uemura T, Nakamura M, Sakamoto A, Suzuki T, Dohmae N, Terui Y, Tomitori H, Casero RA Jr, Kashiwagi K, Igarashi K (2016) Decrease in acrolein toxicity based on the decline of polyamine oxidases. Int J Biochem Cell Biol 79:151–157

Chapter 39

Polyamines and Cancer

Elisabetta Damiani and Heather M. Wallace

Abstract

This chapter provides an overview of how the polyamine pathway has been exploited as a target for the treatment and prevention of multiple forms of cancer, since this pathway is disrupted in all cancers. It is divided into three main sections. The first explores how the polyamine pathway has been targeted for *chemotherapy*, starting from the first drug to target it, difluoromethylornithine (DFMO) to the large variety of polyamine analogues that have been synthesised and tested throughout the years with all their potentials and pitfalls. The second section focuses on the use of polyamines as vectors for *drug delivery*. Knowing that the polyamine transport system is upregulated in cancers and that polyamines naturally bind to DNA, a range of polyamine analogues and polyamine-like structures have been synthesised to target epigenetic regulators, with encouraging results. Furthermore, the use of polyamines as transport vectors to introduce toxic/bioactive/fluorescent agents more selectively to the intended target in cancer cells is discussed. The last section concentrates on *chemoprevention*, where the different strategies that have been undertaken to interfere with polyamine metabolism and function for antiproliferative intervention are outlined and discussed.

Key words Polyamines, Cancer, Chemoprevention, Chemotherapy, Drug delivery

1 Introduction

Ever since their discovery in the eighteenth century, spermidine and spermine, and their diamine precursor, putrescine, have been associated with cell growth. They are aliphatic amines which are positively charged at physiological pH known collectively as the polyamines. The link between polyamines and cancer is well-established and dates back to the seminal discoveries in the 1970s when increased levels of urinary polyamines were found in patients with a variety of tumours [1]. The discovery in the 1980s that polyamine concentrations in cancer cells and tissues were also increased compared to the equivalent normal tissue reinforced this association [2, 3]. As critical factors for cell growth and development, it is not surprising to find these increased concentrations of polyamines in tumour cells, tissues and urine of cancer patients. What remains to be proven, however, is what exactly is the nature of

Rubén Alcázar and Antonio F. Tiburcio (eds.), *Polyamines: Methods and Protocols*, Methods in Molecular Biology, vol. 1694, DOI 10.1007/978-1-4939-7398-9_39, © Springer Science+Business Media LLC 2018

the relationship. Are increases in polyamine concentrations a result of cancer development or are they the cause of the uncontrolled growth of cells we recognise as cancers?

Ornithine decarboxylase (ODC), the first enzyme in the polyamine biosynthetic pathway, exhibits increased enzyme activity in several cancers including breast, prostate, skin and colon [4, 5]. Furthermore ODC is itself an oncogene [6] and is a target of several oncogenes such as c-MYC and Ras, which are amplified in multiple cancers [6, 7]. Given the strong association between increased polyamine concentrations, elevated ODC activity and cancer and the need for effective anticancer drugs, the obvious next step was to determine if preventing their synthesis could block cancer development. Having highlighted the biosynthetic pathway as potentially a druggable target, ODC was the primary candidate. The exciting possibility was that modulation of the polyamine pathway may be useful in the treatment and prevention of multiple forms of cancer as this pathway is disrupted in all cancers. The aim of this chapter is to provide an overview of the evidence of association between polyamines and cancer and to discuss how the knowledge accrued to date on these unique and ubiquitous molecules has been exploited as a target for the treatment and prevention of this disease (Fig. 1).

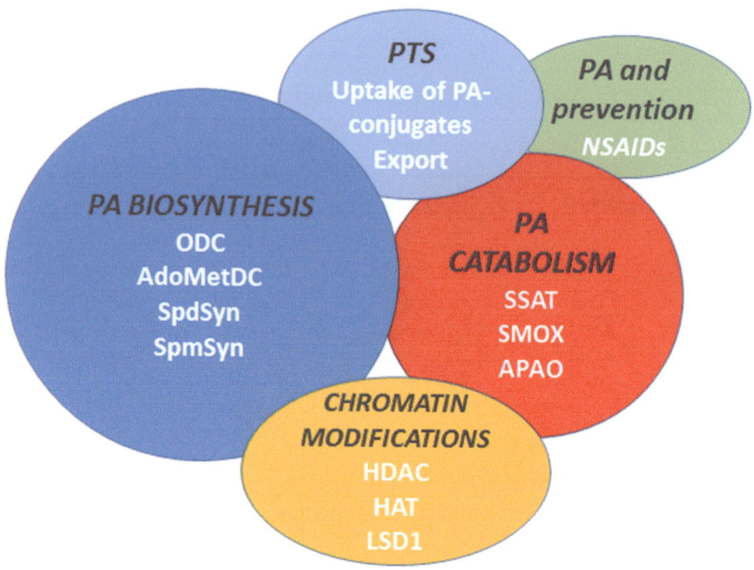

Fig. 1 Targets for intervention in cancer. Abbreviations: *PA* polyamine, *PTS* polyamine transport system, *ODC* ornithine decarboxylase, *AdoMetDC* S-adenosylmethioninedecarboxylase, *SSAT* spermine/spermidine acetyltransferase, *SMOX* spermine oxidase, *SpdSyn* spermidine synthase, *SpmSyn* spermine synthase, *APAO* acetylated polyamine oxidase, *HDAC* histone deacetylase, *HAT* histone acetylase, *LSD1* lysine-specific histone demethylase 1, *NSAIDs* non-steroid anti-inflammatory drugs

2 Targeting the Polyamine Pathway for Chemotherapy

The first anticancer drug to target the polyamine pathway was directed towards inhibiting ODC, a tightly regulated enzyme whose reaction product is the diamine, putrescine. α-Difluoromethylornithine (DFMO) was one of the first inhibitors of this enzyme to be developed in the 1970s specifically to target cancer cell growth [8]. DFMO is an irreversible inhibitor of ODC and while uptake is by diffusion which is not ideal for a drug it did inhibit cancer cell growth in cultured cells and decrease putrescine and spermidine concentrations, but interestingly spermine was most times little affected by DFMO exposure. Unfortunately, despite this success in vitro in several tumour cell types [9], it proved to be less effective in vivo and in clinical trials both alone, and in combination with other agents [10, 11]. The main reason for its ineffectiveness was ascribed to compensatory mechanisms that occur as a result of the depleted polyamine pools. These include increased polyamine transport of preformed polyamines and upregulation of other crucial enzymes in the biosynthetic pathway. The only reported success of DFMO in cancer chemotherapy has been in the treatment of recurrent gliomas [12], although subsequently, in a phase III trial for newly diagnosed patients with glioblastoma multiforme, it proved ineffective compared to standard fractionated irradiation [13]. Despite these disappointments, DFMO is a potentially useful agent as it has little inherent toxicity with only reversible ototoxicity being reported [14]. This low toxicity has kept DFMO 'alive' as an anticancer agent and recently it has gone through a revival and is currently being used as a chemopreventative agent against colorectal cancer and neuroblastoma (discussed in the section on Chemoprevention).

A range of other polyamine biosynthesis blockers, such as the antileukaemic agent, methylglyoxal bis(guanylhydrazone) (MGBG), have been investigated. MGBG is a competitive inhibitor of S-adenosylmethionine decarboxylase (AdoMetDC) which is a second crucial enzyme in the synthesis of spermidine and spermine [15]. MGBG proved to have several drawbacks the most critical of which was its mitochondrial toxicity [16]. This differentiated MGBG from DFMO in terms of potential use and resulted in loss of interest in the former agent. It did however fuel research into synthesising less toxic AdoMetDC inhibitors, the most successful being 4-amidinoindan-1-one-2′-amidinhydrazone (SAM486A/CGP48664) [17]. SAM486A is more potent and specific than MGBG and that has been tested in phase I and phase II clinical trials both as a single agent and in combination with other antiproliferative drugs for multiple cancers with mixed success [18, 19]. Attempts to target the biosynthetic synthases of the polyamine pathway were also explored, and small libraries of inhibitors were synthesised and tested (for review *see* [20]). Compounds such as S-adenosyl-3-thio-1,8-diaminooctane (AdoDATO) and S-adenosyl-1,12-diamino-3-thio-9-azadodecane (AdoDATAD) inhibited spermidine and spermine synthases respectively, but both showed little promise as chemotherapeutic agents.

As a result of the poor outcomes outlined above with compounds aimed at blocking individual enzymes in polyamine biosynthesis, focus shifted to finding other ways of exploiting the polyamine pathway for anticancer therapy. The polyamine pathway is well-known to be self-regulating in nature, therefore one idea was to modulate it by using polyamine-like compounds able to interrupt polyamine biosynthesis and compete for uptake, thus reducing normal polyamine content required for cell growth. A range of polyamine analogues were therefore synthesised. The key requirements of the analogues were: (1) to mimic the polyamines so as to compete with the natural extracellular amines for recognition and subsequent intracellular uptake by the polyamine transporter. In this way, the compensatory mechanisms seen with DFMO treatment which were mainly responsible for its failure would be prevented, (2) to induce negative feedback inhibition of ODC and polyamine biosynthesis, (3) to be sufficiently distinct so as to not replace the function of natural polyamines in cells and should not undergo rapid catabolism by the polyamine catabolic enzymes. The theory is that the resultant compounds with all these key features will significantly deplete all three natural polyamines and produce cell growth arrest and exhibit tumour-selective activity [20].

Based on the different substitutions on the backbone of spermine and spermidine, the polyamine analogues can be subdivided into three groups (for reviews *see* [21, 22]). The first to be developed were the symmetrically substituted ones which have their primary amino termini protected by symmetrical bis(ethyl) groups. This inhibits their oxidation to toxic compounds by multiple amine oxidases [23]. Besides meeting the above requirements, the symmetrically substituted analogues were also found to significantly induce the catabolic pathway, mainly SSAT [24]. This led to much excitement in the field since super-induction of SSAT means increased polyamine catabolism hence greater polyamine depletion [25]. Concomitantly, these analogues also lead to an increase in reactive oxygen species (ROS), which was later discovered to be due to the induction of spermine oxidase (SMOX) [26]. The most successful among these first polyamine analogues was N^1, N^{11}-bis(ethyl)norspermine (BENSpm). This analogue enters the cell via the polyamine transport system and downregulates both ODC and AdoMetDC while upregulating SSAT and SMOX. It proved highly successful against pancreatic and melanoma cells and against bladder, ovarian and lung carcinoma xenografts [27], such that it entered phase I and II clinical trials directed towards a number of solid tumours [28, 29]. However, due to poor activity in addition to unacceptable toxicity at the level of the central nervous system, it did not progress. Despite its failure as a single agent, BENSpm combined with standard chemotherapeutic agents such as oxaliplatin and 5 fluorouracil (5FU), showed synergistic effects in different tumour cell lines, even when low doses of BENSpm were used [26, 30, 31]. These results are encouraging for future clinical combination studies using standard chemotherapeutics and polyamine analogues.

Following this, the second and third generation of polyamine analogues emerged in the hope of improving on the first generation. The second generation are the unsymmetrical substituted alkyl polyamines where cyclopropyl, cycloheptyl or isopropyl substituents were placed on the spermine or norspermine backbone. The third generation are conformationally restricted analogues, characterised by being rotationally restricted at the central carbons of the polyamine chain including the addition of double bonds or cyclic moieties [32, 33]. Although the unsymmetrical analogues showed increased antiproliferative potential and reduced toxicity over the symmetrically substituted ones in a variety of tumour cells [34, 35], it was with the conformationally restricted analogues that more encouraging results were achieved. In fact, two of these, PG-11093 and PG-11047 went as far as being clinically tested. PG-11047 inhibited growth in both small cell and non-small cell lung cancer cell lines as well as in breast and colon cancers in vitro. In phase I clinical trials it was studied both as a single agent and in combination with other antitumour agents such as cisplatin and bevacizumab [36]. PG-11093 was also shown to enhance the anti-myeloma activity of bortezomib and was evaluated in phase I clinical trials [37]. Unfortunately, their further development was halted by the choices made following the merging of pharmaceutical companies and the results of these trials have not been disclosed.

Since polyamines have long been known to interact with nucleic acids and chromatin [21], this fact was exploited to produce another class of polyamine analogues, the oligoamines. These compounds have an increased number of protonable nitrogens (8–14 amines in both saturated and conformationally restricted, unsaturated forms) that increase their affinity for DNA thus increasing their potential as anti-tumour agents. The group of Frydman synthesised a variety of these compounds, the most representative of which was CGC-11144. It exhibited significant antiproliferative activity against human prostate cancer cell lines which correlated well with its ability to aggregate DNA, and was also effective in both in vivo and in vitro models of breast cancer [38]. Interestingly, this compound along with other oligoamines, was found to have anti-oestrogenic effects which could be exploited for breast-cancer therapy and/or prevention [39]. CGC-11144 also possessed other important features that became apparent after the discovery of the histone demethylase, LSD1 in 2004 [40]. The enzyme LSD1 demethylates mono- and dimethyl lysine 4 of histone H3 in the promoter region of active genes, and is responsible in a complex with specific transcription factors, for the transcriptional repression of certain important tumour suppressor genes like p53 and SFRP1. In cancer, LSD1 is frequently over-expressed therefore its inhibition should result in the re-expression of these genes, thereby inhibiting cancer growth [41]. Incidentally, LSD1 shares 60% homology with SMOX and it is this oxidase activity which is responsible for demethylation of lysine residues. It follows that polyamine-like compounds bearing moieties known to inhibit SMOX would also inhibit LSD1, hence a series of previously synthesised (bis)guanidines, (bis)biguanidines, octamines and decamines were screened, some of which indeed were able to re-express aberrantly silenced tumour suppressor genes in colon cancer cells in vitro [42]. CGC-11144 in combination with one such polyaminobiguanidine derivative was particularly effective in inhibiting LSD1 in a human breast cancer cell line [43]. Interestingly, CGC-11144 also showed synergism with DFMO in two different colon cancer cell lines, HCT116 and HT-29. DFMO-induced depletion of polyamines, stimulated the uptake of extracellular polyamines including the polyamine analogue, which in turn inhibited LSD1 leading to the re-expression of previously silenced tumour suppressor genes such as p16 and SFRP2 [44]. Recently, other oligoamines have been synthesised and explored as LSD1 inhibitors in the form of (bis)urea and (bis)thiourea derivatives [45]. Structurally related to these, are another series of compounds, the amidoximes, which induced profound changes to the methylation status of histone H3 at lysine 4, the target for LSD1, leading to the re-expression of genes believed to play a role in tumourigenesis in the same cell lines: SFRP2 (a soluble modulator of Wnt signalling secreted frizzle-related protein), HCAD (the membrane-bound mediator of Ca-dependent cell–cell adhesion H-cadherin) and GATA4 (Zn-finger transcription factor) [46]. Certainly, all these new polyamine derivatives that target LSD1 involved in the complex regulation of gene expression, appear to have strong potential as future anti-tumour drugs.

A number of other biologically interesting polyamine analogues have been synthesised over the years such as the aminoxy analogues of spermidine by Khomutov and colleagues which reversibly inhibit enzymes of the polyamine biosynthetic pathway and inhibit proliferation in colon cells in vitro [47]. Additional macrocyclic polyamine analogues have also been developed that are related to the Budmunchiamine family of natural plant alkaloids. These relatively simple polyamine analogues have been shown to deplete intracellular ATP through their ATPase like activity and inhibit cell growth in prostate cancer cells. They also act as potential artificial nucleases by binding to and hydrolysing DNA. Hence these compounds may have potential as chemotherapeutic agents (reviewed in [48]).

Overall, although all the above polyamine analogues have been shown to have varying degrees of potential as anti-tumour agents being profoundly anti-proliferative and/or causing apoptosis, none have actually been successful when tested in the clinical setting. Does the lack of clinical efficacy mean the end for the analogues? Absolutely not, there is still much to be done with the large variety of analogues available (as outlined above) and those that are still to be synthesised. Indeed, the link between polyamine analogues and the interaction with DNA is of interest from the possibility of using these as inactivators of chromatin-modifier enzymes. They have also been, and continue to be, useful and successful research tools that have helped advance the understanding of the regulation and function of natural polyamines.

3 Polyamines in Drug Delivery

In several cancers, an increase in expression of specific oncogenes or a decrease in tumour-suppressor gene expression has been observed which is linked to altered chromatin modifications, also known as 'epigenetic' modifications. These are typically DNA methylation and histone modifications that alter DNA accessibility and chromatin structure, thereby regulating the patterns of gene expression. A more condensed chromatin structure mediated by DNA and histone methyl transferases and histone deacetylases (HDACs) as well as other complexes, is normally associated with gene silencing. A more relaxed chromatin structure mediated by histone acetylases (HATs) and demethylases, is instead generally associated with gene transcription and activation. Because the dysregulation of these chromatin-modifier enzymes has been observed in several cancers, targeting them has been one of the main goals of recent anti-neoplastic therapies, mainly through the advent of various HDAC inhibitors (HDACi). Polyamines have been shown to alter the expression and activity of both HATs and HDACs and therefore can modulate chromatin arrangement and transcriptional regulation of gene expression (reviewed in [49]). As polyamines are important downstream targets of HDAC activity using an HDACi to modify the polyamine content could be an alternative strategy with considerable therapeutic potential. Polyamines naturally bind to DNA, therefore polyamine analogues and polyamine-like structures could target epigenetic regulators. With this in mind, Varghese et al. synthesised a range of polyaminohydroxamic acid derivatives (PAHAs) [50]. These contain a spermine or spermidine side chain linked to a hydroxamic moiety which is common amongst the potent HDACi, such as Trichostatin A or suberoylanilide hydroxamic acid. Such compounds, thanks to their polyamine moiety, could enter the cell via the polyamine transport system (PTS), since this is often upregulated in tumour compared to normal cells [51]. In addition, they could target compressed and transcriptionally repressed chromatin in the promoter region of tumour suppressor genes of cancer cells with the pharmacologically active component, thereby reactivating gene expression. Indeed, these compounds have been shown to decrease HDAC activity in vitro by, at least, 60% in ML-1 mouse leukaemia cells and in the colon cancer cell line, HCT116. This is equivalent or better than the HDACi used in clinical trials [50]. In addition, the cell cycle regulator p21 was reactivated, promoting apoptosis of these transformed cells and HDAC6 responsible for metastatic invasion by cancer was selectively inhibited [52]. These findings prompted further research and development into these compounds and new derivatives containing the polyaminobenzamide moiety were synthesised (PABAs) [53]. Due to differential charge distribution, these new PABA analogues proved to be more effective substrates for the PTS than their earlier PAHA derivatives. They also proved to be very effective in several cancer cell lines in promoting apoptosis, inhibiting HDAC selectively and activating p21 [54]. These encouraging results stimulate the need for research into finding new polyamine-based chromatin-targeting molecules. The polyamine moiety is amenable to substitution from the synthetic point of view as the amine groups can be conjugated readily to many reactive moieties. Targeting the 11 specific isoforms of the zinc-dependent HDACs, may not only improve their therapeutic activity, but may also help identify which of the HDAC isoforms are actually involved in the carcinogenic phenotype. Thus, these classes of polyamine derivatives with HDAC inhibitory activity could be promising new molecules that could one day be added to current therapeutic cancer protocols [54].

The idea of using polyamines as vectors to introduce inhibitors of chromatin-modifier enzymes more selectively to cancer cells, should have prompted exploration into other types of inhibitors, for example those that would target the HATs too. To date, there is only one report in the literature directed at this. Bandyopdhyay and coworkers [55] synthesised such inhibitors, consisting of spermidine (Spd) linked to the S-terminus of CoA through a thioglycolic acid linkage (Spd-CoA) and variations of this (coenzyme A is the acetyl donor of HATs). These molecules caused rapid inhibition of histone acetylation that correlated with arrest of DNA synthesis and inhibition of DNA repair in several cancer cell lines compared to normal cells. Furthermore, Spd-CoA was shown to have synergistic effects with a variety of DNA damaging treatments, including several commonly used chemotherapeutic agents, to induce cancer cell killing. Therefore this class of polyamine-based chromatin inhibitors may be further fine-tuned for clinical relevance, particularly to reduce therapy toxicity and reverse therapy resistance.

However, the use of polyamines as transport vectors to introduce toxic/bioactive/fluorescent agents more selectively to the intended target in cancer cells does not stop here and is an area of growing interest with great potential for drug delivery in this field, (reviewed in [56, 57]). Targeting tumour cells more selectively than normal cells to overcome the off-target adverse effects associated with most anti-tumour agents is a major challenge facing cancer therapy and in this respect, polyamines appear to be suitable vectors for drug delivery. This is because the PTS which, as mentioned earlier, is upregulated in cancer cells [51], is not restricted to natural polyamines, but extends to various polyamine analogues and polyamine-drug conjugates that structurally mimic the natural polyamine to be transported. Thus polyamine-conjugates would enhance the activity of already established chemotherapeutic agents via the preferential uptake and diminish the secondary effects on healthy cells. In addition to the chromatin-modifier polyamine conjugates, several polyamines with a variety of N-substitutions have been synthesised as delivery vehicles for known anti-tumour agents. The most recent example, and the first compound to be clinically tested, is that of a PA-vectorised inhibitor of topoisomerase II that exhibited potent anti-tumour efficacy in highly PTS-positive and platinum-resistant ovarian cancer-bearing patients [58]. This compound, F14512, bears an epipodophyllotoxin core, structurally related to etoposide, combined with a spermine chain able to specifically target cancer cells with an active PTS and bind more tightly to DNA than the original anticancer drug, etoposide. The efficacy of F14512 in ovarian cancer patients was predicted by using a spermine-linked fluorescent probe, F17073, which was first used to identify the patients eligible for the clinical trials. This phase I clinical trial with F14512 is still ongoing in platinum-resistant ovarian cancer-bearing patients and the outcomes could pave the way for future chemotherapy delivery via the PTS. The concept of using a labelled PA fluoro-probe to identify tumour cells with a highly active PTS that could hence be potentially more sensitive to F14512, is however not new. The same research group had previously used another spermine-based fluorescent marker F96982, in pre-clinical investigations, to select patients with acute myeloid leukaemia (AML) who could benefit from treatment with F14512 [59]. Their encouraging results have opened new-insights to develop F14512 treatment coupled with the use of fluorescent probes both in the onco-haematology field and in solid tumours [60].

In order to determine the molecular recognition and transport requirements involved in the PTS [56], a series of PA-anthracene conjugates were synthesised with different PA skeletons. Subsequent studies showed that the most promising was the homo-spermidine-anthracene conjugate, Ant 4,4. This compound proved to be particularly effective in a number of different cancer cell lines, despite the fact that the effects in HL-60 cells were not long-lasting [61]. Other conjugates that show potent activity and tumour selectivity are the naphtalimide-polyamine ones [62]. One in particular, 3-amino-naphtalimide-spermine, exhibited selectivity between hepatoma HepG2 cells and normal QSG7701 hepatocytes, and its anti-proliferative action was ascribed to cell apoptosis via the PI3K/Akt and Akt/mTOR signalling pathways, a common characteristic of both naphtalimide and PA analogues. Furthermore, preliminary toxicological evaluations showed that it had potent anti-tumour activity in vivo, with no obvious system toxicity [63]. Hence this conjugate merits further investigation as a potential chemotherapeutic candidate against hepatocellular carcinoma. Other interesting conjugates were also developed by linking spermine to acidic retinoids [64]. The potential use of retinoids in several types of cancer and anti-therapeutic regimes has been reported, particularly since they inhibit angiogenesis, which is a significant requirement for tumour growth and progression [65]. However, they have also shown to have multiple side effects. Hence to improve their therapeutic efficacy, biological profile and selectivity in order to decrease their side effects, attempts were made to conjugate them to spermine. One such compound, N^1,N^{12}-bis(all-trans-retinoyl)spermine, was studied on angiogenesis in vivo and on the viability in vitro of human endothelial and prostate cancer cells [66]. It was shown to inhibit angiogenesis dose-dependently in the chicken embryo chorioallantoic membrane model and to decrease the number of endothelial and prostate cancer cells. The inhibitory effect here was attributed to up-regulation of the tumour repressing gene, retinoic acid receptor beta.

Other variations on the theme of polyamines as drug delivery systems for cancer treatment come from the recent advances in the nanotechnology field. Superparamagnetic nanoparticles and advanced nanosystems based on directed nucleic acid assemblies, polyamine-induced DNA condensation, and bovine serum amine oxidase have been proposed for futuristic anticancer gene therapy (for review *see* [67]). In this context, Cui et al. have recently developed a polyamine-based nanoparticle gene-delivery system that is coated with bovine serum albumin to improve in vivo circulation stability and tumour targeting [68]. They used agmatine as an anticancer agent since it inhibits ODC activity and competes for the polyamine transport system, thus reducing cellular polyamine content. The pH-responsive cationic poly(agmatine) nanoparticles function as both a gene delivery vector for AdoMetDC siRNA as well as an anticancer agent after intracellular degradation. These nanoparticles were shown to be taken up by liver cancer cells via endocytosis and after degradation, agmatine and AdoMetDC siRNA are released that synergistically inhibit PA biosynthesis, thereby inhibiting tumour proliferation.

In summary, the investigations carried out so far on the use of polyamines as vectors for targeted drug delivery is convincing, versatile and very attractive, and this approach could, in the future, be at the forefront of strategies being advanced to treat cancer, either alone, or in combination with other drugs and techniques, such as irradiation and immunotherapy.

4 Polyamines in Chemoprevention

Probably, the field where interfering with polyamine metabolism and function as a strategy for antiproliferative intervention has had the most success so far, is in chemoprevention. This is aimed at individuals with a known increased risk to certain cancers due to genetic predisposition or prior history of dysplasia or early-stage cancers. Here, the most widely used clinical drug is DFMO, the irreversible, suicide inhibitor of ODC, which is used either alone or in combination with agents such as non-steroidal anti-inflammatory drugs (reviewed in [9, 69]). The use of DFMO, which generally results in cytostasis rather than apparent toxicity, has been shown to have profound cancer preventative effects, particularly in colon, prostate and skin cancers as well as neuroblastomas, although other cancers are also being explored. The doses of DFMO used in chemoprevention trials are substantially lower than those used in the earlier therapeutic trials and are well tolerated even at gram doses (1–3 g/m^2/day), which is partly why DFMO has been survived so long as an agent for targeting polyamine metabolism. Also, the observed toxicities of thrombocytopaenia, gastrointestinal effects and reversible hearing loss observed in phase I and II trials are considered to be minor compared with other cancer drugs [14]. One other favourable feature of this drug is that its well-described chemical synthesis and long-term stability make it reasonably priced for daily consumption. For the above reasons, DFMO meets many of the criteria for being an effective chemopreventive agent. Additionally, it is already an FDA- and European Commission-approved drug, used for the treatment of African sleeping sickness and for hirsutism, which makes its testing for chemoprevention less complex [70].

The group of Gerner and Myskens has been particularly active in promoting the chemopreventive effects of DFMO both through basic research and by undertaking several clinical trials. The first report of DFMO in clinical trials dates back to 1986 in advanced small cell lung cancer and colon cancer [10] and since then, a number of others have been carried out against several epithelial cancers and solid tumours [9]. In colorectal cancer tissue, polyamines are consistently increased such that their measurement in bowel mucosa is a good indicator of colorectal cancer and familial adenomatous polyposis, the related hereditary condition [4]. Due to the suppressive effects of both PA inhibitors and non-steroidal anti-inflammatory drugs (NSAIDs) on colon carcinogenesis, it is logical to assume that combination of these types of agents might lead to positive outcomes. The clinical trials examining the reoccurrence of colonic polyps have shown that prolonged treatment with DFMO combined with the NSAID, sulindac, caused a 70% reduction in advanced and/or multiple adenomas [71]. Clinical trials are still on going with this combinatorial treatment to further establish whether DFMO and sulindac prevent the recurrence of high-risk adenomas and second primary colorectal cancers in patients up to stage 3 colon or rectal cancers [72]. The chemoprevention effects of DFMO in colorectal cancers, appears to be, besides the diminished levels of polyamines and decreased cell turnover, associated with reduced levels of folate-dependent metabolites, including S-adenosylmethionione (SAM), thymidine pools and related pathway intermediates [73]. DFMO leads to perturbations in single-carbon transfer reactions through the increased consumption and regeneration of SAM required for polyamine biosynthesis, which results in decreased availability of tetrahydrofolate (THF) for the synthesis of thymidine that is required for DNA synthesis. THF becomes depleted upon DFMO treatment—hence this appears to be a fundamental mechanism of DFMO cytostatic activity. In the future, improved success with DFMO for colorectal cancer prevention could benefit from identifying the genetic features that may be markers for both treatment benefit and toxicity, since patients bearing the ODC1A allele have been shown to respond differently to DFMO and sulindac compared with GG genotype patients [74].

One recent area in which the use of DFMO is expanding is in the treatment of neuroblastomas (reviewed in [75]). Neuroblastoma is the most common cancer in children up to age 5, and a major risk factor in this disease is the amplification of the MYCN oncogene or deregulation of MYC signalling which correlates with aggressive disease and poor outcomes [76]. Therefore the rationale behind the use of DFMO in neuroblastoma is based in part on its role as an inhibitor downstream of MYCN. Polyamine depletion in the presence of DFMO was shown to decrease cell proliferation by activating the $p27^{Kip1}$/retinoblastoma (Rb) signalling axis and by inducing cell cycle arrest in the G1 phase in several neuroblastoma cell lines [77]. Recently, it was shown to synergise with the anti-inflammatory and immunomodulatory drug, sulfasalazine, in different neuroblastoma tumour cells, suggesting that this combination may hold promise as a novel therapy for the treatment of neuroblastoma patients [78]. These encouraging results have paved the way to the first phase I chemoprevention trial of DFMO in patients with relapsed/refractory neuroblastoma [79]. The study showed that besides no toxicity, patients with a specific single nucleotide polymorphism (SNP), specifically, the minor T allele at rs2302616 of the ODC gene, had higher levels of urinary polyamine markers and responded better to DFMO therapy compared to those with the major G allele at the same locus. This suggests that this patient subset could be uniquely susceptible to therapies targeting the polyamine pathway. Once again, the success of DFMO chemoprevention appears to be associated with specific genetic patterns, hence tailoring treatment using pharmacogenetics information is worthy of greater attention and comprehension for making DFMO-based chemoprevention even more successful.

DFMO has also given promising results in skin cancer both topically and orally in human trials. Topical application reduced by 25% the number of pre-malignant actinic keratosis compared to controls, while treatment with oral doses (0.5 mg/m^2/day) over 4–5 years in phase III clinical trials, was seen to protect against all non-melanoma skin cancers in subjects with a previous history of this disease and significantly decreased the incidence of basal cell carcinomas [80, 81]. However, in a phase IIB randomised study of topical DFMO treatment in combination with the NSAID, diclofenac, for chemoprevention against non-melanoma skin cancers on sun-damaged skin of the forearms, no benefit with either agent alone or in combination was observed. This apparent lack of benefit was likely due to inflammation in the treatment groups [82]. Overall, the results in both humans and animal models substantiate the use of DFMO as a chemopreventive agent for non-melanoma skin cancers and emphasise the importance of the polyamine pathway in skin tumour development [83]. Other indications that support the evaluation of DFMO as a chemopreventive agent are in Barrett's esophageal cancer and prostate cancers where clinical trials have been undertaken with successful patient outcomes [84, 85]. However, it is worth mentioning that DFMO chemopreventive treatment does not work for all cancers since it has had no success so far in other human trials looking at causing regression of cervical, bladder and breast cancer, despite all the extensive in vitro studies with DFMO on the respective tumour cell lines [86–88].

Targeting the biosynthetic polyamine pathway for chemoprevention is not the only route that has been explored—the catabolic pathway is also being pursued. In particular, the enzyme SMOX has come under close scrutiny because of its involvement in inflammatory-associated cancers, especially during the earliest stages of the carcinogenic process in conditions of chronic inflammation. This enzyme catalyses the decomposition of spermine to produce high levels of H_2O_2 and the toxic aldehyde, 3-aminopropanal, that lead to DNA damage, oxidative stress, mutations and/or altered gene expression which are preludes to cancer. Interestingly, it appears to be highly inducible by both bacterial infections and inflammatory cytokines [89, 90]. One such trigger of SMOX expression is infection with *Helicobacter pylori* which is linked to peptic ulcer disease, chronic gastritis and gastric cancer

[91]. By inducing SMOX-driven ROS production, *H. pylori* evades the immune system by killing the immune cells responsible for eliminating the infection. These effects in both gastrointestinal macrophages and epithelial cells, were abrogated by either inhibition of SMOX with the powerful polyamine oxidase inhibitor *N*, *N′*-bis(2,3-butadienyl)-1,4-butanediamine (MDL72527), by knockdown of SMOX via siRNA, or by catalase that decomposes H_2O_2 [90]. The hypothesis that SMOX expression is directly linked to the increased risk of gastric cancer by *H. pylori* was additionally corroborated by a recent study in the Andean region of Colombia [92]. The authors showed that when clinical strains of *H. pylori* from the high-risk Andean mountain region were co-cultured with gastric epithelial cells, they induced more SMOX expression and oxidative DNA damage, and less apoptosis than low-risk clinical isolates from the Pacific coast, and that this was linked to the increased risk of gastric cancer in the Andean population of Colombia. Recently, it has been suggested that epigenetic factors may be involved in the link between SMOX and carcinogenesis. Murray-Stewart et al. [93] found that the tumour suppressor micro RNA, miR-124, which becomes silenced by DNA methylation, negatively regulates SMOX expression through a recognition site in its 3′-untranslated region of mRNA. In gastric adenocarcinoma cells harbouring highly methylated and silenced miR-124 gene loci, 5-azacytidine treatment restored miR-124 expression and decreased SMOX expression. Furthermore, in gastric biopsies from *H. pylori* infected individuals, it was found that the extent of miR-124 gene methylation correlated with the expression of SMOX. These findings suggest yet another strategy for chemoprevention, in that specific epigenetic therapies designed to re-express aberrantly silenced genes, which are common in cancers, could be combined with SMOX inhibitors to more effectively reduce the onset of cancerous cells.

Induction of SMOX has also been observed in human colonic epithelial cells following infection by another pathogen, the enterotoxigenic *Bacteroides fragilis* (ETBF), and in a mouse model of colon tumourigenesis that was triggered by ETBF, inflammation of the distal colon and subsequent formation of numerous polyps significantly diminished upon treatment with MDL72527 [94]. In this case, the relationship between ETBF induction of polyamine metabolism and colon cancer, has been linked to the recruitment of epigenetic chromatin modifiers to the site of oxidised DNA [95] which could result in silencing of important tumour suppressor genes. In a human study of 69 patients with ulcerative colitis which is a chronic, inflammatory disease of the gastrointestinal tract that increases the risk for colorectal cancer, SMOX mRNA expression and SMOX protein were increased in tissue biopsies obtained from diseased versus normal patients [96]. The initiation of prostate cancer also appears to be associated with persistent inflammation followed by a hyperproliferative state, and once again, SMOX expression was found to be increased in pre-neoplastic intraepithelial prostatic lesions [97]. All the above findings appear to indicate that a correlation exists between SMOX expression in early inflammatory diseases linked to cancers, making it a rational target for chemoprevention. By resorting to various mouse models, the importance of another catabolic enzyme, SSAT, in the induction of intestinal tumourigenesis and skin carcinogenesis was demonstrated. In the former case, SSAT knockout mice had a significantly reduced number of intestinal tumours, and in the latter case, the use of MDL72527 diminished the number of skin tumours [98, 99].

Hence, polyamine oxidase inhibitors, such as MDL72527 that irreversibly inhibits both SMOX and acetylated polyamine oxidase (APAO), may potentially find use as chemopreventive drugs. However, at present the use of MDL72527 is limited due to possible toxicity following long-term use, therefore greater efforts should be made for finding more specific and selective inhibitors that target the polyamine catabolic pathway for preventing neoplastic lesions. In addition, because polyamines have been linked to inflammatory-associated cancers, an interesting strategy to reduce risk factors associated with their development, is that of using anti-inflammatory drugs such as NSAIDs since they also target polyamine enzymes [100, 101], and synergise them with other drugs that target this pathway (for review *see* [102]). Indeed sulindac combined with DFMO was shown to be effective in reducing the number of polyps in at-risk patients. In another study, Xie et al. showed that the combination of celecoxib with the SSAT-inducing polyamine napthalamide conjugate NPC-16, induces cell death in several human colorectal cancer cell lines [103].

5 Concluding Remarks

Although the knowledge of polyamine concentrations and metabolism in cells is very ancient, and both are known to be tightly regulated at distinct stages, there are still many unanswered questions and their precise role in cells evades precise definition. Targeting this important cellular pathway to treat cancer has several obstacles, especially since polyamines appear to be involved in a myriad of cellular processes which are all highly fine-tuned. Indeed, inducing optimal suppression of polyamines in vivo is highly challenging because of the multiple compensatory pathways present regulating polyamine metabolism. It may be that only when the precise role of polyamines is clearly defined that exploiting this pathway to treat or prevent cancer may become truly successful. Until then, the most promising prospect for the use of polyamines and their pathway is, in our view, to develop polyamines as vectors to target specific drugs to cancer cells thus leaving normal cells intact, and to use combinations of DFMO with other agents, especially NSAIDs, for chemoprevention. Likewise, the combination of NSAIDs with inhibitors of SMOX/SSAT is also worthy of further investigation. Other directions that could be revisited include the dosing schedules of the analogues that showed promise in treating some forms of cancer, such as BENSpm, in combination with standard cytotoxic drugs and further development of other agents that inhibit pathways linked into or from polyamine metabolism.

References

1. Russell DH, Levy CC, Schimpff SC et al (1971) Urinary polyamines in cancer patients. Cancer Res 31:1555–1558
2. Kingsnorth AN, Wallace HM, Bundred NJ et al (1984) Polyamines in breast cancer. Br J Surg 71:352–356
3. Kingsnorth AN, Lumsden AB, Wallace HM (1984) Polyamines in colorectal cancer. Br J Surg 71:791–794
4. Giardiello FM, Hamilton SR, Hylind LM et al (1997) Ornithine decarboxylase and polyamines in familial adenomatous polyposis. Cancer Res 57:199–201
5. Manni A, Grove R, Kunselman S et al (1995) Involvement of the polyamine pathway in breast cancer progression. Cancer Lett 92:49–57
6. Bello-Fernandez C, Packham G, Cleveland JL (1993) The ornithine decarboxylase gene is a transcriptional target of c-Myc. Proc Natl Acad Sci U S A 90:7804–7808
7. Shantz LM, Levin VA (2007) Regulation of ornithine decarboxylase during oncogenic transformation: mechanisms and therapeutic potential. Amino Acids 33:213–223
8. Prakash NJ, Schechter PJ, Grove J et al (1978) Effect of alpha-difluoromethylornithine, an enzyme-activated irreversible inhibitor of ornithine decarboxylase, on L1210 leukemia in mice. Cancer Res 38:3059–3062
9. Meyskens FL Jr, Gerner EW (1999) Development of difluoromethylornithine (DFMO) as a chemoprevention agent. Clin Cancer Res 5:945–951
10. Abeloff MD, Rosen ST, Luk GD et al (1986) Phase II trials of alpha-difluoromethylornithine, an inhibitor of polyamine synthesis, in advanced small cell lung cancer and colon cancer. Cancer Treat Rep 70:843–845
11. Horn Y, Schechter PJ, Marton LJ (1987) Phase I–II clinical trial with alpha-difluoromethylornithine—an inhibitor of polyamine biosynthesis. Eur J Cancer Clin Oncol 23:1103–1107
12. Levin VA, Prados MD, Yung WK et al (1992) Treatment of recurrent gliomas with eflornithine. J Natl Cancer Inst 84:1432–1437
13. Prados MD, Wara WM, Sneed PK et al (2001) Phase III trial of accelerated hyperfractionation with or without difluromethylornithine (DFMO) versus standard fractionated radiotherapy with or without DFMO for newly diagnosed patients with glioblastoma multiforme. Int J Radiat Oncol Biol Phys 49:71–77
14. Meyskens FL Jr, Gerner EW, Emerson S et al (1998) Effect of alpha-difluoromethylornithine on rectal mucosal levels of polyamines in a randomized, double-blinded trial for colon cancer prevention. J Natl Cancer Inst 90:1212–1218
15. Williams-Ashman HG, Schenone A (1972) Methyl glyoxal bis(guanylhydrazone) as a potent inhibitor of mammalian and yeast S-adenosylmethionine decarboxylases. Biochem Biophys Res Commun 46:288–295
16. Nass MM (1984) Analysis of methylglyoxal bis(guanylhydrazone)-induced alterations of hamster tumor mitochondria by correlated studies of selective rhodamine binding, ultrastructural damage, DNA replication, and reversibility. Cancer Res 44:2677–2688
17. Regenass U, Mett H, Stanek J et al (1994) CGP 48664, a new S-adenosylmethionine decarboxylase inhibitor with broad spectrum antiproliferative and antitumor activity. Cancer Res 54:3210–3217
18. Pless M, Belhadj K, Menssen HD et al (2004) Clinical efficacy, tolerability, and safety of SAM486A, a novel polyamine biosynthesis inhibitor, in patients with relapsed or refractory non-Hodgkin's lymphoma: results from a phase II multicenter study. Clin Cancer Res 10:1299–1305

19. van Zuylen L, Bridgewater J, Sparreboom A et al (2004) Phase I and pharmacokinetic study of the polyamine synthesis inhibitor SAM486A in combination with 5-fluorouracil/leucovorin in metastatic colorectal cancer. Clin Cancer Res 10:1949–1955
20. Wallace HM, Niiranen K (2007) Polyamine analogues – an update. Amino Acids 33:261–265
21. Wallace HM, Fraser AV (2003) Polyamine analogues as anticancer drugs. Biochem Soc Trans 31:393–396
22. Casero RA Jr, Marton LJ (2007) Targeting polyamine metabolism and function in cancer and other hyperproliferative diseases. Nat Rev Drug Discov 6:373–390
23. Porter CW, Bergeron RJ (1988) Regulation of polyamine biosynthetic activity by spermidine and spermine analogs—a novel antiproliferative strategy. Adv Exp Med Biol 250:677–690
24. Porter CW, Ganis B, Libby PR et al (1991) Correlations between polyamine analogue-induced increases in spermidine/spermine N1-acetyltransferase activity, polyamine pool depletion, and growth inhibition in human melanoma cell lines. Cancer Res 51:3715–3720
25. Wallace HM, Mackarel AJ (1998) Regulation of polyamine acetylation and efflux in human cancer cells. Biochem Soc Trans 26:571–575
26. Pledgie-Tracy A, Billam M, Hacker A et al (2010) The role of the polyamine catabolic enzymes SSAT and SMO in the synergistic effects of standard chemotherapeutic agents with a polyamine analogue in human breast cancer cell lines. Cancer Chemother Pharmacol 65:1067–1081
27. Casero RA Jr, Celano P, Ervin SJ et al (1989) Differential induction of spermidine/spermine N1-acetyltransferase in human lung cancer cells by the bis(ethyl)polyamine analogues. Cancer Res 49:3829–3833
28. Wolff AC, Armstrong DK, Fetting JH et al (2003) A Phase II study of the polyamine analog N1,N11-diethylnorspermine (DENSpm) daily for five days every 21 days in patients with previously treated metastatic breast cancer. Clin Cancer Res 9:5922–5928
29. Streiff RR, Bender JF (2001) Phase 1 study of N1-N11-diethylnorspermine (DENSPM) administered TID for 6 days in patients with advanced malignancies. Investig New Drugs 19:29–39
30. Hector S, Porter CW, Kramer DL et al (2004) Polyamine catabolism in platinum drug action: Interactions between oxaliplatin and the polyamine analogue N1,N11-diethylnorspermine at the level of spermidine/spermine N1-acetyltransferase. Mol Cancer Ther 3:813–822
31. Choi W, Gerner EW, Ramdas L et al (2005) Combination of 5-fluorouracil and N1,N11-diethylnorspermine markedly activates spermidine/spermine N1-acetyltransferase expression, depletes polyamines, and synergistically induces apoptosis in colon carcinoma cells. J Biol Chem 280:3295–3304
32. Casero RA Jr, Mank AR, Saab NH et al (1995) Growth and biochemical effects of unsymmetrically substituted polyamine analogues in human lung tumor cells 1. Cancer Chemother Pharmacol 36:69–74
33. Hacker A, Marton LJ, Sobolewski M et al (2008) In vitro and in vivo effects of the conformationally restricted polyamine analogue CGC-11047 on small cell and non-small cell lung cancer cells. Cancer Chemother Pharmacol 63:45–53
34. Saab NH, West EE, Bieszk NC et al (1993) Synthesis and evaluation of unsymmetrically substituted polyamine analogues as modulators of human spermidine/spermine-N1-acetyltransferase (SSAT) and as potential antitumor agents. J Med Chem 36:2998–3004
35. Fraser AV, Woster PM, Wallace HM (2002) Induction of apoptosis in human leukaemic cells by IPENSpm, a novel polyamine analogue and anti-metabolite. Biochem J 367:307–312
36. Dredge K, Kink JA, Johnson RM et al (2009) The polyamine analog PG11047 potentiates the antitumor activity of cisplatin and bevacizumab in preclinical models of lung and prostate cancer. Cancer Chemother Pharmacol 65:191–195
37. Carew JS, Nawrocki ST, Reddy VK et al (2008) The novel polyamine analogue CGC-11093 enhances the antimyeloma activity of bortezomib. Cancer Res 68:4783–4790
38. Huang Y, Hager ER, Phillips DL et al (2003) A novel polyamine analog inhibits growth and induces apoptosis in human breast cancer cells. Clin Cancer Res 9(7):2769–2777
39. Huang Y, Keen JC, Pledgie A et al (2006) Polyamine analogues down-regulate estrogen receptor alpha expression in human breast cancer cells. J Biol Chem 281:19055–19063
40. Shi Y, Lan F, Matson C et al (2004) Histone demethylation mediated by the nuclear amine oxidase homolog LSD1. Cell 119:941–953
41. Zheng YC, Ma J, Wang Z et al (2015) A systematic review of histone lysine-specific

demethylase 1 and its inhibitors. Med Res Rev 35:1032–1071
42. Huang Y, Stewart TM, Wu Y et al (2009) Novel oligoamine analogues inhibit lysine-specific demethylase 1 and induce reexpression of epigenetically silenced genes. Clin Cancer Res 15:7217–7228
43. Zhu Q, Huang Y, Marton LJ et al (2012) Polyamine analogs modulate gene expression by inhibiting lysine-specific demethylase 1 (LSD1) and altering chromatin structure in human breast cancer cells. Amino Acids 42:887–898
44. Wu Y, Steinbergs N, Murray-Stewart T et al (2012) Oligoamine analogues in combination with 2-difluoromethylornithine synergistically induce re-expression of aberrantly silenced tumour-suppressor genes. Biochem J 442:693–701
45. Nowotarski SL, Pachaiyappan B, Holshouser SL et al (2015) Structure-activity study for (bis)ureidopropyl- and (bis)thioureidopropyl-diamine LSD1 inhibitors with 3-5-3 and 3-6-3 carbon backbone architectures. Bioorg Med Chem 23:1601–1612
46. Hazeldine S, Pachaiyappan B, Steinbergs N et al (2012) Low molecular weight amidoximes that act as potent inhibitors of lysine-specific demethylase 1. J Med Chem 55:7378–7391
47. Milovica V, Turchanowa L, Khomutov AR et al (2001) Hydroxylamine-containing inhibitors of polyamine biosynthesis and impairment of colon cancer cell growth. Biochem Pharmacol 61:199–206
48. Liang F, Wan S, Li Z et al (2006) Medical applications of macrocyclic polyamines. Curr Med Chem 13:711–727
49. Pasini A, Caldarera CM, Giordano E (2014) Chromatin remodeling by polyamines and polyamine analogs. Amino Acids 46:595–603
50. Varghese S, Gupta D, Baran T et al (2005) Alkyl-substituted polyaminohydroxamic acids: a novel class of targeted histone deacetylase inhibitors. J Med Chem 48:6350–6365
51. Palmer AJ, Wallace HM (2010) The polyamine transport system as a target for anticancer drug development. Amino Acids 38:415–422
52. Aldana-Masangkay GI, Sakamoto KM (2011) The role of HDAC6 in cancer. J Biomed Biotechnol 2011:875824
53. Varghese S, Senanayake T, Murray-Stewart T et al (2008) Polyaminohydroxamic acids and polyaminobenzamides as isoform selective histone deacetylase inhibitors. J Med Chem 51:2447–2456
54. Sharma SK, Hazeldine S, Crowley ML et al (2012) Polyamine-based small molecule epigenetic modulators. Medchemcomm 3:14–21
55. Bandyopadhyay K, Baneres JL, Martin A et al (2009) Spermidinyl-CoA-based HAT inhibitors block DNA repair and provide cancer-specific chemo- and radiosensitization. Cell Cycle 8:2779–2788
56. Phanstiel O 4th, Kaur N, Delcros JG (2007) Structure-activity investigations of polyamine-anthracene conjugates and their uptake via the polyamine transporter. Amino Acids 33:305–313
57. Xie S, Wang J, Zhang Y et al (2010) Antitumor conjugates with polyamine vectors and their molecular mechanisms. Expert Opin Drug Deliv 7:1049–1061
58. Thibault B, Clement E, Zorza G et al (2016) F14512, a polyamine-vectorized inhibitor of topoisomerase II, exhibits a marked anti-tumor activity in ovarian cancer. Cancer Lett 370:10–18
59. Annereau JP, Brel V, Dumontet C et al (2010) A fluorescent biomarker of the polyamine transport system to select patients with AML for F14512 treatment. Leuk Res 34:1383–1389
60. Kruczynski A, Pillon A, Creancier L et al (2013) F14512, a polyamine-vectorized anti-cancer drug, currently in clinical trials exhibits a marked preclinical anti-leukemic activity. Leukemia 27:2139–2148
61. Traquete R, Ghani RA, Phanstiel O et al (2013) Ant 4,4, a polyamine-anthracene conjugate, induces cell death and recovery in human promyelogenous leukemia cells (HL-60). Amino Acids 44:1193–1203
62. Tian ZY, Xie SQ, Du YW et al (2009) Synthesis, cytotoxicity and apoptosis of naphthalimide polyamine conjugates as antitumor agents. Eur J Med Chem 44:393–399
63. Li M, Li Q, Zhang YH et al (2013) Antitumor effects and preliminary systemic toxicity of ANISpm in vivo and in vitro. Anti-Cancer Drugs 24:32–42
64. Magoulas G, Papaioannou D, Papadimou E et al (2009) Preparation of spermine conjugates with acidic retinoids with potent ribonuclease P inhibitory activity. Eur J Med Chem 44:2689–2695
65. Tang XH, Gudas LJ (2011) Retinoids, retinoic acid receptors, and cancer. Annu Rev Pathol 6:345–364
66. Vourtsis D, Lamprou M, Sadikoglou E et al (2013) Effect of an all-trans-retinoic acid conjugate with spermine on viability of human

prostate cancer and endothelial cells in vitro and angiogenesis in vivo. Eur J Pharmacol 698:122–130

67. Agostinelli E, Vianello F, Magliulo G et al (2015) Nanoparticle strategies for cancer therapeutics: nucleic acids, polyamines, bovine serum amine oxidase and iron oxide nanoparticles (review). Int J Oncol 46:5–16
68. Cui PF, Xing L, Qiao JB et al (2016) Polyamine metabolism-based dual functional gene delivery system to synergistically inhibit the proliferation of cancer. Int J Pharm 506:79–86
69. Jeter JM, Alberts DS (2012) Difluoromethylornithine: the proof is in the polyamines. Cancer Prev Res (Phila) 5:1341–1344
70. Shapiro J, Lui H (2001) Vaniqa—eflornithine 13.9% cream. Skin Therapy Lett 6:1–3. 5
71. Gerner EW, Meyskens FL Jr (2009) Combination chemoprevention for colon cancer targeting polyamine synthesis and inflammation. Clin Cancer Res 15:758–761
72. Burke CA, Dekker E, Samadder NJ et al (2016) Efficacy and safety of eflornithine (CPP-1X)/sulindac combination therapy versus each as monotherapy in patients with familial adenomatous polyposis (FAP): design and rationale of a randomized, double-blind, Phase III trial. BMC Gastroenterol 16:87
73. Witherspoon M, Chen Q, Kopelovich L et al (2013) Unbiased metabolite profiling indicates that a diminished thymidine pool is the underlying mechanism of colon cancer chemoprevention by alpha-difluoromethylornithine. Cancer Discov 3:1072–1081
74. Zell JA, McLaren CE, Chen WP et al (2010) Ornithine decarboxylase-1 polymorphism, chemoprevention with eflornithine and sulindac, and outcomes among colorectal adenoma patients. J Natl Cancer Inst 102:1513–1516
75. Bassiri H, Benavides A, Haber M et al (2015) Translational development of difluoromethylornithine (DFMO) for the treatment of neuroblastoma. Transl Pediatr 4:226–238
76. Westermann F, Muth D, Benner A et al (2008) Distinct transcriptional MYCN/c-MYC activities are associated with spontaneous regression or malignant progression in neuroblastomas. Genome Biol 9:R150
77. Wallick CJ, Gamper I, Thorne M et al (2005) Key role for p27Kip1, retinoblastoma protein Rb, and MYCN in polyamine inhibitor-induced G1 cell cycle arrest in MYCN-amplified human neuroblastoma cells. Oncogene 24:5606–5618
78. Yco LP, Geerts D, Mocz G et al (2015) Effect of sulfasalazine on human neuroblastoma: analysis of sepiapterin reductase (SPR) as a new therapeutic target. BMC Cancer 15:477–488
79. Saulnier Sholler GL, Gerner EW, Bergendahl G et al (2015) A phase I trial of DFMO targeting polyamine addiction in patients with relapsed/refractory neuroblastoma. PLoS One 10:e0127246
80. Alberts DS, Dorr RT, Einspahr JG et al (2000) Chemoprevention of human actinic keratoses by topical 2-(difluoromethyl)-dl-ornithine. Cancer Epidemiol Biomark Prev 9:1281–1286
81. Kreul SM, Havighurst T, Kim K et al (2012) A phase III skin cancer chemoprevention study of DFMO: long-term follow-up of skin cancer events and toxicity. Cancer Prev Res (Phila) 5:1368–1374
82. Jeter JM, Curiel-Lewandrowski C, Stratton SP et al (2016) Phase IIB randomized study of topical difluoromethylornithine and topical diclofenac on sun-damaged skin of the forearm. Cancer Prev Res (Phila) 9:128–134
83. Elmets CA, Athar M (2010) Targeting ornithine decarboxylase for the prevention of nonmelanoma skin cancer in humans. Cancer Prev Res (Phila) 3:8–11
84. Meyskens FL Jr, Simoneau AR, Gerner EW (2014) Chemoprevention of prostate cancer with the polyamine synthesis inhibitor difluoromethylornithine. Recent Results Cancer Res 202:115–120
85. Sinicrope FA, Broaddus R, Joshi N et al (2011) Evaluation of difluoromethylornithine for the chemoprevention of Barrett's esophagus and mucosal dysplasia. Cancer Prev Res (Phila) 4:829–839
86. Vlastos AT, West LA, Atkinson EN et al (2005) Results of a phase II double-blinded randomized clinical trial of difluoromethylornithine for cervical intraepithelial neoplasia grades 2 to 3. Clin Cancer Res 11:390–396
87. Messing E, Kim KM, Sharkey F et al (2006) Randomized prospective phase III trial of difluoromethylornithine vs placebo in preventing recurrence of completely resected low risk superficial bladder cancer. J Urol 176:500–504
88. Fabian CJ, Kimler BF, Brady DA et al (2002) A phase II breast cancer chemoprevention trial of oral alpha-difluoromethylornithine: breast tissue, imaging, and serum and urine biomarkers. Clin Cancer Res 8:3105–3117
89. Babbar N, Gerner EW, Casero RA Jr (2006) Induction of spermidine/spermine N1-

acetyltransferase (SSAT) by aspirin in Caco-2 colon cancer cells. Biochem J 394:317–324

90. Xu H, Chaturvedi R, Cheng Y et al (2004) Spermine oxidation induced by Helicobacter pylori results in apoptosis and DNA damage: implications for gastric carcinogenesis. Cancer Res 64:8521–8525
91. Chaturvedi R, de Sablet T, Peek RM et al (2012) Spermine oxidase, a polyamine catabolic enzyme that links Helicobacter pylori CagA and gastric cancer risk. Gut Microbes 3:48–56
92. Chaturvedi R, de Sablet T, Asim M et al (2015) Increased Helicobacter pylori-associated gastric cancer risk in the Andean region of Colombia is mediated by spermine oxidase. Oncogene 34:3429–3440
93. Murray-Stewart T, Sierra JC, Piazuelo MB et al (2016) Epigenetic silencing of miR-124 prevents spermine oxidase regulation: implications for Helicobacter pylori-induced gastric cancer. Oncogene 35:5480–5488
94. Goodwin AC, Destefano Shields CE, Wu S et al (2011) Polyamine catabolism contributes to enterotoxigenic Bacteroides fragilis-induced colon tumorigenesis. Proc Natl Acad Sci U S A 108:15354–15359
95. O'Hagan HM, Wang W, Sen S et al (2011) Oxidative damage targets complexes containing DNA methyltransferases, SIRT1, and polycomb members to promoter CpG Islands. Cancer Cell 20:606–619
96. Hong SK, Chaturvedi R, Piazuelo MB et al (2010) Increased expression and cellular localization of spermine oxidase in ulcerative colitis and relationship to disease activity. Inflamm Bowel Dis 16:1557–1566
97. Goodwin AC, Jadallah S, Toubaji A et al (2008) Increased spermine oxidase expression in human prostate cancer and prostatic intraepithelial neoplasia tissues. Prostate 68:766–772
98. Tucker JM, Murphy JT, Kisiel N et al (2005) Potent modulation of intestinal tumorigenesis in Apcmin/+ mice by the polyamine catabolic enzyme spermidine/spermine N1-acetyltransferase. Cancer Res 65:5390–5398
99. Wang X, Feith DJ, Welsh P et al (2007) Studies of the mechanism by which increased spermidine/spermine N1-acetyltransferase activity increases susceptibility to skin carcinogenesis. Carcinogenesis 28:2404–2411
100. Hughes A, Smith NI, Wallace HM (2003) Polyamines reverse non-steroidal anti-inflammatory drug-induced toxicity in human colorectal cancer cells. Biochem J 374:481–488
101. Saunders FR, Wallace HM (2007) Polyamine metabolism and cancer prevention. Biochem Soc Trans 35:364–368
102. Babbar N, Gerner EW (2011) Targeting polyamines and inflammation for cancer prevention. Recent Results Cancer Res 188:49–64
103. Xie SQ, Zhang YH, Li Q et al (2012) COX-2-independent induction of apoptosis by celecoxib and polyamine naphthalimide conjugate mediated by polyamine depression in colorectal cancer cell lines. Int J Color Dis 27:861–868

Chapter 40

Potential Applications of Polyamines in Agriculture and Plant Biotechnology

Antonio F. Tiburcio and Rubén Alcázar

Abstract

The polyamines putrescine, spermidine and spermine have been implicated in a myriad of biological functions in many organisms. Research done during the last decades has accumulated a large body of evidence demonstrating that polyamines are key modulators of plant growth and development. Different experimental approaches have been employed including the measurement of endogenous polyamine levels and the activities of polyamine metabolic enzymes, the study of the effects resulting from exogenous polyamine applications and chemical or genetic manipulation of endogenous polyamine titers. This chapter reviews the role of PAs in seed germination, root development, plant architecture, in vitro plant regeneration, flowering and plant senescence. Evidence presented here indicates that polyamines should be regarded as plant growth regulators with potential applications in agriculture and plant biotechnology.

Key words Putrescine, Spermidine, Spermine, Polyamines, Plant growth and development, Seed germination, Vegetative and reproductive development, Senescence, Agriculture, Biotechnology

1 Introduction

Polyamines (PAs) are low-molecular-mass organic polycations derived from amino acids. Structurally, PAs are mostly aliphatic chains containing two or more amine groups. In plants, the best studied PAs are the diamine putrescine (Put; 1,4-diaminobutane), the triamine spermidine (Spd) and the tetraamine spermine (Spm). Chemical or genetic depletion of Put and Spd results in plant lethality, whereas Spm is not essential for growth but is needed for plant responses to abiotic stresses. Plants also produce an isomer of Spm, thermospermine (Tspm), that appears to have an important role in vascular development [1, 2]. Cadaverine (Cad; 1,5-diaminopentane) is another diamine that is produced from lysine, which also plays physiological roles in many plants [3]. The biological functions of PAs were initially associated with their ability to bind anionic macromolecules with unique structural roles. However, later investigations revealed that PAs also act as regulatory

Rubén Alcázar and Antonio F. Tiburcio (eds.), *Polyamines: Methods and Protocols*, Methods in Molecular Biology, vol. 1694, DOI 10.1007/978-1-4939-7398-9_40, © Springer Science+Business Media LLC 2018

molecules in fundamental cellular processes, enclosing cell division, differentiation, gene expression, and DNA and protein synthesis as they do in many organisms [4]. In plants, PAs occur not only as free forms but also conjugated to phenolic compounds that act on defence responses [5] or bound to proteins like the translation initiation factor 5A (eIF5A) to form the amino acid hypusine which is involved in plant senescence [6]. PAs are modulators of many developmental processes including seed germination, root and shoot development, in vitro plant regeneration, flower and fruit development, senescence, and plant responses to abiotic and biotic stresses [1, 2]. Most of these processes involved in the ontogeny of plants are reviewed in this chapter. Fruit ripening and polyamine's involvement during stress will not be addressed here because they have been discussed in other chapters of this book. The potential applications in agriculture and plant biotechnology of several experimental approaches are discussed in the different sections of this chapter.

2 Seed Germination

Germination in plants is the process by which a dormant seed begins to sprout and grows into a seedling under the appropriate growing conditions. Seed germination and seedling growth are the two critical stages for the establishment of crops. Successful seed germination depends on the individual seed variety and it is closely linked to the ecological conditions of a plant's natural habitat [7, 8]. Thus, germination response of some seeds is affected by environmental conditions, which means that seed germination may be delayed or prevented by various abiotic stresses [7, 8].

2.1 Promotive Effects of Seed Priming

Seed priming is a pre-sowing treatment that exposes seeds to a certain solution during a particular time-period that allows partial hydration, but radicle emergence does not occur [7]. Priming allows the initiation of many of the physiological processes associated with early phase of germination (pre-germination metabolism), but prevents transition towards full germination [7]. There are several priming techniques, which enclose hydropriming, osmopriming, halopriming or chemical priming among the most commonly used [7, 9]. After removal of the seeds from the priming solution, seeds are re-dried to the initial moisture content, thus maintaining the beneficial effects of the priming treatment without loss of quality caused by rapid seed deterioration [7]. When the primed seeds are sown, the swelling of the embryo inside the primed seed speeds up germination by facilitating water absorption. It stimulates pre-germination metabolic processes, so that the seedlings emerge faster, grow more vigorously and perform better in adverse conditions, thus protecting seeds from abiotic and biotic stresses during the critical phase of seedling establishment [7]. Furthermore, seed priming synchronizes emergence leading to uniform stand and improves yield [7].

A number of natural or synthetic compounds have been used for chemical priming [9]. Naturally occurring metabolites, such as some vitamins, hormones and plant growth regulators have attracted current attention. During recent years there is a great interest on the use of PAs for chemical priming [9]. For instance, pre-sowing rice seeds with Put resulted in earlier, synchronized and enhanced seed germination leading to the improvement of shoot and root lenght, seedling fresh and dry weight, root and leaf score [8]. Similar effects have been observed in other plants such as maize, wheat, chamomile and sweet majoran, in which Put was used as seed priming agent [10–12].

Exogenous Spd also improves seed germination in many plant systems. For example, investigations performed with white clover revealed that seeds primed with Spd not only improved germination percentages and shortened mean germination but also resulted in enhanced seed vigor as indicated by longer root lenght, seedling fresh and dry weights compared with control [13]. It was proposed that seed priming with Spd improves starch metabolism presumabily due to elevated α- and β-amylase activities [13]. The positive effects in seed germination by Spd seed priming were observed in other studies [14, 15]. Recently, it has been reported that Spd soaking pre-treatment significantly improves seed germination of corn, while exogenous application of cyclohexylamine (CHA; an inhibitor of Spd biosynthesis) significantly inhibited seed germination and declined seed vigour [16]. The results indicated that enhancement of seed germination by Spd priming was closely related with the metabolism of hormones such as gibberelins, ABA and ethylene [16]. Pre-treatments of rice seeds with Spd and gibberelic acid improved seedling growth, being Spd the most effective seed priming agent [17]. The beneficial effects of seed priming with Spm in wheat have also been attributed to altered hormonal balance [18].

In conclusion, seed priming with PAs is a promising technological approach to be used in the improvement of crop seed germination especially under adverse growth conditions.

2.2 Chemical Inhibition

Guazatine is an inhibitor of polyamine oxidase (PAO) activity which is currently used as fungicide against several plant diseases. We studied the effects of guazatine treatment on seed germination and suprisingly observed that guazatine inhibited seed germination of a weed plant (i.e. *Arabidopsis thaliana*), while the same concentration of the PAO inhibitor did not affect seed germination of a crop plant like oat [19]. These results indicate that targeted inhibition of PAO activity has the potential to control weed growth due to its selective herbicidal properties, thus avoiding the competition of weeds with crops for water, sunlight and nutrients in the soil, which results in improved crop yield [19].

3 Root Development

Roots are important to plants for a wide range of processes, enclosing nutrient and water uptake, anchoring and mechanical support. Roots serve as the major interface between the plant and various biotic and abiotic factors present in the soil by both sensing and responding to environmental cues [20]. Thus, plants can dramatically alter their root architecture to optimize growth in a broad variety of environmental and soil nutrient conditions [20]. Primary roots are the first root to emerge and are derived from embryonically formed meristematic tissue. Primary and mature roots contain meristematic tissue at the tip (the root apical meristem), which forms the basic stem cell pool for other cell types in the root. Root branching is essential to increase the surface area of the root system, allowing the plant to get more distant reserves of water and nutrients and improve soil anchorage. In contrast to primary roots, lateral roots are post-embryonically formed. The later stages of lateral root emergence require de novo creation of a new apical meristem within the emerging lateral root [20]. Moreover, plants develop new roots from stems, petioles or leaves through adventitious processes [21].

It is well established that root development is controlled by hormonal signals, especially auxins [22], and environmental factors that act on the genetic programs of root development and their hormonal and metabolic control [23]. In this complex network of interactions resulting in root developmental plasticity, PAs, like other plant growth regulators seem to be involved [24]. It is well documented that PAs play an important role in primary, lateral and adventitious root development [21, 25].

Several studies have shown that depletion of PA pools is linked with root growth inhibition. For example, DL-α-difluormethylarginine (DFMA), a specific inhibitor of plant arginine decarboxylase (ADC) [26] was shown to decrease root elongation in lateral roots in hairy root cultures of *Hyosciamus muticus* [27] and to inhibit adventitious root growth in grapevine microcuttings [28]. Our work showed that DFMA inhibits root initiation in tobacco thin cell layers, while exogenous Put supply reverses this effect [29]. The Arabidopsis mutant *bud2* (encoding an isoform of *S*-adenosylmethionine decarboxylase; SAMDC) contains more Put and has enhanced root growth, thus giving further support to the promotive effect of this diamine on root growth [30].

On the other hand, exogenous application of Put (0.05 mM) to trifoliate oranges seedlings produced higher total root lenght, tap root lenght, projected surface areas and root volume than non-treated controls [31]. Moreover, the same concentration of Put was also optimal for promoting the number of lateral roots [31]. The promotive effect of Put in lateral root number may be attributed to stimulation of meristematic activity, induction of nitric oxide (NO) signal and regulation of endogenous auxins and cytokinins levels [31].

Exogenous Spd and Spm applied at micromolar concentrations increased root elongation and growth in Virginia pine by increasing root cell division [32]. In citrus seedlings, Spd application to plants symbiotically associated to arbuscular mycorrhizal fungi optimized root system architecture (RSA), thus improving plant growth [33]. Our work has also shown that exogenous applied Spm promotes root branching in *Arabidopsis thaliana* seedlings (Zarza et al. unpublished), thus corroborating the role of the higher PAs on RSA.

Cad is another diamine that is formed by a pathway catalysed by lysine decarboxylase (LDC), which is distinct from the well characterized ODC and ADC Put biosynthetic pathways [3, 34]. Cad has received little attention, in spite of the fact that it has been detected in many plants. For example, the *Leguminoseae* produce Cad and Cad-derived secondary metabolites (eg. quinolizidine alkaloids) some of them involved in insect plant defence or presenting terapheutical properties of interest for the pharmaceutical industry. This diamine is also produced by rizhospere and phyllosphere microbes suggesting that exogenous Cad may affect RSA [3, 34]. In fact, early work showed that exogenous Cad decreased primary root growth, while increased lateral root branching in soybean seedlings [35]. More recently, the inducing effects of Cad in RSA have been observed in other plants [3, 34]. Loci that contribute to root-growth response to Cad were identified through exploration of plant natural vatiation. Thus, several Arabidopsis accesions were tested on Cad-containing media, showing accesion-specific Cad response for primary root growth, skewing, waving and lateral root formation [36]. By using a QTL approach, an ORGANIC CATION TRANSPORTER 1 (OCT1) was identified, which apparently contributes to the variation between different Arabidopsis accesions. Null *oct1* mutants exhibited a hypersensitive root-growth response to Cad, thus suggesting a role for OCT1 in Cad efflux [36].

Interestingly, it seems that Cad regulates root development by inducing Spm accumulation [37]. Thus, *spms* and *pao4–1* Arabidopsis mutants displayed resistant and hypersensitive root-growth responses to Cad, respectively, as compared with wild type, whereas exogenous Cad increased Put and Spd but reduced Spm content. On the other hand, it has also been shown that Spm modulates plant sensitivity to Cad [38]. Overall, these studies suggest a possible cross-talk between Cad and Put-derived pathways in the modulation of root growth [34]. This hypothesis is sustained by our work in which it was observed that specific combinations of Put and Cad promoted root branching in a major extent than individual diamine treatments in Arabidopsis (Zarza et al. unpublished). This is in concordance with the synergistic effect of Put and Cad on the promotion of plant growth and development observed in oats and other plants by using patented formulations [39].

Many plants use roots for storage functions and some root structures serve as human food sources, including root vegetables (such as carrots, parsnips, turnips and others) and spices (such as for example ginger, arrowroot and liquorice) [20]. It seems clear that further understanding of the role of PAs and their cross-talks with other growth regulators on development and architecture of roots holds potential for the exploitation and manipulation of root characteristics to both increase plant yield and optimize agricultural land use.

4 Plant Architecture

Plant architecture can be defined as the three-dimensional organisation of the plant body, which in the case of aireal parts of the plant includes the branching pattern, as well as the size, shape and position of leaves and flower organs. Plant architecture is of major agronomic importance, since influences the suitability of a plant for cultivation, its yield and the efficiency with which it can be harvested. One of the great successes of the Green Revolution, which led to major increases in crop productivity, was based on the modification of plant architecture. For instance, the selection of wheat varieties with shorter and robust stems resulted in plants with improved yield while still resisting damage from wind and rain [40].

In most plants, the growth of axillary meristems is initially suppressed by the shoot tip, a phenomenon known as apical dominance [40]. Increased apical dominance, which is mediated primarily by the gene *TEOSINTE BRANCHED1* (*TB1*), has been one of the major traits to be selected for the domestication of maize from its ancestor teosinte [41]. In the *tb1* mutant, all axillary meristems grow out leading to highly branched maize plants, whereas in the tomato mutant *lateral suppressor* (*ls*) the vegetative axillary meristems are suppressed and no lateral branches are formed, which is of great interest to tomato breeders because manual pruning is labour intensive [40, 42].

Auxin is regarded as a direct regulator of plant architecture, whereas cytokinin acts as a second messenger that mediates the action of auxin in controlling the apical dominance, since it promotes the outgrowth of lateral buds [30]. There is evidence that PAs also play a key regulatory role in the control of plant architecture. Thus, it has been shown that PAs control the outgrowth of axillary buds in the Arabidopsis *bushy and dwarf 2* (*bud2*) mutant, which shows a bushy and dwarf phenotype [43]. The *BUD2* gene encodes SAMDC4 that is required for the biosynthesis of PAs in Arabidopsis. In the *bud2* mutant, the homeostasis of PAs is altered, leading to an increase in Put and decreases in Spd and Spm, which in turn leads to an altered plant morphology [43]. This in concordance with previous work suggesting that PAs may be involved in the formation of plant architecture including shoot apical dominance [44]. Cui et al. showed that deficiency in PA biosynthesis (due to SAMDC mutation) causes hyposensitivity to auxin and hypersensitivity to cytokinin [30], thus suggesting that PAs may play a role in regulating plant morphology by affecting plant response to the ratio of auxin and cytokinin in a similar way to that observed during in vitro plant morphogenesis (*see* below). However, a challenge in the future is to understand how PAs affect the sensitivities of plants to auxins and cytokinins.

5 In Vitro Plant Regeneration

The regenerative capacity of plant cells can be enhanced in vitro when explants are cultured on nutrient media supplemented with plant hormones [45]. The explants can regenerate an entire plant by following two alternative mechanisms, namely organogenesis and somatic embryogenesis (SE) [46]. In the first case, shoots and roots form sequentially and in response to appropriate culture conditions. This type of development is also characterized by the presence of vascular connections between the mother tissue and the regenerating section [46]. On the other hand, SE can be described as the process by which somatic cells develop into structures that resemble zygotic embryos (i.e., bipolar structures and without any vascular connection with the parental tissue) through an orderly series of characteristic embryological stages without fusion of gametes [46].

5.1 Organogenesis

Plant regeneration via organogenesis depends on the balance of exogenous applied auxin and cytokinin, which determines the fate of regenerating tissue. Generally, high ratios of auxin to cytokinin lead to root regeneration, whereas high ratios of cytokinin to auxin promote shoot regeneration [45]. This regeneration process occurs either directly from parental tissues or indirectly via the formation of a callus [46]. Thin cell layers (TCL) explants obtained from internodes of floral branches of tobacco can regenerate directly from the explant either vegetative buds (shoots), floral buds or roots by control of the culture conditions [47]. TCL explants grown with equimolar concentrations (1 μM) of auxin and cytokinin yield floral buds, but rising the cytokinin level tenfold leads to differentiation of shoots instead [47]. It has been shown that explants regenerating shoots have a high ratio of Put to Spd, whereas floral explants are characterized for having the opposite ratio. Exogenous application of the Spd synthase inhibitor CHA to cultures programmed to produce flowers regenerate shoots instead, whereas application of Spd to cultures programmed to produce only vegetative buds regenerate flowers [48]. These findings demonstrated that de novo organ regeneration from tobacco TCL explants depends on the ratio of Put to Spd, independently of the auxin to cytokinin ratio [48].

More recent studies made on callus systems showed that DFMO and DFMA suppressed shoot proliferation in *Achras sapota* cultures, whereas the inhibitory effect was reversed by the application of Put [49]. The influence of different forms of cytokinins, auxins and PAs were tested for mass multiplication and regeneration of cotton and it was found that exogenous Put was the best treatment in terms of shoot proliferation [50]. Similarly, exogenous Spd and Spm treatments to *Withania somnifera* callus cultures also produced higher number of shoots than in the presence of BA or BA/IAA combinations [51]. These results clearly support the essential role of PAs on de novo shoot differentiation and proliferation.

It is known that organogenetic potential decreases with age of the callus and it was suggested that adequate PA levels and their ratios might be important for optimum morphogenesis [52]. In this regard, we have shown that maize callus pretreated with DFMA during 3 months lost their morphogenetic capacity probably due to a senescence-like effect (as suggested by loss of callus greening) [53]. However, when these calli were subcultured on media lacking the inhibitor a four-fold increase in the number of regenerated shoots was observed as compared to non DFMA-pretreated cultures. Furthermore, the regeneration frequency enhanced threefold relative to the controls [53]. We proposed that a rejuvenation effect was produced on the callus after DFMA removal, which could explain the improvement of morphogenetic capacity [53]. The results suggest that pretreatment with DFMA or other inhibitors of PA metabolism may be a good approach to significantly improve in vitro plant regeneration potential of cultured crop plants, which is very important for agricultural and plant biotechnological sectors.

5.2 Somatic Embryogenesis

As indicated above, plants can also regenerate in vitro through SE, by developing cellular structures similar to zygotic embryos and subsequently generating whole plant bodies. SE has many advantages over organogenesis since: (a) it permits the culture of large numbers of somatic embryos with the presence of both root and shoot meristems in the same element; (b) it permits easy scale-up transfers with low labor inputs because embryos can be grown individually and freely floating in liquid medium; (c) they originate from single cells and the cultures can be synchronized and purified so that one can deal with homogeneous material; and, (d) somatic embryos are less variable in terms of mutations than those derived via organogenesis [54]. However, SE has limitations because it cannot be accomplished in all plants since it depends on the genotype and culture conditions [55].

Auxin is considered to be the most important hormone in regulating SE, since many plant species undergo SE when they are cultured on auxin-containing medium and then transferred to auxin-free medium [46]. Among several synthetic auxin-like substances, it seems that 2,4-dichlorophenoxyacetic acid (2,4-D) is the most effective inducer of somatic embryos in many plants [46].

There is evidence that PAs play a critical role in SE, which has been extensively studied in carrot [56]. It was firsly shown that there is a rise in ADC activity and Put levels when carrot callus cultures are transferred to "embryogenesis" medium [57]. DFMA inhibited the formation of somatic embryos, while addition of Put and DFMA restored carrot embryogenesis [58]. In contrast, DFMA did not inhibit the growth of the carrot cells placed on callus media, thus indicating the requirement of Put in the transition from callus to somatic embryos [58]. The importance of Put for SE was also confirmed by transgenic expression of mouse ODC, in carrot, in which a significant increase of Put levels in the transgenic cultures was accompanied by increased number of somatic embryos [56]. More recently, it has been reported that exogenous application of PAs to suspension cultures of different plant species including *Ocotea catharinensis*, *Araucaria angustifolia* and *Pinus taeda* altered the endogenous levels of NO in somatic embryos, which suggests that interactions between PAs and NO may play an important role in SE (*see* [59] for review).

In summary, PAs are good candidates for enhancing plant regeneration potential from in vitro cultures. Propagation of plants through multiplication of embryogenic propagules is the most commercially attractive application [46]. Unfortunately, SE cannot be achieved in many plants [55], which should be alternatively micropropagated via organogenesis. Remarkably, PAs have been found to be key regulators of both types of morphogenetic processes.

6 Flowering

The coordinated transition from vegetative growth to reproductive development is triggered by external cues, such as photoperiod, temperature and nutrient conditions, as well as by endogenous factors [60]. External signals are perceived in leaves and roots and long-distance transport of signaling endogenous compounds to the shoot meristem then triggers its switch from vegetative to inflorescence meristem identity [60]. These changes in meristem identities result in alteration of internode growth, branching and the identities of the floral organs that are produced by these meristems [61].

Gibberellin is probably the best studied hormone in flowering but other hormones including abscisic acid, jasmonate, salicylic acid, brassinosteroids, cytokinins, ethylene and NO have been reported to play a role in regulating the flowering [62].

The is a large body of evidence suggesting a close connection between PAs and the physiological events leading to flowering [63, 64]. Changes in the levels of PAs accompanies floral development in many plants [63, 64]. For example, in *Pharbitis nil*—a short-day (SD) plant—photoperiodic induction causes increases in foliar PA levels as a prelude to their increase in the stem apex, where floral initiation ultimately occurs [65]. In the SD plant *Xanthium strumarium* [66] and the long-day (LD) plant *Sinapis alba* [67], large increases in foliar PA content appear after only a single photoinductive cycle. A direct morphogenetic effect of applied PAs is seen in the SD plant *Pharbitis nil*, where the application of exogenous Put leads to increases in endogenous PA levels and to flowering under LD conditions [68]. In Arabidopsis, exogenous application of Spd to plants grown under SD conditions, where flowering is naturally delayed, increased Spd content and augmented the rate and extent of flowering [69].

PA biosynthetic inhibitors have also been used to investigate the relationship between PAs and flower development. For example, we have found that the ADC inhibitor DFMA prevents floral bud initiation in tobacco TCL explants, while DL-α-difluoromethylornithine (DFMO), the analogous inhibitor of ornithine decarboxylase (ODC), does not [70]. On the other hand, DFMO inhibits the subsequent development of newly regenerated floral buds, while DFMA does not. It thus appears that PAs derived through ADC may be involved in floral bud initiation, while PAs derived through ODC are required for subsequent growth and development of such buds [70]. We have already mentioned that the Spd synthase inhibitor CHA added to these TCL tobacco explants inhibits flowering, whereas exogenous Spd induces flowering in vegetative explants [48]. Similarly, the SAMDC inhibitor methylglyoxal-bis (guanylhydrazone) (MGBG) caused inhibition of flowering in *Spirodela punctuata*, whereas the inhibitory effect was abolished by applying Spd [71].

PA-deficient mutants have also been used to study the implication of PAs in floral development. For example, tobacco mutants deficient in PA metabolism exhibited aberrant morphology in anthers and ovules [72, 73]. Petunia mutants with abnormal PA levels exhibited irregular development of the floral organs [74]. Similarly, high levels of PAs in flowers contributes to abnormal stament development in tomato mutants [75]. In Arabidopsis, it seems that Spd promotes flowering in late-flowering *CS* 3123 mutant, but not in rapidly flowering 35S::APETALA1 [69]. The authors suggested the possibility that there is an optimum level of PAs with respect to bolting and flowering, and that the addition of exogenous Spd promotes these processes when the concentration is suboptimal, but inhibits them when the endogenous level is already optimal [69]. Taking into account that PAs are abundant wherever cell division occurs it was thought that different groups of cells in the apex are differentially sensitive to PAs, and that raising or lowering the concentration of Spd might affect the relative rates of cell division in the different group of cells so as to alter morphogenetic patterns [69].

Taken together, these results point to an important role of PAs in the transition from vegetative growth to reproductive activity and highlight in particular the biological significance of Spd in flowering [64, 76].

7 Leaf Senescence

During the course of their lifespan, leaves undergo a series of developmental, physiological and metabolic transitions that culminate in senescence and death. Leaf senescence allows the degradation of the nutrients that are produced during the growth phase of the leaf and their redistribution to developing seeds or other parts of the plant, which is a strategy that has evolved to maximize the fitness of the plant [77]. During leaf senescence, there are dramatic metabolic changes such as degradation of essential macromolecules (nucleic acids, proteins and lipids) and photosynthetic pigments. Thus, leaf yellowing is one of the first visible morphological symptoms of senescence. There is also an increase of lipid peroxidation and membrane leakiness associated with increased production of reactive oxygen species (ROS), which contribute to ultra-structural modifications including decay of the cytoskeleton, fragmentation of the endoplasmic reticulum, degradation of ribosomes, and structural changes within the chloroplast [78]. Leaf senescence is a complex process that is controlled by multiple layers of regulation involving the function of multiple genes and signaling pathways that integrate age with various endogenous and exogenous signals [77].

Although all classical plant hormones have been described as playing a role in the regulation of leaf senescence, ethylene is widely acknowledged as the senescence-promoting hormone [79]. Ethylene and PAs use a common precursor S-adenosyl-methionine (SAM) for their biosynthesis, but they act in opposite directions [1]. Thus, PAs play a crucial role in suppressing the onset of leaf senescence, while ethylene promotes this event [80]. The involvement of PAs during leaf senescence was firsly demonstrated in the laboratory of Arthur W. Galston (Yale University) in oat leaf protoplasts. It was shown that upon isolation protoplasts start to senesce and do not undergo sustained cell division. Treatment with PAs (particularly, Spd and Spm) stabilized protoplasts against lysis, retarded senescence and even induced limited mitosis [81]. In our work, DFMA-pretreatment of oat leaves before incubation in the dark in the presence of 0.6 M sorbitol—which is the osmotica used routinely in protoplast isolation—resulted in a decrease of Put, but a dramatic increase of Spd and Spm levels due to activation of Spd synthase activity, which was accompanied with a reduction of chlorophyll loss [82]. Moreover, protoplasts isolated from DFMA-pretreated oat leaves exhibited significant improvement of viability, cell division and cell proliferation to the extent that they were capable of forming microcallus [82]. We also showed that DFMA pretreatment stabilized the composition of thylakoid membranes, adding further support that PAs play a key role in preserving membrane integrity [83].

To prevent the degradation of endogenous Spd and Spm levels in osmotically stressed oat leaves, we also used the PAO inhibitor guazatine [84]. Pretreatment with this inhibitor decreased the levels of 1,3-diaminopropane (Dap) (a product of PA catabolism), with a concomitant increase in the levels of Spd and Spm [84], which was correlated with prevention of (Chl) loss and inhibition of lipid peroxidation [85].

It was initially proposed that the antagonistic effects of PAs and ethylene during leaf senescence were due to metabolic competition for SAM, which is its common precursor (*see* [80] for review). However, more recent work has shown that SAMDC-transgenic tomato fruits with elevated Spd and Spm levels and displaying attenuated fruit ripening also produced more ethylene than the non-transgenic controls [86]. This indicates that the two metabolic pathways can operate simultaneously, suggesting that the levels of the SAM precursor are not a limiting factor for either pathway. The authors proposed that PA accumulation may affect ethylene perception [86]. The same group also studied the effect of overexpression of Spd synthase in tomato and found that Spd not only increased fruit shelf live but also delayed the senescence of vegetative parts of the plant including the leaves [87]. This demonstrates a role of Spd on in vivo prevention of leaf senescence.

As indicated above, PA catabolism is also involved in changes produced during dark-induced senescence [84]. In our recent work, we have provided a global view of metabolic changes affected by *PAO4* mutation in Arabidopsis, which are associated with delayed entry into dark-induced senescence [88]. Our findings suggest that the delayed *pao4* senescence may be associated with high Spm levels, reduced ROS production and increased NO levels. Furthermore, our results pointed to an important role of Spm as a 'signaling' metabolite promoting stress protection through metabolic connections involving ASC/GSH redox state modifications, changes in sugar and nitrogen metabolism, cross-talk with ethylene biosynthesis and mitochondrial electron transport chain modulation, all of which are involved in the nitro-oxidative response after stress imposition [88].

PAs can be covalently incorporated into proteins through the transglutaminase reaction [89] and by hypusine biosynthesis [90]. In the latter, Spd is utilized for the post-translational modification of the inactive precursor protein eIF5A [6]. This post-translational modification occurs in a two-step reaction. The first reaction, catalyzed by deoxyhypusine synthase (DHS) consits in adding a butylamine group derived from Spd to the conserved Lys residue to form deoxyhypusine. Subsequently, the deoxyhypusine is converted into hypusine in a second reaction catalyzed by deoxyhypusine hydrolase [6].

In Arabidopsis it has been reported that antisense suppression of *DHS* gene resulted in a delay in the onset of natural rosette leaf senescence ranging from 2 to at least 6 weeks depending on the degree of DHS suppression [91, 92]. Antisense suppression of *DHS* also delayed premature leaf senescence induced by drought stress resulting in enhanced survival in comparison with wild-type plants. In addition, detached leaves from *DHS*-suppressed plants exhibited delayed post-harvest senescence [91, 92]. Antisense suppression of *DHS* in tomato delays the onset of fruit ripening, which further supports the view that suppression of DHS plays a central role in the regulation of plant senescence [93].

Overall, the experimental evidences presented here demonstrate a key role of PAs in leaf senescence and that application of exogenous PAs results in an extension of plant longevity. This is in concordance with recent findings demonstrating that PAs can enhance longevity in many organisms including mammals, in which administration of PA-rich food results in lifespan extension [94].

8 Flower Senescence

Post-harvest senescence is a major limitation to the marketing of many species of cut flowers and considerable efforts have been devoted to develop treatments to extend the marketing period. Silver thiosulfate (STS) is widespread used to delay senescence in ethylene-sensitive cut flowers [95]. However, concerns have been raised over the use of silver as it is a heavy toxic metal for the environment. Therefore, many countries are actively working towards its elimination from commercial use [95]. 1-Methylcyclopropene (1-MCP) has been shown to prevent the action of ethylene and has been demonstrated to extend the storage life of a range of cut flowers and most importantly is non-toxic to humans [96]. More recently, post-harvest application of NO has been shown to be effective in extending the shelf life of a range of flowers, fruits and vegetables when applied as a short-term fumigation treatment at low concentrations [97]. Thus, fumigation with either 1-MCP or NO represent competitive alternatives to STS [96, 97]. However, the gaseous nature of both compounds is a barrier to their use as a commercial treatment because of the need to develop and construct the infrastructure to undertake large-scale fumigation.

There are evidences that PAs are capable of extending the shelf life of a range of flowers. For example, rose flower (one of most important cut-flowers) has a very short vase life, which is characterized by early wilting and bending of the pedicels. Morphological responses and vase life of *Rosa hybrida* to exogenous PAs were studied in plants grown in hydroponic system [98]. The results indicated that spraying rose plants with either Put, Spd or Spm at different concentrations extended the shelf life and significantly increased flower quality characteristics [98]. Similar effects produced by exogenous application of PAs have been reported in other cultivars of rosa [99, 100]. and other ornamental plants like gladiolus [101], carnation [102], dahlia [103], periwinkle [104], and *Chrysanthemum indicum* L. [105]. We have work with *Phalaenopsis* because the cut flower clusters, as well as the intact clusters, of some species of this orchid are considered among the longest lasting [106]. High levels of endogenous Spd and Spm were correlated with longest vase live of the flowers. We suggested that these PAs may be included among the antisenescence factors responsible for the maintenance of the quality of sepals and petals of unpollinated *Phalaenopsis* flowers [107].

Ornamental crop industry is an important component of specialty crop agriculture since the total value of sales of these industries exceed $15 billion and make up over 15% of the total value of sales of all production crops and 36 percent of the value of sales of all specialty crops [108]. We envisage that the improvement of flower set, quality and vase life by PAs will have a great impact in the ornamental and floriculture industry.

9 Concluding Remarks

In this chapter, we have discussed about the use of chemical or genetic approaches aiming at the manipulation of endogenous PA levels and their effects on different aspects of plant development. Thus, we have seen that pre-soaking seeds with PAs significantly improves seed germination and seedling performace specially under adverse environmental conditions, and that chemical inhibition of PAO activity by guazatine has the potential to control weed growth due to its selective herbicidal properties. Interestingly, the diamines Put and Cad act sinergistically in the promotion of root branching and plant growth and development, which may be due at least in part to cross-talks of the corresponding biosynthetic pathways. We have also discussed that PAs regulate plant morphology in vivo and plant organogenesis in vitro depending on the ratio of Put to Spd, independently of auxin to cytokinin ratios. We envisage that further understanding of hormonal cross-talks with PAs holds potential for the exploitation and manipulation of plant architecture characteristics to increase crop yield and optimize agricultural land use. Treatment with DFMA produces a senescence-like effect of callus cultured in vitro, but upon inhibitor removal from the culture medium a rejuvenation effect is produced in the callus, which significantly improves plant regeneration potential. Similarly, a rejuvenation effect leading to the induction of somatic embryos also results from exogenous Spd application to cell lines that had lost embryogenic potential through repeated subculture. Spraying ornamental plants with PAs delays flower vase life and significantly improves flower quality characteristics. We have also discussed that pre-treatments with inhibitors of PA biosynthesis or catabolism are good approaches for delaying plant senescence, whereas genetic depletion of hypusine (a Spd derivative) also delays plant senescence. Overall, these experimental evidences, together with PA effects on fruit ripening and PA responses to abiotic and biotic stresses (discussed in other chapters of this book) indicate that PAs should be regarded as plant growth regulators with potential applications in agriculture and plant biotechnology.

Acknowledgements

A.F.T. acknowledges funding support from Spanish Ministerio de Ciencia e Innovación (BIO2011-29683). R.A. acknowledges further funding support from the Ramón y Cajal Program (RYC-2011-07847) of the Ministerio de Ciencia e Innovación (Spain), the BFU2013-41337-P grant of the Programa Estatal de Fomento de la Investigación Científica y Técnica de Excelencia (Ministerio de Economía y Competitividad, Spain) and a Marie Curie Career Integration Grant (DISEASENVIRON, PCIG10-GA-2011-303568) of the European Union. R.A. and A.F.T. are members of the Group de Recerca Consolidat 2014 SGR-920 of the Generalitat de Catalunya.

References

1. Tiburcio AF, Altabella T, Bitrián M, Alcázar R (2014) The roles of polyamines during the lifespan of plants: from development to stress. Planta 240:1–18
2. Alcázar R, Altabella T, Marco F, Bortolotti C, Reymond M, Koncz C, Carrasco P, Tiburcio AF (2010) Polyamines: molecules with regulatory functions in plant abiotic stress tolerance. Planta 231:1237–1249
3. Jancewicz AL, Gibbs NM, Masson PH (2016) Cadaverine's functional role in plant development and environmental response. Front Plant Sci 7:870

4. Igarashi K, Kashiwagi K (2010) Modulation of cellular function by polyamines. Int J Biochem Cell Biol 42:39–51
5. Walters D, Meurer-Grimes B, Rovira I (2001) Antifungal activity of three spermidine conjugates. FEMS Microbiol Lett 201:255–258
6. Feng H, Chen Q, Feng J, Zhang J, Yang X, Zuo J (2007) Functional characterization of the Arabidopsis eukaryotic translation initiation factor 5A-2 that plays a crucial role in plant growth and development by regulating cell division, cell growth, and cell death. Plant Physiol 144:1531–1545
7. Ibrahim EA (2016) Seed priming to alleviate salinity stress in germinating seeds. J Plant Physiol 192:38–46
8. Farooq M, Basra SMA, Rehman H, Hussain M (2008) Seed priming with polyamines improves the germination and early seedling growth in fine rice. J New Seeds 9:145–155
9. Savvides A, Ali S, Tester M, Fotopoulos V (2016) Chemical priming of plants against multiple abiotic stresses: mission possible? Trends Plant Sci 21:329–340
10. Cao DD, Hu J, Gao CH, Guan YJ, Zang S, Xiao JF (2008) Chilling tolerance of maize can be improved by seed soaking in putrescine. Seed Sci Technol 36:191–197
11. Yang L, Hong XU, Xiao-xia W, Yun-cheng L (2016) Effect of polyamines on wheat under drought stress is related to changes in hormones and carbohydrates. J Integr Agric 15:60345–60347
12. Ali RM, Abbas HM, Kamal RK (2009) The effects of treatment with polyamines on dry matter and some metabolites in salinity-stressed chamomile and sweet majoram seedlings. Plant Soil Environ 55:477–483
13. Li Z, Peng Y, Zhang XQ, Ma X, Huang LK, Yan YH (2014) Exogenous spermidine improves seed germination of white clover under water stress via involvement in starch metabolism, antioxidant defenses and relevant gene expression. Molecules 19:18003–18024
14. Rebecca LJ, Das S, Dhanalakshmi V, Anbuselvi S (2010) Effect of exogenous spermidine on salinity tolerance with respect to seed germination. Int J Appl Agric Res 5:163–169
15. Sedagahat S, Rahemi M (2011) Effect of pre-soaking seeds in polyamines on seed germination and seedling growth of *Pistacia vera* L. cv. Ghazvini. Int J f Nuts Relat Sci 2:7–14
16. Huang Y, Lin C, He F, Li Z, Guan Y, Hu Q, Hu J (2017) Exogenous spermidine improves seed germination of sweet corn via involvement in phytohormone interactions, H_2O_2 and relevant gene expression. BMC Plant Biol 17:1
17. Chunthaburee S, Sanichon J, Pattanagul W, Theerakulpisut (2014) Alleviation of salt stress in seedlings of Black glutinous rice by seed priming with spermidine and gibberelic acid. Not Bot Horti Agrobot 42:405–413
18. Iqbal M, Ashraf M, Rehman S-U, Rha ES (2006) Does polyamine seed pretreatment modulate growth and levels of some plant growth regulators in hexaploid wheat plants under salt stress? Bot Stud 47:239–250
19. Ferrando A, Carrasco P, Tiburcio AF (2009) Modulation of seed growth and development by inhibition of polyamine catabolism. Patent WO2009074700
20. Smith S, De Smet I (2012) Root system architecture: insights from Arabidopsis and cereal crops. Philos Trans R Soc Lond Ser B Biol Sci 367:1441–1452
21. Couée I, Hummel I, Sulmon C, Gouesbert G, El Amrani A (2004) Involvement of polyamines in root development. Plant Cell Tissue Organ Cult 76:1–10
22. Celenza JL Jr, Grisafi PL, Fink GR (1995) A pathway for lateral root formation in Arabidopsis thaliana. Genes Dev 9:2131–2142
23. Zhang H, Jennings A, Barlow PW, Forde BG (1999) Dual pathways for regulation of root branching by nitrate. Proc Natl Acad Sci U S A 96:6529–6534
24. Kende H, Zeevaart J (1997) The five "classical" plant hormones. Plant Cell 9:1197–1210
25. Martin-Tanguy J (2001) Metabolism and function of polyamines in plants: recent development (new approaches). Plant Growth Regul 34:135–148
26. Flores HE, Galston AW (1982) Polyamines and plant stress: activation of putrescine biosynthesis by osmotic shock. Science 217:1259–1261
27. Biondi S, Mengoli M, Mott D, Bagni N (1993) Hairy root cultures of *Hyosciamus muticus*-effect of polyamine biosynthesis inhibitors. Plant Physiol Biochem 31:51–58
28. Martin-Tanguy J, Carré M (1993) Polyamines in grapevine microcuttings cultivated *in vitro*-effects of amines and inhibitors of polyamine biosynthesis on polyamine levels and microcutting growth and development. Plant Growth Regul 13:269–280
29. Tiburcio AF, Amin Gendy C, Tran Than Van K (1989) Morphogenesis in tobacco subepidermal cells: putrescine as a marker of root differentiation. Plant Cell Tissue Organ Cult 19:43–54

30. Cui X, Ge C, Wang R, Wang H, Chen W, Fu Z, Jiang X, Li J, Wang Y (2010) The *BUD2* mutation affects plant architecture through altering cytokinin and auxin responses in Arabidopsis. Cell Res 20:576–586
31. Wu Q-S, Zou Y-N, Liu C-Y, Cheng K (2012) Effects of exogenous putrescine on mycorrhiza, root system architecture, and physiological traits of *Glomus mosseae*-colonized trifoliate orange seedlings. Not Bot Horti Agrobot 40:80–85
32. Tang W, Newton RJ (2005) Polyamines promote root elongation and growth by increasing root cell division in regenerated Virginia pine (*Pinus virginiana* Mill.) plantlets. Plant Cell Rep 24:581–589
33. Wu Q-S, Zou Y-N, Liu C-Y, Lu T (2010) Interacted effect of arbuscular mycorrizal fungi and polyamines on root system architecture of citrus seedlings. J Integr Biol 11:1675–1681
34. Tomar PC, Lakra N, Mishra SN (2013) Cadaverine: a lysine catabolite involved in plant growth and development. Plant Signal Behav 8(10)
35. Gamarnik A, Frydman RB (1991) Cadaverine, an essential diamine for the normal root development of germinating soybean (Glycine max) seeds. Plant Physiol 97:778–785
36. Strohm AK, Vaughn LM, Masson PH (2015) Natural variation in the expression of ORGANIC CATION TRANSPORTER1 affects root length responses to cadaverine in Arabidopsis. J Exp Bot 66:853–862
37. Liu T, Dobashi H, Kim DW, Sagor GH, Niitsu M, Berberich T, Kusano T (2014) Arabidopsis mutant plants with diverse defects in polyamine metabolism show unequal sensitivity to exogenous cadaverine probably based on their spermine content. Physiol Mol Biol Plants 20:151–159
38. Sagor GH, Berberich T, Kojima S, Niitsu M, Kusano T (2016) Spermine modulates the expression of two probable polyamine transporter genes and determines growth responses to cadaverine in Arabidopsis. Plant Cell Rep 35:1247–1257
39. Salabert A (1995) Obtaining and use of diamines, polyamines and other complementary active elements from treated natural products. Patent EP0726240A1
40. Reinhardt D, Kuhlemeier C (2002) Plant architecture. EMBO Rep 3:846–851
41. Doebley J, Stec A, Hubbard L (1997) The evolution of apical dominance in maize. Nature 386:485–488
42. Schumacher K, Schmitt T, Rossberg M, Schmitz G, Theres K (1999) The *Lateral suppressor* (*Ls*) gene of tomato encodes a new member of the VHIID protein family. Proc Natl Acad Sci U S A 96:290–295
43. Ge C, Cui X, Wang Y, Hu Y, Fu Z, Zhang D, Cheng Z, Li J (2006) *BUD2*, encoding an S-adenosylmethionine decarboxylase, is required for Arabidopsis growth and development. Cell Res 16:446–456
44. Geuns JM, Smets R, Struyf T, Prinsen E, Valcke R, Van Onckelen H (2001) Apical dominance in Pssu-ipt-transformed tobacco. Phytochemistry 58:911–921
45. Murashige T (1974) Plant propagation through tissue cultures. Annu Rev Plant Physiol 25:135–166
46. Ikeuchi M, Ogawa Y, Iwase A, Sugimoto K (2016) Plant regeneration: cellular origins and molecular mechanisms. Development 143:1442–1451
47. Tran Thanh Van M (1973) Direct flower neoformation from superficial tissue of small explants of *Nicotiana tabacum* L. Planta 115:87–92
48. Kaur-Sawhney R, Tiburcio AF, Galston AW (1988) Spermidine and flower-bud differentiation in thin-layer explants of tobacco. Planta 173:282–284
49. Purohit SD, Singhvi A, Nagori R, Vyas S (2007) Polyamines stimulate shoot bud proliferation in *Achras sapota* grown in culture. Indian J Biotechnol 6:85–90
50. Ganesan M, Jayabalan N (2006) Influence of cytokinins, auxins and polyamines on in vitro mass multiplication of cotton (*Gossypium hirsutum* L. cv. SVPR2). Indian J Exp Biol 44:506–513
51. Sivanandhan G, Mariashibu TS, Arun M, Kasthurirengan S, Selvaraj N, Ganapathi A (2011) The effect of polyamines on the efficiency of multiplication and rooting of *Eithania somnifera* (L.) Dunal and content of some withanolides in obtained plants. Acta Physiol Plant 33:2279–2288
52. Bajaj S, Rajam MV (1996) Polyamine accumulation and near loss of morphogenesis in long-term callus cultures of rice (restoration of plant regeneration by manipulation of cellular polyamine levels). Plant Physiol 112:1343–1348
53. Tiburcio AF, Figueras X, Claparols I, Santos M, Torné JM (1991) Improved plant regeneration in maize callus cultures after pretreatment with DL-alpha difluoro-methylarginine. Plant Cell Tissue Organ Cult 27:27–32

54. Ammirato PV (1984) Induction, maintenance, and manipulation of development in embryogenic cell suspension cultures. In: Vasil IK (ed) Cell culture and somatic cell genetics of plants, vol 1. Academic Press, New York, pp 139–151
55. Jiménez VM, Bangerth F (2001) Endogenous hormone levels in explants and in embryogenic and non-embryogenic cultures of carrot. Physiol Plant 111:389–395
56. Bastola DR, Minocha SC (1995) Increased putrescine biosynthesis through transfer of mouse ornithine decarboxylase cDNA in carrot promotes somatic embryogenesis. Plant Physiol 109:63–71
57. Montague MJ, Armstrong TA, Jaworski EG (1979) Polyamine metabolism in embryogenic cells of daucus carota: II. Changes in arginine decarboxylase activity. Plant Physiol 63:341–345
58. Feirer RP, Mignon G, Litvay JD (1984) Arginine decarboxylase and polyamines required for embryogenesis in the wild carrot. Science 223:1433–1435
59. Wimalasekera R, Tebartz F, Scherer GFE (2011) Polyamines, polyamine oxidase and nitric oxid in development, abiotic and biotic stresses. Plant Sci 181:593–603
60. Andres F, Coupland G (2012) The genetic basis of flowering responses to seasonal cues. Nat Rev Genet 13:627–639
61. Wils CR, Kaufmann K (2017) Gene-regulatory networks controlling inflorescence and flower development in *Arabidopsis thaliana*. Biochim Biophys Acta 1860:95–105
62. Davis SJ (2009) Integrating hormones into the floral-transition pathway of *Arabidopsis thaliana*. Plant Cell Environ 32:1201–1210
63. Galston AW, Sawhney RK (1990) Polyamines in plant physiology. Plant Physiol 94:406–410
64. Walden R, Cordeiro A, Tiburcio AF (1997) Polyamines: small molecules triggering pathways in plant growth and development. Plant Physiol 113:1009–1013
65. Dai YR, Wang J (1987) Relation of polyamine titer to photoperiodic induction of flowering in *Pharbitis*. Plant Sci 51:137–139
66. Hamasaki N, Galston AW (1990) The polyamines of *Xanthium strumarium* and their response to photoperiod. Photochem Photobiol 52:181–186
67. Havelange A, Lejeune P, Bernier G, Kaur-Sawhney R, Galston AW (1996) Putrescine export from leaves in relation to floral transition in *Sinapis alba*. Physiol Plant 96:59–65
68. Wada N, Shinozaki M, Iwamura H (1994) Flower induction by polyamines and related compounds in seedlings of morning glory (*Pharbitis nil* cv. Kidachi). Plant Cell Physiol 35:469–472
69. Applewhite PB, Kaur-Sawhney R, Galston AW (2000) A role for spermidine in the bolting and flowering of Arabidopsis. Physiol Plant 108:314–320
70. Tiburcio AF, Kaur-Sawhney R, Galston AW (1988) Polyamine biosynthesis during vegetative-and floral-bud differentiation in thin-layer tobacco tissues cultures. Plant Cell Physiol 29:1241–1249
71. DeCantu LB, Kandeler R (1989) Significance of polyamines for flowering in *Spirodela punctata*. Plant Cell Physiol 30:455–458
72. Malmberg RL, McIndoo J (1983) Abnormal floral development of a tobacco mutant with elevated polyamine levels. Nature 305:623–625
73. Malmberg RL, McIndoo J (1988) Nicotiana plants with altered polyamine levels and floral organs. Patent US4751348
74. Gerats AG, Kaye C, Collins C, Malmberg RL (1988) Polyamine levels in petunia genotypes with normal and abnormal floral morphologies. Plant Physiol 86:390–393
75. Rastogi R, Sawhney VK (1990) Polyamines and flower development in the male sterile stamenless-2 mutant of tomato (*Lycopersicon esculentum* Mill.): II. Effects of polyamines and their biosynthetic inhibitors on the development of normal and mutant floral buds cultured *in vitro*. Plant Physiol 93:446–452
76. Lu J-H, Honda C, Moriguchi T (2006) Involvement of polyamines in floral and fruit development. Jpn Agric Res Q 40:51–58
77. Woo HR, Kim HJ, Nam HG, Lim PO (2013) Plant leaf senescence and death – regulation by multiple layers of control and implications for aging in general. J Cell Sci 126:4823–4833
78. Thomas H, Ougham HJ, Wagstaff C, Stead AD (2003) Defining senescence and death. J Exp Bot 54:1127–1132
79. Jibran R, A Hunter D, P Dijkwel P (2013) Hormonal regulation of leaf senescence through integration of developmental and stress signals. Plant Mol Biol 82:547–561
80. Bais HP, Ravishankar GA (2002) Role of polyamines in the ontogeny of plants and their biotechnological applications. Plant Cell Tissue Organ Cult 69:1–34
81. Kaur-Sawhney R, Flores HE, Galston AW (1980) Polyamine-induced DNA synthesis and mitosis in oat leaf protoplasts. Plant Physiol 65:368–371

82. Tiburcio AF, Kaur-Sawhney R, Galston AW (1986) Polyamine metabolism and osmotic stress. II. Improvement of oat protoplasts by an inhibitor of arginine decarboxylase. Plant Physiol 82:375–378
83. Besford RT, Richardson CM, Campos JL, Tiburcio AF (1993) Effect of polyamines on stabilization of molecular complexes in thylakoid membranes of osmotically-stressed oat leaves. Planta 189:201–206
84. Capell T, Campos JL, Tiburcio AF (1993) Antisenescence properties of guazatine in osmotically-stressed oat leaves. Phytochemistry 33:785–788
85. Borrell A, Carbonell R, Farràs R, Puig-Parellada P, Tiburcio AF (1997) Polyamines inhibit lipid peroxidation in senescing oat leaves. Physiol Plant 99:385–390
86. Mehta RA, Cassol T, Li N, Ali N, Handa AK, Mattoo AK (2002) Engineered polyamine accumulation in tomato enhances phytonutrient content, juice quality, and vine life. Nat Biotechnol 20:613–618
87. Nambeesan S, Datsenka T, Ferruzzi MG, Malladi A, Mattoo AK, Handa AK (2010) Overexpression of yeast spermidine synthase impacts ripening, senescence and decay symptoms in tomato. Plant J 63:836–847
88. Sequera-Mutiozabal MI, Erban A, Kopka J, Atanasov KE, Bastida J, Fotopoulos V, Alcázar R, Tiburcio AF (2016) Global metabolic profiling of Arabidopsis polyamine oxidase 4 (AtPAO4) loss-of-function mutants exhibiting delayed dark-induced senescence. Front Plant Sci 7:173
89. Del Duca S, Serafini-Fracassini D, Cai G (2014) Senescence and programmed cell death in plants: polyamine action mediated by transglutaminase. Front Plant Sci 5:120
90. Park MH (2006) The post-translational synthesis of a polyamine-derived amino acid, hypusine, in the eukaryotic translation initiation factor 5A (eIF5A). J Biochem 139:161–169
91. Thompson JE, Tzann-Wei Wang T-W, Lu DL (2003) DNA encoding a plant deoxyhypusine synthase, a plant eukaryotic initiation factor 5A, transgenic plants and a method for controlling senescence programmed and cell death in plants. Patent US 6538182 B1
92. Wang TW, Lu L, Zhang CG, Taylor C, Thompson JE (2003) Pleiotropic effects of suppressing deoxyhypusine synthase expression in *Arabidopsis thaliana*. Plant Mol Biol 52:1223–1235
93. Wang TW, Zhang CG, Wu W, Nowack LM, Madey E, Thompson JE (2005) Antisense suppression of deoxyhypusine synthase in tomato delays fruit softening and alters growth and development. Plant Physiol 138:1372–1382
94. Kibe R, Kurihara S, Sakai Y, Suzuki H, Ooga T, Sawaki E, Muramatsu K, Nakamura A, Yamashita A, Kitada Y, Kakeyama M, Benno Y, Matsumoto M (2014) Upregulation of colonic luminal polyamines produced by intestinal microbiota delays senescence in mice. Sci Rep 4:4548
95. Stabyl GL, Basel RM, Michael S, Reid MS, Dodge LL (1993) Efficacies of commercial anti-ethylene products for fresh cut flowers. HortTechnology 3:199–202
96. Watkins CB (2006) The use of 1-methylcyclopropene (1-MCP) on fruits and vegetables. Biotechnol Adv 24:389–409
97. Leshem YY, Wills RBH, Veng-Va Ku V (1998) Evidence for the function of the free radical gas – nitric oxide (NO) – as an endogenous maturation and senescence regulating factor in higher plants. Plant Physiol Biochem 36:825–833
98. Farahi MH, Khalighi A, Kholdbarin B, Akbarboojar MM, Eshghi S (2013) Morphological responses and vase life of *Rosa hybrida* cv. dolcvita to polyamines spray in hydroponic system. World Appl Sci J 21:1681–1686
99. Ling X, ZhongShen W, Zifa D (2007) Effects of polyamines and penicillin on preservation of cut roses. J Nanjing For Univ Nat Sci Ed 31:53–56
100. Nada K, Kawaguchi T, Tachibana S (2004) Effects of polyamines in the vase water on the vase life of cut rose flowers. Hortic Res (Japan) 3:101–104
101. Dantuluri VSR, Misra RL, Singh VP (2008) Effect of polyamines on post harvest life of gladiolus spikes. J Ornam Hort 11:66–68
102. Upfold SJ, Van Staden J (1991) Polyamines and carnation flower senescence: endogenous levels and the effect of applied polyamines on senescence. Plant Growth Regul 10:355–362
103. Mahgoub MH, Abd El Aziz NG, Mazhar MA (2011) Response of *Dahlia pinnata* L plant to foliar spray with putrescine and thiamine on growth, flowering and photosynthetic pigments. American-Eurasian J Agric Environ Sci 10:769–775
104. Iman Talaat M, Bekheta MA, Mahgoub MM (2005) Physiological response of periwinkle plants (*Catharanthus roseus* L.) to tryptophan and putrescine. Int J Agric Biol 7:210–213
105. Mahros KM, El-Saady MB, Mahgoub MH, Afaf MH, El-Sayed MI (2011) Effect of putrescine and uniconazole treatments on

flower characters and photosynthetic pigments of *Chrysanthemum indicum* L. Plant J Am Sci 7:399–408

106. Gelein C (1984) Catalogue: cut flowers-pot plants-bedding plants. Verenige Bloemenveilingen Aalsmeer, The Netherlands, pp 105–115
107. Tiburcio AF, Campos JL, Figueras X, Marce M, Capell T, Riera R, Bestford RT (1993) Polyamines and morphogenesis in monocots: experimental systems and mechanisms of action. In: Roubelakis-Angelakis KA, Tran Thanh Van K (eds) Morphogenesis in plants. Plenum Press, New York, pp 113–135
108. Miller SR, Abdulkadri A (2008) The U.S. economic impact of the IR-4 ornamental horticulture project. Dec 4, pp 1–18

Correction to: Analysis of Cotranslational Polyamine Sensing During Decoding of ODC Antizyme mRNA

R. Palanimurugan, Daniela Gödderz, Leo Kurian, and R. J. Dohmen

Correction to:
Chapter 26 in: Rubén Alcázar and Antonio F. Tiburcio (eds.), *Polyamines: Methods and Protocols*, Methods in Molecular Biology, vol. 1694, https://doi.org/10.1007/978-1-4939-7398-9_26

The author's first and last name was incorrectly captured in Pubmed. This has been corrected to read as follows:

Dohmen R.J.

The updated online version of this chapter can be found at https://doi.org/10.1007/978-1-4939-7398-9_26

Rubén Alcázar and Antonio F. Tiburcio (eds.), *Polyamines: Methods and Protocols*, Methods in Molecular Biology, vol. 1694, https://doi.org/10.1007/978-1-4939-7398-9_41,

Index

Rubén Alcázar and Antonio F. Tiburcio (eds.), *Polyamines: Methods and Protocols*, Methods in Molecular Biology, vol. 1694, DOI 10.1007/978-1-4939-7398-9, © Springer Science+Business Media LLC 2018

MIX
Papier aus verantwortungsvollen Quellen
Paper from responsible sources
FSC® C105338

If you have any concerns about our products,
you can contact us on
ProductSafety@springernature.com

In case Publisher is established outside the EU,
the EU authorized representative is:
Springer Nature Customer Service Center GmbH
Europaplatz 3, 69115 Heidelberg, Germany

Printed by Libri Plureos GmbH
in Hamburg, Germany